Toxicological Chemistry and Biochemistry

Third Edition

Toxicological Chemistry and Biochemistry

THIRD EDITION

Stanley E. Manahan

LEWIS PUBLISHERS

A CRC Press Company
Boca Raton   London   New York   Washington, D.C.

Cover image courtesy of the North Carolina State University Center for Applied Aquatic Ecology.

Library of Congress Cataloging-in-Publication Data

Manahan, Stanley E.
Toxicological chemistry and biochemistry / by Stanley E. Manahan.-- 3rd ed.
   p. cm.
   Includes bibliographical references and index.
   ISBN 1-56670-618-1
   1. Toxicological chemistry. 2. Environmental chemistry. 3. Biochemical toxicology. I.
Title.

RA1219.3 .M36 2002
815.9′001′54--dc21   2002072486

This book contains information obtained from authentic and highly regarded sources. Reprinted material is quoted with permission, and sources are indicated. A wide variety of references are listed. Reasonable efforts have been made to publish reliable data and information, but the author and the publisher cannot assume responsibility for the validity of all materials or for the consequences of their use.

Neither this book nor any part may be reproduced or transmitted in any form or by any means, electronic or mechanical, including photocopying, microfilming, and recording, or by any information storage or retrieval system, without prior permission in writing from the publisher.

The consent of CRC Press LLC does not extend to copying for general distribution, for promotion, for creating new works, or for resale. Specific permission must be obtained in writing from CRC Press LLC for such copying.

Direct all inquiries to CRC Press LLC, 2000 N.W. Corporate Blvd., Boca Raton, Florida 33431.

Trademark Notice: Product or corporate names may be trademarks or registered trademarks, and are used only for identification and explanation, without intent to infringe.

Visit the CRC Press Web site at www.crcpress.com

© 2003 by CRC Press LLC
Lewis Publishers is an imprint of CRC Press LLC

No claim to original U.S. Government works
International Standard Book Number 1-56670-618-1
Library of Congress Card Number 2002072486
Printed in the United States of America 1 2 3 4 5 6 7 8 9 0
Printed on acid-free paper

Preface

The first edition of *Toxicological Chemistry* (1989) was written to bridge the gap between toxicology and chemistry. It defined toxicological chemistry as the science that deals with the chemical nature and reactions of toxic substances, their origins and uses, and the chemical aspects of their exposure, transformation, and elimination by biological systems. It emphasized the chemical formulas, structures, and reactions of toxic substances. The second edition of *Toxicological Chemistry* (1992) was significantly enlarged and increased in scope compared to the first edition. In addition to toxicological chemistry, it addressed the topic of environmental biochemistry, which pertains to the effects of environmental chemical substances on living systems and the influence of life-forms on such chemicals. It did so within a framework of environmental chemistry, defined as that branch of chemistry that deals with the origins, transport, reactions, effects, and fates of chemical species in the water, the air, and terrestrial and living environments.

The third edition has been thoroughly updated and expanded into areas important to toxicological chemistry based upon recent advances in several significant fields. In recognition of the increased emphasis on the genetic aspects of toxicology, the toxic effects to various body systems, and xenobiotics analysis, the title has been changed to *Toxicological Chemistry and Biochemistry*. The new edition has been designed to be useful to a wide spectrum of readers with various interests and a broad range of backgrounds in chemistry, biochemistry, and toxicology. For readers who have had very little exposure to chemistry, Chapter 1, "Chemistry and Organic Chemistry," outlines the basic concepts of general chemistry and organic chemistry needed to understand the rest of the material in the book. The second chapter, "Environmental Chemistry," is an overview of that topic, presented so that the reader may understand the remainder of the book within a framework of environmental chemistry. Chapter 3, "Biochemistry," gives the fundamentals of the chemistry of life processes essential to understanding toxicological chemistry and biochemistry. Chapter 4, "Metabolic Processes," covers the basic principles of metabolism needed to understand how toxicants interact with organisms. Chapter 5, "Environmental Biological Processes and Ecotoxicology," is a condensed and updated version of three chapters from the second edition dealing with microbial processes, biodegradation and bioaccumulation, and biochemical processes that occur in aquatic and soil environments; the major aspects of ecotoxicology are also included. Chapter 6, "Toxicology," defines and explains toxicology as the science of poisons. Chapter 7, "Toxicological Chemistry," bridges the gap between toxicology and chemistry, emphasizing chemical aspects of toxicological phenomena, including fates and effects of xenobiotic chemicals in living systems. Chapter 8, "Genetic Aspects of Toxicology," is new; it recognizes the importance of considering the crucial role of nucleic acids, the basic genetic material of life, in toxicological chemistry. It provides the foundation for understanding the important ways in which chemical damage to DNA can cause mutations, cancer, and other toxic effects. It also considers the role of genetics in determining genetic susceptibilities to various toxicants. Also new is Chapter 9, "Toxic Responses," which considers toxicities to various systems in the body, such as the endocrine and reproductive systems. It is important for understanding the specific toxic effects of various toxicants on certain body organs, as discussed in later chapters. Chapters 10 to 18 discuss toxicological chemistry within an organizational structure based on classes of chemical substances, and Chapter 19 deals with toxicants from natural sources. Another new addition is Chapter 20, "Analysis of Xenobiotics," which deals with the determination of toxicants and their metabolites in blood and other biological materials.

Every effort has been made to retain the basic information and structure that have made the first two editions of this book popular among and useful to students, faculty, regulatory agency personnel, people working with industrial hygiene aspects, and any others who need to understand toxic effects of chemicals from a chemical perspective. The chapters that have been added are designed to enhance the usefulness of the book and to modernize it in important areas such as genetics and xenobiotics analysis.

This book is designed to be both a textbook and a general reference book. Questions at the end of each chapter are written to summarize and review the material in the chapter. References are given for specific points covered in the book, and supplementary references are cited at the end of each chapter for additional reading about the topics covered.

The assistance of David Packer, Publisher, CRC Press, in developing the third edition of *Toxicological Chemistry and Biochemistry* is gratefully acknowledged. The author would also like to acknowledge the excellent work of Judith Simon, Project Editor, and the staff of CRC Press in the production of this book.

The Author

Stanley E. Manahan is a professor of chemistry at the University of Missouri–Columbia, where he has been on the faculty since 1965, and is president of ChemChar Research, Inc., a firm developing nonincinerative thermochemical waste treatment processes. He received his A.B. in chemistry from Emporia State University in 1960 and his Ph.D. in analytical chemistry from the University of Kansas in 1965. Since 1968, his primary research and professional activities have been in environmental chemistry, toxicological chemistry, and waste treatment. He teaches courses on environmental chemistry, hazardous wastes, toxicological chemistry, and analytical chemistry. He has lectured on these topics throughout the United States as an American Chemical Society local section tour speaker, in Puerto Rico, at Hokkaido University in Japan, at the National Autonomous University in Mexico City, and at the University of the Andes in Merida, Venezuela. He was the recipient of the Year 2000 Award of the environmental chemistry division of the Italian Chemical Society.

Professor Manahan is the author or coauthor of approximately 100 journal articles in environmental chemistry and related areas. In addition to *Fundamentals of Environmental Chemistry*, 2nd ed., he is the author of *Environmental Chemistry*, 7th ed. (Lewis Publishers, 2000), which has been published continuously in various editions since 1972. Other books that he has written include *Industrial Ecology: Environmental Chemistry and Hazardous Waste* (Lewis Publishers, 1999), *Environmental Science and Technology* (Lewis Publishers, 1997), *Toxicological Chemistry*, 2nd ed. (Lewis Publishers, 1992), *Hazardous Waste Chemistry, Toxicology, and Treatment* (Lewis Publishers, 1992), *Quantitative Chemical Analysis* (Brooks/Cole, 1986), and *General Applied Chemistry*, 2nd ed. (Willard Grant Press, 1982).

Contents

Chapter 1 Chemistry and Organic Chemistry
1.1 Introduction 1
1.2 Elements 1
    1.2.1 Subatomic Particles and Atoms 2
    1.2.2 Subatomic Particles 2
    1.2.3 Atom Nucleus and Electron Cloud 3
    1.2.4 Isotopes 3
    1.2.5 Important Elements 3
    1.2.6 The Periodic Table 4
        1.2.6.1 Features of the Periodic Table 4
    1.2.7 Electrons in Atoms 6
        1.2.7.1 Lewis Symbols of Atoms 6
    1.2.8 Metals, Nonmetals, and Metalloids 6
1.3 Chemical Bonding 7
    1.3.1 Chemical Compounds 7
    1.3.2 Molecular Structure 8
    1.3.3 Ionic Bonds 8
    1.3.4 Summary of Chemical Compounds and the Ionic Bond 9
    1.3.5 Molecular Mass 9
    1.3.6 Oxidation State 10
1.4 Chemical Reactions and Equations 10
    1.4.1 Reaction Rates 11
1.5 Solutions 11
    1.5.1 Solution Concentration 11
    1.5.2 Water as a Solvent 12
    1.5.3 Solutions of Acids and Bases 12
        1.5.3.1 Acids, Bases, and Neutralization Reactions 12
        1.5.3.2 Concentration of H^+ Ion and pH 13
        1.5.3.3 Metal Ions Dissolved in Water 13
        1.5.3.4 Complex Ions Dissolved in Water 14
    1.5.4 Colloidal Suspensions 14
1.6 Organic Chemistry 14
    1.6.1 Molecular Geometry in Organic Chemistry 15
1.7 Hydrocarbons 15
    1.7.1 Alkanes 15
        1.7.1.1 Formulas of Alkanes 16
        1.7.1.2 Alkanes and Alkyl Groups 17
        1.7.1.3 Names of Alkanes and Organic Nomenclature 17
        1.7.1.4 Summary of Organic Nomenclature as Applied to Alkanes ...18
        1.7.1.5 Reactions of Alkanes 19
    1.7.2 Alkenes and Alkynes 20
        1.7.2.1 Addition Reactions 20
    1.7.3 Alkenes and *Cis–trans* Isomerism 21
    1.7.4 Condensed Structural Formulas 21
    1.7.5 Aromatic Hydrocarbons 21
        1.7.5.1 Benzene and Naphthalene 23
        1.7.5.2 Polycyclic Aromatic Hydrocarbons 23

    1.8   Organic Functional Groups and Classes of Organic Compounds 23
            1.8.1   Organooxygen Compounds 24
            1.8.2   Organonitrogen Compounds 25
            1.8.3   Organohalide Compounds 27
                    1.8.3.1   Alkyl Halides 27
                    1.8.3.2   Alkenyl Halides 28
                    1.8.3.3   Aryl Halides 28
                    1.8.3.4   Halogenated Naphthalene and Biphenyl 28
                    1.8.3.5   Chlorofluorocarbons, Halons, and Hydrogen-Containing
                                  Chlorofluorocarbons 29
                    1.8.3.6   Chlorinated Phenols 30
            1.8.4   Organosulfur Compounds 30
                    1.8.4.1   Thiols and Thioethers 30
                    1.8.4.2   Nitrogen-Containing Organosulfur Compounds 32
                    1.8.4.3   Sulfoxides and Sulfones 32
                    1.8.4.4   Sulfonic Acids, Salts, and Esters 32
                    1.8.4.5   Organic Esters of Sulfuric Acid 32
            1.8.5   Organophosphorus Compounds 32
                    1.8.5.1   Alkyl and Aromatic Phosphines 32
                    1.8.5.2   Organophosphate Esters 33
                    1.8.5.3   Phosphorothionate Esters 33
1.9   Optical Isomerism 34
1.10   Synthetic Polymers 34
Supplementary References 36
Questions and Problems 36

Chapter 2   Environmental Chemistry
2.1   Environmental Science and Environmental Chemistry 39
        2.1.1   The Environment 39
        2.1.2   Environmental Chemistry 41
2.2   Water 42
2.3   Aquatic Chemistry 44
        2.3.1   Oxidation–Reduction 44
        2.3.2   Complexation and Chelation 45
        2.3.3   Water Interactions with Other Phases 45
        2.3.4   Water Pollutants 45
        2.3.5   Water Treatment 46
2.4   The Geosphere 46
        2.4.1   Solids in the Geosphere 46
2.5   Soil 47
2.6   Geochemistry and Soil Chemistry 49
        2.6.1   Physical and Chemical Aspects of Weathering 49
        2.6.2   Soil Chemistry 50
2.7   The Atmosphere 51
2.8   Atmospheric Chemistry 52
        2.8.1   Gaseous Oxides in the Atmosphere 53
        2.8.2   Hydrocarbons and Photochemical Smog 54
        2.8.3   Particulate Matter 54
2.9   The Biosphere 55

2.10 The Anthrosphere and Green Chemistry 55
        2.10.1 Green Chemistry..56
References 56
Supplementary References...57
Questions and Problems...57

Chapter 3 Biochemistry

3.1 Biochemistry...59
        3.1.1 Biomolecules 59
3.2 Biochemistry and the Cell..60
        3.2.1 Major Cell Features 60
3.3 Proteins 61
        3.3.1 Protein Structure 64
        3.3.2 Denaturation of Proteins 65
3.4 Carbohydrates...65
3.5 Lipids 66
3.6 Enzymes..69
3.7 Nucleic Acids..72
        3.7.1 Nucleic Acids in Protein Synthesis...75
        3.7.2 Modified DNA 75
3.8 Recombinant DNA and Genetic Engineering..76
3.9 Metabolic Processes 76
        3.9.1 Energy-Yielding Processes 76
Supplementary References...77
Questions and Problems...77

Chapter 4 Metabolic Processes

4.1 Metabolism in Environmental Biochemistry 79
        4.1.1 Metabolism Occurs in Cells..79
        4.1.2 Pathways of Substances and Their Metabolites in the Body 79
4.2 Digestion...80
        4.2.1 Carbohydrate Digestion...81
        4.2.2 Digestion of Fats 82
        4.2.3 Digestion of Proteins...83
4.3 Metabolism of Carbohydrates, Fats, and Proteins...83
        4.3.1 An Overview of Catabolism...83
        4.3.2 Carbohydrate Metabolism 85
        4.3.3 Metabolism of Fats 85
        4.3.4 Metabolism of Proteins 85
4.4 Energy Utilization by Metabolic Processes...87
        4.4.1 High-Energy Chemical Species 87
        4.4.2 Glycolysis 88
        4.4.3 Citric Acid Cycle 89
        4.4.4 Electron Transfer in the Electron Transfer Chain 90
        4.4.5 Electron Carriers...91
        4.4.6 Overall Reaction for Aerobic Respiration 91
        4.4.7 Fermentation 92
4.5 Using Energy to Put Molecules Together: Anabolic Reactions 92

4.6   Metabolism and Toxicity 94
    4.6.1   Stereochemistry and Xenobiotics Metabolism 94
Supplementary References 95
Questions and Problems 95

Chapter 5   Environmental Biological Processes and Ecotoxicology
5.1   Introduction 97
5.2   Toxicants 98
5.3   Pathways of Toxicants into Ecosystems 99
    5.3.1   Transfers of Toxicants between Environmental Spheres 100
    5.3.2   Transfers of Toxicants to Organisms 101
5.4   Bioconcentration 102
    5.4.1   Variables in Bioconcentration 103
    5.4.2   Biotransfer from Sediments 103
5.5   Bioconcentration and Biotransfer Factors 103
    5.5.1   Bioconcentration Factor 103
    5.5.2   Biotransfer Factor 105
    5.5.3   Bioconcentration by Vegetation 105
5.6   Biodegradation 105
    5.6.1   Biochemical Aspects of Biodegradation 106
    5.6.2   Cometabolism 107
    5.6.3   General Factors in Biodegradation 107
    5.6.4   Biodegradability 108
5.7   Biomarkers 108
5.8   Endocrine Disrupters and Developmental Toxicants 109
5.9   Effects of Toxicants on Populations 110
5.10 Effects of Toxicants on Ecosystems 110
Supplementary References 112
Questions and Problems 113

Chapter 6   Toxicology
6.1   Introduction 115
    6.1.1   Poisons and Toxicology 115
    6.1.2   History of Toxicology 115
    6.1.3   Future of Toxicology 116
    6.1.4   Specialized Areas of Toxicology 116
    6.1.5   Toxicological Chemistry 116
6.2   Kinds of Toxic Substances 117
6.3   Toxicity-Influencing Factors 117
    6.3.1   Classification of Factors 117
    6.3.2   Form of the Toxic Substance and Its Matrix 118
    6.3.3   Circumstances of Exposure 119
    6.3.4   The Subject 119
6.4   Exposure to Toxic Substances 120
    6.4.1   Percutaneous Exposure 121
        6.4.1.1   Skin Permeability 121
    6.4.2   Barriers to Skin Absorption 121
        6.4.2.1   Measurement of Dermal Toxicant Uptake 122
        6.4.2.2   Pulmonary Exposure 122

	6.4.3 Gastrointestinal Tract	123
	6.4.4 Mouth, Esophagus, and Stomach	123
	6.4.5 Intestines	123
	6.4.6 The Intestinal Tract and the Liver	123
6.5	Dose–Response Relationships	124
	6.5.1 Thresholds	125
6.6	Relative Toxicities	125
	6.6.1 Nonlethal Effects	125
6.7	Reversibility and Sensitivity	127
	6.7.1 Hypersensitivity and Hyposensitivity	127
6.8	Xenobiotic and Endogenous Substances	128
	6.8.1 Examples of Endogenous Substances	128
6.9	Kinetic and Nonkinetic Toxicology	129
	6.9.1 Kinetic Toxicology	129
6.10	Receptors and Toxic Substances	129
	6.10.1 Receptors	129
6.11	Phases of Toxicity	130
6.12	Toxification and Detoxification	131
	6.12.1 Synergism, Potentiation, and Antagonism	132
6.13	Behavioral and Physiological Responses	132
	6.13.1 Vital Signs	132
	6.13.2 Skin Symptoms	134
	6.13.3 Odors	134
	6.13.4 Eyes	135
	6.13.5 Mouth	135
	6.13.6 Gastrointestinal Tract	135
	6.13.7 Central Nervous System	135
6.14	Reproductive and Developmental Effects	135
References		136
Supplementary References		136
Questions and Problems		137

Chapter 7 Toxicological Chemistry

7.1	Introduction	139
	7.1.1 Chemical Nature of Toxicants	139
	7.1.2 Biochemical Transformations	140
7.2	Metabolic Reactions of Xenobiotic Compounds	141
	7.2.1 Phase I and Phase II Reactions	142
7.3	Phase I Reactions	143
	7.3.1 Oxidation Reactions	143
	7.3.2 Hydroxylation	144
	7.3.3 Epoxide Hydration	144
	7.3.4 Oxidation of Noncarbon Elements	144
	7.3.5 Alcohol Dehydrogenation	145
	7.3.6 Metabolic Reductions	147
	7.3.7 Metabolic Hydrolysis Reactions	148
	7.3.8 Metabolic Dealkylation	148
	7.3.9 Removal of Halogen	149
7.4	Phase II Reactions of Toxicants	149
	7.4.1 Conjugation by Glucuronides	150
	7.4.2 Conjugation by Glutathione	152

	7.4.3 Conjugation by Sulfate	153
	7.4.4 Acetylation	154
	7.4.5 Conjugation by Amino Acids	155
	7.4.6 Methylation	156
7.5	Biochemical Mechanisms of Toxicity	157
7.6	Interference with Enzyme Action	158
	7.6.1 Inhibition of Metalloenzymes	159
	7.6.2 Inhibition by Organic Compounds	159
7.7	Biochemistry of Mutagenesis	159
7.8	Biochemistry of Carcinogenesis	161
	7.8.1 Alkylating Agents in Carcinogenesis	163
	7.8.2 Testing for Carcinogens	163
7.9	Ionizing Radiation	163
References		164
Questions and Problems		164

Chapter 8 Genetic Aspects of Toxicology

8.1	Introduction	167
	8.1.1 Chromosomes	167
	8.1.2 Genes and Protein Synthesis	169
	8.1.3 Toxicological Importance of Nucleic Acids	169
8.2	Destructive Genetic Alterations	170
	8.2.1 Gene Mutations	170
	8.2.2 Chromosome Structural Alterations, Aneuploidy, and Polyploidy	170
	8.2.3 Genetic Alteration of Germ Cells and Somatic Cells	171
8.3	Toxicant Damage to DNA	171
8.4	Predicting and Testing for Genotoxic Substances	173
	8.4.1 Tests for Mutagenic Effects	173
	8.4.2 The Bruce Ames Test and Related Tests	175
	8.4.3 Cytogenetic Assays	175
	8.4.4 Transgenic Test Organisms	176
8.5	Genetic Susceptibilities and Resistance to Toxicants	176
8.6	Toxicogenomics	177
	8.6.1 Genetic Susceptibility to Toxic Effects of Pharmaceuticals	178
References		180
Supplementary Reference		180
Questions and Problems		180

Chapter 9 Toxic Responses

9.1	Introduction	183
9.2	Respiratory System	184
9.3	Skin	186
	9.3.1 Toxic Responses of Skin	187
	9.3.2 Phototoxic Responses of Skin	188
	9.3.3 Damage to Skin Structure and Pigmentation	188
	9.3.4 Skin Cancer	189
9.4	The Liver	189
9.5	Blood and the Cardiovascular System	192
	9.5.1 Blood	192
	9.5.2 Hypoxia	194
	9.5.3 Leukocytes and Leukemia	195

	9.5.4	Cardiotoxicants	195
	9.5.5	Vascular Toxicants	196
9.6	Immune System		196
9.7	Endocrine System		198
9.8	Nervous System		200
9.9	Reproductive System		203
9.10	Developmental Toxicology and Teratology		205
	9.10.1	Thalidomide	206
	9.10.2	Accutane	206
	9.10.3	Fetal Alcohol Syndrome	206
9.11	Kidney and Bladder		206
References			207
Supplementary References			208
Questions and Problems			208

Chapter 10 Toxic Elements

10.1	Introduction		211
10.2	Toxic Elements and the Periodic Table		211
10.3	Essential Elements		212
10.4	Metals in an Organism		212
	10.4.1	Complex Ions and Chelates	212
	10.4.2	Metal Toxicity	213
	10.4.3	Lithium	214
	10.4.4	Beryllium	215
	10.4.5	Vanadium	215
	10.4.6	Chromium	216
	10.4.7	Cobalt	216
	10.4.8	Nickel	217
	10.4.9	Cadmium	217
	10.4.10	Mercury	218
		10.4.10.1 Absorption and Transport of Elemental and Inorganic Mercury	219
		10.4.10.2 Metabolism, Biologic Effects, and Excretion	219
		10.4.10.3 Minimata Bay	220
	10.4.11	Lead	220
		10.4.11.1 Exposure and Absorption of Inorganic Lead Compounds	220
		10.4.11.2 Transport and Metabolism of Lead	221
		10.4.11.3 Manifestations of Lead Poisoning	221
		10.4.11.4 Reversal of Lead Poisoning and Therapy	222
	10.4.12	Defenses Against Heavy Metal Poisoning	222
10.5	Metalloids: Arsenic		223
	10.5.1	Sources and Uses	223
	10.5.2	Exposure and Absorption of Arsenic	224
	10.5.3	Metabolism, Transport, and Toxic Effects of Arsenic	224
10.6	Nonmetals		225
	10.6.1	Oxygen and Ozone	225
	10.6.2	Phosphorus	228
	10.6.3	The Halogens	228
		10.6.3.1 Fluorine	229
		10.6.3.2 Chlorine	229
		10.6.3.3 Bromine	229
		10.6.3.4 Iodine	230

  10.6.4 Radionuclides..230
    10.6.4.1 Radon 230
    10.6.4.2 Radium 230
    10.6.4.3 Fission Products 231
References 231
Supplementary Reference 231
Questions and Problems..231

Chapter 11 Toxic Inorganic Compounds
11.1 Introduction...235
  11.1.1 Chapter Organization..235
11.2 Toxic Inorganic Carbon Compounds 235
  11.2.1 Cyanide 235
    11.2.1.1 Biochemical Action of Cyanide..................................236
  11.2.2 Carbon Monoxide 237
  11.2.3 Biochemical Action of Carbon Monoxide 237
  11.2.4 Cyanogen, Cyanamide, and Cyanates 238
11.3 Toxic Inorganic Nitrogen Compounds 238
  11.3.1 Ammonia..238
  11.3.2 Hydrazine...238
  11.3.3 Nitrogen Oxides...239
  11.3.4 Effects of NO_2 Poisoning 239
  11.3.5 Nitrous Oxide 239
11.4 Hydrogen Halides..240
  11.4.1 Hydrogen Fluoride...240
  11.4.2 Hydrogen Chloride 240
  11.4.3 Hydrogen Bromide and Hydrogen Iodide.....................................240
11.5 Interhalogen Compounds and Halogen Oxides 240
  11.5.1 Interhalogen Compounds 241
  11.5.2 Halogen Oxides 241
  11.5.3 Hypochlorous Acid and Hypochlorites 242
  11.5.4 Perchlorates..242
11.6 Nitrogen Compounds of the Halogens 242
  11.6.1 Nitrogen Halides..242
  11.6.2 Azides 243
  11.6.3 Monochloramine and Dichloramine..243
11.7 Inorganic Compounds of Silicon 243
  11.7.1 Silica 243
  11.7.2 Asbestos...244
  11.7.3 Silanes..244
  11.7.4 Silicon Halides and Halohydrides 245
11.8 Inorganic Phosphorus Compounds..245
  11.8.1 Phosphine...245
  11.8.2 Phosphorus Pentoxide..245
  11.8.3 Phosphorus Halides 246
  11.8.4 Phosphorus Oxyhalides 246
11.9 Inorganic Compounds of Sulfur..246
  11.9.1 Hydrogen Sulfide...246
  11.9.2 Sulfur Dioxide and Sulfites 247
  11.9.3 Sulfuric Acid..248

11.9.4	Carbon Disulfide	248
11.9.5	Miscellaneous Inorganic Sulfur Compounds	249

References 249
Questions and Problems..250

Chapter 12 Organometallics and Organometalloids

12.1	The Nature of Organometallic and Organometalloid Compounds	253
12.2	Classification of Organometallic Compounds	253
	12.2.1 Ionically Bonded Organic Groups	254
	12.2.2 Organic Groups Bonded with Classical Covalent Bonds	254
	12.2.3 Organometallic Compounds with Dative Covalent Bonds	256
	12.2.4 Organometallic Compounds Involving π-Electron Donors	257
12.3	Mixed Organometallic Compounds	257
12.4	Organometallic Compound Toxicity	258
12.5	Compounds of Group 1A Metals	258
	12.5.1 Lithium Compounds	258
	12.5.2 Compounds of Group 1A Metals Other Than Lithium	259
12.6	Compounds of Group 2A Metals	260
	12.6.1 Magnesium	260
	12.6.2 Calcium, Strontium, and Barium	261
12.7	Compounds of Group 2B Metals	261
	12.7.1 Zinc	261
	12.7.2 Cadmium	263
	12.7.3 Mercury	263
12.8	Organotin and Organogermanium Compounds	264
	12.8.1 Toxicology of Organotin Compounds	265
	12.8.2 Organogermanium Compounds	266
12.9	Organolead Compounds	266
	12.9.1 Toxicology of Organolead Compounds	266
12.10	Organoarsenic Compounds	267
	12.10.1 Organoarsenic Compounds from Biological Processes	267
	12.10.2 Synthetic Organoarsenic Compounds	268
	12.10.3 Toxicities of Organoarsenic Compounds	269
12.11	Organoselenium and Organotellurium Compounds	270
	12.11.1 Organoselenium Compounds	270
	12.11.2 Organotellurium Compounds	270

References 270
Supplementary References...271
Questions and Problems..271

Chapter 13 Toxic Organic Compounds and Hydrocarbons

13.1	Introduction	273
13.2	Classification of Hydrocarbons	273
	13.2.1 Alkanes	273
	13.2.2 Unsaturated Nonaromatic Hydrocarbons	275
	13.2.3 Aromatic Hydrocarbons	275
13.3	Toxicology of Alkanes	276
	13.3.1 Methane and Ethane	276
	13.3.2 Propane and Butane	277
	13.3.3 Pentane through Octane	277

	13.3.4	Alkanes above Octane	278
	13.3.5	Solid and Semisolid Alkanes	278
	13.3.6	Cyclohexane	278
13.4	Toxicology of Unsaturated Nonaromatic Hydrocarbons		279
	13.4.1	Propylene	280
	13.4.2	1,3-Butadiene	280
	13.4.3	Butylenes	282
	13.4.4	Alpha-Olefins	282
	13.4.5	Cyclopentadiene and Dicyclopentadiene	282
	13.4.6	Acetylene	283
13.5	Benzene and Its Derivatives		283
	13.5.1	Benzene	283
		13.5.1.1 Acute Toxic Effects of Benzene	283
		13.5.1.2 Chronic Toxic Effects of Benzene	284
		13.5.1.3 Metabolism of Benzene	284
	13.5.2	Toluene, Xylenes, and Ethylbenzene	285
	13.5.3	Styrene	286
13.6	Naphthalene		287
	13.6.1	Metabolism of Naphthalene	288
	13.6.2	Toxic Effects of Naphthalene	288
13.7	Polycyclic Aromatic Hydrocarbons		288
	13.7.1	PAH Metabolism	289
References			290
Questions and Problems			290

Chapter 14 Organooxygen Compounds

14.1	Introduction		293
	14.1.1	Oxygen-Containing Functional Groups	293
14.2	Alcohols		293
	14.2.1	Methanol	293
	14.2.2	Ethanol	295
	14.2.3	Ethylene Glycol	296
	14.2.4	The Higher Alcohols	296
14.3	Phenols		297
	14.3.1	Properties and Uses of Phenols	297
	14.3.2	Toxicology of Phenols	298
14.4	Oxides		298
14.5	Formaldehyde		299
	14.5.1	Properties and Uses of Formaldehyde	300
	14.5.2	Toxicity of Formaldehyde and Formalin	300
14.6	Aldehydes and Ketones		300
	14.6.1	Toxicities of Aldehydes and Ketones	301
14.7	Carboxylic Acids		302
	14.7.1	Toxicology of Carboxylic Acids	302
14.8	Ethers		303
	14.8.1	Examples and Uses of Ethers	303
	14.8.2	Toxicities of Ethers	304
14.9	Acid Anhydrides		304
	14.9.1	Toxicological Considerations	305
14.10	Esters		305
	14.10.1	Toxicities of Esters	306

References 306
Questions and Problems 307

Chapter 15   Organonitrogen Compounds
15.1   Introduction 309
15.2   Nonaromatic Amines 309
        15.2.1   Lower Aliphatic Amines 309
        15.2.2   Fatty Amines 310
        15.2.3   Alkyl Polyamines 310
        15.2.4   Cyclic Amines 311
15.3   Carbocyclic Aromatic Amines 311
        15.3.1   Aniline 311
        15.3.2   Benzidine 313
        15.3.3   Naphthylamines 313
15.4   Pyridine and Its Derivatives 314
15.5   Nitriles 314
15.6   Nitro Compounds 315
        15.6.1   Nitro Alcohols and Nitro Phenols 316
        15.6.2   Dinoseb 316
15.7   Nitrosamines 317
15.8   Isocyanates and Methyl Isocyanate 318
15.9   Pesticidal Compounds 320
        15.9.1   Carbamates 320
        15.9.2   Bipyridilium Compounds 321
15.10 Alkaloids 322
References 323
Questions and Problems 324

Chapter 16   Organohalide Compounds
16.1   Introduction 327
        16.1.1   Biogenic Organohalides 327
16.2   Alkyl Halides 328
        16.2.1   Toxicities of Alkyl Halides 328
        16.2.2   Toxic Effects of Carbon Tetrachloride on the Liver 329
        16.2.3   Other Alkyl Halides 330
        16.2.4   Hydrochlorofluorocarbons 330
        16.2.5   Halothane 331
16.3   Alkenyl Halides 332
        16.3.1   Uses of Alkenyl Halides 332
        16.3.2   Toxic Effects of Alkenyl Halides 333
        16.3.3   Hexachlorocyclopentadiene 335
16.4   Aryl Halides 336
        16.4.1   Properties and Uses of Aryl Halides 336
        16.4.2   Toxic Effects of Aryl Halides 337
16.5   Organohalide Insecticides 338
        16.5.1   Toxicities of Organohalide Insecticides 338
        16.5.2   Hexachlorocyclohexane 340
        16.5.3   Toxaphene 340
16.6   Noninsecticidal Organohalide Pesticides 341
        16.6.1   Toxic Effects of Chlorophenoxy Herbicides 342
        16.6.2   Toxicity of TCDD 342

  16.6.3 Alachlor 342
  16.6.4 Chlorinated Phenols 343
  16.6.5 Hexachlorophene 344
References 344
Questions and Problems..344

Chapter 17 Organosulfur Compounds

17.1 Introduction 347
  17.1.1 Classes of Organosulfur Compounds......................................347
  17.1.2 Reactions of Organic Sulfur..347
17.2 Thiols, Sulfides, and Disulfides 349
  17.2.1 Thiols 349
  17.2.2 Thiols as Antidotes for Heavy Metal Poisoning 350
  17.2.3 Sulfides and Disulfides..350
  17.2.4 Organosulfur Compounds in Skunk Spray.............................351
  17.2.5 Carbon Disulfide and Carbon Oxysulfide...............................351
17.3 Organosulfur Compounds Containing Nitrogen or Phosphorus 352
  17.3.1 Thiourea Compounds 352
  17.3.2 Thiocyanates 353
  17.3.3 Disulfiram 354
  17.3.4 Cyclic Sulfur and Nitrogen Organic Compounds 354
  17.3.5 Dithiocarbamates 355
  17.3.6 Phosphine Sulfides 355
  17.3.7 Phosphorothionate and Phosphorodithioate Esters..................356
17.4 Sulfoxides and Sulfones..356
17.5 Sulfonic Acids, Salts, and Esters 357
17.6 Organic Esters of Sulfuric Acid..358
17.7 Miscellaneous Organosulfur Compounds 359
  17.7.1 Sulfur Mustards 359
  17.7.2 Sulfur in Pesticides..359
  17.7.3 Sulfa Drugs 360
17.8 Organically Bound Selenium 360
References 360
Questions and Problems..360

Chapter 18 Organophosphorus Compounds

18.1 Introduction 363
  18.1.1 Phosphine..363
18.2 Alkyl and Aryl Phosphines 363
18.3 Phosphine Oxides and Sulfides...365
18.4 Phosphonic and Phosphorous Acid Esters 365
18.5 Organophosphate Esters..366
  18.5.1 Orthophosphates and Polyphosphates.....................................366
  18.5.2 Orthophosphate Esters...367
  18.5.3 Aromatic Phosphate Esters 368
  18.5.4 Tetraethylpyrophosphate...368
18.6 Phosphorothionate and Phosphorodithioate Esters.........................368
18.7 Organophosphate Insecticides 369
  18.7.1 Chemical Formulas and Properties 369
  18.7.2 Phosphate Ester Insecticides 369
  18.7.3 Phosphorothionate Insecticides 371

		18.7.4	Phosphorodithioate Insecticides	372

        18.7.4   Phosphorodithioate Insecticides...372
        18.7.5   Toxic Actions of Organophosphate Insecticides...374
                18.7.5.1   Inhibition of Acetylcholinesterase...374
                18.7.5.2   Metabolic Activation..375
                18.7.5.3   Mammalian Toxicities...375
                18.7.5.4   Deactivation of Organophosphates..376
18.8   Organophosphorus Military Poisons...377
References...378
Supplementary Reference..379
Questions and Problems...379

Chapter 19   Toxic Natural Products

19.1   Introduction...383
19.2   Toxic Substances from Bacteria..384
        19.2.1   *In Vivo* Bacterial Toxins..384
                19.2.1.1   Toxic Shock Syndrome...385
        19.2.2   Bacterial Toxins Produced Outside the Body..385
19.3   Mycotoxins..385
        19.3.1   Aflatoxins..386
        19.3.2   Other Mycotoxins..386
        19.3.3   Mushroom Toxins...387
19.4   Toxins from Protozoa..387
19.5   Toxic Substances from Plants...388
        19.5.1   Nerve Toxins from Plants...389
                19.5.1.1   Pyrethrins and Pyrethroids..390
        19.5.2   Internal Organ Plant Toxins...390
        19.5.3   Eye and Skin Irritants...391
        19.5.4   Allergens...392
        19.5.5   Mineral Accumulators..392
        19.5.6   Toxic Algae..393
19.6   Insect Toxins...393
        19.6.1   Bee Venom..393
        19.6.2   Wasp and Hornet Venoms..394
        19.6.3   Toxicities of Insect Venoms...394
19.7   Spider Toxins..394
        19.7.1   Brown Recluse Spiders...394
        19.7.2   Widow Spiders...394
        19.7.3   Other Spiders..395
19.8   Reptile Toxins...395
        19.8.1   Chemical Composition of Snake Venoms..395
        19.8.2   Toxic Effects of Snake Venom..395
19.9   Nonreptile Animal Toxins...396
References...397
Supplementary References..398
Questions and Problems...398

Chapter 20   Analysis of Xenobiotics

20.1   Introduction...401
20.2   Indicators of Exposure to Xenobiotics...401
20.3   Determination of Metals...402

        20.3.1 Direct Analysis of Metals 402
        20.3.2 Metals in Wet-Ashed Blood and Urine 403
        20.3.3 Extraction of Metals for Atomic Absorption Analysis 403
20.4 Determination of Nonmetals and Inorganic Compounds 403
20.5 Determination of Parent Organic Compounds 404
20.6 Measurement of Phase I and Phase II Reaction Products 404
        20.6.1 Phase I Reaction Products 404
        20.6.2 Phase II Reaction Products 406
        20.6.3 Mercapturates 407
20.7 Determination of Adducts 407
20.8 The Promise of Immunological Methods 408
References 409
Supplementary References 409
Questions and Problems 410

Index 413

CHAPTER 1

Chemistry and Organic Chemistry

1.1 INTRODUCTION

This book is about toxicological chemistry, the branch of chemical science dealing with the toxic effects of substances. In order to understand this topic, it is essential to have an understanding of **chemistry**, the science of matter. The nature of toxic substances depends upon their chemical characteristics, how they are bonded together, and how they react. Mechanisms of toxicity are basically chemical in nature. Chemical processes carried out by organisms play a strong role in determining the fates of toxic substances. In some cases, chemical modification of toxicants by organisms reduces the toxicity of chemical substances or makes them entirely nontoxic. In other cases, chemical activation of foreign compounds makes them more toxic. For example, benzo(a)pyrene, a substance produced by the partial combustion of organic matter, such as that which occurs when smoking cigarettes, is not itself toxic, but it reacts with oxygen through the action of enzymes in the body to produce a species that can bind with DNA and cause cancer.

The chemical processes that occur in organisms are addressed by biochemistry, which is discussed in Chapter 3. In order to understand biochemistry, however, it is essential to have a basic understanding of chemistry. Since most substances in living organisms, as well as most toxic substances, are organic materials containing carbon, it is also essential to have an understanding of organic chemistry in order to consider toxicological chemistry. Therefore, this chapter starts with a brief overview of chemistry and includes the basic principles of organic chemistry as well.

It is important to consider the effects of toxic substances within the context of the environment through which exposure of various organisms occurs. Furthermore, toxic substances are created, altered, or detoxified by environmental chemical processes in water, in soil, and when substances are exposed to the atmosphere. Therefore, Chapter 2 deals with environmental chemistry and environmental chemical processes. The relationship of toxic substances and the organisms that they affect in the environment is addressed specifically by ecotoxicology in Chapter 5.

1.2 ELEMENTS

All substances are composed of only about a hundred fundamental kinds of matter called **elements**. Elements themselves may be of environmental and toxicological concern. The heavy metals, including lead, cadmium, and mercury, are well recognized as toxic substances in the environment. Elemental forms of otherwise essential elements may be very toxic or cause environmental damage. Oxygen in the form of ozone, O_3, is the agent most commonly associated with atmospheric smog pollution and is very toxic to plants and animals. Elemental white phosphorus is highly flammable and toxic.

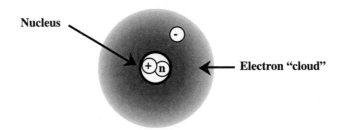

Figure 1.1 Representation of a deuterium atom. The nucleus contains one proton (+) and one neutron (n). The electron (−) is in constant, rapid motion around the nucleus, forming a cloud of negative electrical charge, the density of which drops off with increasing distance from the nucleus.

Table 1.1 Properties of Protons, Neutrons, and Electrons

Subatomic Particle	Symbol[a]	Unit Charge	Mass Number	Mass in μ	Mass in Grams
Proton	p	+1	1	1.007277	1.6726×10^{-24}
Neutron	n	0	1	1.008665	1.6749×10^{-24}
Electron	e	−1	0	0.000549	9.1096×10^{-28}

[a] The mass number and charge of each of these kinds of particles can be indicated by a superscript and subscript, respectively, in the symbols $^{1}_{1}p$, $^{1}_{0}n$, $^{0}_{-1}e$.

Each element is made up of very small entities called **atoms**; all atoms of the same element behave identically chemically. The study of chemistry, therefore, can logically begin with elements and the atoms of which they are composed. Each element is designated by an atomic number, a name, and a **chemical symbol**, such as carbon, C; potassium, K (for its Latin name kalium); or cadmium, Cd. Each element has a characteristic **atomic mass** (atomic weight), which is the average mass of all atoms of the element.

1.2.1 Subatomic Particles and Atoms

Figure 1.1 represents an atom of deuterium, a form of the element hydrogen. As shown, such an atom is made up of even smaller **subatomic particles**: positively charged **protons**, negatively charged **electrons**, and uncharged (neutral) **neutrons**.

1.2.2 Subatomic Particles

The subatomic particles differ in mass and charge. Their masses are expressed by the **atomic mass unit**, u (also called the **dalton**), which is also used to express the masses of individual atoms, and molecules (aggregates of atoms). The atomic mass unit is defined as a mass equal to exactly 1/12 that of an atom of carbon-12, the isotope of carbon that contains six protons and six neutrons in its nucleus.

The proton, p, has a mass of 1.007277 u and a unit charge of +1. This charge is equal to 1.6022×10^{-19} coulombs; a coulomb is the amount of electrical charge involved in a flow of electrical current of 1 ampere for 1 sec. The neutron, n, has no electrical charge and a mass of 1.008665 u. The proton and neutron each have a mass of essentially 1 u and are said to have a *mass number* of 1. (Mass number is a useful concept expressing the total number of protons and neutrons, as well as the approximate mass, of a nucleus or subatomic particle.) The electron, e, has an electrical charge of −1. It is very light, however, with a mass of only 0.000549 u, about 1/1840 that of the proton or neutron. Its mass number is 0. The properties of protons, neutrons, and electrons are summarized in Table 1.1.

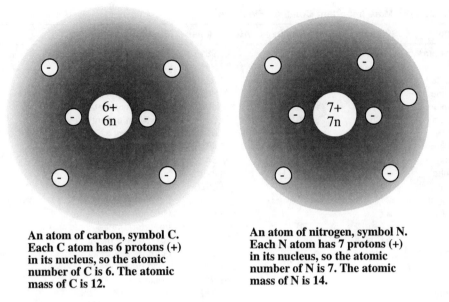

Figure 1.2 Atoms of carbon and nitrogen.

Although it is convenient to think of the proton and neutron as having the same mass, and each is assigned a mass number of 1, Table 1.1 shows that their exact masses differ slightly from each other. Furthermore, the mass of an atom is not exactly equal to the sum of the masses of subatomic particles composing the atom. This is because of the energy relationships involved in holding the subatomic particles together in an atom so that the masses of the atom's constituent subatomic particles do not add up to exactly the mass of the atom.

1.2.3 Atom Nucleus and Electron Cloud

Protons and neutrons are contained in the positively charged **nucleus** of the atom. Protons and neutrons have relatively high masses compared to electrons. Therefore, the nucleus has essentially all of the mass, but occupies virtually none of the volume, of the atom. An uncharged atom has the same number of electrons as protons. The electrons in an atom are contained in a cloud of negative charge around the nucleus that occupies most of the volume of the atom. These concepts are illustrated in Figure 1.2.

1.2.4 Isotopes

Atoms with the *same* number of protons, but *different* numbers of neutrons in their nuclei are chemically identical atoms of the same element, but have different masses and may differ in their nuclear properties. Such atoms are **isotopes** of the same element. Some isotopes are **radioactive isotopes**, or **radionuclides**, which have unstable nuclei that give off charged particles and gamma rays in the form of **radioactivity**. Radioactivity may have detrimental, or even fatal, health effects; a number of hazardous substances are radioactive, and they can cause major environmental problems. The most striking example of such contamination resulted from a massive explosion and fire at a power reactor in the Ukrainian city of Chernobyl in 1986.

1.2.5 Important Elements

An abbreviated list of a few of the most important elements, which the reader may find useful, is given in Table 1.2. A complete list of the well over 100 known elements is given on the inside front

Table 1.2 The More Important Common Elements

Element	Symbol	Atomic Number	Atomic Mass	Significance
Aluminum	Al	13	26.9815	Abundant in Earth's crust
Argon	Ar	18	39.948	Noble gas
Arsenic	As	33	74.9216	Toxic metalloid
Bromine	Br	35	79.904	Toxic halogen
Cadmium	Cd	48	112.40	Toxic heavy metal
Calcium	Ca	20	40.08	Abundant essential element
Carbon	C	6	12.011	Life element
Chlorine	Cl	17	35.453	Halogen
Copper	Cu	29	63.54	Useful metal
Fluorine	F	9	18.998	Halogen
Helium	He	2	4.0026	Lightest noble gas
Hydrogen	H	1	1.008	Lightest element
Iodine	I	53	126.904	Halogen
Iron	Fe	26	55.847	Important metal
Lead	Pb	82	207.19	Toxic heavy metal
Magnesium	Mg	12	24.305	Light metal
Mercury	Hg	80	200.59	Toxic heavy metal
Neon	Ne	10	20.179	Noble gas
Nitrogen	N	7	14.0067	Important nonmetal
Oxygen	O	8	15.9994	Abundant, essential nonmetal
Phosphorus	P	15	30.9738	Essential nonmetal
Potassium	K	19	39.0983	Alkali metal
Silicon	Si	14	28.0855	Abundant metalloid
Silver	Ag	47	107.87	Valuable, reaction-resistant metal
Sodium	Na	11	22.9898	Essential, abundant alkali metal
Sulfur	S	16	32.064	Essential element, occurs in air pollutant sulfur dioxide, SO_2
Tin	Sn	50	118.69	Useful metal
Uranium	U	92	238.03	Fissionable metal used for nuclear fuel
Zinc	Zn	30	65.37	Useful metal

cover of this book. Fortunately, most of the chemistry covered in this book requires familiarity only with the shorter list of elements in Table 1.2.

1.2.6 The Periodic Table

The properties of elements listed in order of increasing atomic number repeat in a periodic manner. For example, elements with atomic numbers 2, 10, and 18 are gases that do not undergo chemical reactions and consist of individual atoms, whereas those with atomic numbers larger by 1 — elements with atomic numbers 3, 11, and 19 — are unstable, highly reactive metals. An arrangement of the elements reflecting this recurring behavior is the **periodic table** (Figure 1.3). This table is extremely useful in understanding chemistry and predicting chemical behavior because it organizes the elements in a systematic manner related to their chemical behavior as a consequence of the structures of the atoms that compose the elements. As shown in Figure 1.3, the entry for each element in the periodic table gives the element's atomic number, symbol, and atomic mass. More detailed versions of the table include other information as well.

1.2.6.1 Features of the Periodic Table

Groups of elements having similar chemical behavior are contained in vertical columns in the periodic table. **Main group** elements may be designated as A groups (IA and IIA on the left, IIIA through VIIIA on the right). **Transition elements** are those between main groups IIA and IIIA. **Noble gases** (group VIIIA), a group of gaseous elements that are virtually chemically unreactive,

CHEMISTRY AND ORGANIC CHEMISTRY

Noble gases

Period	IA 1	IIA 2	IIIB 3	IVB 4	VB 5	VIB 6	VIIB 7	VIIIB 8	VIIIB 9	VIIIB 10	IB 11	IIB 12	IIIA 13	IVA 14	VA 15	VIA 16	VIIA 17	VIII 18
1	1 H 1.008																	2 He 4.003
2	3 Li 6.941	4 Be 9.012											5 B 10.81	6 C 12.01	7 N 14.01	8 O 16.00	9 F 19.00	10 Ne 20.18
3	11 Na 22.99	12 Mg 24.3											13 Al 26.98	14 Si 28.09	15 P 30.97	16 S 32.07	17 Cl 35.45	18 Ar 39.95
4	19 K 39.10	20 Ca 40.08	21 Sc 44.96	22 Ti 47.88	23 V 50.94	24 Cr 52.00	25 Mn 54.94	26 Fe 55.85	27 Co 58.93	28 Ni 58.69	29 Cu 63.55	30 Zn 65.39	31 Ga 69.72	32 Ge 72.59	33 As 74.92	34 Se 78.96	35 Br 79.9	36 Kr 83.8
5	37 Rb 85.47	38 Sr 87.62	39 Y 88.91	40 Zr 91.22	41 Nb 92.91	42 Mo 95.94	43 Tc 98.91	44 Ru 101.1	45 Rh 102.9	46 Pd 106.4	47 Ag 107.9	48 Cd 112.4	49 In 114.8	50 Sn 118.7	51 Sb 121.8	52 Te 127.6	53 I 126.9	54 Xe 131.3
6	55 Cs 132.9	56 Ba 137.3	57* La 138.9	72 Hf 178.5	73 Ta 180.9	74 W 183.8	75 Re 186.2	76 Os 190.2	77 Ir 192.2	78 Pt 195.1	79 Au 197.0	80 Hg 200.6	81 Tl 204.4	82 Pb 207.2	83 Bi 209.0	84 Po (210)	85 At (210)	86 Rn (222)
7	87 Fr (223)	88 Ra (226)	89** Ac (227)	104 Rf (261)	105 Ha (262)	106 Sg (263)	107 Ns (262)	108 Ha (265)	109 Mt (266)									

Inner Transition Elements

Lanthanide series *6

58 Ce 140.1	59 Pr 140.9	60 Nd 144.2	61 Pm 144.9	62 Sm 150.4	63 Eu 152.0	64 Gd 157.2	65 Tb 158.9	66 Dy 162.5	67 Ho 164.9	68 Er 167.3	69 Tm 168.9	70 Yb 173.0	71 Lu 175.0

Actinide series **7

90 Th 232.0	91 Pa 231.0	92 U 238.0	93 Np 237.0	94 Pu 239.1	95 Am 243.1	96 Cm 247.1	97 Bk 247.1	98 Cf 252.1	99 Es 252.1	100 Fm 257.1	101 Md 256.1	102 No 259.1	103 Lr 260.1

Figure 1.3 The periodic table of elements.

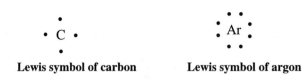

Figure 1.4 Lewis symbols of carbon and argon.

are in the far right column. The chemical similarities of elements in the same group are especially pronounced for groups IA, IIA, VIIA, and VIIIA.

Horizontal rows of elements in the periodic table are called **periods**, the first of which consists of only hydrogen (H) and helium (He). The second period begins with atomic number 3 (lithium) and terminates with atomic number 10 (neon), whereas the third goes from atomic number 11 (sodium) through atomic number 18 (argon). The fourth period includes the first row of transition elements, whereas lanthanides and actinides, which occur in the sixth and seventh periods, respectively, are listed separately at the bottom of the table.

1.2.7 Electrons in Atoms

Although the placement of electrons in atoms determines how the atoms behave chemically and, therefore, the chemical properties of each element, it is beyond the scope of this book to discuss electronic structure in detail. Several key points pertaining to this subject are mentioned here.

Electrons in atoms occupy **orbitals** in which electrons have different energies, orientations in space, and average distances from the nucleus. Each orbital may contain a maximum of two electrons. The chemical behavior of an atom is determined by the placement of electrons in its orbitals; in this respect, the outermost orbitals and the electrons contained in them are the most important. These **outer electrons** are the ones beyond those of the immediately preceding noble gas in the periodic table. They are of particular importance because they become involved in the sharing and transfer of electrons through which chemical bonding occurs, resulting in the formation of huge numbers of different substances from only a few elements.

1.2.7.1 Lewis Symbols of Atoms

Outer electrons are called **valence electrons** and are represented by dots in **Lewis symbols**, as shown for carbon and argon in Figure 1.4.

The four electrons shown for the carbon atom are those added beyond the electrons possessed by the noble gas that immediately precedes carbon in the periodic table (helium, atomic number 2). Eight electrons are shown around the symbol of argon. This is an especially stable electron configuration for noble gases known as an **octet**. (Helium is the exception among noble gases in that it has a stable shell of only two electrons.) When atoms interact through the sharing, loss, or gain of electrons to form molecules and chemical compounds (see Section 1.3), many attain an octet of outer-shell electrons. This tendency is the basis of the **octet rule** of chemical bonding. (Two or three of the lightest elements, most notably hydrogen, attain stable helium-like electron configurations containing two electrons when they become chemically bonded.)

1.2.8 Metals, Nonmetals, and Metalloids

Elements are divided between metals and nonmetals; a few elements with an intermediate character are called metalloids. **Metals** are elements that are generally solid, shiny in appearance, electrically conducting, and malleable — that is, they can be pounded into flat sheets without disintegrating. They tend to have only one to three outer electrons, which they may lose in forming chemical compounds. Examples of metals are iron, copper, and silver. Most metallic objects that

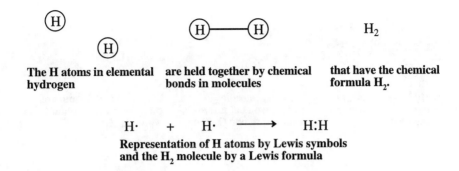

Figure 1.5 Molecule and Lewis formula of H_2.

are commonly encountered are not composed of just one kind of elemental metal, but are alloys consisting of homogeneous mixtures of two or more metals. **Nonmetals** often have a dull appearance, are not at all malleable, and frequently occur as gases or liquids. Colorless oxygen gas, green chlorine gas (transported and stored as a liquid under pressure), and brown bromine liquid are common nonmetals. Nonmetals tend to have close to a full octet of outer-shell electrons, and in forming chemical compounds, they gain or share electrons. **Metalloids**, such as silicon or arsenic, may have properties of both, in some respects behaving like metals, in other respects behaving like nonmetals.

1.3 CHEMICAL BONDING

Only a few elements, particularly the noble gases, exist as individual atoms; most atoms are joined by chemical bonds to other atoms. This can be illustrated very simply by elemental hydrogen, which exists as **molecules**, each consisting of two H atoms linked by a **chemical bond,** as shown in Figure 1.5. Because hydrogen molecules contain two H atoms, they are said to be diatomic and are denoted by the **chemical formula** H_2. The H atoms in the H_2 molecule are held together by a **covalent bond** made up of two electrons, each contributed by one of the H atoms and shared between the atoms. (Bonds formed by transferring electrons between atoms are described later in this section.) The shared electrons in the covalent bonds holding the H_2 molecule together are represented by two dots between the H atoms in Figure 1.5. By analogy with Lewis symbols defined in the preceding section, such a representation of molecules showing outer-shell and bonding electrons as dots is called a **Lewis formula**.

1.3.1 Chemical Compounds

Most substances consist of two or more elements joined by chemical bonds. For example, consider the chemical combination of the elements hydrogen and oxygen shown in Figure 1.6. Oxygen, chemical symbol O, has an atomic number of 8 and an atomic mass of 16, and it exists in the elemental form as diatomic molecules of O_2. Hydrogen atoms combine with oxygen atoms to form molecules in which two H atoms are bonded to one O atom in a substance with a chemical formula of H_2O (water). A substance such as H_2O that consists of a chemically bonded combination of two or more elements is called a **chemical compound**. In the chemical formula for water the letters H and O are the chemical symbols of the two elements in the compound and the subscript 2 indicates that there are two H atoms per one O atom. (The absence of a subscript after the O denotes the presence of just one O atom in the molecule.)

As shown in Figure 1.6, each of the hydrogen atoms in the water molecule is connected to the oxygen atom by a chemical bond composed of two electrons shared between the hydrogen and oxygen atoms. For each bond, one electron is contributed by the hydrogen and one by oxygen. The

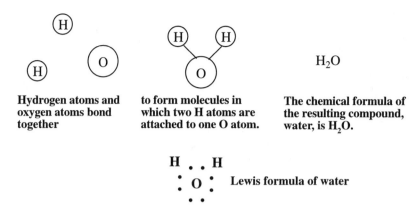

Figure 1.6 Formation and Lewis formula of a chemical compound, water.

two dots located between each H and O in the Lewis formula of H_2O represent the two electrons in the covalent bond joining these atoms. Four of the electrons in the octet of electrons surrounding O are involved in H–O bonds and are called bonding electrons. The other four electrons shown around the oxygen that are not shared with H are nonbonding outer electrons.

1.3.2 Molecular Structure

As implied by the representations of the water molecule in Figure 1.6, the atoms and bonds in H_2O form an angle somewhat greater than 90°. The shapes of molecules are referred to as their **molecular geometry**, which is crucial in determining the chemical and toxicological activity of a compound and structure-activity relationships.

1.3.3 Ionic Bonds

As shown by the example of magnesium oxide in Figure 1.7, the transfer of electrons from one atom to another produces charged species called **ions**. Positively charged ions are called **cations**, and negatively charged ions are called **anions**. Ions that make up a solid compound are held together by **ionic bonds** in a **crystalline lattice** consisting of an ordered arrangement of the ions in which each cation is largely surrounded by anions and each anion by cations. The attracting forces of the oppositely charged ions in the crystalline lattice constitute ionic bonds in the compound.

The formation of magnesium oxide is shown in Figure 1.7. In naming this compound, the cation is simply given the name of the element from which it was formed, magnesium. However, the ending of the name of the anion, *oxide*, is different from that of the element from which it was formed, *oxygen*.

Rather than individual atoms that have lost or gained electrons, many ions are groups of atoms bonded together covalently and have a net charge. A common example of such an ion is the ammonium ion, NH_4^+:

It consists of four hydrogen atoms covalently bonded to a single nitrogen (N) atom, and it has a net electrical charge of +1 for the whole cation, as shown by its Lewis formula above.

CHEMISTRY AND ORGANIC CHEMISTRY

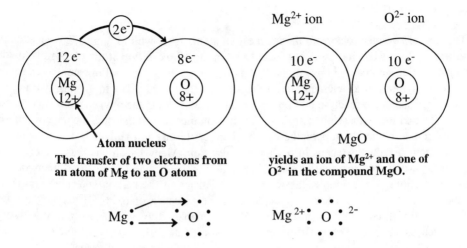

Figure 1.7 Ionic bonds are formed by the transfer of electrons and the mutual attraction of oppositely charged ions in a crystalline lattice.

1.3.4 Summary of Chemical Compounds and the Ionic Bond

The preceding several pages have just covered material on chemical compounds and bonds that is essential for understanding chemistry. To summarize:

- Atoms of two or more different elements can form *chemical bonds* with each other to yield a product that is entirely different from the elements. Such a substance is called a *chemical compound*.
- The *formula* of a chemical compound gives the symbols of the elements and uses subscripts to show the relative numbers of atoms of each element in the compound.
- *Molecules* of some compounds are held together by *covalent bonds* consisting of shared electrons.
- Another kind of compound is composed of *ions* consisting of electrically charged atoms or groups of atoms held together by *ionic bonds* that exist because of the mutual attraction of oppositely charged ions.

1.3.5 Molecular Mass

The average mass of all molecules of a compound is its **molecular mass** (formerly called molecular weight). The molecular mass of a compound is calculated by multiplying the atomic mass of each element by the relative number of atoms of the element, then adding all the values obtained for each element in the compound. For example, the molecular mass of NH_3 is $14.0 + 3 \times 1.0 = 17.0$. For another example, consider the following calculation of the molecular mass of ethylene, C_2H_4:

1. The chemical formula of the compound is C_2H_4.
2. Each molecule of C_2H_4 consists of two C atoms and four H atoms.
3. From the periodic table or Table 1.2, the atomic mass of C is 12.0 and that of H is 1.0.
4. Therefore, the molecular mass of C_2H_4 is

$$12.0 + 12.0 + 1.0 + 1.0 + 1.0 + 1.0 = 28.0$$

$\underbrace{\hphantom{12.0 + 12.0}}_{\text{From 2 C atoms}} \quad \underbrace{\hphantom{1.0 + 1.0 + 1.0 + 1.0}}_{\text{From 4 H atoms}}$

1.3.6 Oxidation State

The loss of two electrons from the magnesium atom, as shown in Figure 1.7, is an example of **oxidation**, and the Mg^{2+} ion product is said to be in the +2 **oxidation state**. (A positive oxidation state or oxidation number is conventionally denoted by a roman numeral in parentheses following the name or symbol of an element, as in magnesium(II) and Mg(II).) In gaining two negatively charged electrons in the reaction that produces magnesium oxide, the oxygen atom is **reduced** and is in the –2 oxidation state. (Unlike positive oxidation numbers, negative ones are not conventionally shown by roman numerals in parentheses.) In chemical terms, an **oxidizer** is a species that takes electrons from a reducing agent in a chemical reaction. Many hazardous waste substances are oxidizers or strong reducers, and oxidation–reduction is the driving force behind many dangerous chemical reactions. For example, the reducing tendencies of the carbon and hydrogen atoms in propane cause it to burn violently or explode in the presence of oxidizing oxygen in air. The oxidizing ability of concentrated nitric acid, HNO_3, enables it to react destructively with organic matter, such as cellulose or skin. Strong oxidants such as 30% hydrogen peroxide, H_2O_2, are classified as corrosive poisons because of their ability to attack exposed tissue.

Covalently bonded atoms that have not actually lost or gained electrons to produce ions are also assigned oxidation states. This can be done because in covalent compounds electrons are not shared equally. Therefore, an atom of an element with a greater tendency to attract electrons is assigned a negative oxidation state, compared to the positive oxidation state assigned to an element with a lesser tendency to attract electrons. For example, Cl atoms attract electrons more strongly than H atoms do, so in hydrogen chloride gas, HCl, the Cl atom is in the –1 oxidation state and the H atoms are in the +1 oxidation state. **Electronegativity** values are assigned to elements on the basis of their tendencies to attract electrons.

The oxidation state (oxidation number) of an element in a compound may have a strong influence on the hazards and toxicities posed by the compound. For example, chromium from which each atom has lost three electrons to form a chemical compound, designated as chromium(III) or Cr(III), is not toxic, whereas chromium in the +6 oxidation state (Cr(VI), chromate) is regarded as a cancer-causing chemical when inhaled.

1.4 CHEMICAL REACTIONS AND EQUATIONS

Chemical reactions occur when substances are changed to other substances through the breaking and formation of chemical bonds. For example, water is produced by the chemical reaction of hydrogen and oxygen:

Hydrogen plus oxygen yields water

Chemical reactions are written as **chemical equations**. The chemical reaction between hydrogen and water is written as the **balanced chemical equation**

$$2H_2 + O_2 \rightarrow 2H_2O \qquad (1.4.1)$$

in which the arrow is read as "yields" and separates the hydrogen and oxygen **reactants** from the water **product**. Note that because elemental hydrogen and elemental oxygen occur as *diatomic molecules* of H_2 and O_2, respectively, it is necessary to write the equation in a way that reflects these correct chemical formulas of the elemental form. All correctly written chemical equations are **balanced**. For a chemical equation to be properly balanced, *the same number of each kind of atom must be shown on both sides of the equation*. The equation above is balanced because of the following:

On the left:
- There are two H_2 *molecules*, each containing two H *atoms*, for a total of four H atoms on the left.
- There is one O_2 *molecule* containing two O *atoms*, for a total of two O atoms on the left.

On the right:
- There are two H_2O *molecules*, each containing two H *atoms* and one O atom, for a total of four H atoms and two O atoms on the right.

1.4.1 Reaction Rates

Most chemical reactions give off heat and are classified as exothermic reactions. The rate of a reaction may be calculated by the Arrhenius equation, which contains absolute temperature, K, equal to the Celsius temperature plus 273, in an exponential term. As a general rule, the speed of a reaction doubles for each 10°C increase in temperature. Reaction rates are important in fires or explosions involving hazardous chemicals. A remarkable aspect of biochemical reactions is that they occur rapidly at very mild conditions, typically at body temperature in humans (see Chapter 3). For example, industrial fixation of atmospheric elemental nitrogen to produce chemically bound nitrogen in ammonia requires very high temperatures and pressures, whereas *Rhizobium* bacteria accomplish the same thing under ambient conditions.

1.5 SOLUTIONS

A **solution** is formed when a substance in contact with a liquid becomes dispersed homogeneously throughout the liquid in a molecular form. The substance, called a **solute**, is said to **dissolve**. The liquid is called a **solvent**. There may be no readily visible evidence that a solute is present in the solvent; for example, a deadly poisonous solution of sodium cyanide in water looks like pure water. The solution may have a strong color, as is the case for intensely purple solutions of potassium permanganate, $KMnO_4$. It may have a strong odor, such as that of ammonia, NH_3, dissolved in water. Solutions may consist of solids, liquids, or gases dissolved in a solvent. Technically, it is even possible to have solutions in which a solid is a solvent, although such solutions are not discussed in this book.

1.5.1 Solution Concentration

The quantity of solute relative to that of solvent or solution is called the **solution concentration**. Concentrations are expressed in numerous ways. Very high concentrations are often given as percent by weight. For example, commercial concentrated hydrochloric acid is 36% HCl, meaning that 36% of the weight has come from dissolved HCl and 64% from water solvent. Concentrations of very dilute solutions, such as those of hazardous waste leachate containing low levels of contaminants, are expressed as weight of solute per unit volume of solution. Common units are milligrams per liter (mg/L) or micrograms per liter (μg/L). Since a liter of water weighs essentially 1000 g, a concentration of 1 mg/L is equal to 1 part per million (ppm) and a concentration of 1 μg/L is equal to 1 part per billion (ppb).

Chemists often express concentrations in moles per liter, or molarity, M. Molarity is given by the relationship

$$M = \frac{\text{Number of moles of solute}}{\text{Number of liters of solution}} \quad (1.5.1)$$

The number of moles of a substance is its mass in grams divided by its molar mass. For example, the molecular mass of ammonia, NH_3, is 14 + 1 + 1 + 1, so a mole of ammonia has a mass of 17 g. Therefore, 17 g of NH_3 in 1 L of solution has a value of M equal to 1 mole/L.

1.5.2 Water as a Solvent

Most liquid wastes are solutions or suspensions of waste materials in water. Water has some unique properties as a solvent that arise from its molecular structure, as represented by the following Lewis formula of water:

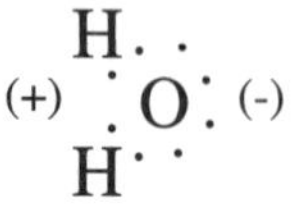

$$\text{(+)} \quad \overset{H}{\underset{H}{\,\,\ddot{\,}\,\ddot{O}\ddot{\,}\,}} \quad \text{(-)}$$

The H atoms are not on opposite sides of the O atom, and the two H–O bonds form an angle of 105°. Furthermore, the O atom (–2 oxidation state) is able to attract electrons more strongly than the two H atoms (each in the +1 oxidation state) so that the molecule is **polar**, with the O atom having a partial negative charge and the end of the molecule with the two H atoms having a partial positive charge. This means that water molecules can cluster around ions with the positive ends of the water molecules attracted to negatively charged anions and the negative end to positively charged cations. This kind of interaction is part of the general phenomenon of **solvation**. It is specifically called **hydration** when water is the solvent, and it is partially responsible for water's excellent ability to dissolve ionic compounds, including acids, bases, and salts.

Water molecules form a special kind of bond called a **hydrogen bond** with each other and with solute molecules that contain O, N, or S atoms. As its name implies, a hydrogen bond involves a hydrogen atom held between two other atoms of O, N, or S. Hydrogen bonding is partly responsible for water's ability to solvate and dissolve chemical compounds capable of hydrogen bonding.

As noted above, the water molecule is a polar species, which affects its ability to act as a solvent. Solutes may likewise have polar character. In general, solutes with polar molecules are more soluble in water than nonpolar ones. The polarity of an impurity solute in wastewater is a factor in determining how it may be removed from water. Nonpolar organic solutes are easier to take out of water by an adsorbent species, such as activated carbon, than are more polar solutes.

1.5.3 Solutions of Acids and Bases

1.5.3.1 Acids, Bases, and Neutralization Reactions

A substance that produces H^+ ion in water is an **acid**. A substance that reacts with H^+ ion or that produces OH^- ion, which can react with H^+ to produce H_2O, is a **base**. A common example of an acid is an aqueous solution of hydrogen chloride, which is completely ionized to H^+ and Cl^- ions in water to produce a solution of hydrochloric acid. Although it does not contain H^+ ion, carbon dioxide acts as an acid in water because it undergoes the following reaction, producing H^+:

$$CO_2 + H_2O \rightarrow H^+ + HCO_3^- \qquad (1.5.2)$$

A typical base is sodium hydroxide, NaOH, which exists as Na^+ and OH^- ions in water. Ammonia gas, NH_3, dissolves in water to produce a basic solution:

$$NH_3 + H_2O \rightarrow NH_4^+ + OH^- \qquad (1.5.3)$$

The reaction between H^+ ion from an acid and OH^- ion from a base is a **neutralization reaction**. For a specific example, consider the reaction of H^+ from a solution of sulfuric acid, H_2SO_4, and OH^- from a solution of calcium hydroxide:

$$H_2SO_4 + Ca(OH)_2 \longrightarrow 2H_2O + CaSO_4 \quad (1.5.4)$$

where H_2SO_4 is the acid (source of H^+ ion), $Ca(OH)_2$ is the base (source of OH^- ion), H_2O is water, and $CaSO_4$ is the salt.

In addition to water, which is always the product of a neutralization reaction, the other product is calcium sulfate, $CaSO_4$. This compound is a **salt** composed of Ca^{2+} ions and SO_4^{2-} ions held together by ionic bonds. A salt, consisting of a cation other than H^+ and an anion other than OH^-, is the other product produced in addition to water when an acid and base react. Some salts are hazardous substances and environmental pollutants because of their dangerous or harmful properties. Some examples include the following:

- Ammonium perchlorate, NH_4ClO_4 (reactive oxidant)
- Barium cyanide, $Ba(CN)_2$ (toxic)
- Lead acetate, $Pb(C_2H_3O_2)_2$ (toxic)
- Thallium(I) carbonate, Tl_2CO_3 (toxic)

1.5.3.2 Concentration of H⁺ Ion and pH

Molar concentrations of hydrogen ion, $[H^+]$, range over many orders of magnitude and are conveniently expressed by pH, defined as

$$pH = -\log[H^+] \quad (1.5.5)$$

In absolutely pure water, the value of $[H^+]$ is exactly 1×10^{-7} mole/L, the pH is 7.00, and the solution is **neutral** (neither acidic nor basic). **Acidic** solutions have pH values of less than 7, and **basic** solutions have pH values of greater than 7.

Strong acids and strong bases are **corrosive** substances that exhibit extremes of pH. They are destructive to materials and flesh. Strong acids can react with cyanide and sulfide compounds to release highly toxic hydrogen cyanide (HCN) or hydrogen sulfide (H_2S) gases, respectively. Bases liberate noxious ammonia gas (NH_3) from solid ammonium compounds.

1.5.3.3 Metal Ions Dissolved in Water

Metal ions dissolved in water have some unique characteristics that influence their properties as natural water constituents and heavy metal pollutants and in biological systems. The formulas of metal ions are usually represented by the symbol for the ion followed by its charge. For example, iron(II) ion (from a compound such as iron(II) sulfate, $FeSO_4$) dissolved in water is represented as Fe^{2+}. Actually, in water solution each iron(II) ion is strongly solvated and bonded to water molecules, so that the formula is more correctly shown as $Fe(H_2O)_6^{2+}$. Many metal ions have a tendency to lose hydrogen ions from the solvating water molecules, as shown by the following:

$$Fe(H_2O)_6^{2+} \rightarrow Fe(OH)(H_2O)_5^+ + H^+ \quad (1.5.6)$$

Ions of the next higher oxidation state, iron(III), have such a tendency to lose H^+ ion in aqueous solution that, except in rather highly acidic solutions, they precipitate out as solid hydroxides, such as iron(III) hydroxide, $Fe(OH)_3$:

$$Fe(H_2O)_6^{3+} \rightarrow Fe(OH)_3(s) + 3H_2O + 3H^+ \quad (1.5.7)$$

1.5.3.4 Complex Ions Dissolved in Water

It was noted above that metal ions are solvated (hydrated) by binding to water molecules in aqueous solution. Some species in solution have a stronger tendency than water to bond to metal ions. An example of such a species is cyanide ion, CN⁻, which displaces water molecules from some metal ions in solution, as shown below:

$$Ni(H_2O)_4^{2+} + 4CN^- \rightarrow Ni(CN)_4^{2-} + 4H_2O \qquad (1.5.8)$$

(The tendency of cyanide ion to bond with iron(III) is responsible for its toxicity in that it bonds with iron(III) in one of the enzymes involved in the utilization of molecular oxygen in respiration processes.) This prevents utilization of oxygen with potentially fatal results, as discussed in Chapter 11.) The species that bonds to the metal ion, cyanide in this case, is called a **ligand**, and the product of the reaction is a **complex ion** or metal complex. The overall process is called **complexation**.

1.5.4 Colloidal Suspensions

Very small particles on the order of 1 μm or less in size, called **colloidal particles**, may stay suspended in a liquid for an indefinite period of time. Such a mixture is a **colloidal suspension**, and it behaves in many respects like a solution. Colloidal suspensions are used in many industrial applications. Many waste materials are colloidal and are often emulsions consisting of colloidal liquid droplets suspended in another liquid, usually wastewater. One of the challenges in dealing with colloidal wastes is removing a relatively small quantity of colloidal material from a large quantity of water by precipitating the colloid. This process is called **coagulation** or **flocculation** and is often brought about by the addition of chemical agents.

1.6 ORGANIC CHEMISTRY

Most carbon-containing compounds are **organic chemicals** and are addressed by the subject of **organic chemistry**. Organic chemistry is a vast, diverse discipline because of the enormous number of organic compounds that exist as a consequence of the versatile bonding capabilities of carbon. Such diversity is due to the ability of carbon atoms to bond to each other through single bonds (two shared electrons), double bonds (four shared electrons), and triple bonds (six shared electrons), in a limitless variety of straight chains, branched chains, and rings.

Among organic chemicals are included the majority of important industrial compounds, synthetic polymers, agricultural chemicals, biological materials, and most substances that are of concern because of their toxicities and other hazards. Pollution of the water, air, and soil environments by organic chemicals is an area of significant concern.

Chemically, most organic compounds can be divided among hydrocarbons, oxygen-containing compounds, nitrogen-containing compounds, sulfur-containing compounds, organohalides, phosphorus-containing compounds, or combinations of these. Each of these classes of organic compounds is discussed briefly here.

All organic compounds, of course, contain carbon. Virtually all also contain hydrogen and have at least one C-H bond. The simplest organic compounds, and those easiest to understand, are those that contain only hydrogen and carbon. These compounds are called **hydrocarbons** and are addressed first among the organic compounds discussed in this chapter. Hydrocarbons are used here to illustrate some of the most fundamental points of organic chemistry, including organic formulas, structures, and names.

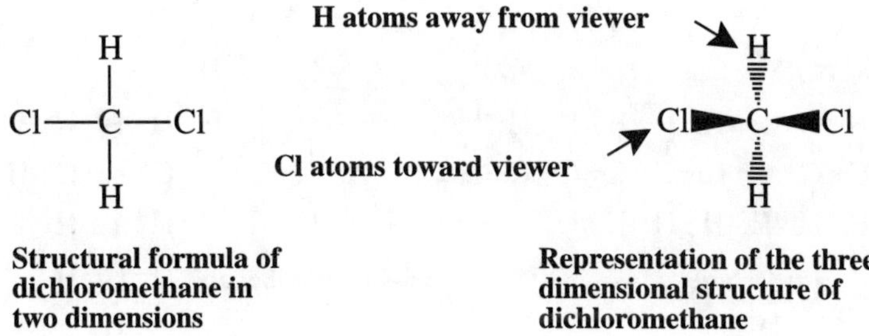

CHEMISTRY AND ORGANIC CHEMISTRY 15

Structural formula of dichloromethane in two dimensions

Representation of the three-dimensional structure of dichloromethane

H atoms away from viewer
Cl atoms toward viewer

Figure 1.8 Structural formulas of dichloromethane, CH_2Cl_2; the formula on the right provides a three-dimensional representation.

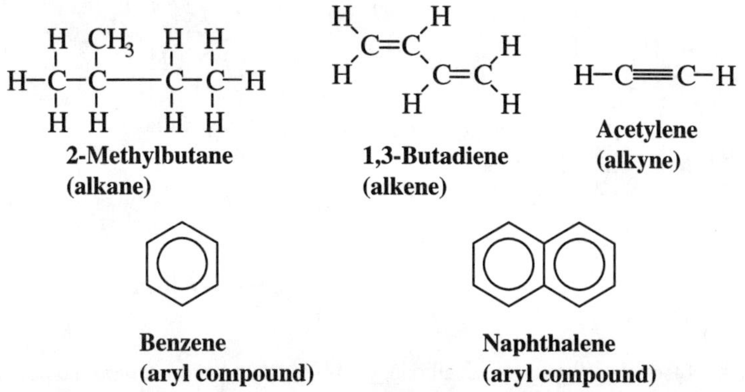

2-Methylbutane (alkane)

1,3-Butadiene (alkene)

Acetylene (alkyne)

Benzene (aryl compound)

Naphthalene (aryl compound)

Figure 1.9 Examples of major types of hydrocarbons.

1.6.1 Molecular Geometry in Organic Chemistry

The three-dimensional shape of a molecule, that is, its molecular geometry, is particularly important in organic chemistry. This is because its molecular geometry determines in part the properties of an organic molecule, particularly its interactions with biological systems. Shapes of molecules are represented in drawings by lines of normal, uniform thickness for bonds in the plane of the paper, broken lines for bonds extending away from the viewer, and heavy lines for bonds extending toward the viewer. These conventions are shown by the example of dichloromethane, CH_2Cl_2, an important organochloride solvent and extractant, illustrated in Figure 1.8.

1.7 HYDROCARBONS

As noted above, hydrocarbon compounds contain only carbon and hydrogen. The major types of hydrocarbons are alkanes, alkenes, alkynes, and aromatic compounds. Examples of each are shown in Figure 1.9.

1.7.1 Alkanes

Alkanes, also called **paraffins** or **aliphatic hydrocarbons**, are hydrocarbons in which the C atoms are joined by single covalent bonds (sigma bonds) consisting of two shared electrons (see

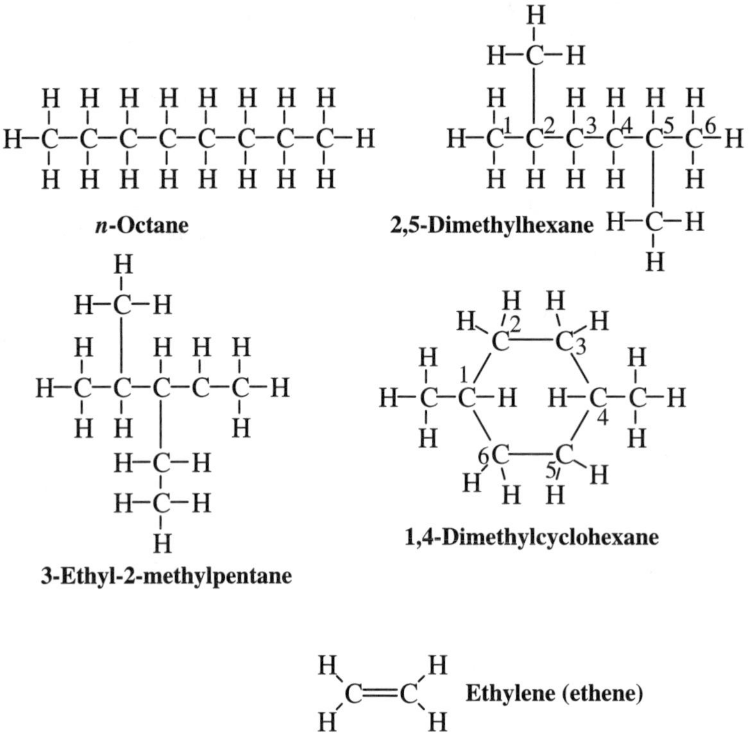

Figure 1.10 Structural formulas of four hydrocarbons, each containing eight carbon atoms, that illustrate the structural diversity possible with organic compounds. Numbers used to denote locations of atoms for purposes of naming are shown on two of the compounds.

Section 1.3). Some examples of alkanes are shown in Figure 1.10. As with other organic compounds, the carbon atoms in alkanes may form straight chains or branched chains. The three kinds of alkanes are **straight-chain alkanes, branched-chain alkanes,** and **cycloalkanes,** respectively. A typical branched chain alkane is 2-methylbutane, a volatile, highly flammable liquid, the structural formula of which is shown in Figure 1.9. It is a component of gasoline, which may explain why it has been detected as an air pollutant in urban air. The general molecular formula for straight- and branched-chain alkanes is C_nH_{2n+2}, and that of cyclic alkanes is C_nH_{2n}. The four hydrocarbon molecules in Figure 1.10 contain eight carbon atoms each. In one of the molecules, all of the carbon atoms are in a straight chain; in two, they are in branched chains; and in the fourth, six of the carbon atoms are in a ring.

1.7.1.1 Formulas of Alkanes

Formulas of organic compounds present information at several different levels of sophistication. **Molecular formulas,** such as that of octane (C_8H_{18}), give the number of each kind of atom in a molecule of a compound. As shown in Figure 1.10, however, the molecular formula of C_8H_{18} may apply to several alkanes, each one of which has unique chemical, physical, and toxicological properties. These different compounds are designated by **structural formulas** showing the order in which the atoms in a molecule are arranged. Compounds that have the same molecular, but different structural formulas are called **structural isomers.** Of the compounds shown in Figure 1.10, n-octane, 2,5-dimethylhexane, and 3-ethyl-2-methylpentane are structural isomers, all having the formula C_8H_{18}, whereas 1,4-dimethylcyclohexane is not a structural isomer of the other three compounds because its molecular formula is C_8H_{16}.

CHEMISTRY AND ORGANIC CHEMISTRY

Table 1.3 Some Alkanes and Substituent Groups Derived from Them

Alkane	Substituent Group(s) Derived from Alkane
H–CH₃ (Methane)	H–CH₂–* (Methyl group)
CH₃–CH₃ (Ethane)	CH₃–CH₂–* (Ethyl group)
CH₃–CH₂–CH₃ (Propane)	CH₃–CH₂–CH₂–* (n-Propyl group) ; CH₃–CH(*)–CH₃ (Isopropyl group)
CH₃–CH₂–CH₂–CH₃ (n-Butane)	CH₃–CH₂–CH₂–CH₂–* (n-Butyl group) ; CH₃–CH(*)–CH₂–CH₃ (sec-Butyl group) ; (CH₃)₃C–* (tert-Butyl group)
CH₃–CH₂–CH₂–CH₂–CH₃ (n-Pentane)	CH₃–CH₂–CH₂–CH₂–CH₂–* (n-Pentyl group)

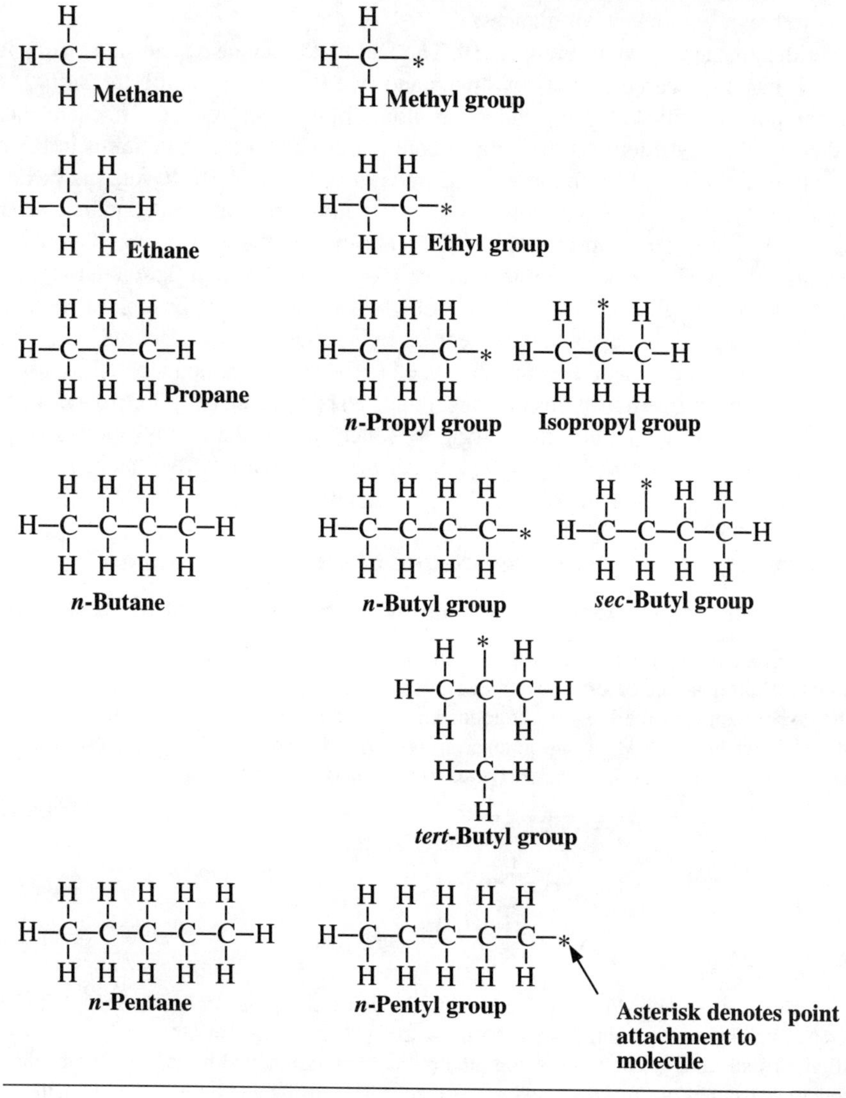

Asterisk denotes point attachment to molecule

1.7.1.2 Alkanes and Alkyl Groups

Most organic compounds can be derived from alkanes. In addition, many important parts of organic molecules contain one or more alkane groups, minus a hydrogen atom, bonded as substituents onto the basic organic molecule. As a consequence of these factors, the names of many organic compounds are based on alkanes. It is useful to know the names of some of the more common alkanes and substituent groups derived from them, as shown in Table 1.3.

1.7.1.3 Names of Alkanes and Organic Nomenclature

Systematic names, from which the structures of organic molecules can be deduced, have been assigned to all known organic compounds. The more common organic compounds, including many toxic and hazardous organic sustances, likewise have **common names** that have no structural implications. Although it is not possible to cover organic nomenclature in any detail in this chapter, the basic approach to nomenclature is presented, along with some pertinent examples. The simplest approach is to begin with names of alkanes.

Consider the alkanes shown in Figure 1.10. The fact that *n*-octane has no side chains is denoted by "*n*," that it has eight carbon atoms by "oct," and that it is an alkane by the suffix "ane." The names of compounds with branched chains or atoms other than H or C attached make use of numbers that stand for positions on the longest continuous chain of carbon atoms in the molecule. This convention is illustrated by the second compound in Figure 1.10. It gets the hexane part of its name from the fact that it is an alkane with six carbon atoms in its longest continuous chain ("hex" stands for six). However, it has a methyl group (CH_3) attached on the second carbon atom of the chain and another on the fifth. Hence the full systematic name of the compound is 2,5-dimethylhexane, where "di" indicates two methyl groups. In the case of 3-ethyl-2-methylpentane, the longest continuous chain of carbon atoms contains five carbon atoms, denoted by *pent*ane; an ethyl group, C_2H_5, is attached to the third carbon atom; and a methyl group is attached to the second carbon atom. The last compound shown in the figure has six carbon atoms in a ring, indicated by the prefix "cyclo," so it is a cyclo*hex*ane compound. Furthermore, the carbon in the ring to which one of the methyl groups is attached is designated by 1, and another methyl group is attached to the fourth carbon atom around the ring. Therefore, the full name of the compound is 1,4-dimethylcyclohexane.

1.7.1.4 Summary of Organic Nomenclature as Applied to Alkanes

Naming relatively simple alkanes is a straightforward process. The basic rules to be followed are:

1. The name of the compound is based on the longest continuous chain of carbon atoms. (The structural formula may be drawn such that this chain is not immediately obvious.)
2. The carbon atoms in the longest continous chain are numbered sequentially from one end. The end of the chain from which the numbering is started is chosen to give the lower numbers for substituent groups in the final name. For example, the compound

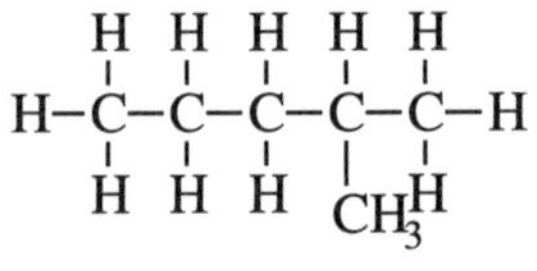

$$\begin{array}{c} H\ H\ H\ H\ H \\ |\ \ |\ \ |\ \ |\ \ | \\ H-C-C-C-C-C-H \\ |\ \ |\ \ |\ \ |\ \ | \\ H\ H\ H\ CH_3\ H \end{array}$$

   could be named 4-methylpentane (numbering the five-carbon chain from the left), but should be named 2-methylpentane (numbering the five-carbon chain from the right).
3. All groups attached to the longest continuous chain are designated by the number of the carbon atom to which they are attached and by the name of the substituent group (2-methyl in the example cited in step 2 above).
4. A prefix is used to denote multiple substitutions by the same kind of group. This is illustrated by 2,2,3-trimethylpentane,

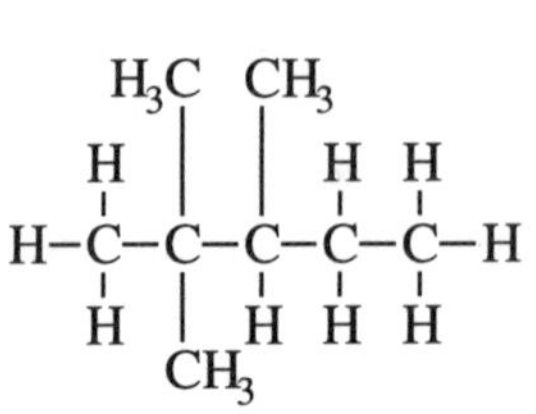

$$\begin{array}{c} H_3C\ \ CH_3 \\ H\ \ |\ \ \ \ |\ \ H\ H \\ |\ \ \ \ \ \ \ \ \ \ \ \ \ \ |\ \ | \\ H-C-C-C-C-C-H \\ |\ \ \ \ \ \ |\ \ \ \ \ |\ \ |\ \ | \\ H\ \ \ \ CH_3\ H\ H\ H \end{array}$$ **2,2,3-Trimethylpentane**

   in which the prefix *tri* is used to show that three methyl groups are attached to the pentane chain.
5. The complete name is assigned such that it denotes the longest continuous chain of carbon atoms and the name and location on this chain of each substituent group.

CHEMISTRY AND ORGANIC CHEMISTRY

1.7.1.5 Reactions of Alkanes

Alkanes contain only C–C and C–H bonds, both of which are relatively strong. For that reason, they have little tendency to undergo many kinds of reactions common to some other organic chemicals, such as acid–base reactions or low-temperature oxidation–reduction reactions. However, at elevated temperatures alkanes readily undergo oxidation — more specifically combustion — with molecular oxygen in air, as shown by the following reaction of propane:

$$C_3H_8 + 5O_2 \rightarrow 3CO_2 + 4H_2O + \text{heat} \tag{1.7.1}$$

Common alkanes are highly flammable, and the more volatile lower molecular mass alkanes form explosive mixtures with air. Furthermore, combustion of alkanes in an oxygen-deficient atmosphere or in an automobile engine produces significant quantities of carbon monoxide, CO, the toxic properties of which are discussed in Section 11.2.2.

In addition to combustion, alkanes undergo **substitution reactions** in which one or more H atoms on an alkane are replaced by atoms of another element. The most common such reaction is the replacement of H by chlorine, to yield **organochlorine** compounds. For example, methane reacts with chlorine to give chloromethane. This reaction begins with the dissociation of molecular chlorine, usually initiated by ultraviolet electromagnetic radiation:

$$Cl_2 + UV \text{ energy} \rightarrow Cl\cdot + Cl\cdot \tag{1.7.2}$$

The Cl· product is a **free radical** species in which the chlorine atom has only seven outer-shell electrons, as shown by the Lewis symbol,

$$:\ddot{\text{Cl}}\cdot$$

instead of the favored octet of eight outer-shell electrons. In gaining the octet required for chemical stability, the chlorine atom is very reactive. It abstracts a hydrogen from methane,

$$Cl\cdot + CH_4 \rightarrow HCl + CH_3\cdot \tag{1.7.3}$$

to yield HCl gas and another reactive species with an unpaired electron, $CH_3\cdot$, called methyl radical. The methyl radical attacks molecular chlorine,

$$CH_3\cdot + Cl_2 \rightarrow CH_3Cl + Cl\cdot \tag{1.7.4}$$

to give the chloromethane (CH_3Cl) product and regenerate Cl·, which can attack additional methane, as shown in Reaction 1.7.3. The reactive Cl· and $CH_3\cdot$ species continue to cycle through the two preceding reactions.

The reaction sequence shown above illustrates three important aspects of chemistry that will be shown to be very important in the discussion of atmospheric chemistry in Section 2.8. The first of these is that a reaction may be initiated by a **photochemical process** in which a photon of "light" (electromagnetic radiation) energy produces a reactive species, in this case the Cl· atom. The second point illustrated is the high chemical reactivity of **free radical species** with unpaired electrons and incomplete octets of valence electrons. The third point illustrated is that of **chain reactions**, which can multiply manyfold the effects of a single reaction-initiating event, such as the photochemical dissociation of Cl_2.

1.7.2 Alkenes and Alkynes

Alkenes, or **olefins**, are hydrocarbons that have double bonds consisting of four shared electrons. The simplest and most widely manufactured alkene is ethylene,

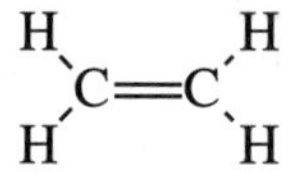

$$\underset{H}{\overset{H}{\diagdown}}C=C\underset{H}{\overset{H}{\diagup}} \quad \text{Ethylene (ethene)}$$

used for the production of polyethylene polymer. Another example of an important alkene is 1,3-butadiene (Figure 1.9), widely used in the manufacture of polymers, particularly synthetic rubber. The lighter alkenes, including ethylene and 1,3-butadiene, are highly flammable and form explosive mixtures with air. There have been a number of tragic industrial explosions and fires involving ethylene and other lighter alkenes.

Acetylene (Figure 1.9) is an **alkyne**, a class of hydrocarbons characterized by carbon–carbon triple bonds consisting of six shared electrons. Highly flammable acetylene is used in large quantities as a chemical raw material and fuel for oxyacetylene torches. It forms dangerously explosive mixtures with air.

1.7.2.1 Addition Reactions

The double and triple bonds in alkenes and alkynes have "extra" electrons capable of forming additional bonds. Therefore, the carbon atoms attached to these bonds can add atoms without losing any atoms already bonded to them; the multiple bonds are said to be **unsaturated**. Therefore, alkenes and alkynes both undergo **addition reactions**, in which pairs of atoms are added across unsaturated bonds, as shown in the reaction of ethylene with hydrogen to give ethane:

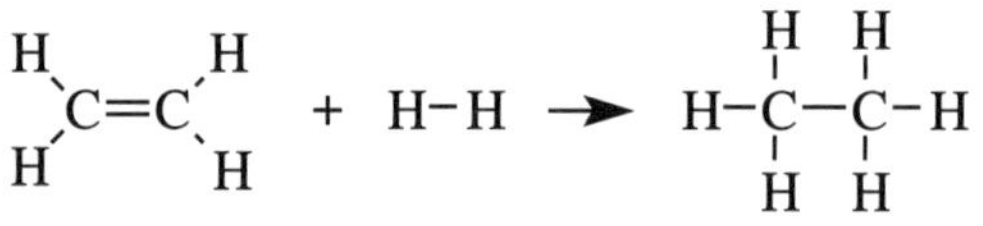

$$\underset{H}{\overset{H}{\diagdown}}C=C\underset{H}{\overset{H}{\diagup}} + H-H \rightarrow H-\underset{H}{\overset{H}{\underset{|}{C}}}-\underset{H}{\overset{H}{\underset{|}{C}}}-H \qquad (1.7.5)$$

This is an example of a **hydrogenation reaction**, a very common reaction in organic synthesis, food processing (manufacture of hydrogenated oils), and petroleum refining. Another example of an addition reaction is the biologically mediated reaction of vinyl chloride with oxygen,

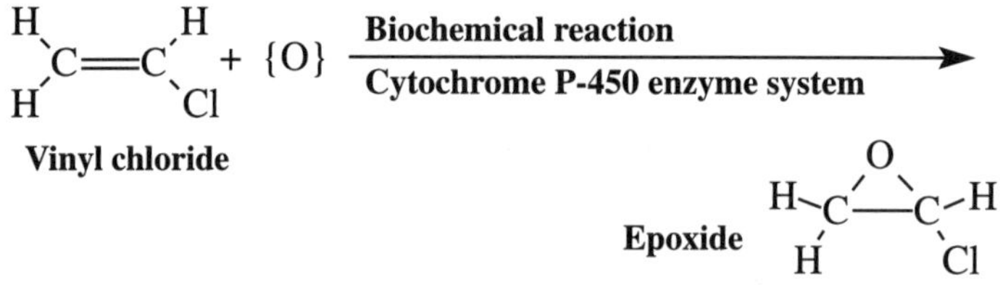

$$\underset{H}{\overset{H}{\diagdown}}C=C\underset{Cl}{\overset{H}{\diagup}} + \{O\} \xrightarrow[\text{Cytochrome P-450 enzyme system}]{\text{Biochemical reaction}} \underset{H}{\overset{H}{\diagdown}}C\overset{O}{\overline{\quad\quad}}C\underset{Cl}{\overset{H}{\diagup}} \qquad (1.7.6)$$

Vinyl chloride

Epoxide

to produce a reactive epoxide that can bind to biomolecules in the body to produce liver cancer. This kind of reaction, which is not possible with alkanes, adds to the chemical and metabolic versatility of compounds containing unsaturated bonds and is a factor contributing to their generally higher toxicities. It makes unsaturated compounds much more chemically reactive, more hazardous to handle in industrial processes, and more active in atmospheric chemical processes, such as smog formation (see Section 2.8).

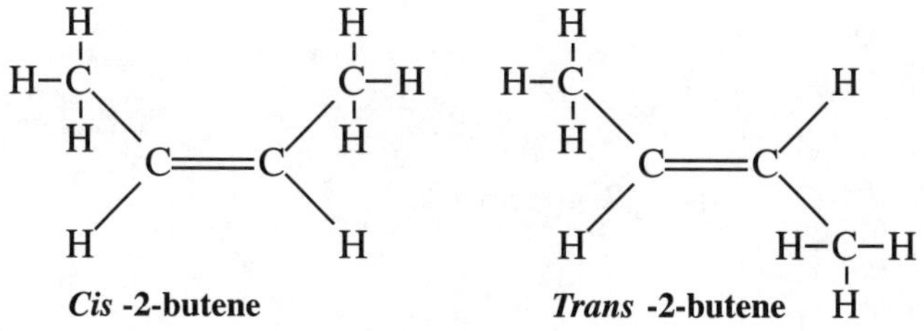

Figure 1.11 Cis and trans isomers of the alkene 2-butene.

1.7.3 Alkenes and *Cis–trans* Isomerism

As shown by the two simple compounds in Figure 1.11, the two carbon atoms connected by a double bond in alkenes cannot rotate relative to each other. For this reason, another kind of isomerism, called **cis–trans**, is possible for alkenes. *Cis–trans* isomers have different parts of the molecule oriented differently in space, although these parts occur in the same order. Both alkenes illustrated in Figure 1.11 have a molecular formula of C_4H_8. In the case of *cis*-2-butene, the two CH_3 (methyl) groups attached to the C=C carbon atoms are on the same side of the molecule, whereas in *trans*-2-butene they are on opposite sides.

1.7.4 Condensed Structural Formulas

To save space, structural formulas are conveniently abbreviated as **condensed structural formulas**. The condensed structural formula of 3-ethyl-2-methylpentane is $CH_3CH(CH_3)CH(C_2H_5)CH_2CH_3$, where the CH_3 (methyl) and C_2H_5 (ethyl) groups are placed in parentheses to show that they are branches attached to the longest continuous chain of carbon atoms, which contains five carbon atoms. It is understood that each of the methyl and ethyl groups is attached to the carbon immediately preceding it in the condensed structural formula (methyl attached to the second carbon atom, ethyl to the third).

As illustrated by the examples in Figure 1.12, the structural formulas of organic molecules may be represented in a very compact form by lines and figures such as hexagons. The ends and intersections of straight-line segments in these formulas indicate the locations of carbon atoms. Carbon atoms at the terminal ends of lines are understood to have three H atoms attached. C atoms at the intersections of two lines are understood to have two H atoms attached to each. One H atom is attached to a carbon represented by the intersection of three lines. No hydrogen atoms are bonded to C atoms where four lines intersect. Other atoms or groups of atoms, such as the Cl atom or OH group, that are substituted for H atoms are shown by their symbols attached to a C atom with a line.

1.7.5 Aromatic Hydrocarbons

Benzene (Figure 1.13) is the simplest of a large class of **aromatic**, or **aryl**, hydrocarbons. Many important aryl compounds have substituent groups containing atoms of elements other than hydrogen and carbon and are called **aromatic compounds**. Most aromatic compounds discussed in this book contain six-carbon-atom benzene rings, as shown for benzene, C_6H_6, in Figure 1.13. Aromatic compounds have ring structures and are held together in part by particularly stable bonds that contain delocalized clouds of so-called π (pi, pronounced "pie") electrons. In an oversimplified sense, the structure of benzene can be visualized as resonating between the two equivalent structures shown on the left in Figure 1.13 by the shifting of electrons in chemical bonds. This structure can be shown more simply and accurately by a hexagon with a circle in it.

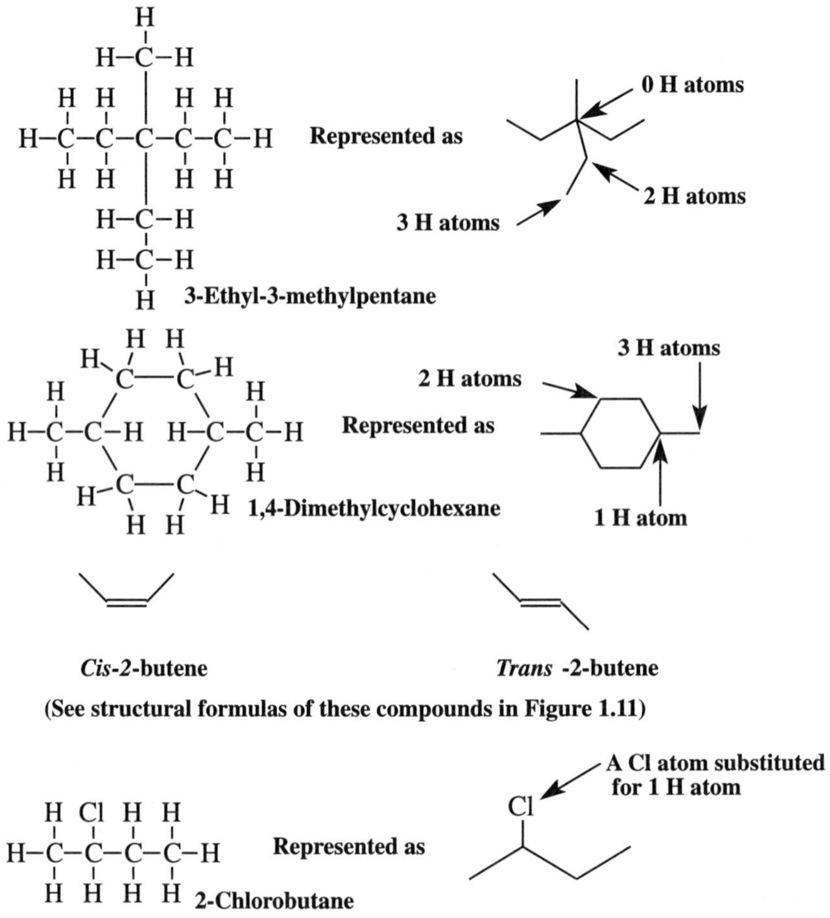

Figure 1.12 Representation of structural formulas with lines. A carbon atom is understood to be at each corner and at the end of each line. The numbers of hydrogen atoms attached to carbons at several specific locations are shown with arrows.

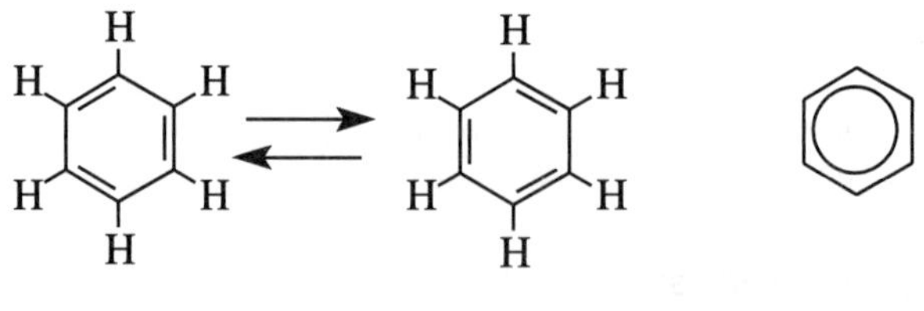

Figure 1.13 Representation of the aromatic benzene molecule with two resonance structures (left) and, more accurately, as a hexagon with a circle in it (right). Unless shown by symbols of other atoms, it is understood that a C atom is at each corner and that 1 H atom is bonded to each C atom.

Aromatic compounds have special characteristics of **aromaticity**, which include a low hydrogen:carbon atomic ratio, C–C bonds that are quite strong and of intermediate length between such bonds in alkanes and those in alkenes, tendency to undergo substitution reactions rather than the addition reactions characteristic of alkenes, and delocalization of π electrons over several carbon atoms. The last phenomenon adds substantial stability to aromatic compounds and is known as **resonance stabilization**.

Many toxic substances, environmental pollutants, and hazardous waste compounds, such as benzene, toluene, naphthalene, and chlorinated phenols, are aromatic compounds (see Figure 1.14).

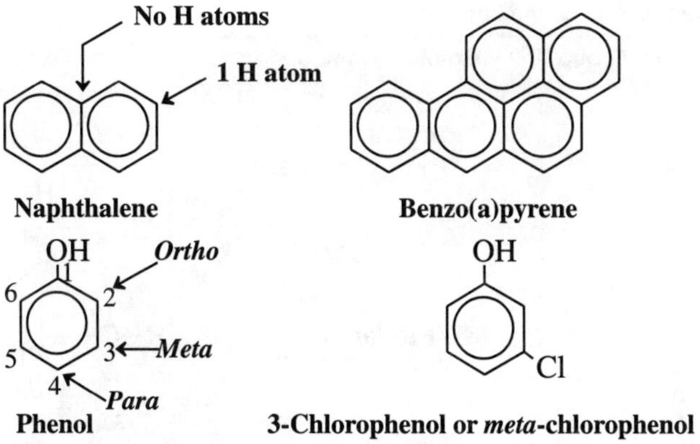

Figure 1.14 Aromatic compounds containing fused rings (top) and showing the numbering of carbon atoms for purposes of nomenclature.

As shown in Figure 1.14, some aromatic compounds, such as naphthalene and the polycyclic aromatic compound, benzo(a)pyrene, contain fused rings.

1.7.5.1 Benzene and Naphthalene

Benzene is a volatile, colorless, highly flammable liquid that is consumed as a raw material for the manufacture of phenolic and polyester resins, polystyrene plastics, alkylbenzene surfactants, chlorobenzenes, insecticides, and dyes. It is hazardous both for its ignitability and toxicity (exposure to benzene causes blood abnormalities that may develop into leukemia). Naphthalene is the simplest member of a large number of multicyclic aromatic hydrocarbons having two or more fused rings. It is a volatile white crystalline solid with a characteristic odor and has been used to make mothballs. The most important of the many chemical derivatives made from naphthalene is phthalic anhydride, from which phthalate ester plasticizers are synthesized.

1.7.5.2 Polycyclic Aromatic Hydrocarbons

Benzo(a)pyrene (Figure 1.14) is the most studied of the polycyclic aromatic hydrocarbons (PAHs), which are characterized by condensed ring systems ("chicken wire" structures). These compounds are formed by the incomplete combustion of other hydrocarbons, a process that consumes hydrogen in preference to carbon. The carbon residue is left in the thermodynamically favored condensed aromatic ring system of the PAH compounds.

Because there are so many partial-combustion and pyrolysis processes that favor production of PAHs, these compounds are encountered abundantly in the atmosphere, soil, and elsewhere in the environment from sources that include engine exhausts, wood stove smoke, cigarette smoke, and charbroiled food. Coal tars and petroleum residues such as road and roofing asphalt have high levels of PAHs. Some PAH compounds, including benzo(a)pyrene, are of toxicological concern because they are precursors to cancer-causing metabolites. As discussed in more detail in Chapters 7 and 13, benzo(a)pyrene and some other PAHs are partially oxidized by enzyme systems in humans and other animals to produce species that can bind with DNA, resulting in cancer in some cases.

1.8 ORGANIC FUNCTIONAL GROUPS AND CLASSES OF ORGANIC COMPOUNDS

The presence of elements other than hydrogen and carbon in organic molecules greatly increases the diversity of their chemical behavior. **Functional groups** consist of specific bonding configurations

Table 1.4 Examples of Some Important Functional Groups

Type of Functional Group	Example Compound	Structural Formula of Group[a]
Alkene (olefin)	Propene (propylene)	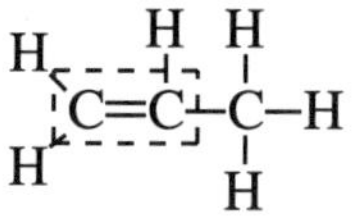 H₂C=CH–CH₃ (with H's shown)
Alkyne	Acetylene	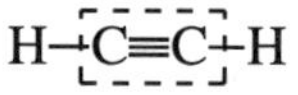 H–C≡C–H
Alcohol (-OH attached to alkyl group)	2-Propanol	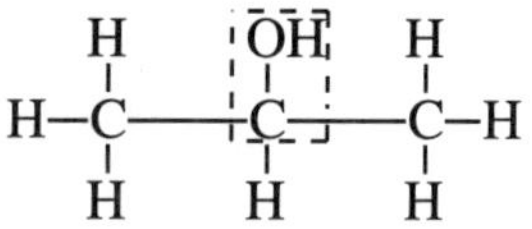 H₃C–CH(OH)–CH₃
Phenol (-OH attached to aryl group)	Phenol	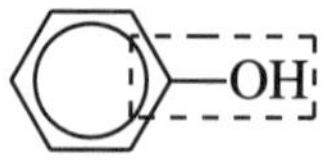 C₆H₅–OH
Ketone (When $-\overset{O}{\underset{\parallel}{C}}-H$ group is on end carbon, compound is an aldehyde)	Acetone	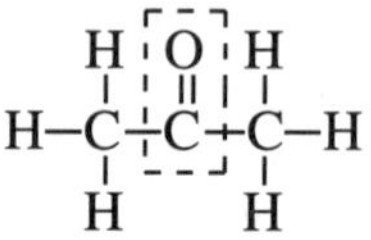 H₃C–C(=O)–CH₃
Amine	Methylamine	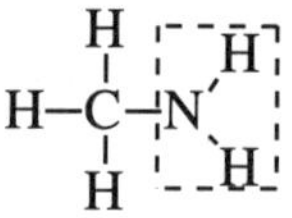 H₃C–NH₂
Nitro compounds	Nitromethane	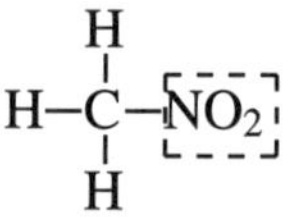 H₃C–NO₂
Sulfonic acids	Benzenesulfonic acid	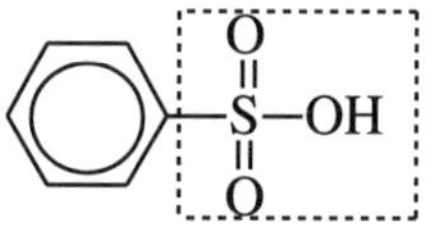 C₆H₅–S(=O)₂–OH
Organohalides	1,1–Dichloro-ethane	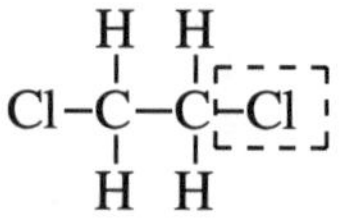 Cl₂CH–CH₃

[a] Functional group outlined by dashed line.

of atoms in organic molecules. Most functional groups contain at least one element other than carbon or hydrogen, although two carbon atoms joined by a double bond (alkenes) or triple bond (alkynes) are likewise considered to be functional groups. Table 1.4 shows some of the major functional groups that determine the nature of organic compounds.

1.8.1 Organooxygen Compounds

The most common types of compounds with oxygen-containing functional groups are epoxides, alcohols, phenols, ethers, aldehydes, ketones, and carboxylic acids. The functional groups characteristic of these compounds are illustrated by the examples of oxygen-containing compounds shown in Figure 1.15.

Ethylene oxide is a moderately to highly toxic sweet-smelling, colorless, flammable, explosive gas used as a chemical intermediate, sterilant, and fumigant. It is a mutagen and a carcinogen to

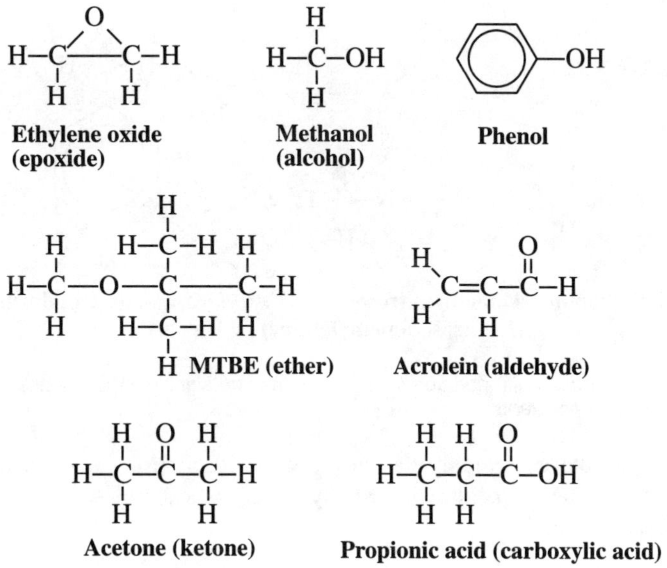

Figure 1.15 Examples of oxygen-containing organic compounds that may be significant as wastes, toxic substances, or environmental pollutants.

experimental animals. It is classified as hazardous for both its toxicity and ignitability. **Methanol** is a clear, volatile, flammable liquid alcohol used for chemical synthesis, as a solvent, and as a fuel. It has been advocated strongly in some quarters as an alternative to gasoline that would result in significantly less photochemical smog formation than currently used gasoline formulations. At least one major automobile manufacturer has been advocating methanol as a fuel that can be catalytically decomposed to produce hydrogen for fuel cells to propel automobiles. Ingestion of methanol can be fatal, and blindness can result from sublethal doses. **Phenol** is a dangerously toxic aromatic alcohol widely used for chemical synthesis and polymer manufacture. **Methyltertiary-butyl ether** (MTBE) is an ether that replaced tetraethyllead as an octane booster in gasoline. More recently, however, it has been found to be a common water pollutant and is now being phased out of gasoline. **Acrolein** is an alkenic aldehyde and a volatile, flammable, highly reactive chemical. It forms explosive peroxides upon prolonged contact with O_2. An extreme lachrimator and strong irritant, acrolein is quite toxic by all routes of exposure. **Acetone** is the lightest of the ketones. Like all ketones, acetone has a carbonyl (C=O) group that is bonded to two carbon atoms (that is, it is somewhere in the middle of a carbon atom chain). Acetone is a good solvent and is chemically less reactive than the aldehydes, which all have the functional group

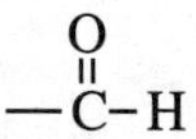

in which binding of the C=O to H makes the molecule significantly more reactive. **Propionic acid** is a typical organic carboxylic acid. The oxidized $-CO_2H$ group characteristic of carboxylic acids may be produced by oxidizing aldehydes and alcohols that have an $-OH$ group or C=O group on an end carbon atom.

1.8.2 Organonitrogen Compounds

Figure 1.16 shows examples of three classes of the many kinds of compounds that contain N (amines, nitrosamines, and nitro compounds). Nitrogen occurs in many functional groups in organic compounds, some of which contain nitrogen in ring structures, or along with oxygen.

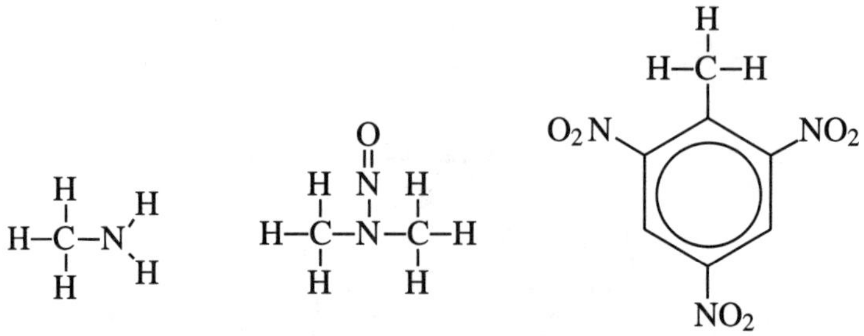

Methylamine   **Dimethylnitrosamine**   **2,4,6-trinitrotoluene**
              (N-nitrosodimethylamine)   **(TNT)**

Figure 1.16 Examples of organonitrogen compounds that may be significant as wastes, toxic substances, or environmental pollutants.

Methylamine is a colorless, highly flammable gas with a strong odor. It is a severe irritant affecting eyes, skin, and mucous membranes. Methylamine is the simplest of the **amine** compounds, which have the general formula

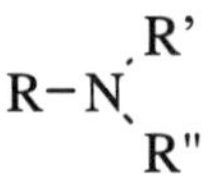

$$R-N\begin{matrix}R'\\R''\end{matrix}$$

where the Rs are hydrogen or hydrocarbon groups, at least one of which is the latter.

Dimethylnitrosamine is an N-nitroso compound, all of which contain the N–N=O functional group. It was once widely used as an industrial solvent, but was observed to cause liver damage and jaundice in exposed workers. Subsequently, numerous other N-nitroso compounds, many produced as by-products of industrial operations and food and alcoholic beverage processing, were found to be carcinogenic.

Solid **trinitrotoluene** (TNT) has been widely used as a military explosive. TNT is moderately to very toxic and has caused toxic hepatitis or aplastic anemia in exposed individuals, a few of whom have died from its toxic effects. It belongs to the general class of nitro compounds characterized by the presence of $-NO_2$ groups bonded to a hydrocarbon structure.

Some organonitrogen compounds are chelating agents that bind strongly to metal ions and play a role in the solubilization and transport of heavy metal wastes. Prominent among these are salts of the aminocarboxylic acids, which, in the acid form, have $-CH_2CO_2H$ groups bonded to nitrogen atoms. An important example of such a compound is the monohydrate of trisodium nitrilotriacetate (NTA):

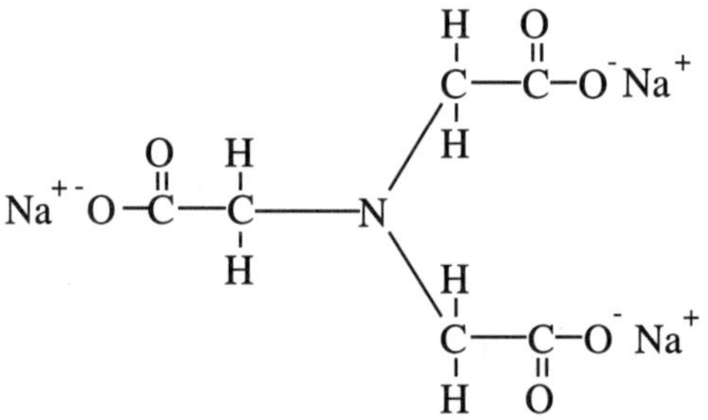

This compound has been widely used in Canada as a substitute for detergent phosphates to bind to calcium ion and make the detergent solution basic. NTA is used in metal plating formulations.

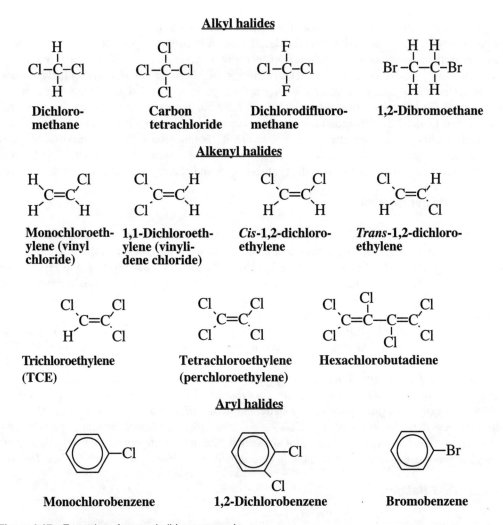

Figure 1.17 Examples of organohalide compounds.

It is highly water soluble and quickly eliminated with urine when ingested. It has a low acute toxicity, and no chronic effects have been shown for plausible doses. However, concern does exist over its interaction with heavy metals in waste treatment processes and in the environment.

1.8.3 Organohalide Compounds

Organohalides (Figure 1.17) exhibit a wide range of physical and chemical properties. These compounds consist of halogen-substituted hydrocarbon molecules, each of which contains at least one atom of F, Cl, Br, or I. They may be saturated (**alkyl halides**), unsaturated (**alkenyl halides**), or aromatic (**aryl halides**). The most widely manufactured organohalide compounds are chlorinated hydrocarbons, many of which are regarded as environmental pollutants or hazardous wastes.

1.8.3.1 Alkyl Halides

Substitution of halogen atoms for one or more hydrogen atoms on alkanes gives **alkyl halides**; example structural formulas are given in Figure 1.17. Most of the commercially important alkyl halides are derivatives of alkanes of low molecular mass. A brief discussion of the uses of the compounds listed in Figure 1.17 is given here to provide an idea of the versatility of the alkyl halides.

Dichloromethane is a volatile liquid with excellent solvent properties for nonpolar organic solutes. It has been used as a solvent for the decaffeination of coffee, in paint strippers, as a blowing agent in urethane polymer manufacture, and to depress vapor pressure in aerosol formulations. Once commonly sold as a solvent and stain remover, highly toxic **carbon tetrachloride** is now largely restricted to uses as a chemical intermediate under controlled conditions, primarily to manufacture chlorofluorocarbon refrigerant fluid compounds, which are also discussed in this section. Insecticidal **1,2-dibromoethane** has been consumed in large quantities as a lead scavenger in leaded gasoline and to fumigate soil, grain, and fruit. (Fumigation with this compound has been discontinued because of toxicological concerns.) An effective solvent for resins, gums, and waxes, it serves as a chemical intermediate in the syntheses of some pharmaceutical compounds and dyes.

1.8.3.2 Alkenyl Halides

Viewed as hydrocarbon-substituted derivatives of alkenes, the **alkenyl**, or **olefinic**, **organohalides** contain at least one halogen atom and at least one carbon–carbon double bond. The most significant of these are the lighter chlorinated compounds, such as those illustrated in Figure 1.17.

Vinyl chloride is consumed in large quantities as a raw material to manufacture pipe, hose, wrapping, and other products fabricated from polyvinylchloride plastic. This highly flammable, volatile, sweet-smelling gas is a known human carcinogen.

As shown in Figure 1.17, there are three possible dichloroethylene compounds, all clear, colorless liquids. Vinylidene chloride forms a copolymer with vinyl chloride used in some kinds of coating materials. The geometrically isomeric 1,2-dichloroethylenes are used as organic synthesis intermediates and as solvents. **Trichloroethylene** is a clear, colorless, nonflammable, volatile liquid. It is an excellent degreasing and dry-cleaning solvent and has been used as a household solvent and for food extraction (for example, in decaffeination of coffee). Colorless, nonflammable liquid **tetrachloroethylene** has properties and uses similar to those of trichloroethylene. **Hexachlorobutadiene**, a colorless liquid with an odor somewhat like that of turpentine, is used as a solvent for higher hydrocarbons and elastomers, as a hydraulic fluid, in transformers, and for heat transfer.

1.8.3.3 Aryl Halides

Aryl halide derivatives of benzene and toluene have many uses in chemical synthesis: as pesticides and raw materials for pesticides manufacture, solvents, and a diverse variety of other applications. These widespread uses over many decades have resulted in substantial human exposure and environmental contamination. Three example aryl halides are shown in Figure 1.17. Monochlorobenzene is a flammable liquid boiling at 132°C. It is used as a solvent, heat transfer fluid, and synthetic reagent. Used as a solvent, 1,2-dichlorobenzene is employed for degreasing hides and wool. It also serves as a synthetic reagent for dye manufacture. Bromobenzene is a liquid boiling at 156°C that is used as a solvent, motor oil additive, and intermediate for organic synthesis.

1.8.3.4 Halogenated Naphthalene and Biphenyl

Two major classes of halogenated aryl compounds containing two benzene rings are made by the chlorination of naphthalene and biphenyl, once marketed as mixtures with varying degrees of chlorine content. Examples of chlorinated naphthalenes and polychlorinated biphenyls (PCBs) (discussed later) are shown in Figure 1.18. The less highly chlorinated of these compounds are liquids; those with higher chlorine contents are solids. Because of their physical and chemical stabilities and other qualities, these compounds have had many uses, including heat transfer fluids, hydraulic fluids, and dielectrics. Polybrominated biphenyls (PBBs) have served as flame retardants. However, because chlorinated naphthalenes, PCBs, and PBBs are environmentally extremely persistent, their uses have been severely curtailed.

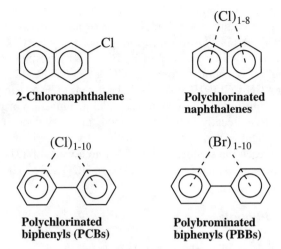

Figure 1.18 Halogenated naphthalenes and biphenyls.

1.8.3.5 Chlorofluorocarbons, Halons, and Hydrogen-Containing Chlorofluorocarbons

Chlorofluorocarbons (CFCs) are volatile one- and two-carbon compounds that contain Cl and F bonded to carbon. These compounds are notably stable and nontoxic. They were once widely used in the fabrication of flexible and rigid foams and as fluids for refrigeration and air conditioning. The most widely manufactured of these compounds are CCl_3F (CFC-11), CCl_2F_2 (CFC-12), $C_2Cl_3F_3$ (CFC-113), $C_2Cl_2F_4$ (CFC-114), and C_2ClF_5 (CFC-115).

Halons are related compounds that contain bromine and are used in fire extinguisher systems. The most commonly produced commercial halons are $CBrClF_2$ (Halon-1211), $CBrF_3$ (Halon-1301), and $C_2Br_2F_4$ (Halon-2402), where the sequence of numbers denotes the number of carbon, fluorine, chlorine, and bromine atoms, respectively, per molecule. Halons are particularly effective fire-extinguishing agents because of the way in which they stop combustion. Some fire suppressants, such as carbon dioxide, act by depriving the flame of oxygen by a smothering effect, whereas water cools a burning substance to a temperature below which combustion is supported. Halons act by chain reactions (discussed briefly in Section 1.7.1.5) that destroy hydrogen atoms that sustain combustion. The basic sequence of reactions involved is outlined in the following reactions:

$$CBrClF_2 + H\cdot \longrightarrow CClF_2\cdot + HBr \qquad (1.8.1)$$

$$HBr + H\cdot \longrightarrow Br\cdot + H_2 \qquad (1.8.2)$$

$$Br\cdot + H\cdot \longrightarrow HBr \qquad (1.8.3)$$

(Chain reaction)

Halons are used in automatic fire-extinguishing systems, such as those located in flammable solvent storage areas, and in specialty fire extinguishers, such as those on aircraft. It has proven difficult to find substitutes for halons that have the same excellent performance characteristics.

All of the chlorofluorocarbons and halons discussed above have been implicated in the halogen atom-catalyzed destruction of atmospheric ozone. As a result of U.S. Environmental Protection Agency regulations imposed in accordance with the 1986 Montreal Protocol on Substances that

Deplete the Ozone Layer, production of CFCs and halocarbons in the United States was curtailed starting in 1989. The most common substitutes for these halocarbons are hydrogen-containing chlorofluorocarbons (HCFCs) and hydrogen-containing fluorocarbons (HFCs). The substitute compounds produced commercially first were CH_2FCF_3 (HFC-134a, substitute for CFC-12 in automobile air conditioners and refrigeration equipment), $CHCl_2CF_3$ (HCFC-123, substitute for CFC-11 in plastic foam blowing), CH_3CCl_2F (HCFC-141b, substitute for CFC-11 in plastic foam blowing), and $CHClF_2$ (HCFC-22, substitute in air conditioners and manufacture of plastic foam food containers). Because of the more readily broken H–C bonds that they contain, these compounds are more easily destroyed by atmospheric chemical reactions (particularly with hydroxyl radical, see Section 2.8) before they reach the stratosphere. Relative to a value of 1.0 for CFC-11, the ozone-depletion potentials of these substitutes are HFC-134a, 0; HCFC-123, 0.016; HCFC-141b, 0.081; and HCFC-22, 0.053.

1.8.3.6 Chlorinated Phenols

The chlorinated phenols, particularly **pentachlorophenol**,

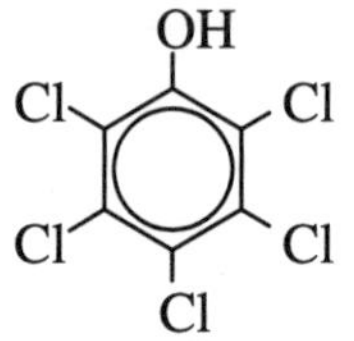

and the trichlorophenol isomers are significant hazardous wastes. These compounds are biocides that were used to treat wood to prevent rot by fungi and to prevent termite infestation. They are toxic, causing liver malfunction and dermatitis; contaminant polychlorinated dibenzodioxins may be responsible for some of the observed effects. Wood preservative chemicals such as pentachlorophenol have been encountered at many hazardous waste sites in wastewaters and sludges.

1.8.4 Organosulfur Compounds

Sulfur is chemically similar to, but more diverse than, oxygen. Whereas, with the exception of peroxides, most chemically combined organic oxygen is in the –2 oxidation state, sulfur occurs in the –2, +4, and +6 oxidation states. Many organosulfur compounds are noted for their foul, "rotten egg," or garlic odors, which makes them very unpleasant environmental contaminants, but warns of their presence even at very low levels. A number of example organosulfur compounds are shown in Figure 1.19.

1.8.4.1 Thiols and Thioethers

Substitution of alkyl or aryl hydrocarbon groups, such as phenyl and methyl for H on hydrogen sulfide, H_2S, leads to a number of different organosulfur **thiols** (mercaptans, R–SH) and **sulfides**, also called thioethers (R–S–R). Structural formulas of examples of these compounds are shown in Figure 1.19.

Methanethiol and other, lighter alkyl thiols are fairly common air pollutants that have "ultragarlic" odors; both 1- and 2-butanethiol are associated with skunk odor. Gaseous methanethiol is used as an odorant leak-detecting additive for natural gas, propane, and butane; it is also employed as an intermediate in pesticide synthesis. A toxic, irritating volatile liquid with a strong garlic odor, 2-propene-1-thiol (allyl mercaptan) is a typical alkenyl mercaptan. Benzenethiol (phenyl mercaptan) is the simplest of the aryl thiols. It is a toxic liquid with a severely repulsive odor.

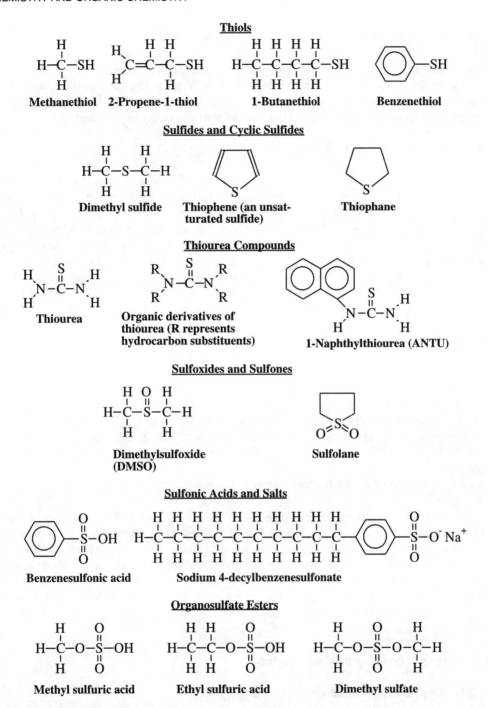

Figure 1.19 Examples of organosulfur compounds.

Alkyl sulfides or thioethers contain the C–S–C functional group. The lightest of these compounds is dimethyl sulfide, a volatile liquid (boiling point (bp) 38°C) that is moderately toxic by ingestion. It is released to the atmosphere by marine organisms and is a major source of atmospheric sulfur. Cyclic sulfides contain the C–S–C group in a ring structure. The most common of these compounds is thiophene, a heat-stable liquid (bp 84°C) with a solvent action much like that of benzene that is used in the manufacture of pharmaceuticals, dyes, and resins. Its saturated analog is tetrahydrothiophene, or thiophane.

1.8.4.2 Nitrogen-Containing Organosulfur Compounds

Many important organosulfur compounds also contain nitrogen. One such compound is **thiourea**, the sulfur analog of urea. Its structural formula is shown in Figure 1.19. Thiourea and **phenylthiourea** have been used as rodenticides. Commonly called ANTU, **1-naphthylthiourea** is an excellent rodenticide that is virtually tasteless and has a very high rodent:human toxicity ratio.

1.8.4.3 Sulfoxides and Sulfones

Sulfoxides and **sulfones** (Figure 1.19) contain both sulfur and oxygen. **Dimethylsulfoxide** (DMSO) is a liquid with numerous uses and some very interesting properties. It is used to remove paint and varnish, as a hydraulic fluid, mixed with water as an antifreeze solution, and in pharmaceutical applications as an anti-inflammatory and bacteriostatic agent. A polar aprotic (no ionizable H) solvent with a relatively high dielectric constant, **sulfolane** dissolves both organic and inorganic solutes. It is the most widely produced sulfone because of its use in an industrial process called BTX processing, in which it selectively extracts benzene, toluene, and xylene from aliphatic hydrocarbons; as the solvent in the Sulfinol process, by which thiols and acidic compounds are removed from natural gas; as a solvent for polymerization reactions; and as a polymer plasticizer.

1.8.4.4 Sulfonic Acids, Salts, and Esters

Sulfonic acids and sulfonate salts contain the $-SO_3H$ and $-SO_3^-$ groups, respectively, attached to a hydrocarbon moiety. The structural formulas of benzene sulfonic acids and of sodium 1-(*p*-sulfophenyl)decane, a biodegradable detergent surfactant, are shown in Figure 1.19. The common sulfonic acids are water-soluble strong acids that lose virtually all ionizable H^+ in aqueous solution. They are used commercially to hydrolyze fat and oil esters to produce fatty acids and glycerol used in chemical synthesis.

1.8.4.5 Organic Esters of Sulfuric Acid

Replacement by a hydrocarbon group of one H on sulfuric acid, H_2SO_4, yields an acid ester, and replacement of both yields an ester. Examples of these esters are shown in Figure 1.19. Sulfuric acid esters are used as alkylating agents, which act to attach alkyl groups (such as methyl) to organic molecules, in the manufacture of agricultural chemicals, dyes, and drugs. **Methylsulfuric acid** and **ethylsulfuric acid** are oily water-soluble liquids that are strong irritants to skin, eyes, and mucous tissue. Dimethylsulfate in which both H atoms of sulfuric acid are substituted by methyl groups is of particular concern because of some evidence it is a primary carcinogen that does not require bioactivation to cause cancer.

1.8.5 Organophosphorus Compounds

1.8.5.1 Alkyl and Aromatic Phosphines

The first two examples in Figure 1.20 illustrate that the structural formulas of alkyl and aromatic phosphine compounds may be derived by substituting organic groups for the H atoms in phosphine (PH_3), the hydride of phosphorus, discussed as a toxic inorganic compound in Section 12.10. **Methylphosphine** is a colorless, reactive gas. Crystalline, solid **triphenylphosphine** has a low reactivity and moderate toxicity when inhaled or ingested.

As shown by the reaction,

$$4C_3H_9P + 26O_2 \rightarrow 12CO_2 + 18H_2O + P_4O_{10} \qquad (1.8.4)$$

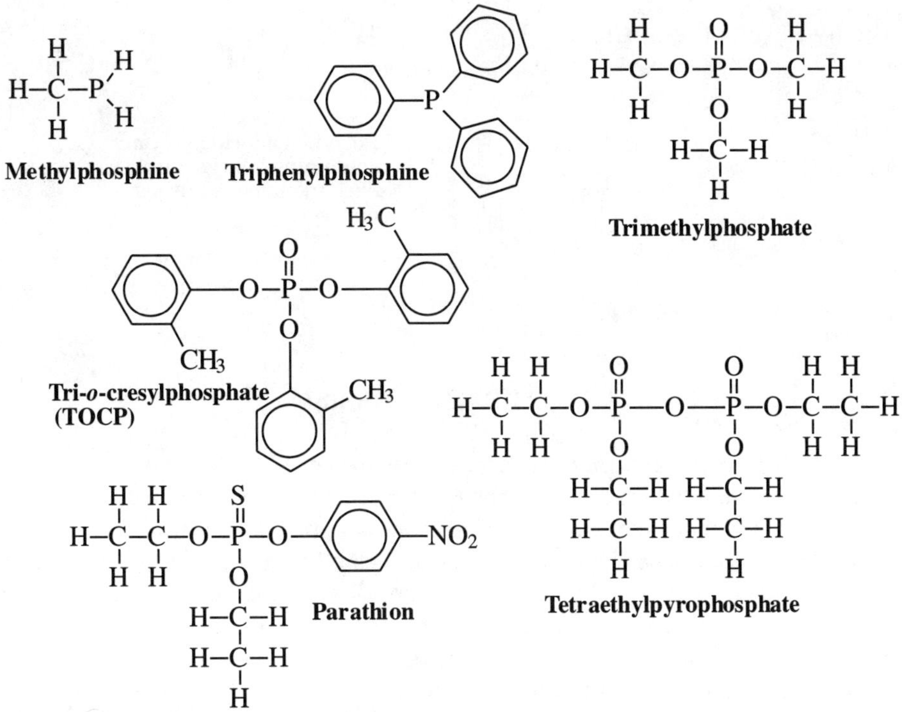

Figure 1.20 Some representative organophosphorus compounds.

combustion of aromatic and alkyl phosphines produces P_4O_{10}, a corrosive irritant toxic substance that reacts with moisture in the air to generate a visibility-obscuring fog of aerosol droplets consisting of corrosive orthophosphoric acid, H_3PO_4.

1.8.5.2 Organophosphate Esters

The structural formulas of three esters of orthophosphoric acid (H_3PO_4) and an ester of pyrophosphoric acid ($H_4P_2O_6$) are shown in Figure 1.20. Although **trimethylphosphate** is considered to be only moderately toxic, **tri-o-cresyl-phosphate** (TOCP) has a notorious record of poisonings. **Tetraethylpyrophosphate** (TEPP) was developed in Germany during World War II as a substitute for insecticidal nicotine. Although it is a very effective insecticide, its use in that application was of very short duration because it kills almost everything else, too.

1.8.5.3 Phosphorothionate Esters

Parathion, shown in Figure 1.20, is an example of a **phosphorothionate** ester. These compounds are used as insecticidal acetylcholinesterase inhibitors. They contain the P=S (thiono) group, which increases their insect:mammal toxicity ratios. Since the first organophosphate insecticides were developed in Germany during the 1930s and 1940s, many insecticidal organophosphate compounds have been synthesized. One of the earliest and most successful of these is **parathion**, *O,O*-diethyl-*O-p*-nitrophenylphosphorothionate (banned from use in the United States in 1991 because of its acute toxicity to humans). From a long-term environmental standpoint, organophosphate insecticides are superior to the organohalide insecticides that they largely displaced because the organophosphates readily undergo biodegradation and do not bioaccumulate.

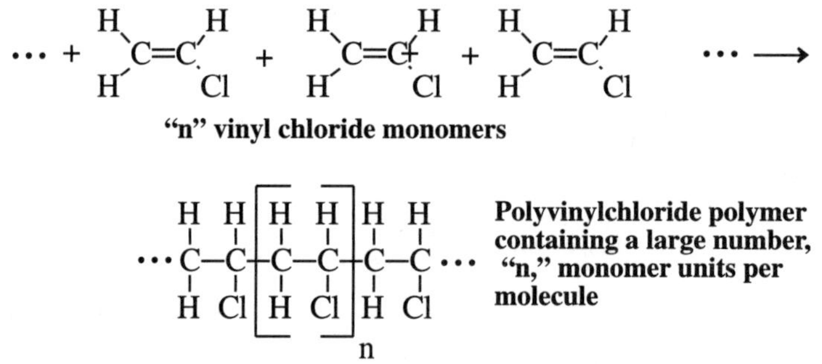

$$\cdots + \underset{H}{\overset{H}{\diagdown}}C=C\underset{Cl}{\overset{H}{\diagup}} + \underset{H}{\overset{H}{\diagdown}}C=C\underset{Cl}{\overset{H}{\diagup}} + \underset{H}{\overset{H}{\diagdown}}C=C\underset{Cl}{\overset{H}{\diagup}} \cdots \longrightarrow$$

"n" vinyl chloride monomers

Polyvinylchloride polymer containing a large number, "n," monomer units per molecule

Figure 1.21 Polyvinylchloride polymer.

1.9 OPTICAL ISOMERISM

Figure 1.11 showed *cis* and *trans* isomers of the alkene 2-butene. This is an example of **geometrical isomerism** in which the same bonds are present in a molecule, but are arranged in a different way. Another, more subtle form of isomerism is **optical isomerism**, which is based on the fact that some molecules can be mirror images of each other that are not superimposable, just as left and right gloves, both placed palm down, cannot be superimposed on each other, although placed palm to palm they are mirror images. Such isomers are known as optical isomers because in the pure form they rotate plane-polarized light either left or right. The *d* isomer rotates such light right and is said to be **dextrorotatory**, whereas the *l* isomer rotates plane-polarized light left and is labeled **levorotatory**. The *d* and *l* isomers of the same compound are called **enantiomers**, and the compounds that form enantiomers are said to be **chiral**.

Except for the property of rotating plane-polarized light in opposite directions, the physical properties of enantiomers of the same compound are identical. In addition, their chemical properties are identical, except when they are acted upon by another chiral molecule. One such kind of molecule consists of enzymes, large molecules of proteins that catalyze biochemical reactions. Therefore, many biochemical reactions involve chiral molecules.

Chirality has a strong influence on toxicity. This was tragically illustrated in the case of thalidomide (see Section 6.14) produced as a mixture of equal parts of *d* and *l* isomers, called a **racemic mixture**, and administered to counteract the nausea of morning sickness in early stages of pregnancy. One of the enantiomers was effective in this application and the other, tragically, caused devastating birth defects when the drug was used in Europe and Japan in the 1960s. Recently, using chiral synthetic procedures, the "good" isomer has been synthesized in a pure form and is now a well-regarded pharmaceutical for some applications.

1.10 SYNTHETIC POLYMERS

A large fraction of the chemical industry worldwide is devoted to polymer manufacture, which is very important in the area of hazardous wastes, as a source of environmental pollutants, in toxicology, and in the manufacture of materials used to alleviate environmental and waste problems. Synthetic **polymers** are produced when small molecules called **monomers** bond together to form a much smaller number of very large molecules. Many natural products are polymers; for example, cellulose in wood, paper, and many other materials is a polymer of the sugar glucose. Synthetic polymers form the basis of many industries, such as rubber, plastics, and textiles manufacture.

An important example of a polymer is polyvinylchloride, shown in Figure 1.21. This polymer is synthesized in large quantities for the manufacture of water and sewer pipes, water-repellant

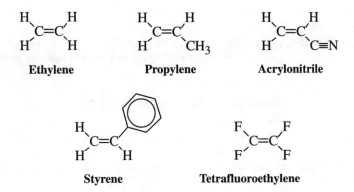

Figure 1.22 Monomers from which commonly used polymers are synthesized.

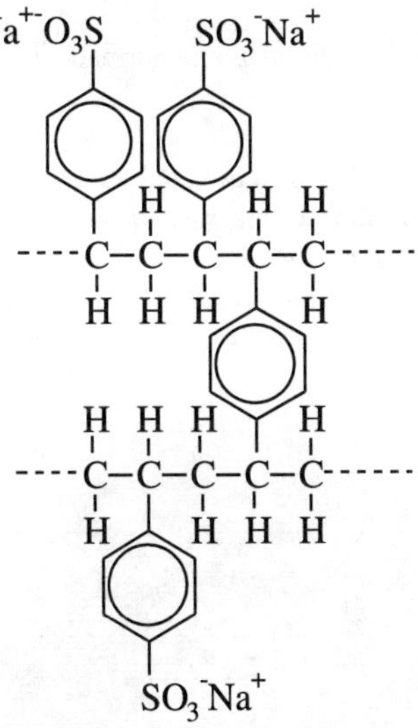

Figure 1.23 Polymeric cation exchanger in the sodium form.

liners, and other plastic materials. Other major polymers include polyethylene (plastic bags, milk cartons), polypropylene (impact-resistant plastics, indoor–outdoor carpeting), polyacrylonitrile (Orlon, carpets), polystyrene (foam insulation), and polytetrafluoroethylene (Teflon coatings, bearings); the monomers from which these substances are made are shown in Figure 1.22.

Polymers have a number of applications in waste treatment and disposal. Waste disposal landfill liners are made from synthetic polymers, as are the fiber filters that remove particulate pollutants from flue gas in baghouses. Membranes used for ultrafiltration and reverse osmosis treatment of water are composed of very thin sheets of synthetic polymers. Organic solutes can be removed from water by sorption onto hydrophobic (water-repelling) organophilic beads of Amberlite XAD resin. Heavy metal pollutants are removed from wastewater by cation exchange resins made of polymers with anionic functional groups. Typically, these resins exchange harmless sodium ion, Na^+, on the solid resin for toxic heavy metal ions in water. Figure 1.23 shows a segment of the polymeric structure of a cation exchange resin in the sodium form.

Many of the hazards from the polymer industry arise from the monomers used as raw materials. Many monomers are reactive and flammable, with a tendency to form explosive vapor mixtures with air. All have a certain degree of toxicity; vinyl chloride is a known human carcinogen. The combustion of many polymers may result in the evolution of toxic gases, such as hydrogen cyanide

(HCN) from polyacrylonitrile or hydrogen chloride (HCl) from polyvinylchloride. Another hazard presented by plastics results from the presence of **plasticizers** added to provide essential properties, such as flexibility. The most widely used plasticizers are phthalates (see Chapter 14), which are environmentally persistent, resistant to treatment processes, and prone to undergo bioaccumulation.

SUPPLEMENTARY REFERENCES

Manahan, S.E., *Fundamentals of Environmental Chemistry*, 2nd ed., CRC Press/Lewis Publishers, Boca Raton, FL, 2000.

Manahan, S.E., *Green Chemistry: Fundamentals of Chemical Science and Technology*, ChemChar, Columbia, MO, 2002.

QUESTIONS AND PROBLEMS

1. What distinguishes a radioactive isotope from a "normal" stable isotope?
2. Why is the periodic table so named?
3. Match the following:
    1. O_2 (a) Element consisting of individual atoms
    2. NH_3 (b) Element consisting of chemically bonded atoms
    3. Ar   (c) Ionic compound
    4. NaCl   (d) Covalently bound compound
4. Consider the following atom:

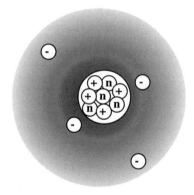

How many electrons, protons, and neutrons does it have? What is its atomic number? Give the name and chemical symbol of the element of which it is composed.
5. After examining Figure 1.7, consider what might happen when an atom of sodium (Na), atomic number 11, loses an electron to an atom of fluorine, (F), atomic number 9. What kinds of particles are formed by this transfer of a negatively charged electron? Is a chemical compound formed? What is it called?
6. Give the chemical formula and molecular mass of the molecule represented below:

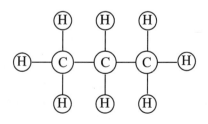

7. Calculate the molecular masses of (a) C_2H_2, (b) N_2H_4, (c) Na_2O, (d) O_3 (ozone), (e) PH_3, and (f) CO_2.
8. Is the equation $H_2 + O_2 \rightarrow H_2O$ a balanced chemical equation? Explain. Point out the reactants and products in the equation.
9. Define and distinguish the differences between environmental chemistry, environmental biochemistry, and toxicological chemistry.
10. An uncharged atom has the same number of _____ as protons. The electrons in an atom are contained in a cloud of _____ around the nucleus that occupies most of the _____ atom.
11. Match:
    A. Argon   1. A halogen
    B. Hydrogen   2. Fissionable element
    C. Uranium   3. Noble gas
    D. Chlorine   4. Has an isotope with a mass number of 2
    E. Mercury   5. Toxic heavy metal
12. The entry for each element in the periodic table gives the element's _____ _____, and the periodic table is arranged horizontally in _____ and vertically in _____ .
13. Electrons in atoms occupy _____ in which electrons have different _____. Each of these may contain a maximum of _____ electrons.
14. The Lewis symbol of carbon is _____, in which each dot represents a _____.
15. Elements that are generally solid, shiny in appearance, electrically conducting, and malleable are _____, whereas elements that tend to have a dull appearance, are not at all malleable, and frequently occur as gases or liquids are _____. Elements with intermediate properties are _____.
16. In the Lewis formula

    $$H : H$$

    the two dots represent _____.
17. Explain why H_2 is not a chemical compound, whereas H_2O is a chemical compound.
18. Using examples, distinguish between covalent and ionic bonds.
19. In terms of c, h, o, and the appropriate atomic masses, write a formula for the molecular mass of a compound with a general formula of $C_cH_hO_o$.
20. Considering oxidation and reduction phenomena, when Al reacts with O_2 to produce Al_2O_3, which contains the Al^{3+} ion, the Al is said to have been and is in the _____ oxidation state.
21. Calculate the concentration in moles per liter of (a) a solution that is 27.0% H_2SO_4 by mass and that has a density of 1198 g/L, and (b) of a solution that is 1 mg/L H_2SO_4 having a density of 1000 g/L.
22. Calculate the pH of the second solution described in the preceding problem, keeping in mind that each H_2SO_4 molecule yields two H^+ ions.
23. Write a balanced neutralization reaction between NaOH and H_2SO_4.
24. Distinguish between solutions and colloidal suspensions.
25. What is the nature of the Fe^{3+} ion? Why are solutions containing this ion acidic?
26. What kind of species is $Ni(CN)_4^{2-}$?
27. Explain the bonding properties of carbon that make organic chemistry so diverse.
28. Distinguish among alkanes, alkenes, alkynes, and aromatic compounds. To which general class of organic compounds do all belong?
29. In what sense are alkanes saturated? Why are alkenes more reactive than alkanes?
30. Name the compound below:

$$\begin{array}{c} H_3C CH_3 \\ H | | H H H \\ H-C-C-C-C-C-C-H \\ H H | H H H \\ CH_3 \end{array}$$

31. What is indicated by *n* in a hydrocarbon name?
32. Discuss the chemical reactivity of alkanes. Why are they chemically reactive or unreactive?
33. Discuss the chemical reactivity of alkenes. Why are they chemically reactive or unreactive?
34. What are the characteristics of aromaticity? What are the chemical reactivity characteristics of aromatic compounds?
35. Describe chain reactions, discussing what is meant by free radicals and photochemical processes.
36. Define, with examples, what is meant by isomerism.
37. Describe how the two forms of 1,2-dichloroethylene can be used to illustrate *cis–trans* isomerism.
38. Give the structural formula corresponding to the condensed structural formula of $CH_3CH(C_2H_5)CH(C_2H_5)CH_2CH_3$.
39. Discuss how organic functional groups are used to define classes of organic compounds.
40. Give the functional groups corresponding to (a) alcohols, (b) aldehydes, (c) carboxylic acids, (d) ketones, (e) amines, (f) thiol compounds, and (g) nitro compounds.
41. Give an example compound of each of the following: epoxides, alcohols, phenols, ethers, aldehydes, ketones, and carboxylic acids.
42. Which functional group is characteristic of N-nitroso compounds, and why are these compounds toxicologically significant?
43. Give an example of each of the following: alkyl halides, alkenyl halides, and aromatic halides.
44. Give an example compound of a chlorinated naphthalene and of a PCB.
45. What explains the tremendous chemical stability of CFCs? What kinds of compounds are replacing CFCs? Why?
46. How does a thio differ from a thioether?
47. How does a sulfoxide differ from a sulfone?
48. Which inorganic compound is regarded as the parent compound of alkyl and aromatic phosphines? Give an example of each of these.
49. What are organophosphate esters and what are their toxicological significance?
50. Define what is meant by a polymer and give an example of one.

CHAPTER 2

Environmental Chemistry

2.1 ENVIRONMENTAL SCIENCE AND ENVIRONMENTAL CHEMISTRY

In order to understand toxicological chemistry, it is necessary to have some understanding of the environmental context in which toxicological chemical phenomena occur. This in turn requires an understanding of the broader picture of environmental science and environmental chemistry, which are addressed in this chapter. Also needed is an understanding of how environmental chemicals interact with organisms and their ecosystems, as addressed by the topic of ecotoxicology, covered in Chapter 5.

Environmental science can be defined as *the study of the earth, air, water, and living environments, and the effects of technology thereon.*[1] To a significant degree, environmental science has evolved from investigations of the ways by which, and places in which, living organisms carry out their life cycles. This is the discipline of **natural history**, which in recent times has evolved into **ecology**, the study of environmental factors that affect organisms and how organisms interact with these factors and with each other.

2.1.1 The Environment

Traditionally, environmental science has been divided among the study of the atmosphere, the hydrosphere, the geosphere, and the biosphere. To an increasing extent during their brief time on earth, humans have used their ingenuity and technology to cause enormous perturbations in the natural environment. This has occurred to such a degree that it is now necessary to recognize a fifth sphere of the environment that is constructed and operated by humans, the **anthrosphere**. The five spheres of the environment are shown in Figure 2.1.

The **atmosphere** is the thin layer of gases that cover Earth's surface. In addition to its role as a reservoir of gases, the atmosphere moderates Earth's temperature, absorbs energy and damaging ultraviolet radiation from the sun, transports energy away from equatorial regions, and serves as a pathway for vapor-phase movement of water in the hydrologic cycle. The **hydrosphere** contains Earth's water. Over 97% of Earth's water is in oceans, and most of the remaining fresh water is in the form of ice. Therefore, only a relatively small percentage of the total water on Earth is actually involved with terrestrial, atmospheric, and biological processes. Exclusive of seawater, the water that circulates through environmental processes and cycles occurs in the atmosphere, underground as ground water, and as surface water in streams, rivers, lakes, ponds, and reservoirs. The **geosphere** consists of the solid earth, including soil, which supports most plant life. The part of the geosphere that is directly involved with environmental processes through contact with the atmosphere, the hydrosphere, and living things is the solid **lithosphere**. The lithosphere varies from 50 to 100 km in thickness. The most important part of it insofar as interactions with the other spheres of the environment are concerned is its thin outer skin, composed largely of lighter, silicate-based minerals

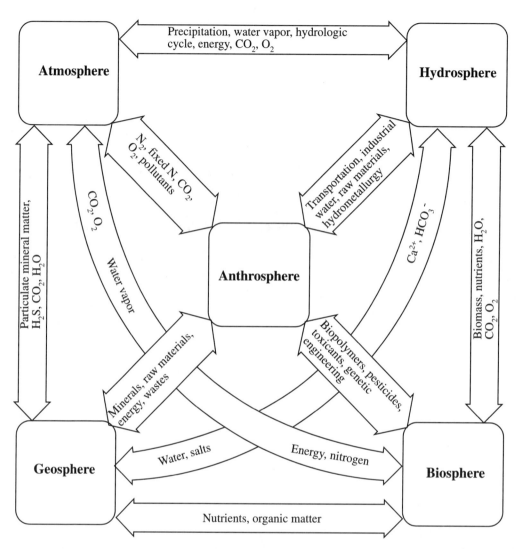

Figure 2.1 Illustration of the close relationships of the air, water, and earth environments with each other and with living systems, as well as the tie-in with technology (the anthrosphere).

and called the **crust**. All living entities on Earth compose the **biosphere**. Living organisms and the aspects of the environment pertaining directly to them are called **biotic**, and other portions of the environment are **abiotic**. The **anthrosphere** consists of all the structures and devices made and operated by humans. The anthrosphere is composed of buildings, highways, parking lots, railroads, vehicles, aircraft, and other things that people make and do in Earth's environment. It obviously has a major influence on all environmental phenomena, and any realistic treatment of environmental science must consider the anthrosphere along with the other four environmental spheres.

To a large extent, the strong interactions among living organisms and the various spheres of the abiotic environment are best described by cycles of matter that involve biological, chemical, and geological processes and phenomena. Such cycles are called **biogeochemical cycles**. Organisms participate in biogeochemical cycles, which describe the circulation of matter, particularly plant and animal nutrients, through ecosystems. As part of the carbon cycle, atmospheric carbon in CO_2 is fixed as biomass; as part of the nitrogen cycle, atmospheric N_2 is fixed in organic matter. The reverse of these kinds of processes is **mineralization**, in which biologically bound elements are returned to inorganic states. A typical biogeochemical cycle, the nitrogen cycle, is shown in Figure 2.2.

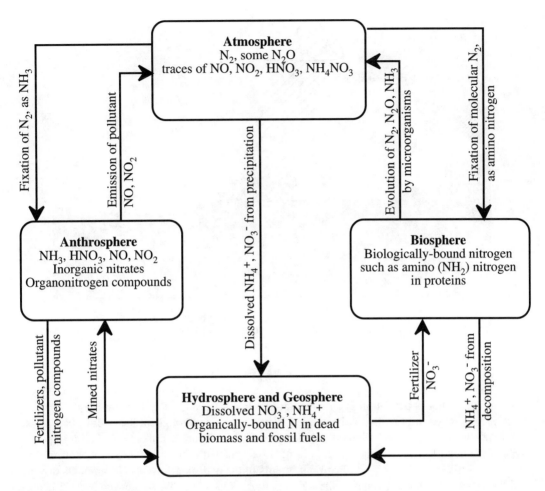

Figure 2.2 The nitrogen cycle.

2.1.2 Environmental Chemistry

Environmental chemistry is *the study of the sources, reactions, transport, effects, and fates of chemical species in the water, air, terrestrial, and living environments and the effects of human activities thereon.*[2] Some idea of the complexity of environmental chemistry as a discipline may be realized by examining Figure 2.1, which indicates the interchange of chemical species among various environmental spheres. Throughout an environmental system there are variations in temperature, mixing, intensity of solar radiation, input of materials, and various other factors that strongly influence chemical conditions and behavior. Because of its complexity, environmental chemistry must be approached with simplified models. This chapter presents an overview of environmental chemistry

The definition of environmental chemistry given above is illustrated for a typical environmental pollutant in Figure 2.3. Pollutant sulfur dioxide is generated in the anthrosphere by combustion of sulfur in coal, which has been extracted from the geosphere. The SO_2 is transported to the atmosphere with flue gas and oxidized by chemical and photochemical processes in the atmosphere to sulfuric acid. The sulfuric acid, in turn, falls as acidic precipitation, where it may have detrimental effects, such as toxic effects, on trees and other plants in the biosphere. Eventually the sulfuric acid is carried by stream runoff in the hydrosphere to a lake or ocean, where its ultimate fate is to be stored in solution in the water or precipitated as solid sulfates and returned to the geosphere.

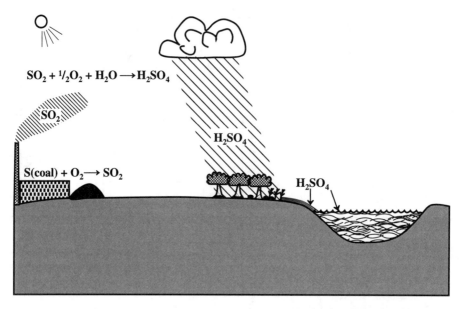

Figure 2.3 Illustration of the definition of environmental chemistry by the example of pollutant sulfuric acid formed by the oxidation of sulfur dioxide generated during the combustion of sulfur-containing coal.

2.2 WATER

Although water is part of all environmental spheres, it is convenient to regard portions of the environment as constituting the **hydrosphere**. Water circulates throughout the hydrosphere in one of nature's great cycles, the **hydrologic cycle**. As shown in Figure 2.4, there are several important parts of the hydrosphere, and these are, in turn, closely related to the other environmental spheres. Earth's water is contained in several **compartments** of the hydrologic cycle. The amounts of water and the **residence times** of water in these compartments vary greatly. By far, the largest of these compartments consists of the **oceans**, containing about 97% of all Earth's chemically unbound water, with a residence time of about 3000 years. Oceans serve as a huge reservoir for water and as the source of most water vapor that enters the hydrologic cycle. As vast heat sinks, oceans have a tremendous moderating effect on climate. A relatively large amount of water is also contained in the solid state as ice, snowpacks, glaciers, and the polar ice caps. Surface water is found in lakes, streams, and reservoirs. Groundwater is located in aquifers underground. Another fraction of water is present as water vapor in the atmosphere (clouds). There is a strong connection between the *hydrosphere*, where water is found, and the *geosphere*, or land; human activities affect both.

Water has a number of unique properties that are essential to life, due largely to its molecular structure and bonding properties. Among the special characteristics of water are the fact that it is an excellent solvent, it has a temperature–density relationship that results in bodies of water becoming stratified in layers, it is transparent, and it has an extraordinary capacity to absorb, retain, and release heat per unit mass of ice, liquid water, or water vapor.

Water's unique temperature–density relationship results in the formation of distinct layers within nonflowing bodies of water, as shown in Figure 2.5. During the summer, a surface layer (**epilimnion**) is heated by solar radiation and, because of its lower density, floats on the bottom layer, or **hypolimnion**. This phenomenon is called **thermal stratification**. When an appreciable temperature difference exists between the two layers, they do not mix, but behave independently and have very different chemical and biological properties. The epilimnion, which is exposed to light, may have a heavy growth of algae. As a result of exposure to the atmosphere and (during daylight hours) because of the photosynthetic activity of algae, the epilimnion contains relatively higher levels of

ENVIRONMENTAL CHEMISTRY

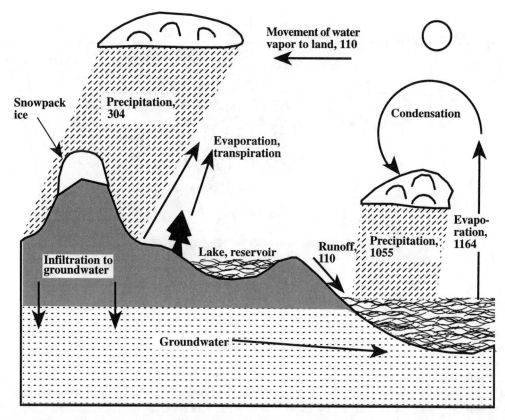

Figure 2.4 The hydrologic cycle, quantities of water in trillions of liters per day.

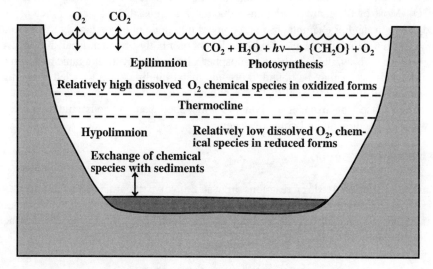

Figure 2.5 Stratification of a body of water.

dissolved oxygen and is said to be *aerobic*. In the hypolimnion, bacterial action on biodegradable organic material consumes oxygen and may cause the water to become *anaerobic*, that is, essentially free of oxygen. As a consequence, chemical species in a relatively reduced form tend to predominate in the hypolimnion.

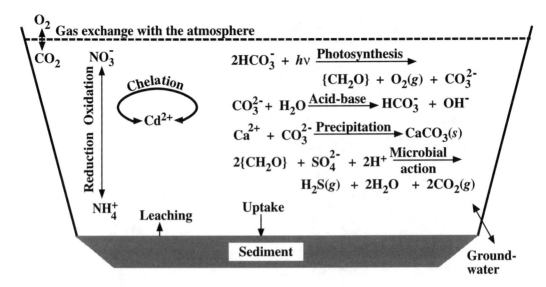

Figure 2.6 Major aquatic chemical processes.

2.3 AQUATIC CHEMISTRY

Figure 2.6 summarizes the more important environmental chemical aspects of **aquatic chemistry**. As shown in this figure, a number of chemical phenomena occur in water. Many aquatic chemical processes are influenced by the action of algae and bacteria in water. For example, it is shown that algal photosynthesis fixes inorganic carbon from HCO_3^- ion in the form of biomass (represented as $\{CH_2O\}$), in a process that also produces carbonate ion, CO_3^{2-}. Carbonate undergoes an acid–base reaction to produce OH^- ion and raise the pH, or it reacts with Ca^{2+} ion to precipitate solid $CaCO_3$. Most of the many oxidation–reduction reactions that occur in water are mediated (catalyzed) by bacteria. For example, bacteria convert inorganic nitrogen largely to ammonium ion, NH_4^+, in the oxygen-deficient (anaerobic) lower layers of a body of water. Near the surface, which is aerobic because O_2 is available from the atmosphere, bacteria convert inorganic nitrogen to nitrate ion, NO_3^-. Metals in water may be bound to organic metal-binding species called chelating agents, such as pollutant nitrilotriacetic acid (NTA) or naturally occurring fulvic acids produced by decay of plant matter. Gases are exchanged with the atmosphere, and various solutes are exchanged between water and sediments in bodies of water.

2.3.1 Oxidation–Reduction

Oxidation–reduction (redox) reactions in water involve the transfer of electrons between chemical species, usually through the action of bacteria. The relative oxidation–reduction tendencies of a chemical system depend on the **activity of the electron** e^-. When the electron activity is relatively high, chemical species, including water, tend to accept electrons,

$$2H_2O + 2e^- \rightleftharpoons H_2(g) + 2OH^- \tag{2.3.1}$$

and are said to be **reduced**. When the electron activity is relatively low, the medium is **oxidizing**, and chemical species such as H_2O may be **oxidized** by the loss of electrons,

$$2H_2O \rightleftharpoons O_2(g) + 4H^+ + 4e^- \tag{2.3.2}$$

The relative tendency toward oxidation or reduction may be expressed by the electrode potential, E, which is more positive in an oxidizing medium and more negative in a reducing medium.

2.3.2 Complexation and Chelation

Metal ions in water are always bonded to water molecules in the form of hydrated ions represented by the general formula $M(H_2O)_x^{n+}$, from which H_2O is often omitted for simplicity. Other species may be present that bond to the metal ion more strongly than does water. For example, cadmium ion dissolved in water, Cd^{2+}, reacts with cyanide ion, CN^-, as follows:

$$Cd^{2+} + CN^- \rightarrow CdCN^+ \qquad (2.3.3)$$

The product of the reaction is called a **complex** (complex ion), and the cyanide ion is called a **ligand**. Some ligands, called chelating agents, can bond with a metal ion in two or more places, forming particularly stable complexes. Chelating agents are almost always organic molecules or anions.

In addition to metal complexes and chelates, another major type of environmentally important metal species consists of **organometallic compounds**. These differ from complexes and chelates in that the organic portion is bonded to the metal by a carbon–metal bond and the organic ligand is frequently not capable of existing as a stable separate species.

Complexation, chelation, and organometallic compound formation have strong effects on metals in the environment. For example, complexation with negatively charged ligands may convert a soluble metal cation, which is readily bound and immobilized by ion exchange processes in soil, to an anion, such as $Ni(CN)_4^{2-}$, that is not strongly held by soil. On the other hand, some chelating agents are used for the treatment of heavy metal poisoning in chelation therapy (see chelation therapy for lead poisoning in Section 10.4). Furthermore, insoluble chelating agents, such as chelating resins, can be used to remove metals from waste streams.

2.3.3 Water Interactions with Other Phases

Most of the important chemical phenomena associated with water do not occur in solution, but rather through interaction of solutes in water with other phases. Such interactions may involve exchange of solute species between water and sediments, gas exchange between water and the atmosphere, and effects of organic surface films. Substances dissolve in water from other phases, and gases are evolved and solids precipitated as the result of chemical and biochemical phenomena in water.

Sediments are repositories of a wide variety of chemical species and are the media in which many chemical and biochemical processes occur. Sediments are sinks for many hazardous organic compounds and heavy metal salts that have entered into water.

Colloids, which consist of very small particles ranging from 0.001 to 1 µm in diameter, have a strong influence on aquatic chemistry. Colloids have very high surface-to-volume ratios, so their physical, chemical, and biological activities may be high. Colloids may be very difficult to remove from water during water treatment.

2.3.4 Water Pollutants

Natural waters are afflicted with a wide variety of inorganic, organic, and biological pollutants. In some cases, such as that of highly toxic cadmium, a pollutant is directly toxic at a relatively low level. In other cases, the pollutant itself is not toxic, but its presence results in conditions detrimental to water quality. For example, biodegradable organic matter in water is often not toxic, but the consumption of oxygen during its degradation prevents the water from supporting fish life.

Some contaminants, such as NaCl, are normal constituents of water at low levels, but harmful pollutants at higher levels. The proper design of industrial ecosystems minimizes the release of water pollutants.

2.3.5 Water Treatment

The treatment of water can be considered under two major categories: (1) treatment before use, and (2) treatment of contaminated water after it has passed through a municipal water system or industrial process. In both cases, consideration must be given to potential contamination by pollutants and their removal from water to acceptable levels.

Several operations may be employed to treat water prior to use. Aeration is used to drive off odorous gases, such as H_2S, and to oxidize soluble Fe^{2+} and Mn^{2+} ions to insoluble forms. Lime is added to remove dissolved calcium (water hardness). $Al_2(SO_4)_3$ forms a sticky precipitate of $Al(OH)_3$, which causes very fine particles to settle. Various filtration and settling processes are employed to treat water. Chlorine, Cl_2, is added to kill bacteria. Formation of undesirable byproducts of water chlorination may be avoided by disinfection with chlorine dioxide, ClO_2, or ozone, O_3.

Municipal wastewater may be subjected to primary, secondary, or advanced water treatment. **Primary** water treatment consists of settling and skimming operations that remove grit, grease, and physical objects from water. **Secondary** water treatment is designed to take out biochemical oxygen demand (BOD). This is normally accomplished by introducing air and microorganisms such that waste biomass in the water, $\{CH_2O\}$, is removed by aerobic respiration of microorganisms acting on degradable biomass:

$$\{CH_2O\} + O_2 \rightarrow CO_2 + H_2O \text{ (aerobic respiration)} \tag{2.3.4}$$

2.4 THE GEOSPHERE

The **geosphere**, or solid Earth, is that part of the Earth upon which humans live and from which they extract most of their food, minerals, and fuels. Once thought to have an almost unlimited buffering capacity against the perturbations of humankind, the geosphere is now known to be rather fragile and subject to harm by human activities, such as mining, acid rain, erosion from poor cultivation practices, and disposal of hazardous wastes. It may be readily seen that the preservation of the geosphere in a form suitable for human habitation is one of the greatest challenges facing humankind.

2.4.1 Solids in the Geosphere

The part of Earth's geosphere that is accessible to humans is the **crust**, which, ranging from 5 to 40 km in thickness, is extremely thin compared to the diameter of the earth. Most of the solid earth crust consists of rocks. Rocks are composed of minerals, where a **mineral** is a naturally occurring inorganic solid with a definite internal crystal structure and chemical composition. A **rock** is a solid, cohesive mass of pure mineral or an aggregate of two or more minerals.

At elevated temperatures deep beneath Earth's surface, rocks and mineral matter melt to produce a molten substance called **magma**. Cooling and solidification of magma produces **igneous rock**. Common igneous rocks are granite, basalt, quartz (SiO_2), feldspar (($Ca,Na,K)AlSi_3O_8$), and magnetite (Fe_3O_4). Exposure of igneous rocks formed under water-deficient, chemically reducing conditions of high temperature and high pressure to wet, oxidizing, low-temperature, and low-pressure conditions at the surface causes the rocks to disintegrate by a process called **weathering**.

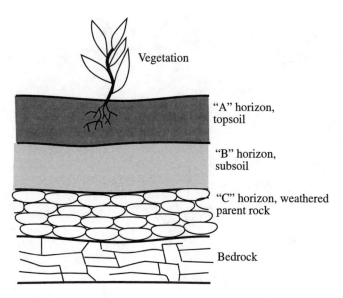

Figure 2.7 Soil profile showing soil horizons.

Erosion from wind, water, or glaciers picks up materials from weathering rocks and deposits them as **sediments** or **soil**. A process called **lithification** describes the conversion of sediments to **sedimentary rocks**. In contrast to the parent igneous rocks, sediments and sedimentary rocks are porous, soft, and chemically reactive. **Metamorphic rock** is formed by the action of heat and pressure on sedimentary, igneous, or other kinds of metamorphic rock that are not in a molten state.

Although over 2000 minerals are known, only about 25 **rock-forming minerals** make up most of the Earth's crust. Oxygen and silicon comprise 49.5 and 25.7% by mass of the Earth's crust, respectively; therefore, most minerals are **silicates** such as quartz, SiO_2, or potassium feldspar, $KAlSi_3O_8$. In descending order of abundance, the other elements in the Earth's crust are aluminum (7.4%), iron (4.7%), calcium (3.6%), sodium (2.8%), potassium (2.6%), magnesium (2.1%), and other (1.6%).

2.5 SOIL

Insofar as the human environment and life on Earth are concerned, the most important part of the Earth's crust is **soil**, which consists of particles that make up a variable mixture of minerals, organic matter, and water, capable of supporting plant life on Earth's surface. It is the final product of the weathering action of physical, chemical, and biological processes on rocks. The organic portion of soil consists of plant biomass in various stages of decay. High populations of bacteria, fungi, and animals such as earthworms may be found in soil. Soil contains air spaces and generally has a loose texture.

Soils usually exhibit distinctive layers with increasing depth (Figure 2.7). These layers, called **horizons**, form as the result of complex interactions among processes that occur during weathering. Rainwater percolating through soil carries dissolved and colloidal solids to lower horizons, where they are deposited. Biological processes, such as bacterial decay of residual plant biomass, produce slightly acidic CO_2, organic acids, and complexing compounds that are carried by rainwater to lower horizons, where they interact with clays and other minerals, altering the properties of the minerals. The top layer of soil, typically several inches in thickness, is known as the A horizon, or **topsoil**. This is the layer of maximum biological activity in the soil; it contains most of the soil's

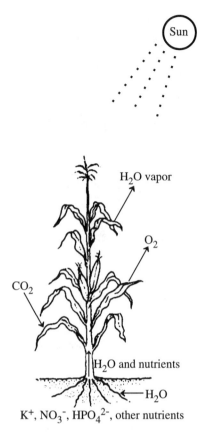

Figure 2.8 Plants transport water from the soil to the atmosphere by transpiration. Nutrients are also carried from the soil to the plant extremities by this process. Plants remove CO_2 from the atmosphere and add O_2 by photosynthesis. The reverse occurs during plant respiration.

organic matter. Metal ions and clay particles in the A horizon are subject to considerable leaching, such that it is sometimes called the zone of leaching. The next layer is the B horizon, or **subsoil**. It receives material such as organic matter, salts, and clay particles leached from the topsoil, so it is called the zone of accumulation. The C horizon is composed of fractured and weathered parent rocks from which the soil originated.

Soils exhibit a large variety of characteristics that are used to classify them for various purposes, including crop production, road construction, and waste disposal. The parent rocks from which soils are formed obviously play a strong role in determining the composition of soils. Other soil characteristics include strength, workability, soil particle size, permeability, and degree of maturity.

Water is a crucial part of the three-phase, solid–liquid–gas system making up soil. It is the solvent of the soil solution (see Section 2.6) and is the basic transport medium for carrying plant nutrients from solid soil particles into plant roots and to the farthest reaches of the plant's leaf structure (Figure 2.8). The water enters the atmosphere from the plant's leaves, a process called **transpiration**. Large quantities of water are required for the production of most plant materials.

Normally, because of the small size of soil particles and the presence of small capillaries and pores in the soil, the water phase is not totally independent of soil solid matter. Water present in larger spaces in soil is relatively more available to plants and readily drains away. Water held in smaller pores or between the unit layers of clay particles is held much more firmly. Water in soil interacts strongly with organic matter and clay minerals.

2.6 GEOCHEMISTRY AND SOIL CHEMISTRY

Geochemistry deals with chemical species, reactions, and processes in the lithosphere and their interactions with the atmosphere and hydrosphere. The branch of geochemistry that explores the complex interactions among the rock–water–air–life (and human) systems that determine the chemical characteristics of the surface environment is **environmental geochemistry**. Obviously, geochemistry and its environmental subdiscipline are very important in environmental science and in considerations of industrial ecology.

Geochemistry addresses a large number of chemical and related physical phenomena. Some of the major areas of geochemistry are the following:

- The chemical composition of major components of the geosphere, including magma and various kinds of solid rocks
- Processes by which elements are mobilized, moved, and deposited in the geosphere through a cycle known as the **geochemical cycle**
- Chemical processes that occur during the formation of igneous rocks from magma
- Chemical processes that occur during the formation of sedimentary rocks
- Chemistry of rock weathering
- Chemistry of volcanic phenomena
- Role of water and solutions in geological phenomena, such as deposition of minerals from hot brine solutions
- The behavior of dissolved substances in concentrated brines

An important consideration in geochemistry is that of the interaction of life-forms with geochemical processes, addressed as biogeochemistry or organic geochemistry. The deposition of biomass and the subsequent changes that it undergoes have caused huge deposits of petroleum, coal, and oil shale. Chemical changes induced by photosynthesis have resulted in massive deposits of calcium carbonate (limestone). Deposition of the biochemically synthesized shells of microscopic animals has led to the development of large formations of calcium carbonate and silica. Biogeochemistry is closely involved with elemental cycles, such as the nitrogen cycle shown in Figure 2.2.

2.6.1 Physical and Chemical Aspects of Weathering

Defined in Section 2.4, *weathering* is discussed here as a geochemical phenomenon. Rocks tend to weather more rapidly when there are pronounced changes over time in physical conditions — alternate freezing and thawing and wet periods alternating with severe drying. Other mechanical aspects are swelling and shrinking of minerals with hydration and dehydration, as well as growth of roots through cracks in rocks. The rates of chemical reactions involved in weathering increase with increasing temperature.

As a chemical phenomenon, weathering can be viewed as the result of the tendency of the rock–water–mineral system to attain equilibrium. This occurs through the usual chemical mechanisms of dissolution and precipitation, acid–base reactions, complexation, hydrolysis, and oxidation–reduction.

Weathering is very slow in dry air. Water increases the rate of weathering by many orders of magnitude for several reasons. Water, itself, is a chemically active substance in the weathering process. Furthermore, water holds weathering agents in solution such that they are transported to chemically active sites on rock minerals and contact the mineral surfaces at the molecular and ionic levels. Prominent among such weathering agents are CO_2, O_2, organic acids, sulfur acids ($SO_2(aq)$, H_2SO_4), and nitrogen acids (HNO_3, HNO_2). Water provides the source of H^+ ion needed for acid-forming gases to act as acids, as shown by the following:

$$CO_2 + H_2O \rightarrow H^+ + HCO_3^- \qquad (2.6.1)$$

$$SO_2 + H_2O \rightarrow H^+ + HSO_3^- \qquad (2.6.2)$$

Rainwater is essentially free of mineral solutes. It is usually slightly acidic due to the presence of dissolved carbon dioxide, or more highly acidic because of acid rain-forming constituents. As a result of its slight acidity and lack of alkalinity and dissolved calcium salts, rainwater is *chemically aggressive* toward some kinds of mineral matter, which it breaks down by a process called **chemical weathering**.

A typical chemical reaction involved in weathering is the dissolution of calcium carbonate (limestone) by water containing dissolved carbon dioxide:

$$CaCO_3(s) + H_2O + CO_2(aq) \rightarrow Ca^{2+}(aq) + 2HCO_3^-\,(aq) \qquad (2.6.3)$$

Weathering may also involve oxidation reactions, such as occurs when pyrite, FeS_2, is oxidized, a process that can produce sulfuric acid (acid mine water):

$$4FeS_2(s) + 15O_2(g) + (8 + 2x)H_2O \rightarrow 2Fe_2O_3 \cdot xH_2O + 8SO_4^{2-}\,(aq) + 16H^+(aq) \qquad (2.6.4)$$

2.6.2 Soil Chemistry

A large variety of chemical and biochemical processes occur in soil. In discussing soil chemistry, it is crucial to consider the **soil solution**, which is the aqueous portion of soil that contains dissolved matter from soil chemical and biochemical processes and from exchange with the hydrosphere and biosphere. This medium transports chemical species to and from soil particles and provides intimate contact between the solutes and the soil particles. In addition to providing water for plant growth, soil solution is an essential pathway for the exchange of plant nutrients between roots and solid soil.

Soil acts as a buffer and resists changes in pH. The buffering capacity depends on the type of soil. The acidity and basicity of soil must be kept within certain ranges to enable the soil to be productive. Usually soil tends to become too acidic through the processes by which plants take up nutrient cations from it (see Reaction 2.6.6 below). Most common plants grow best in soil with a pH near neutrality. If the soil becomes too acidic for optimum plant growth, it may be restored to productivity by liming, ordinarily through the addition of calcium carbonate:

$$\text{Soil}\}\,(H^+)_2 + CaCO_3 \rightarrow \text{Soil}\}\,Ca^{2+} + CO_2 + H_2O \qquad (2.6.5)$$

Chemical processes involving ions in the soil solution and bound to the soil solids are very important. Cation exchange in soil is the mechanism by which potassium, calcium, magnesium, and essential trace-level metals are made available to plants. When nutrient metal ions are taken up by plant roots, hydrogen ion is exchanged for the metal ions. This process, plus the leaching of calcium, magnesium, and other metal ions from the soil by water containing carbonic acid, tends to make the soil acidic:

$$\text{Soil}\}\,Ca^{2+} + 2CO_2 + 2H_2O \rightarrow \text{Soil}\}\,(H^+)_2 + Ca^{2+}(\text{root}) + 2HCO_3^- \qquad (2.6.6)$$

A large number of oxidation–reduction processes occur in soil, almost always mediated by microorganisms. The most common of these processes is the biodegradation of biomass, such as that from crop residues, here represented as $\{CH_2O\}$:

$$\{CH_2O\} + O_2 \rightarrow CO_2 + H_2O \qquad (2.6.7)$$

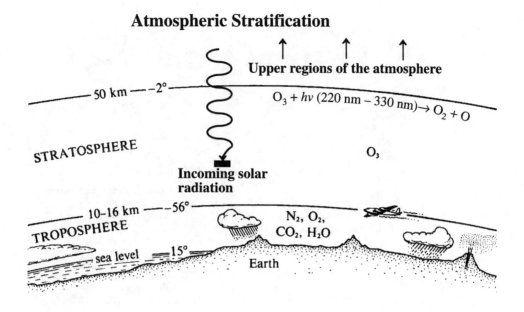

Figure 2.9 The two lower layers of the atmosphere. Above the stratosphere is the mesosphere, a region in which the temperature decreases with increasing altitude, and above that is the thermosphere, a region characterized by increasing temperature with increasing altitude. These two regions play a very important role in filtering out high-energy solar electromagnetic radiation before it reaches Earth's surface.

This process consumes oxygen and produces CO_2. As a result, the oxygen content of air in soil may be as low as 15%, and the carbon dioxide content may be several percent. Thus, the decay of organic matter in soil increases the equilibrium level of dissolved CO_2 in groundwater. This lowers the pH and contributes to weathering of carbonate minerals, particularly calcium carbonate.

Although the organic fraction of solid soil is usually not greater than 5%, organic matter strongly influences the chemical, physical, and biological properties of the soil. **Soil humus** is by far the most significant organic constituent. Humus is the residue left when bacteria and fungi biodegrade plant material and results largely from the microbial alteration of lignin, which is the material that binds plant matter together. Humus is composed of soluble humic and fulvic acids and an insoluble fraction called humin. Humus molecules exhibit variable, complex chemical structures. They have acid–base, ion-exchanging, and metal-chelating properties.

2.7 THE ATMOSPHERE

The **atmosphere** is made up of the thin layer of mixed gases, consisting predominantly of nitrogen, oxygen, and water vapor, that covers the Earth's surface. As shown in Figure 2.9, the atmosphere is divided into several layers on the basis of temperature–density relationships resulting from interrelationships between physical and photochemical (light-induced chemical phenomena) processes in air. Of these, the most significant for human activities are the troposphere, on Earth's surface, and the stratosphere, the next highest layer.

The lowest layer of the atmosphere, extending from sea level to an altitude of 10 to 16 km (depending on time, temperature, latitude, season, climate conditions, and underlying terrestrial features), is the **troposphere**, which is characterized by a generally homogeneous composition of major gases other than water and decreasing temperature with increasing altitude from the heat-radiating surface of the Earth. The temperature of the troposphere ranges from an average of 15°C

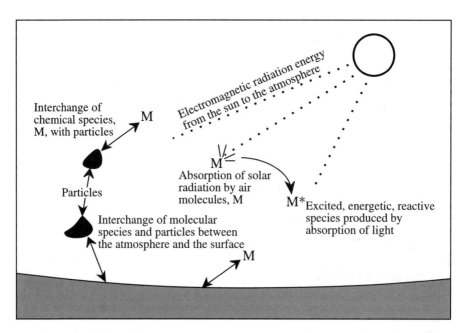

Figure 2.10 Atmospheric chemical processes may occur in the gas phase on or in particles, and on land or water surfaces in contact with the atmosphere. A particular feature of atmospheric chemistry is that of photochemical reactions induced by the absorption of energetic photons of sunlight.

at sea level to an average of −56°C at its upper boundary. The homogeneous composition of the troposphere results from constant mixing by circulating air masses. However, the water vapor content of the troposphere is extremely variable because of cloud formation, precipitation, and evaporation of water from terrestrial water bodies.

The atmospheric layer directly above the troposphere is the **stratosphere**, the average temperature of which increases from −56°C at its boundary with the troposphere to −2°C at its upper boundary, around 50 km. The reason for this increase is absorption of solar ultraviolet energy by ozone (O_3), levels of which may reach around 10 ppm by volume in the midrange of the stratosphere.

Atmospheric air may contain up to 5% water by volume, although the normal range is 1 to 3%. The two major constituents of air are nitrogen and oxygen, present at levels of 78.08 and 20.95% by volume in dry air, respectively. The noble gas argon comprises 0.934% of the volume of dry air. Carbon dioxide makes up about 0.037% by volume of dry air, a figure that fluctuates seasonably because of photosynthesis, and which is increasing steadily as more CO_2 is released to the atmosphere by fossil-fuel combustion. There are numerous trace gases in the atmosphere at levels below 0.002%, including neon, helium, krypton, xenon, sulfur dioxide, ozone, CO, N_2O, NO_2, NH_3, CH_4, SO_2, and CCl_2F_2, a persistent chlorofluorocarbon (Freon compound) released from air conditioners and other sources. Some of the trace gases may have profound effects, such as the role played by chlorofluorocarbons in the depletion of the stratosphere's protective ozone layer and that played by methane in global warming.

2.8 ATMOSPHERIC CHEMISTRY

As indicated in Figure 2.10, many kinds of chemical reactions occur in the atmosphere. These reactions take place in the gas phase, on atmospheric particle surfaces, within particulate water droplets, and on land and water surfaces in contact with the atmosphere. The most significant feature of atmospheric chemistry is the occurrence of **photochemical reactions** that take place when molecules in the atmosphere absorb energy in the form of light photons, designated $h\nu$. (The energy, E, of a photon of visible or ultraviolet light is given by the equation, $E = h\nu$, where h is Planck's constant and ν is the frequency of light, which is inversely proportional to its wavelength. Ultraviolet radiation has a higher frequency than visible light and is, therefore, more energetic and more likely to break chemical bonds in molecules that absorb it.) A molecule that has absorbed a photon of electromagnetic radiation has excess energy and is said to be in an **excited state**. Such a molecule is designated with an asterisk to distinguish it from an unexcited molecule, which is said to be in the **ground state**. For example, a molecule of NO_2 that has absorbed a photon of ultraviolet radiation is denoted NO_2^*. Such a species may be highly reactive; typically NO_2^* dissociates to yield NO and reactive O atoms, the process that initiates photochemical smog formation (see below).

One of the most significant photochemical reactions is the one responsible for the presence of ozone in the stratosphere (see above). It is initiated when O_2 absorbs highly energetic ultraviolet radiation in the wavelength ranges of 135–176 and 240–260 nm in the stratosphere:

$$O_2 + h\nu \rightarrow O + O \qquad (2.8.1)$$

The oxygen atoms produced by the photochemical dissociation of O_2 react with oxygen molecules to produce ozone, O_3:

$$O + O_2 + M \rightarrow O_3 + M \qquad (2.8.2)$$

where M is a third body, such as a molecule of N_2, which absorbs excess energy from the reaction. The ozone that is formed is very effective in absorbing ultraviolet radiation in the 220- to 330-nm wavelength range, which causes the temperature increase observed in the stratosphere. The ozone serves as a very valuable filter to remove ultraviolet radiation from the sun's rays. If this radiation reached the Earth's surface, it would cause skin cancer and other damage to living organisms.

2.8.1 Gaseous Oxides in the Atmosphere

Oxides of carbon, sulfur, and nitrogen are important constituents of the atmosphere and are pollutants at higher levels. Of these, carbon dioxide, CO_2, is the most abundant. It is a natural atmospheric constituent and is required for plant growth. However, the level of carbon dioxide in the atmosphere, now at about 370 ppm by volume, is increasing by about 1 ppm per year. This increase in atmospheric CO_2 may well cause general atmospheric warming — the "greenhouse effect" — with potentially very serious consequences for the global atmosphere and for life on earth. One of the central challenges for humankind and its systems of industrial ecology in the future will be to provide the energy needed to keep operating without adding to the global burden of carbon dioxide.

Though not a global threat, carbon monoxide, CO, can be a serious health threat. This is because carbon monoxide binds very strongly to hemoglobin and prevents blood from transporting oxygen to body tissues (see Chapter 11).

The two most serious nitrogen oxide air pollutants are nitric oxide, NO, and nitrogen dioxide, NO_2, collectively denoted as NO_x. These enter the atmosphere primarily as NO, and photochemical processes in the atmosphere tend to convert NO to NO_2. Further reactions can result in the formation of corrosive nitrate salts or nitric acid, HNO_3. Nitrogen dioxide is particularly significant in atmospheric chemistry because of its photochemical dissociation by light with a wavelength less than 430 nm to produce highly reactive O atoms. This is the first step in the formation of photochemical smog (see below). Sulfur dioxide, SO_2, is a reaction product of the combustion of sulfur-containing fuels, such as high-sulfur coal. Part of this sulfur dioxide is converted in the atmosphere to sulfuric acid, H_2SO_4, normally the predominant contributor to acid precipitation.

2.8.2 Hydrocarbons and Photochemical Smog

The most abundant hydrocarbon in the atmosphere is methane, CH_4. This gas is released from underground sources as natural gas and produced by the fermentation of organic matter. Methane is one of the least reactive atmospheric hydrocarbons and is produced by diffuse sources, so that its participation in the formation of pollutant photochemical reaction products is minimal.

The most significant atmospheric pollutant hydrocarbons are the reactive ones produced from automobile exhaust emissions, as well as from other sources, including even plants such as pine and citrus trees. In the presence of NO, under conditions of temperature inversion (which hold masses of air stationary for several days), low humidity, and sunlight, these hydrocarbons produce undesirable **photochemical smog** manifested by the presence of visibility-obscuring particulate matter, oxidants such as ozone, O_3, and noxious organic species, such as aldehydes. The process of smog formation is initiated by the photochemical dissociation of nitrogen dioxide by ultraviolet solar radiation,

$$NO_2 + h\nu \rightarrow NO + O \qquad (2.8.3)$$

which produces a reactive oxygen atom. This atom readily reacts with hydrocarbons in the atmosphere,

$$R-H + O \rightarrow HO\cdot + R\cdot \qquad (2.8.4)$$

where R–H represents a generic hydrocarbon molecule and one of the hydrogen atoms bound to it. The dots on the products denote unpaired electrons; species with such electrons are very reactive **free radicals**. The hydroxyl radical, $HO\cdot$, is particularly important because of the strong role it plays as a very reactive intermediate in photochemical smog formation and other atmospheric chemical processes. The hydroxyl radical reacts with organic molecules in the atmosphere to initiate a series of **chain reactions** in which $HO\cdot$ is regenerated. The end products of these multiple chain reactions are the noxious materials, including oxidants and aldehydes, that are characteristic of photochemical smog. The hydrocarbon radical, $R\cdot$, reacts with molecular O_2,

$$R\cdot + O_2 \rightarrow RO_2\cdot \qquad (2.8.5)$$

to produce a reactive, strongly oxidizing $RO_2\cdot$ radical. The $RO_2\cdot$ radical in turn reoxidizes NO back to NO_2,

$$NO + RO_2\cdot \rightarrow NO_2 + RO\cdot \qquad (2.8.6)$$

which can undergo photodissociation again (Reaction 2.8.3) to further promote photochemical smog formation.

2.8.3 Particulate Matter

Particles ranging from aggregates of a few molecules to pieces of dust readily visible to the naked eye are commonly found in the atmosphere. Some of these particles, such as sea salt formed by the evaporation of water from droplets of sea spray, are natural and even beneficial atmospheric constituents. Very small particles called **condensation nuclei** serve as bodies for atmospheric water vapor to condense upon and are essential for the formation of precipitation.

Colloidal-sized particles in the atmosphere are called **aerosols**. Those formed by grinding up bulk matter are known as **dispersion aerosols**, whereas particles formed from chemical reactions of gases are **condensation aerosols**; the latter tend to be smaller. Smaller particles are in general the most harmful because they have a greater tendency to scatter light and are the most respirable (tendency to be inhaled into the lungs).

Much of the mineral particulate matter in a polluted atmosphere is in the form of oxides and other compounds produced during the combustion of high-ash fossil fuel. Smaller particles of **fly ash** enter furnace flues and are efficiently collected in a properly equipped stack system. However, some fly ash escapes through the stack and enters the atmosphere. Unfortunately, the fly ash thus released tends to consist of smaller particles that do the most damage to human health, plants, and visibility.

2.9 THE BIOSPHERE

The **biosphere** is the sphere of the environment occupied by living organisms and in which life processes are carried out. **Biology** is the science of life and of living organisms. A **living organism** is constructed of one or more small units called *cells* and has the following characteristics:

1. It is composed in part of large characteristic macromolecules containing carbon, hydrogen, oxygen, and nitrogen along with phosphorus and sulfur.
2. It is capable of **metabolism**; that is, it mediates chemical processes by which it utilizes energy and synthesizes new materials needed for its structure and function.
3. It regulates itself.
4. It interacts with its environment.
5. It reproduces itself.

The biosphere is important in environmental chemistry for two major reasons. The first of these is that organisms carry out many of the chemical transformations that are part of an environmental chemical system. In aquatic systems (see Figure 2.4), algae carry out photosynthesis that produces biomass from inorganic carbon, bacteria degrade biomass, and bacteria mediate transformations between elemental oxidation states, such as the conversion of NO_3^- to NH_4^+ or of SO_4^{2-} to H_2S. The second reason that the biosphere is important in environmental chemistry is that the ultimate concern with respect to any pollutant or environmental phenomenon is the effect that it may have on living organisms in the environment. Such effects are addressed by toxicological chemistry, which is the major topic of this book.

2.10 THE ANTHROSPHERE AND GREEN CHEMISTRY

The anthrosphere is intimately entwined with the other environmental spheres; examples of its interactions with these spheres abound. Arguably the greatest effect of the anthrosphere to date has been through the cultivation of vast areas of the geosphere to produce food for humans and their domesticated animals. Humans have drastically altered the hydrosphere, putting dams on rivers and otherwise altering their normal flow. Vast stores of water left over from the Ice Age in underground aquifers have been mined and depleted to provide water for irrigation. Now humans have the capability of drastically modifying organisms in the biosphere through genetic engineering (see Chapter 8).

In the modern era, the greatest concern with respect to anthrospheric influence on the environment as a whole is the emission of carbon dioxide and other greenhouse gases to the atmosphere that may be causing global warming and an irreversible change in climate that will overshadow the effects of all other human activities. In consideration of this potential effect, Nobel Laureate Paul J. Crutzen of the Max Planck Institute for Chemistry, Mainz, Germany, has speculated that the current halocene epoch, under which humans have lived and generally thrived since the last Ice Age ended about 10,000 years ago, may be replaced by a new age in which climatic conditions on Earth will be largely determined by human influences, particularly the emission of gases that cause global warming. In recognition of anthropogenic effects and the key role of the anthrosphere in determining future conditions on Earth, Dr. Crutzen has called this new epoch the **anthropocene**.

2.10.1 Green Chemistry

Realistically, humans are not going to stop using Earth's resources and return to a "natural" era in which human influences on Earth, its climate, and its resources are minimal. Therefore, the challenge facing humans today is to use their intelligence and growing technological abilities to integrate the anthrosphere with the rest of Earth's environment in a way that minimizes adverse environmental impact and, indeed, enhances the quality of the environment as a whole. One way in which this is done is through the practice of **industrial ecology**, in which industrial enterprises interact in ways analogous to organisms in natural ecosystems, maximizing the efficiency of their utilization of materials and energy, recycling materials, and generally minimizing the impact on the Earth's resources and environment in providing goods and services that humans require for a high standard of living.[3]

Very recently, the exercise of chemical science and technology within the context of the best practice of industrial ecology has given rise to **green chemistry**, defined as *the practice of chemical science and manufacturing in a manner that is sustainable, safe, and nonpolluting and that consumes minimum amounts of materials and energy while producing little or no waste material*.[4] Green chemistry has developed from the realization that the command-and-control approach to pollution reduction based on regulatory processes, though extremely effective in controlling air and water pollution, is inherently unnatural and costly. This approach has used "end-of-pipe" pollution control measures in which pollutants are produced, but are contained and destroyed or not released into the environment. Green chemistry addresses chemical manufacturing and utilization processes from the beginning, with the goal of preventing or minimizing the production and use of chemicals that are toxic or otherwise hazardous. Green chemistry is sustainable and efficient chemistry and, when the costs of pollution control are factored in, the least costly way in which to practice chemical science and engineering.

Toxicological chemistry is of utmost importance in the practice of green chemistry. This is because the influence of toxic substances on chemical plant workers and users of chemical products, as well as potential inadvertent exposure of humans and other organisms to chemicals in the environment, is the foremost concern in the practice of green chemistry. Rather than concentrating on reduction of the exposure of organisms to potentially toxic chemical species, green chemistry concentrates on reduction of the hazard by attempting to avoid the manufacture or use of toxic chemicals.

REFERENCES

1. Manahan, S.E., *Environmental Science and Technology*, CRC Press/Lewis Publishers, Boca Raton, FL, 1997.
2. Manahan, S.E., *Environmental Chemistry*, 7th ed., CRC Press/Lewis Publishers, Boca Raton, FL, 2000.
3. Manahan, S.E., *Industrial Ecology: Environmental Chemistry and Hazardous Waste*, CRC Press/Lewis Publishers, Boca Raton, FL, 1999.
4. Manahan, S.E., *Green Chemistry: Fundamentals of Chemical Science and Technology*, ChemChar, Columbia, MO, 2002.

SUPPLEMENTARY REFERENCES

Andrews, J.E., *Environmental Chemistry*, Blackwell Science Publishers, Cambridge, MA, 1996.
Schlesinger, W.H., *Biogeochemistry*, Academic Press, San Diego, CA, 1991.
Spiro, T.G. and Stigliani, W.M., *Chemistry of the Environment*, Prentice Hall, Upper Saddle River, NJ, 1996.

QUESTIONS AND PROBLEMS

1. Construct a flow diagram that relates environmental science, environmental chemistry, aquatic chemistry, atmospheric chemistry, green chemistry, and toxicological chemistry in a hierarchical form that shows their relationships to each other.
2. Some authorities contend that it is incorrect to categorize the anthrosphere as a fifth sphere of the environment. What are arguments for or against this position.
3. In times past, before it was possible to prepare chemically bound nitrogen from atmospheric N_2, nitrates were mined as Chile saltpeter in arid regions of Chile and used as fertilizer. Where would such nitrogen fall in the nitrogen cycle shown in Figure 2.2?
4. Explain why environmental chemistry must be discussed in terms of simplified models and whether or not this suggests a lack of academic rigor in the discipline.
5. The density of liquid water decreases with decreasing temperature down to 4°C. In the fall season in the vicinity of a large population of lakes and ponds, there is often a period of one to several days in which whole geographic areas are afflicted with alarming foul odors. Based on the information contained in Figures 2.5 and 2.6, give a plausible explanation for this phenomenon.
6. Why might it be accurate to suggest that both the nitrogen cycle and the hydrologic cycle involve transfers of relatively small fractions of total available nitrogen and water at any particular time?
7. Explain how the production of {CH_2O} in water can raise water pH and cause precipitation of calcium carbonate.
8. Explain why the hypolimnion of a body of water may be chemically reducing. Suggest how nitrate ion, NO_3^-, might function in the hypolimnion in the absence of O_2.
9. Why is a notation for a metal ion such as Ca^{2+} not completely accurate for showing a metal ion dissolved in water? What might be a more accurate way of showing a metal ion?
10. The nitrilotriacetate anion,

    $$\begin{array}{c} \text{structure} \end{array}$$

    is a good chelating agent. Considering the nature of the three ionized carboxylic acid groups on the anion and the fact that it has an unshared pair of electrons on the N atom, suggest the structural formula of a metal chelate that it might form, such as with copper(II) ion, Cu^{2+}.
11. What distinguishes organometallic compounds from metal chelates?
12. Suggest how chelation of a metal with a negatively charged chelating agent might influence the metal's behavior. Knowing that organic species such as DDT tend to accumulate in the lipid (fat) tissue of fish and other organisms, suggest how formation of organometallic compounds might influence a metal's behavior in respect to the biosphere.
13. Why are materials in the form of colloidal particles particularly reactive? Since bacterial cells are the size of colloidal particles, how might this small size influence the behavior of bacteria in water and soil?

14. Although biomass represented generally in the environment as {CH_2O} is not normally toxic, how might it act as a water pollutant with toxic effects to fish?
15. How does the reaction {CH_2O} + $O_2 \rightarrow CO_2 + H_2O$ apply to water treatment? Describe the nature of the water pollutant that it eliminates.
16. Aluminum sulfate, $Al_2(SO_4)_3$, is commonly added to water to remove fine colloidal solids. Suggest a reaction of this salt with water to produce a substance that will remove colloidal solids from water.
17. What explains the fact that a very high proportion of rocks in the geosphere are composed of silicate minerals?
18. Contrast the processes of lithification and weathering.
19. Complete the following statement in a reasonable fashion: soil is rock that has been _____ by _____ processes.
20. Explain the role of transpiration in getting required nutrients into plants.
21. Explain how the exposure of mineral pyrite, FeS_2, from coal to air during coal-mining processes can result in acidic pollution.
22. Why is weathering particularly slow in arid desert regions?
23. Explain how CO_2 dissolved in water participates in the weathering of some kinds of minerals.
24. What natural process makes soil acidic? How can acidic soil be treated?
25. How is soil humus formed? Why is it important?
26. Suggest why CCl_2F_2 is now considered to be a permanent trace gas in the atmosphere.
27. By what kind of phenomenon may the species NO_2* be formed? What may happen to this species? Why?
28. What causes a warming effect of the stratosphere?
29. Although carbon dioxide in the atmosphere is not present at toxic levels and is required by plants for photosynthesis, why may it turn out to be the ultimate air pollutant?
30. What are the characteristics of an atmosphere afflicted with photochemical smog?
31. What is the species denoted HO·? What does the dot represent? Why is this species particularly important in the atmosphere?
32. What are condensation nuclei? What is fly ash?
33. What does a living organism do that makes it "alive"?
34. What are two major reasons that the biosphere is particularly important in environmental chemistry?
35. What is the anthropocene? Is Earth now experiencing the anthropocene? How will humans know if the anthropocene develops?
36. What is green chemistry? How does it relate to industrial ecology and toxicological chemistry?

CHAPTER 3

Biochemistry

3.1 BIOCHEMISTRY

Most people have had the experience of looking through a microscope at a single cell. It may have been an amoeba, alive and oozing about like a blob of jelly on the microscope slide, or a cell of bacteria, stained with a dye to make it show up more plainly. Or it may have been a beautiful cell of algae with its bright green chlorophyll. Even the simplest of these cells is capable of carrying out a thousand or more chemical reactions. These life processes fall under the heading of **biochemistry**, the branch of chemistry that deals with the chemical properties, composition, and biologically mediated processes of complex substances in living systems.

Biochemical phenomena that occur in living organisms are extremely sophisticated. In the human body, complex metabolic processes break down a variety of food materials to simpler chemicals, yielding energy and the raw materials to build body constituents, such as muscle, blood, and brain tissue. Impressive as this may be, consider a humble microscopic cell of photosynthetic cyanobacteria only about a micrometer in size, which requires only a few simple inorganic chemicals and sunlight for its existence. This cell uses sunlight energy to convert carbon from CO_2, hydrogen and oxygen from H_2O, nitrogen from NO_3^-, sulfur from SO_4^{2-}, and phosphorus from inorganic phosphate into all the proteins, nucleic acids, carbohydrates, and other materials that it requires to exist and reproduce. Such a simple cell accomplishes what could not be done by human endeavors even in a vast chemical factory costing billions of dollars.

Ultimately, most environmental pollutants and hazardous substances are of concern because of their effects on living organisms. The study of the adverse effects of substances on life processes requires some basic knowledge of biochemistry. Biochemistry is discussed in this chapter, with an emphasis on the aspects that are especially pertinent to environmentally hazardous and toxic substances, including cell membranes, deoxyribonucleic acid (DNA), and enzymes.

Biochemical processes not only are profoundly influenced by chemical species in the environment, but they largely determine the nature of these species, their degradation, and even their syntheses, particularly in the aquatic and soil environments. The study of such phenomena forms the basis of **environmental biochemistry**.[1]

3.1.1 Biomolecules

The biomolecules that constitute matter in living organisms are often polymers with molecular masses of the order of a million or even larger. As discussed later in this chapter, these biomolecules may be divided into the categories of carbohydrates, proteins, lipids, and nucleic acids. Proteins and nucleic acids consist of macromolecules, lipids are usually relatively small molecules, and carbohydrates range from relatively small sugar molecules to high-molecular-mass macromolecules, such as those in cellulose.

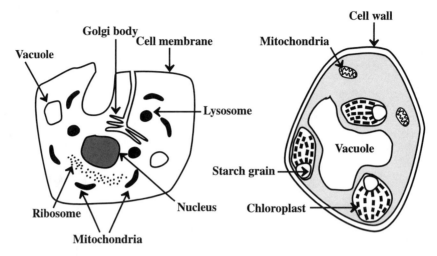

Figure 3.1 Some major features of the eukaryotic cell in animals (left) and plants (right).

The behavior of a substance in a biological system depends to a large extent upon whether the substance is hydrophilic (water-loving) or hydrophobic (water-hating). Some important toxic substances are hydrophobic, a characteristic that enables them to traverse cell membranes readily. Part of the detoxification process carried on by living organisms is to render such molecules hydrophilic, therefore water soluble and readily eliminated from the body.

3.2 BIOCHEMISTRY AND THE CELL

The focal point of biochemistry and biochemical aspects of toxicants is the **cell**, the basic building block of living systems where most life processes are carried. Bacteria, yeasts, and some algae consist of single cells. However, most living things are made up of many cells. In a more complicated organism the cells have different functions. Liver cells, muscle cells, brain cells, and skin cells in the human body are quite different from each other and do different things. Cells are divided into two major categories depending upon whether or not they have a nucleus: **eukaryotic** cells have a nucleus, and **prokaryotic** cells do not. Prokaryotic cells are found in single-celled bacteria. Eukaryotic cells compose organisms other than bacteria.

3.2.1 Major Cell Features

Figure 3.1 shows the major features of the **eukaryotic cell**, which is the basic structure in which biochemical processes occur in multicelled organisms. These features are as follows:

- **Cell membrane**, which encloses the cell and regulates the passage of ions, nutrients, lipid-soluble (fat-soluble) substances, metabolic products, toxicants, and toxicant metabolites into and out of the cell interior because of its varying **permeability** for different substances. The cell membrane protects the contents of the cell from undesirable outside influences. Cell membranes are composed in part of phospholipids that are arranged with their hydrophilic (water-seeking) heads on the cell membrane surfaces and their hydrophobic (water-repelling) tails inside the membrane. Cell membranes contain bodies of proteins that are involved in the transport of some substances through the membrane. One reason the cell membrane is very important in toxicology and environmental biochemistry is because it regulates the passage of toxicants and their products into and out of the cell interior. Furthermore, when its membrane is damaged by toxic substances, a cell may not function properly and the organism may be harmed.
- **Cell nucleus**, which acts as a sort of "control center" of the cell. It contains the genetic directions the cell needs to reproduce itself. The key substance in the nucleus is DNA. **Chromosomes** in the

cell nucleus are made up of combinations of DNA and proteins. Each chromosome stores a separate quantity of genetic information. Human cells contain 46 chromosomes. When DNA in the nucleus is damaged by foreign substances, various toxic effects, including mutations, cancer, birth defects, and defective immune system function may occur.
- **Cytoplasm**, which fills the interior of the cell not occupied by the nucleus. Cytoplasm is further divided into a water-soluble proteinaceous filler called **cytosol**, in which are suspended bodies called **cellular organelles**, such as mitochondria or, in photosynthetic organisms, chloroplasts.
- **Mitochondria**, "powerhouses" that mediate energy conversion and utilization in the cell. Mitochondria are sites in which food materials — carbohydrates, proteins, and fats — are broken down to yield carbon dioxide, water, and energy, which is then used by the cell for its energy needs. The best example of this is the oxidation of the sugar glucose, $C_6H_{12}O_6$:

$$C_6H_{12}O_6 + 6O_2 \rightarrow 6CO_2 + 6H_2O + energy$$

This kind of process is called **cellular respiration**.
- **Ribosomes**, which participate in protein synthesis.
- **Endoplasmic reticulum**, which is involved in the metabolism of some toxicants by enzymatic processes.
- **Lysosome**, a type of organelle that contains potent substances capable of digesting liquid food material. Such material enters the cell through a "dent" in the cell wall, which eventually becomes surrounded by cell material. This surrounded material is called a **food vacuole**. The vacuole merges with a lysosome, and the substances in the lysosome bring about digestion of the food material. The digestion process consists largely of **hydrolysis reactions** in which large, complicated food molecules are broken down into smaller units by the addition of water.
- **Golgi bodies**, which occur in some types of cells. These are flattened bodies of material that serve to hold and release substances produced by the cells.
- **Cell walls** of plant cells. These are strong structures that provide stiffness and strength. Cell walls are composed mostly of cellulose, which will be discussed later in this chapter.
- **Vacuoles** inside plant cells that often contain materials dissolved in water.
- **Chloroplasts** in plant cells that are involved in photosynthesis (the chemical process that uses energy from sunlight to convert carbon dioxide and water to organic matter). Photosynthesis occurs in these bodies. Food produced by photosynthesis is stored in the chloroplasts in the form of **starch grains**.

3.3 PROTEINS

Proteins are nitrogen-containing organic compounds that are the basic units of life systems. Cytoplasm, the jelly-like liquid filling the interior of cells, is made up largely of protein. Enzymes, which act as catalysts of life reactions, are made of proteins; they are discussed later in the chapter. Proteins are composed of **amino acids** (Figure 3.2) joined together in huge chains. Amino acids are organic compounds that contain the carboxylic acid group, $^-CO_2H$, and the amino group, $^-NH_2$. They are sort of a hybrid of carboxylic acids and amines (see Sections 1.8.1 and 1.8.2). Proteins are polymers, or **macromolecules**, of amino acids containing from approximately 40 to several thousand amino acid groups joined by peptide linkages. Smaller molecule amino acid polymers, containing only about 10 to about 40 amino acids per molecule, are called **polypeptides**. A portion of the amino acid left after the elimination of H_2O during polymerization is called a **residue**. The amino acid sequence of these residues is designated by a series of three-letter abbreviations for the amino acid.

Natural amino acids all have the following chemical group:

$$\begin{array}{c} HH \\ \diagdown\diagup \\ NO \\ |\| \\ R-C-C-OH \\ | \\ H \end{array}$$

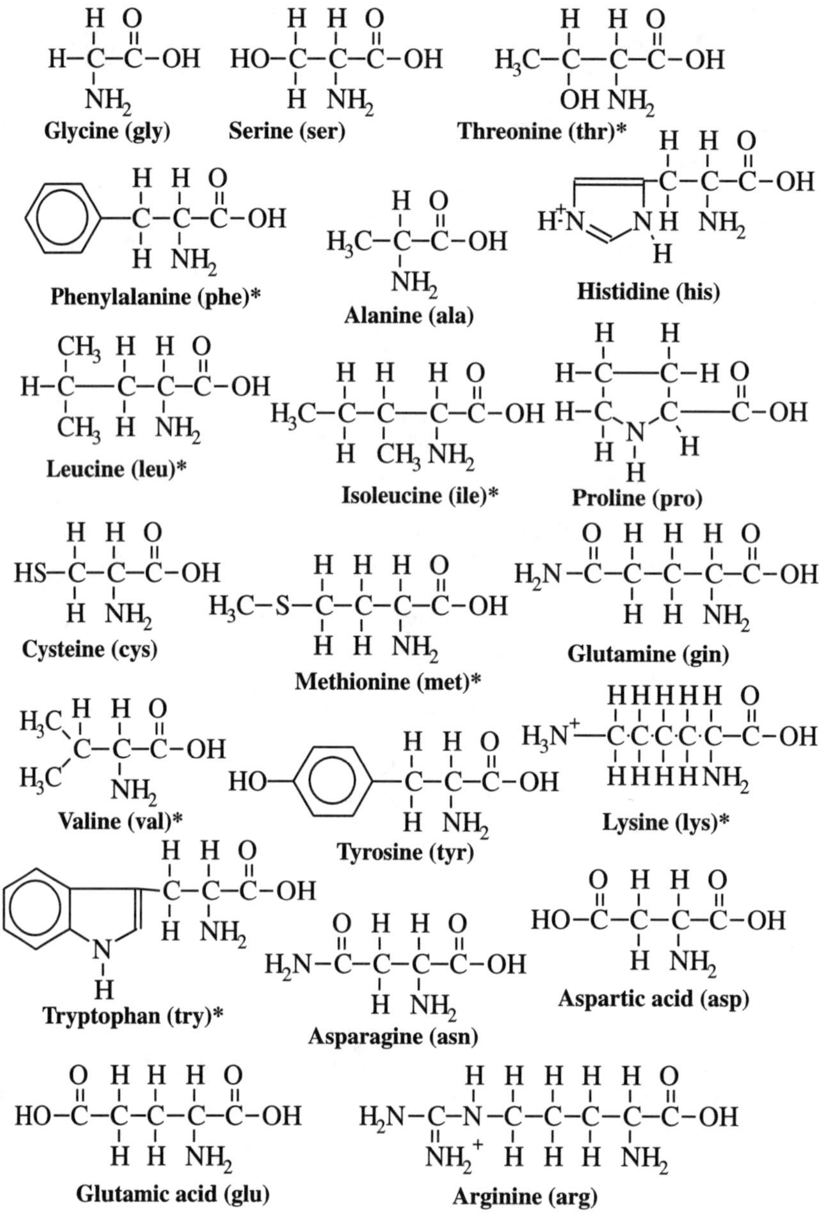

Figure 3.2 Amino acids that occur in proteins. Those marked with an asterisk cannot be synthesized by the human body and must come from dietary sources.

In this structure the –NH$_2$ group is always bonded to the carbon next to the –CO$_2$H group. This is called the "alpha" location, so natural amino acids are alpha-amino acids. Other groups, designated as R, are attached to the basic alpha-amino acid structure. The R groups may be as simple as an atom of H found in glycine, or they may be as complicated as the structure of the R group in tryptophan:

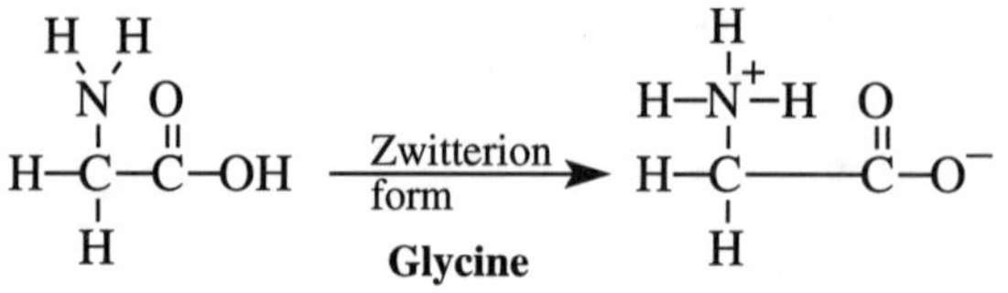

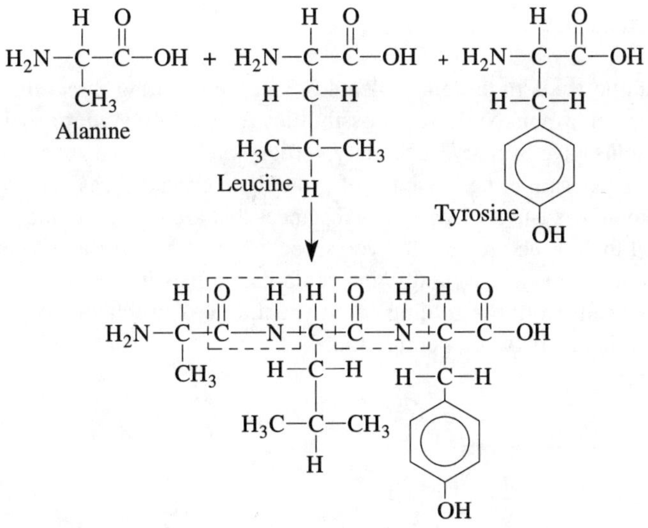

BIOCHEMISTRY

$$H_2N-\underset{CH_3}{\underset{|}{\overset{H}{\overset{|}{C}}}}-\overset{O}{\overset{\|}{C}}-OH + H_2N-\underset{H-C-H}{\underset{|}{\overset{H}{\overset{|}{C}}}}-\overset{O}{\overset{\|}{C}}-OH + H_2N-\underset{H-C-H}{\underset{|}{\overset{H}{\overset{|}{C}}}}-\overset{O}{\overset{\|}{C}}-OH$$

Alanine, Leucine ($H_3C-CH-CH_3$ with H), Tyrosine (phenol with OH)

↓

Tripeptide: $H_2N-\underset{CH_3}{\underset{|}{\overset{H}{\overset{|}{C}}}}-\overset{O}{\overset{\|}{C}}-N-\underset{H-C-H}{\underset{|}{\overset{H}{\overset{|}{C}}}}-\overset{O}{\overset{\|}{C}}-N-\underset{H-C-H}{\underset{|}{\overset{H}{\overset{|}{C}}}}-\overset{O}{\overset{\|}{C}}-OH$ with leucine and tyrosine side chains

Figure 3.3 Condensation of alanine, leucine, and tyrosine to form a tripeptide consisting of three amino acids joined by peptide linkages (outlined by dashed lines).

Table 3.1 Major Types of Proteins

Type of Protein	Example	Function and Characteristics
Nutrient	Casein (milk protein)	Food source; people must have an adequate supply of nutrient protein with the right balance of amino acids for adequate nutrition
Storage	Ferritin	Storage of iron in animal tissues
Structural	Collagen (tendons), keratin (hair)	Structural and protective components in organisms
Contractile	Actin, myosin in muscle tissue	Strong, fibrous proteins that can contract and cause movement to occur
Transport	Hemoglobin	Transport inorganic and organic species across cell membranes, in blood, between organs
Defense	—	Antibodies against foreign agents such as viruses produced by the immune system
Regulatory	Insulin, human growth hormone	Regulate biochemical processes such as sugar metabolism or growth by binding to sites inside cells or on cell membranes
Enzymes	Acetylcholine esterase	Catalysts of biochemical reactions (see Section 3.6)

As shown in Figure 3.2, there are 20 common amino acids in proteins. These are shown with uncharged $-NH_2$ and $-CO_2H$ groups. Actually, these functional groups exist in the charged **zwitterion** form, as shown for glycine above.

Amino acids in proteins are joined together in a specific way. These bonds constitute the **peptide linkage**. The formation of peptide linkages is a condensation process involving the loss of water. For example, consider the condensation of alanine, leucine, and tyrosine shown in Figure 3.3. When these three amino acids join together, two water molecules are eliminated. The product is a *tri*peptide since there are three amino acids involved. The amino acids in proteins are linked as shown for this tripeptide, except that many more monomeric amino acid groups are involved.

Proteins may be divided into several major types that have widely varying functions. These are listed in Table 3.1.

3.3.1 Protein Structure

The order of amino acids in protein molecules, and the resulting three-dimensional structures that form, provide an enormous variety of possibilities for **protein structure**. This is what makes life so diverse. Proteins have primary, secondary, tertiary, and quaternary structures. The structures of protein molecules determine the behavior of proteins in crucial areas such as the processes by which the body's immune system recognizes substances that are foreign to the body. Proteinaceous enzymes depend on their structures for the very specific functions of the enzymes.

The order of amino acids in the protein molecule determines its primary structure. **Secondary protein structures** result from the folding of polypeptide protein chains to produce a maximum number of hydrogen bonds between peptide linkages:

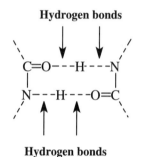

Illustration of hydrogen bonds between N and O atoms in peptide linkages, which constitutes protein secondary structures

Further folding of the protein molecules held in place by attractive forces between amino acid side chains gives proteins a **secondary structure**, which is determined by the nature of the amino acid R groups. Small R groups enable protein molecules to be hydrogen-bonded together in a parallel arrangement, whereas large R groups produce a spiral form known as an **alpha-helix**.

Tertiary structures are formed by the twisting of alpha-helices into specific shapes. They are produced and held in place by the interactions of amino side chains on the amino acid residues constituting the protein macromolecules. Tertiary protein structure is very important in the processes by which enzymes identify specific proteins and other molecules upon which they act. It is also involved with the action of antibodies in blood, which recognize foreign proteins by their shape and react to them. This is what happens in the phenomenon of disease immunity, where antibodies in blood recognize specific proteins from viruses or bacteria and reject them.

Two or more protein molecules consisting of separate polypeptide chains may be further attracted to each other to produce a **quaternary structure**.

Some proteins are **fibrous proteins**, which occur in skin, hair, wool, feathers, silk, and tendons. The molecules in these proteins are long and threadlike and are laid out parallel in bundles. Fibrous proteins are quite tough and do not dissolve in water.

An interesting fibrous protein is keratin, which is found in hair. The cross-linking bonds between protein molecules in keratin are –S–S– bonds formed from two HS– groups in two molecules of the amino acid cysteine. These bonds largely hold hair in place, thus keeping it curly or straight. A "permanent" consists of breaking the bonds chemically, setting the hair as desired, and then reforming the cross-links to hold the desired shape.

Aside from fibrous protein, the other major type of protein form is the **globular protein**. These proteins are in the shape of balls and oblongs. Globular proteins are relatively soluble in water. A typical globular protein is hemoglobin, the oxygen-carrying protein in red blood cells. Enzymes are generally globular proteins.

BIOCHEMISTRY

3.3.2 Denaturation of Proteins

Secondary, tertiary, and quaternary protein structures are easily changed by a process called **denaturation**. These changes can be quite damaging. Heating, exposure to acids or bases, and even violent physical action can cause denaturation to occur. The albumin protein in egg white is denatured by heating so that it forms a semisolid mass. Almost the same thing is accomplished by the violent physical action of an egg beater in the preparation of meringue. Heavy metal poisons such as lead and cadmium change the structures of proteins by binding to functional groups on the protein surface.

3.4 CARBOHYDRATES

Carbohydrates have the approximate simple formula CH_2O and include a diverse range of substances composed of simple sugars such as glucose:

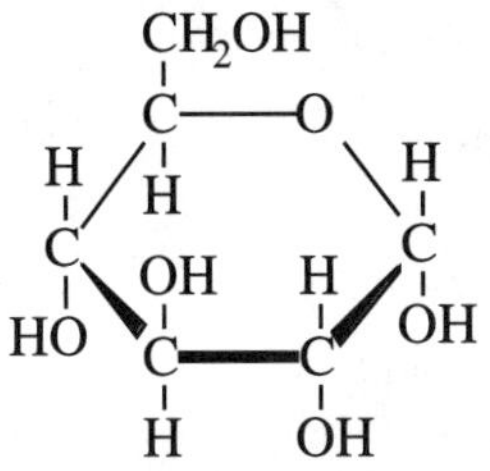

Glucose molecule

High-molecular-mass **polysaccharides**, such as starch and glycogen (animal starch), are biopolymers of simple sugars.

Photosynthesis in a plant cell converts the energy from sunlight to chemical energy in a carbohydrate, $C_6H_{12}O_6$. This carbohydrate may be transferred to some other part of the plant for use as an energy source. It may be converted to a water-insoluble carbohydrate for storage until it is needed for energy. Or it may be transformed to cell wall material and become part of the structure of the plant. If the plant is eaten by an animal, the carbohydrate is used for energy by the animal.

The simplest carbohydrates are the **monosaccharides**. These are also called **simple sugars**. Because they have six carbon atoms, simple sugars are sometimes called *hex*oses. Glucose (formula shown above) is the most common simple sugar involved in cell processes. Other simple sugars with the same formula but somewhat different structures are fructose, mannose, and galactose. These must be changed to glucose before they can be used in a cell. Because of its use for energy in body processes, glucose is found in the blood. Normal levels are from 65 to 110 mg of glucose per 100 ml of blood. Higher levels may indicate diabetes.

Units of two monosaccharides make up several very important sugars known as **disaccharides**. When two molecules of monosaccharides join together to form a disaccharide,

$$C_6H_{12}O_6 + C_6H_{12}O_6 \rightarrow C_{12}H_{22}O_{11} + H_2O \tag{3.4.1}$$

a molecule of water is lost. Recall that proteins are also formed from smaller amino acid molecules by condensation reactions involving the loss of water molecules. Disaccharides include sucrose (cane sugar used as a sweetener), lactose (milk sugar), and maltose (a product of the breakdown of starch).

Polysaccharides consist of many simple sugar units hooked together. One of the most important polysaccharides is **starch**, which is produced by plants for food storage. Animals produce a related material called **glycogen**. The chemical formula of starch is $(C_6H_{10}O_5)_n$, where n may represent a number as high as several hundred. What this means is that the very large starch molecule consists

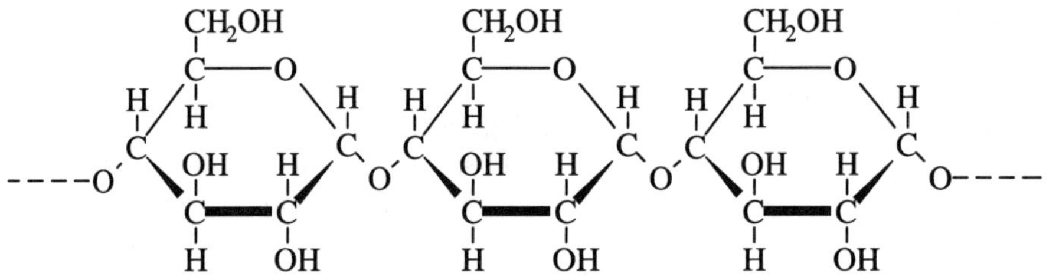

Figure 3.4 Part of a starch molecule showing units of $C_6H_{10}O_5$ condensed together.

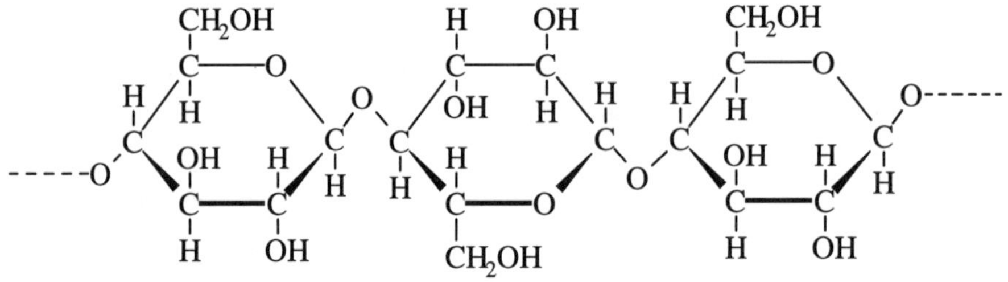

Figure 3.5 Part of the structure of cellulose.

of many units of $C_6H_{10}O_5$ joined together. For example, if n is 100, there are 6 times 100 carbon atoms, 10 times 100 hydrogen atoms, and 5 times 100 oxygen atoms in the molecule. Its chemical formula is $C_{600}H_{1000}O_{500}$. The atoms in a starch molecule are actually present as linked rings, represented by the structure shown in Figure 3.4. Starch occurs in many foods, such as bread and cereals. It is readily digested by animals, including humans.

Cellulose is a polysaccharide that is also made up of $C_6H_{10}O_5$ units. Molecules of cellulose are huge, with molecular weights of around 400,000. The cellulose structure (Figure 3.5) is similar to that of starch. Cellulose is produced by plants and forms the structural material of plant cell walls. Wood is about 60% cellulose, and cotton contains over 90% of this material. Fibers of cellulose are extracted from wood and pressed together to make paper.

Humans and most other animals cannot digest cellulose. Ruminant animals (cattle, sheep, goats, moose) have bacteria in their stomachs that break down cellulose into products that can be used by the animal. Chemical processes are available to convert cellulose to simple sugars by the reaction

$$(C_6H_{10}O_5)_n + nH_2O \rightarrow nC_6H_{12}O_6 \qquad (3.4.2)$$
$$\text{cellulose} \qquad\qquad\qquad \text{glucose}$$

where n may be 2000 to 3000. This involves breaking the linkages between units of $C_6H_{10}O_5$ by adding a molecule of H_2O at each linkage, a hydrolysis reaction. Large amounts of cellulose from wood, sugar cane, and agricultural products go to waste each year. The hydrolysis of cellulose enables these products to be converted to sugars, which can be fed to animals.

Carbohydrate groups are attached to protein molecules in a special class of materials called **glycoproteins**. Collagen is a crucial glycoprotein that provides structural integrity to body parts. It is a major constituent of skin, bones, tendons, and cartilage.

3.5 LIPIDS

Lipids are substances that can be extracted from plant or animal matter by organic solvents, such as chloroform, diethyl ether, or toluene (Figure 3.6). Whereas carbohydrates and proteins are

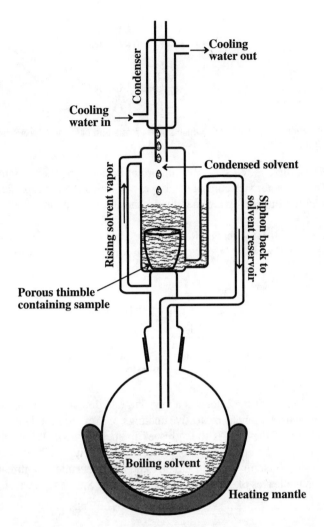

Figure 3.6 Lipids are extracted from some biological materials with a soxhelet extractor (above). The solvent is vaporized in the distillation flask by the heating mantle, rises through one of the exterior tubes to the condenser, and is cooled to form a liquid. The liquid drops onto the porous thimble containing the sample. Siphon action periodically drains the solvent back into the distillation flask. The extracted lipid collects as a solution in the solvent in the flask.

characterized predominately by the monomers (monosaccharides and amino acids) from which they are composed, lipids are defined essentially by their physical characteristic of organophilicity. The most common lipids are fats and oils composed of **triglycerides** formed from alcohol glycerol, $CH_2(OH)CH(OH)CH_2(OH)$, and a long-chain fatty acid such as stearic acid, $CH_3(CH_2)_{16}C(O)OH$ (Figure 3.7). Numerous other biological materials, including waxes, cholesterol, and some vitamins and hormones, are classified as lipids. Common foods, such as butter and salad oils, are lipids. Long-chain fatty acids, such as stearic acid, are also organic soluble and are classified as lipids.

Lipids are toxicologically important for several reasons. Some toxic substances interfere with lipid metabolism, leading to detrimental accumulation of lipids. Many toxic organic compounds are poorly soluble in water, but are lipid soluble, so that bodies of lipids in organisms serve to dissolve and store toxicants.

An important class of lipids consists of **phosphoglycerides** (glycerophosphatides). These compounds may be regarded as triglycerides in which one of the acids bonded to glycerol is ortho-

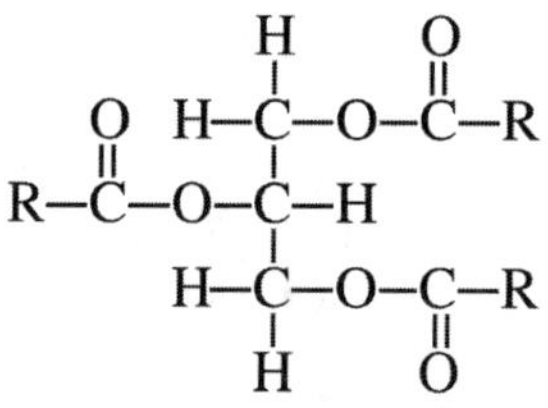

$$
\begin{array}{c}
\overset{\displaystyle H}{|}\overset{\displaystyle O}{\|} \\
\text{O}\text{H--C--O--C--R} \\
\overset{\displaystyle \|}{\text{R--C--O--C--H}} \\
\text{H--C--O--C--R} \\
\underset{\displaystyle H}{|}\underset{\displaystyle O}{\|}
\end{array}
$$

Figure 3.7 General formula of triglycerides, which make up fats and oils. The R group is from a fatty acid and is a hydrocarbon chain, such as $-(CH_2)_{16}CH_3$.

phosphoric acid. These lipids are especially important because they are essential constituents of cell membranes. These membranes consist of bilayers in which the hydrophilic phosphate ends of the molecules are on the outside of the membrane and the hydrophobic "tails" of the molecules are on the inside.

Waxes are also esters of fatty acids. However, the alcohol in a wax is not glycerol; it is often a very long chain alcohol. For example, one of the main compounds in beeswax is myricyl palmitate,

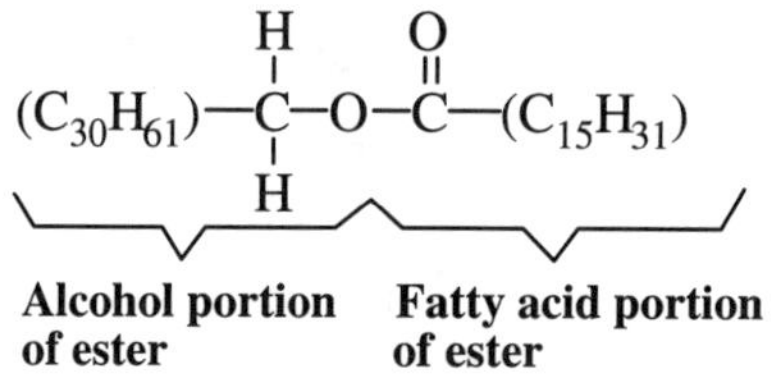

$$(C_{30}H_{61})\!-\!\underset{\underset{\displaystyle H}{|}}{\overset{\overset{\displaystyle H}{|}}{C}}\!-\!O\!-\!\overset{\overset{\displaystyle O}{\|}}{C}\!-\!(C_{15}H_{31})$$

Alcohol portion   **Fatty acid portion**
of ester   **of ester**

in which the alcohol portion of the ester has a very large hydrocarbon chain. Waxes are produced by both plants and animals, largely as protective coatings. Waxes are found in a number of common products. Lanolin is one of these. It is the "grease" in sheep's wool. When mixed with oils and water, it forms stable colloidal emulsions consisting of extremely small oil droplets suspended in water. This makes lanolin useful for skin creams and pharmaceutical ointments. Carnauba wax occurs as a coating on the leaves of some Brazilian palm trees. Spermaceti wax is composed largely of cetyl palmitate,

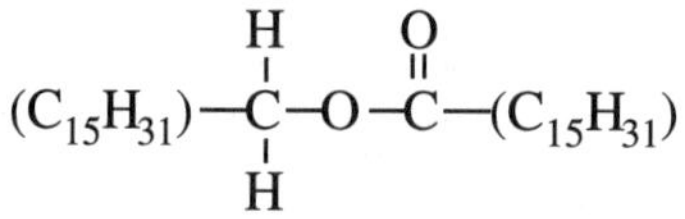

$$(C_{15}H_{31})\!-\!\underset{\underset{\displaystyle H}{|}}{\overset{\overset{\displaystyle H}{|}}{C}}\!-\!O\!-\!\overset{\overset{\displaystyle O}{\|}}{C}\!-\!(C_{15}H_{31}) \quad \text{Cetyl palmitate}$$

which is extracted from the blubber of the sperm whale. It is very useful in some cosmetics and pharmaceutical preparations.

Steroids are lipids found in living systems that all have the ring system shown in Figure 3.8 for cholesterol. Steroids occur in bile salts, which are produced by the liver and then secreted into the intestines. Their breakdown products give feces its characteristic color. Bile salts act on fats in the intestine. They suspend very tiny fat droplets in the form of colloidal emulsions. This enables the fats to be broken down chemically and digested.

Some steroids are **hormones**. Hormones act as "messengers" from one part of the body to another. As such, they start and stop a number of body functions. Male and female sex hormones are examples of steroid hormones. Hormones are given off by glands in the body called **endocrine glands**. The locations of the important endocrine glands are shown in Figure 3.9.

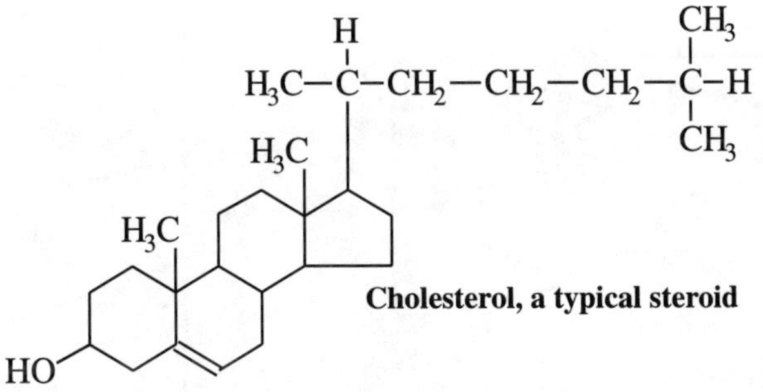

BIOCHEMISTRY

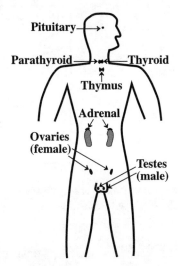

Figure 3.8 Steroids are characterized by the ring structure shown above for cholesterol.

Figure 3.9 Locations of important endocrine glands.

3.6 ENZYMES

Catalysts are substances that speed up a chemical reaction without themselves being consumed in the reaction. The most sophisticated catalysts of all are those found in living systems. They bring about reactions that could not be performed at all, or only with great difficulty, outside a living organism. These catalysts are called **enzymes**. In addition to speeding up reactions by as much as 10- to a 100 million-fold, enzymes are extremely selective in the reactions they promote.

Enzymes are proteinaceous substances with highly specific structures that interact with particular substances or classes of substances called **substrates**. Enzymes act as catalysts to enable biochemical reactions to occur, after which they are regenerated intact to take part in additional reactions. The extremely high specificity with which enzymes interact with substrates results from their "lock and key" action, based on the unique shapes of enzymes, as illustrated in Figure 3.10.

This illustration shows that an enzyme "recognizes" a particular substrate by its molecular structure and binds to it to produce an **enzyme–substrate complex**. This complex then breaks apart to form one or more products different from the original enzyme, regenerating the unchanged enzyme, which is then available to catalyze additional reactions. The basic process for an enzyme reaction is, therefore,

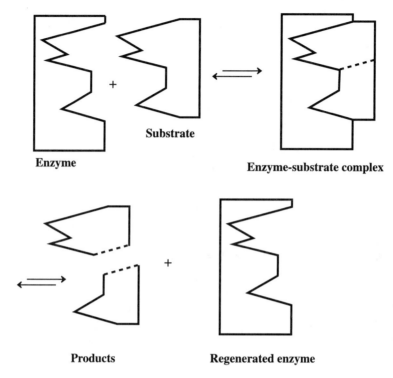

Figure 3.10 Representation of the "lock and key" mode of enzyme action, which enables the very high specificity of enzyme-catalyzed reactions.

$$\text{enzyme} + \text{substrate} \rightleftharpoons \text{enzyme–substrate complex} \rightleftharpoons \text{enzyme} + \text{product} \quad (3.6.1)$$

Several important things should be noted about this reaction. As shown in Figure 3.10, an enzyme acts on a specific substrate to form an enzyme–substrate complex because of the fit between their structures. As a result, something happens to the substrate molecule. For example, it might be split in two at a particular location. Then the enzyme–substrate complex comes apart, yielding the enzyme and products. The enzyme is not changed in the reaction and is now free to react again. Note that the arrows in the formula for enzyme reaction point both ways. This means that the reaction is **reversible**. An enzyme–substrate complex can simply go back to the enzyme and the substrate. The products of an enzymatic reaction can react with the enzyme to form the enzyme–substrate complex again. It, in turn, may again form the enzyme and the substrate. Therefore, the same enzyme may act to cause a reaction to go either way.

Some enzymes cannot function by themselves. In order to work, they must first be attached to **coenzymes**. Coenzymes normally are not protein materials. Some of the vitamins are important coenzymes.

Enzymes are named for what they do. For example, the enzyme given off by the stomach, which splits proteins as part of the digestion process, is called *gastric proteinase*. The "gastric" part of the name refers to the enzyme's origin in the stomach. "Proteinase" denotes that it splits up protein molecules. The common name for this enzyme is pepsin. Similarly, the enzyme produced by the pancreas that breaks down fats (lipids) is called *pancreatic lipase*. Its common name is steapsin. In general, lipase enzymes cause lipid triglycerides to dissociate and form glycerol and fatty acids.

The enzymes mentioned above are **hydrolyzing enzymes**, which bring about the breakdown of high-molecular-weight biological compounds by the addition of water. This is one of the most important reactions involved in digestion. The three main classes of energy-yielding foods that animals eat are carbohydrates, proteins, and fats. Recall that the higher carbohydrates that humans eat are largely disaccharides (sucrose, or table sugar) and polysaccharides (starch). These are formed by the joining together of units of simple sugars, $C_6H_{12}O_6$, with the elimination of an H_2O molecule at the linkage where they join. Proteins are formed by the condensation of amino acids, again with the elimination of a water molecule at each linkage. Fats are esters that are produced when glycerol and fatty acids link together. A water molecule is lost for each of these linkages when a protein, fat, or carbohydrate is synthesized. In order for these substances to be used as a food source, the reverse process must occur to break down large, complicated molecules of protein, fat, or carbohydrate to simple, soluble substances that can penetrate a cell membrane and take part in chemical processes in the cell. This reverse process is accomplished by hydrolyzing enzymes.

Biological compounds with long chains of carbon atoms are broken down into molecules with shorter chains by the breaking of carbon–carbon bonds. This commonly occurs by the elimination of $-CO_2H$ groups from carboxylic acids. For example, *pyruvic decarboxylase* enzyme acts upon pyruvic acid,

$$\underset{\textbf{Pyruvic acid}}{\text{H}-\underset{\underset{\text{H}}{|}}{\overset{\overset{\text{H}}{|}}{\text{C}}}-\overset{\overset{\text{O}}{\|}}{\text{C}}-\overset{\overset{\text{O}}{\|}}{\text{C}}-\text{OH}} \xrightarrow{\text{Pyruvate decarboxylase}} \underset{\textbf{Acetaldehyde}}{\text{H}-\underset{\underset{\text{H}}{|}}{\overset{\overset{\text{H}}{|}}{\text{C}}}-\overset{\overset{\text{O}}{\|}}{\text{C}}-\text{H}} + CO_2 \qquad (3.6.2)$$

to split off CO_2 and produce a compound with one less carbon. It is by such carbon-by-carbon breakdown reactions that long-chain compounds are eventually degraded to CO_2 in the body, or that long-chain hydrocarbons undergo biodegradation by the action of spill bacteria on spilled petroleum. Oxidation and reduction are the major reactions for the exchange of energy in living systems. Cellular respiration, discussed in Section 3.2, is an oxidation reaction in which a carbohydrate, $C_6H_{12}O_6$, is broken down to carbon dioxide and water with the release of energy:

$$C_6H_{12}O_6 + 6O_2 \rightarrow 6CO_2 + 6H_2O + \text{energy} \qquad (3.6.3)$$

Actually, such an overall reaction occurs in living systems by a complicated series of individual steps. Some of these steps involve oxidation. The enzymes that bring about oxidation in the presence of free O_2 are called **oxidases**. In general, biological oxidation–reduction reactions are catalyzed by **oxidoreductase enzymes**.

In addition to the types of enzymes discussed above, there are many enzymes that perform miscellaneous duties in living systems. Typical of these are **isomerases**, which form isomers of particular compounds. For example, there are several simple sugars with the formula $C_6H_{12}O_6$. However, only glucose can be used directly for cell processes. The other isomers are converted to glucose by the action of isomerases. **Transferase enzymes** move chemical groups from one molecule to another, **lyase enzymes** remove chemical groups without hydrolysis and participate in the formation of C=C bonds or addition of species to such bonds, and **ligase enzymes** work in conjunction with adenosine triphosphate (ATP), a high-energy molecule that plays a crucial role in energy-yielding, glucose-oxidizing metabolic processes, to link molecules together with the formation of bonds such as carbon–carbon or carbon–sulfur bonds.

Enzyme action may be affected by many different things. Enzymes require a certain hydrogen ion concentration to function best. For example, gastric proteinase requires the acid environment of the stomach to work well. When it passes into the much less acidic intestines, it stops working. This prevents damage to the intestine walls, which would occur if the enzyme tried to digest them. Temperature is critical. Not surprisingly, the enzymes in the human body work best at around 98.6°F (37°C), which is the normal body temperature. Heating these enzymes to around 140°F permanently destroys them. Some bacteria that thrive in hot springs have enzymes that work best at relatively high temperatures. Other "cold-seeking" bacteria have enzymes adapted to near the freezing point of water.

One of the greatest concerns regarding the effects of surroundings on enzymes is the influence of toxic substances. A major mechanism of toxicity is the alteration or destruction of enzymes by agents such as cyanide, heavy metals, or organic compounds, such as insecticidal parathion. An enzyme that has been destroyed obviously cannot perform its designated function, whereas one that has been altered either may not function at all or may act improperly. Toxicants can affect enzymes in several ways. Parathion, for example, bonds covalently to the nerve enzyme acetylcholinesterase, which can then no longer serve to stop nerve impulses. Heavy metals tend to bind to sulfur atoms in enzymes (such as sulfur from the amino acid cysteine, shown in Figure 3.2), thereby altering the shape and function of the enzyme. Enzymes are denatured by some poisons, causing them to "unravel" so that the enzyme no longer has its crucial specific shape.

3.7 NUCLEIC ACIDS

The essence of life is contained in **deoxyribonucleic acid** (DNA), which stays in the cell nucleus, and **ribonucleic acid** (RNA), which functions in the cell cytoplasm. These substances, which are known collectively as **nucleic acids**, store and pass on essential genetic information that controls reproduction and protein synthesis.

The structural formulas of the monomeric constituents of nucleic acids are given in Figure 3.11. These are pyrimidine or purine nitrogen-containing bases, two sugars, and phosphate. DNA molecules are made up of the nitrogen-containing bases adenine, guanine, cytosine, and thymine; phosphoric acid (H_3PO_4); and the simple sugar 2-deoxy-β-D-ribofuranose (commonly called deoxyribose). RNA molecules are composed of the nitrogen-containing bases adenine, guanine, cytosine, and uracil; phosphoric acid (H_3PO_4); and the simple sugar β-D-ribofuranose (ribose).

The formation of nucleic acid polymers from their monomeric constituents may be viewed as the following steps.

- Monosaccharide (simple sugar) + cyclic nitrogenous base yields **nucleoside**:

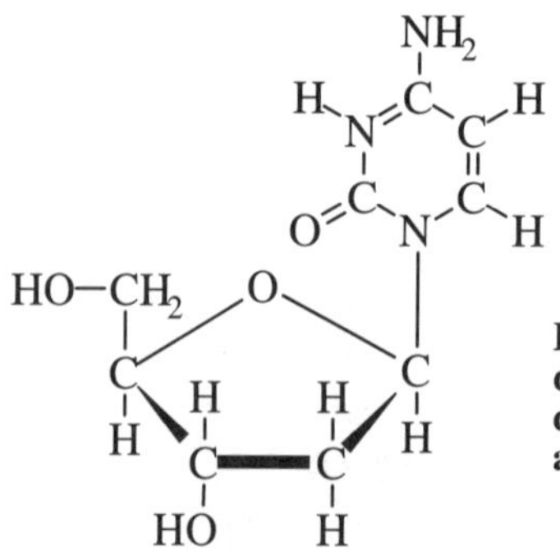

Deoxyctidine formed by the dimerization of cytosine and deoxyribose with the elimination of a molecule of H_2O.

BIOCHEMISTRY

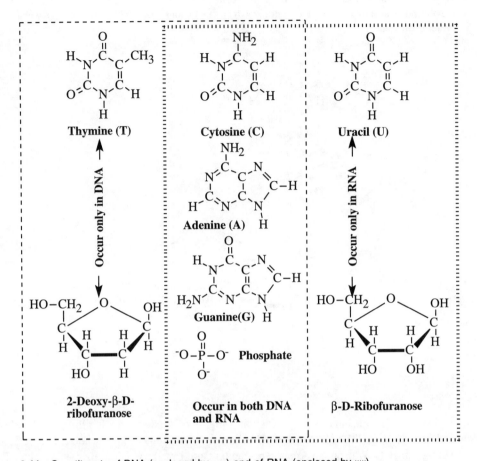

Figure 3.11 Constituents of DNA (enclosed by ----) and of RNA (enclosed by ⅢⅢ).

- Nucleoside + phosphate yields **phosphate ester nucleotide**.

Nucleotide formed by the bonding of a phosphate group to deoxyctidine

- Polymerized nucleotide yields **nucleic acid**, as shown by the structure below. In the nucleic acid the phosphate negative charges are neutralized by metal cations (such as Mg^{2+}) or positively charged proteins (histones).

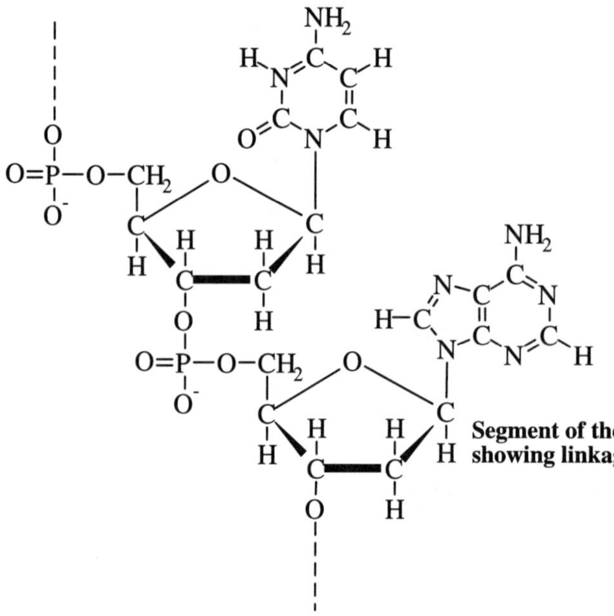

Segment of the DNA polymer showing linkage of two nucleotides

Molecules of DNA are huge, with molecular weights of greater than 1 billion. Molecules of RNA are also quite large. The structure of DNA is that of the famed double helix. It was figured out in 1953 by James D. Watson, an American scientist, and Francis Crick, a British scientist. They received the Nobel Prize for this scientific milestone in 1962. This model visualizes DNA as a so-called double α-helix structure of oppositely wound polymeric strands held together by hydrogen bonds between opposing pyrimidine and purine groups. As a result, DNA has both a primary and a secondary structure; the former is due to the sequence of nucleotides in the individual strands of DNA, and the latter results from the α-helix interaction of the two strands. In the secondary structure of DNA, only cytosine can be opposite guanine and only thymine can be opposite adenine and vice versa. Basically, the structure of DNA is that of two spiral ribbons "counterwound" around each other, as illustrated in Figure 3.12. The two strands of DNA are **complementary**. This means that a particular portion of one strand fits like a key in a lock with the corresponding portion of another strand. If the two strands are pulled apart, each manufactures a new complementary strand, so that two copies of the original double helix result. This occurs during cell reproduction.

The molecule of DNA is like a coded message. This "message," the genetic information contained in and transmitted by nucleic acids, depends on the sequence of bases from which they are composed. It is somewhat like the message sent by telegraph, which consists only of dots, dashes, and spaces in between. The key aspect of DNA structure that enables storage and replication of this information is the famed double helix structure of DNA mentioned above.

Portions of the DNA double helix may unravel, and one of the strands of DNA may produce a strand of RNA. This substance then goes from the cell nucleus out into the cell and regulates the synthesis of new protein. In this way, DNA regulates the function of the cell and acts to control life processes.

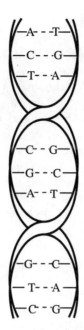

Figure 3.12 Representation of the double helix structure of DNA showing the allowed base pairs held together by hydrogen bonding between the phosphate–sugar polymer "backbones" of the two strands of DNA. The letters stand for adenine (A), cytosine (C), guanine (G), and thymine (T). The dashed lines represent hydrogen bonds.

3.7.1 Nucleic Acids in Protein Synthesis

When a new cell is formed, the DNA in its nucleus must be accurately reproduced from the parent cell. Life processes are absolutely dependent upon accurate protein synthesis, as regulated by cell DNA. The DNA in a single cell must be capable of directing the synthesis of up to 3000 or even more different proteins. The directions for the synthesis of a single protein are contained in a segment of DNA called a **gene**. The process of transmitting information from DNA to a newly synthesized protein involves the following steps:

- The DNA undergoes **replication**. This process involves separation of a segment of the double helix into separate single strands, which then replicate such that guanine is opposite cytosine (and vice versa) and adenine is opposite thymine (and vice versa). This process continues until a complete copy of the DNA molecule has been produced.
- The newly replicated DNA produces **messenger RNA** (mRNA), a complement of the single strand of DNA, by a process called **transcription**.
- A new protein is synthesized using mRNA as a template to determine the order of amino acids in a process called **translation**.

3.7.2 Modified DNA

DNA molecules may be modified by the unintentional addition or deletion of nucleotides or by substituting one nucleotide for another. The result is a **mutation** that is transmittable to offspring. Mutations can be induced by chemical substances. This is a major concern from a toxicological viewpoint because of the detrimental effects of many mutations and because substances that cause mutations often cause cancer as well. DNA malfunction may result in birth defects, and the failure to control cell reproduction results in cancer. Radiation from x-rays and radioactivity also disrupts DNA and may cause mutation.

3.8 RECOMBINANT DNA AND GENETIC ENGINEERING

As noted above, segments of DNA contain information for the specific syntheses of particular proteins. Within the last two decades it has become possible to transfer this information between organisms by means of **recombinant DNA technology**, which has resulted in a new industry based on **genetic engineering**. Most often the recipient organisms are bacteria, which can be reproduced (cloned) over many orders of magnitude from a cell that has acquired the desired qualities. Therefore, to synthesize a particular substance, such as human insulin or growth hormone, the required genetic information can be transferred from a human source to bacterial cells, which then produce the substance as part of their metabolic processes.

The first step in recombinant DNA gene manipulation is to lyze (open up) a donor cell to remove needed DNA material by using enzyme action to cut the sought-after genes from the donor DNA chain. These are next spliced into small DNA molecules. These molecules, called **cloning vehicles**, are capable of penetrating the host cell and becoming incorporated into its genetic material. The modified host cell is then reproduced many times and carries out the desired biosynthesis.

Early concerns about the potential of genetic engineering to produce "monster organisms" or new and horrible diseases have been largely allayed, although caution is still required with this technology. In the environmental area, genetic engineering offers some hope for the production of bacteria engineered to safely destroy troublesome wastes and to produce biological substitutes for environmentally damaging synthetic pesticides.

3.9 METABOLIC PROCESSES

Biochemical processes that involve the alteration of biomolecules fall under the category of **metabolism**. Metabolic processes may be divided into the two major categories of **anabolism** (synthesis) and **catabolism** (degradation of substances). An organism may use metabolic processes to yield energy or to modify the constituents of biomolecules.

3.9.1 Energy-Yielding Processes

Organisms can gain energy by the following three processes:

- **Respiration**, in which organic compounds undergo catabolism that requires molecular oxygen (**aerobic respiration**) or that occurs in the absence of molecular oxygen (**anaerobic respiration**). Aerobic respiration uses the **Krebs cycle** to obtain energy from the following reaction:

$$C_6H_{12}O_6 + 6O_2 \rightarrow 6CO_2 + 6H_2O + \text{energy}$$

- About half of the energy released is converted to short-term stored chemical energy, particularly through the synthesis of ATP nucleoside. For longer-term energy storage, glycogen or starch polysaccharides are synthesized, and for still longer term energy storage, lipids (fats) are generated and retained by the organism.
- **Fermentation**, which differs from respiration in not having an electron transport chain. Yeasts produce ethanol from sugars by fermentation:

$$C_6H_{12}O_6 \rightarrow 2CO_2 + 2C_2H_5OH$$

- **Photosynthesis**, in which light energy captured by plant and algal chloroplasts are used to synthesize sugars from carbon dioxide and water:

$$6CO_2 + 6H_2O + h\nu \rightarrow C_6H_{12}O_6 + 6O_2$$

Plants cannot always get the energy that they need from sunlight. During the dark they must use stored food. Plant cells, like animal cells, contain mitochondria in which stored food is converted to energy by cellular respiration.

Plant cells, which use sunlight for energy and CO_2 for carbon, are said to be **autotrophic**. In contrast, animal cells must depend on organic material manufactured by plants for their food. These are called **heterotrophic** cells. They act as "middlemen" in the chemical reaction between oxygen and food material, using the energy from the reaction to carry out their life processes.

SUPPLEMENTARY REFERENCES

Bettelheim, F.A. and March, J., *Introduction to Organic and Biochemistry*, Saunders College Publishing, Fort Worth, TX, 1998.

Chesworth, J.M., Stuchbury, T., and Scaife, J.R., *An Introduction to Agricultural Biochemistry*, Chapman & Hall, London, 1998.

Garrett, R.H. and Grisham, C.M., *Biochemistry*, Saunders College Publishing, Philadelphia, 1998.

Gilbert, H.F., Ed., *Basic Concepts in Biochemistry*, McGraw-Hill, Health Professions Division, New York, 2000.

Kuchel, P.W., Ed., *Schaum's Outline of Theory and Problems of Biochemistry*, McGraw-Hill, New York, 1998.

Lea, P.J. and Leegood, R.C., Eds., *Plant Biochemistry and Molecular Biology*, 2nd ed., John Wiley & Sons, New York, 1999.

Marks, D.B., *Biochemistry*, Williams & Wilkins, Baltimore, 1999.

Meisenberg, G. and Simmons, W.H., *Principles of Medical Biochemistry*, Mosby, St. Louis, 1998.

Switzer, R.L. and Garrity, L.F., *Experimental Biochemistry*, W.H. Freeman and Co., New York, 1999.

Voet, D., Voet, J.G., and Pratt, C., *Fundamentals of Biochemistry*, John Wiley & Sons, New York, 1998.

Vrana, K.E., *Biochemistry*, Lippincott Williams & Wilkins, Philadelphia, 1999.

Wilson, K. and Walker, J.M., *Principles and Techniques of Practical Biochemistry*, Cambridge University Press, New York, 1999.

QUESTIONS AND PROBLEMS

1. What is the toxicological importance of lipids? How do lipids relate to hydrophobic (water-disliking) pollutants and toxicants?
2. What is the function of a hydrolase enzyme?
3. Match the cell structure on the left with its function on the right:

    A. Mitochondria   1. Toxicant metabolism
    B. Endoplasmic reticulum   2. Fills the cell
    C. Cell membrane   3. DNA
    D. Cytoplasm   4. Mediate energy conversion and utilization
    E. Cell nucleus   5. Encloses the cell and regulates the passage of materials into and out of the cell interior

4. The formula of simple sugars is $C_6H_{12}O_6$. The simple formula of higher carbohydrates is $C_6H_{10}O_5$. Of course, many of these units are required to make a molecule of starch or cellulose. If higher carbohydrates are formed by joining together molecules of simple sugars, why is there a difference in the ratios of C, H, and O atoms in the higher carbohydrates, compared to the simple sugars?
5. Why does wood contain so much cellulose?
6. What would be the chemical formula of a *tri*saccharide made by the bonding together of three simple sugar molecules?

7. The general formula of cellulose may be represented as $(C_6H_{10}O_5)_x$. If the molecular weight of a molecule of cellulose is 400,000, what is the estimated value of x?
8. During 1 month, a factory for the production of simple sugars, $C_6H_{12}O_6$, by the hydrolysis of cellulose processes 1 million kg of cellulose. The percentage of cellulose that undergoes the hydrolysis reaction is 40%. How many kilograms of water are consumed in the hydrolysis of cellulose each month?
9. What is the structure of the largest group of atoms common to all amino acid molecules?
10. Glycine and phenylalanine can join together to form two different dipeptides. What are the structures of these two dipeptides?
11. One of the ways in which two parallel protein chains are joined together, or cross-linked, is by way of an –S–S– link. What amino acid to you think might be most likely to be involved in such a link? Explain your choice.
12. Fungi, which break down wood, straw, and other plant material, have what are called "exoenzymes." Fungi have no teeth and cannot break up plant material physically by force. Knowing this, what do you suppose an exoenzyme is? Explain how you think it might operate in the process by which fungi break down something as tough as wood.
13. Many fatty acids of lower molecular weight have a bad odor. Speculate as to the reasons why rancid butter has a bad odor. What chemical compound is produced that has a bad odor? What sort of chemical reaction is involved in its production?
14. The long-chain alcohol with ten carbons is called decanol. What do you think would be the formula of decyl stearate? To what class of compounds would it belong?
15. Write an equation for the chemical reaction between sodium hydroxide and cetyl stearate. What are the products?
16. What are two endocrine glands that are found only in females? Which of these glands is found only in males?
17. The action of bile salts is a little like that of soap. What function do bile salts perform in the intestine? Look up the action of soaps, and explain how you think bile salts may function somewhat like soap.
18. If the structure of an enzyme is illustrated as

how should the structure of its substrate be represented?
19. Look up the structures of ribose and deoxyribose. Explain where the "deoxy" came from in the name deoxyribose.
20. In what respect are an enzyme and its substrate like two opposite strands of DNA?
21. For what discovery are Watson and Crick noted?
22. Why does an enzyme no longer work if it is denatured?

CHAPTER 4

Metabolic Processes

4.1 METABOLISM IN ENVIRONMENTAL BIOCHEMISTRY

The biochemical changes that substances undergo in a living organism are called metabolism. Metabolism describes the catabolic reactions by which chemical species are broken down by enzymatic action in an organism to produce energy and components for the synthesis of biomolecules required for life processes. It also describes the anabolic reactions in which energy is used to assemble small molecules into larger biomolecules. Metabolism is an essential process for any organism because it provides the two things essential for life — energy and raw materials.

Metabolism is especially important in toxicological chemistry for two reasons: (1) interference with metabolism is a major mode of toxic action, and (2) toxic substances are transformed by metabolic processes to other materials that are usually, though not invariably, less toxic and more readily eliminated from the organism. This chapter introduces the topic of metabolism in general. Specific aspects of the metabolism of toxic substances are discussed in Chapter 7.

4.1.1 Metabolism Occurs in Cells

Metabolic processes occur in cells in organisms. Figure 3.1 shows the general structure of eukaryotic cells in organisms such as animals and fungi. A cell is contained within a *cell membrane* composed of a lipid bilayer that separates the contents of the cell from the aqueous medium around it. Other than the cell nucleus, the material inside the cell is referred to as the *cell cytoplasm*, the fluid part of which is the *cytosol*. The cytosol is an aqueous solution of electrolytes that also contains enzymes that catalyze some important cell functions, including some metabolic processes. Within the cytoplasm are specialized *organelles* that carry out various metabolic functions. Of these, *mitochondria* are of particular importance in metabolism because of their role in synthesizing energetic **adenosine triphosphate** (ATP) using energy-yielding reactions. *Ribosomes* are sites of protein synthesis from mRNA templates (Chapter 8).

4.1.2 Pathways of Substances and Their Metabolites in the Body

In considering metabolic processes, it is important to keep in mind the pathways of nutrients and xenobiotics in organisms. For humans and other vertebrate animals, materials enter into the **gastrointestinal tract**, in which substances are broken down and absorbed into the bloodstream. Most substances enter the bloodstream through the intestinal walls and are transported first to the liver, which is the main organ for metabolic processes in the human body. The other raw material essential for metabolic processes, oxygen from air, enters blood through the lungs. Volatile toxic substances can enter the bloodstream through the lungs, a major pathway for environmental and

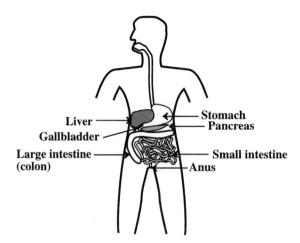

Figure 4.1 Major organs involved in digestion.

occupational exposure to xenobiotics. Toxic substances can also be absorbed through the skin. Undigested food residues and wastes excreted from the liver in bile leave the body through the intestinal tract as feces. The other major pathway for elimination of waste products from metabolic processes consists of the kidneys, which remove such materials from blood, and the bladder and urinary tract through which urine leaves the body. Waste carbon dioxide from the oxidation of food nutrients is eliminated through the lungs.

4.2 DIGESTION

For most food substances and for a very limited number of toxicants, **digestion** is necessary for sorption into the body. Digestion is an enzymatic hydrolysis process by which polymeric macromolecules are broken down with the addition of water into units that can be absorbed from the gastrointestinal tract into blood in the circulatory system; material that cannot be absorbed is excreted as waste, usually after it has been subjected to the action of intestinal bacteria. The digestive tract and organs associated with it are shown in Figure 4.1. A coating of mucus protects the internal surface of the digestive tract from the action of the enzymes that operate in it.

Various enzymes perform digestion by acting on materials in the digestive tract. Carbohydrase, protease, peptidase, lipase, and nuclease enzymes hydrolyze carbohydrates, proteins, peptides, lipids, and nucleic acids, respectively. Digestion begins in the mouth through the action of **amylase** enzyme, which is secreted with saliva and hydrolyzes starch molecules to glucose sugar. The major enzyme that acts in the stomach is **pepsin**, a protein-hydrolyzing enzyme secreted into the stomach as an inactive form (a zymogen) that is activated by a low pH of 1 to 3 in the stomach, resulting from hydrochloric acid secreted into the stomach. In the small intestine, the digestion of carbohydrates and proteins is finished, the digestion of fats is initiated, and the absorption of hydrolysis product nutrients occurs. The first part of the small intestine, the **duodenum**, is where most digestion occurs, whereas nutrient absorption occurs in the lower **jejunum** and **ileum**. The small intestine produces a number of enzymes, including aminopeptidase, which converts peptides to other peptides and amino acids; nuclease; and lactase, which converts lactose (milk sugar) to glactose and glucose.

The liver and the pancreas are not part of the digestive tract as such, but they provide enzymes and secretions required for digestion to occur in the small intestine. The pancreas secretes amylase, lipase, and nuclease enzymes, as well as several enzymes involved in breaking down proteins and peptides. As discussed below with digestion of fats, the liver secretes a substance called **bile** that is stored in the gallbladder and then secreted into the duodenum when needed for digestion of fats.

METABOLIC PROCESSES

By the time that ingested food mass reaches the large intestine or colon, most of the nutrients have been absorbed. Water and ions are absorbed from the mass of material in the colon, concentrating it and converting it to a semisolid state. Much of the material in the colon is converted to bacterial biomass by the action of bacteria, especially *Escherichia coli*, that metabolize food residues not digested by humans or animals. These bacteria produce beneficial vitamins, such as Vitamin K and biotin, that are absorbed through the colon walls and are important in nutrition. The reducing environment maintained by the bacteria in the colon can reduce some xenobiotics (see the discussion of metabolic reductions in Section 7.3). One such product is toxic hydrogen sulfide, H_2S, which is detoxified by special enzymes produced in intestinal wall mucus membranes.

4.2.1 Carbohydrate Digestion

A very simple example of a digestion process is the hydrolysis of sucrose (common table sugar),

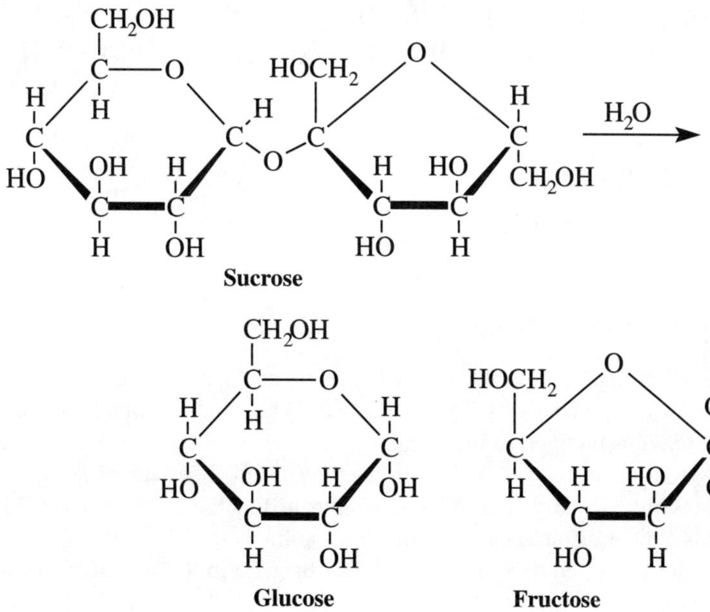

$$\text{Sucrose} + H_2O \longrightarrow \text{Glucose} + \text{Fructose} \tag{4.2.1}$$

to produce glucose and fructose monosaccharides that can be absorbed through intestine walls to undergo metabolism in the body. Each digestive hydrolysis reaction of carbohydrates has its own enzyme. Sucrase enzyme carries out the reaction above, whereas amylase enzyme converts starch to a disaccharide with two glucose molecules called maltose, and maltose in turn is hydrolyzed to glucose by the action of maltase enzyme. A third important disaccharide is **lactose** or "milk sugar," each molecule of which is hydrolyzed by digestive processes to give a molecule of glucose and one of galactose.

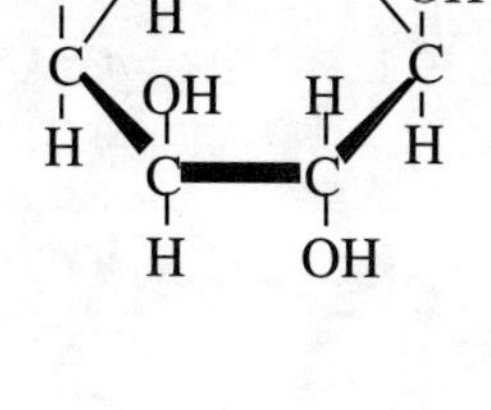

Galactose

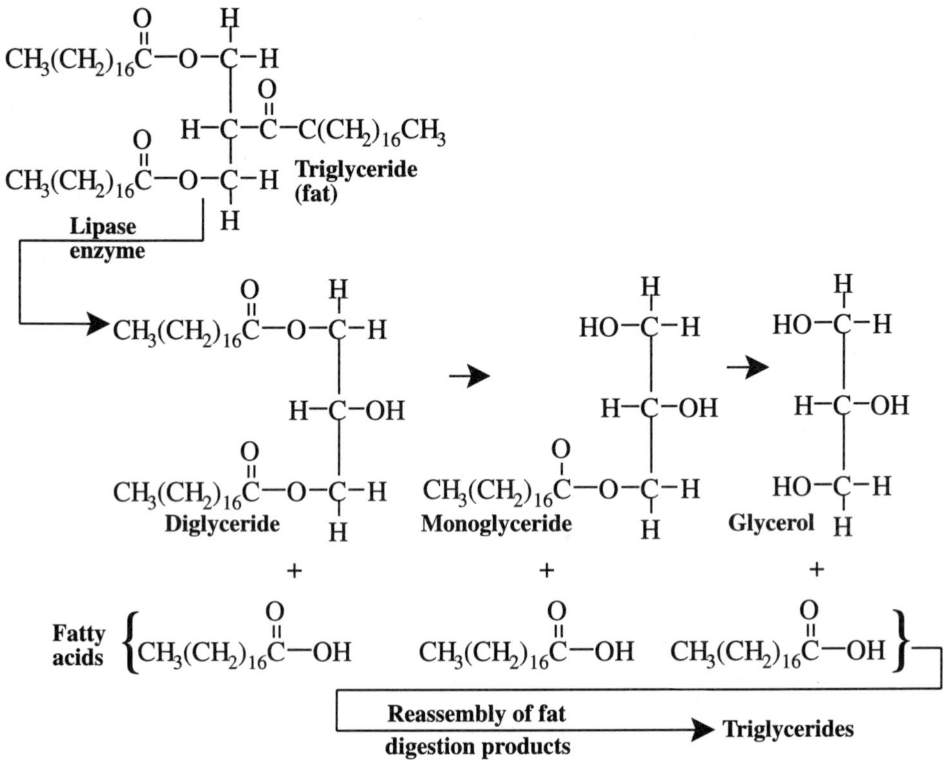

Figure 4.2 Illustration of digestion of fats (triglycerides).

Digestion can be a limiting factor in the ability of organisms to utilize saccharides. Many adults lack the *lactase* enzyme required to hydrolyze lactose. When these individuals consume milk products, the lactose remains undigested in the intestine, where it is acted upon by bacteria. These bacteria produce gas and intestinal pain, and diarrhea may result. The lack of a digestive enzyme for cellulose in humans and virtually all other animals means that these animals cannot metabolize cellulose. The cellulosic plant material eaten by ruminant animals such as cattle is actually digested by the action of enzymes produced by specialized rumen bacteria in the stomachs of such animals.

4.2.2 Digestion of Fats

Fats and oils are the most common lipids that are digested. Digestion breaks fats down from triglycerides to di- and monoglycerides, fatty acids and their salts (soaps) and glycerol, which pass through the intestine wall, where they are resynthesized to triglycerides and transported to the blood through the lymphatic system (see Figure 4.2).

A special consideration in the digestion of fats is that they are not water soluble and cannot be placed in aqueous solution along with the water-soluble lipase digestive enzymes. However, intimate contact is obtained by emulsification of fats through the action of **bile salts** from glycocholic and taurocholic acids produced from cholesterol in the liver:

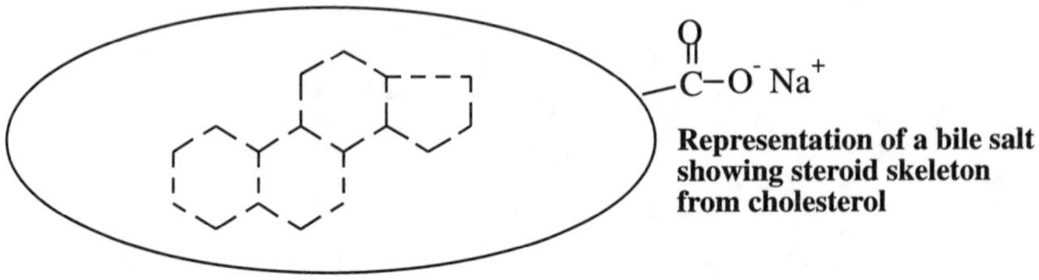

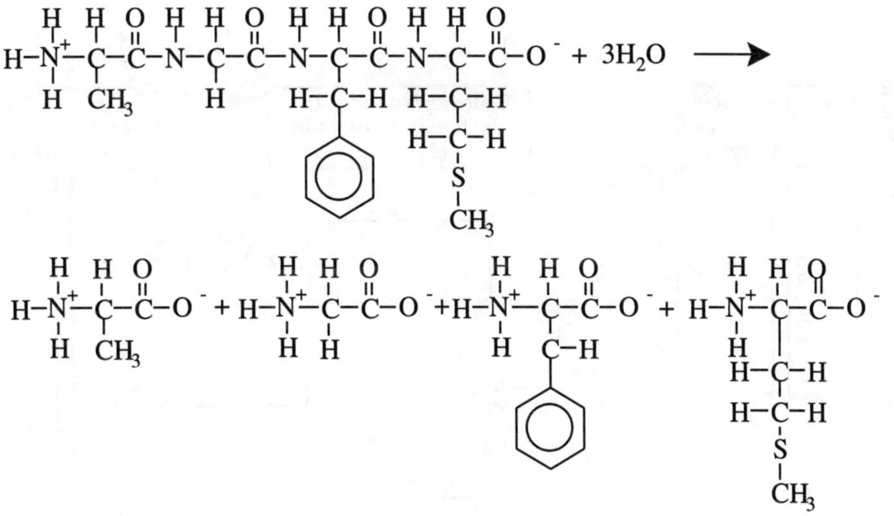

METABOLIC PROCESSES

Figure 4.3 Illustration of the enzymatic hydrolysis of a tetrapeptide such as occurs in the digestion of protein.

4.2.3 Digestion of Proteins

Digestion of proteins occurs by enzymatic hydrolysis in the small intestine (Figure 4.3). The digestion of protein produces single amino acids. These can enter the bloodstream through the small intestine walls. The amino acids circulate in the bloodstream until further metabolized or used for protein synthesis; there is not a "storage depot" for amino acids as there is for lipids, which are stored in "fat depots" in adipose tissue. However, the body does break down protein tissue (muscle) to provide amino acids in the bloodstream.

4.3 METABOLISM OF CARBOHYDRATES, FATS, AND PROTEINS

In the preceding section the digestion of carbohydrates, fats, and proteins by the enzymatic hydrolysis of their molecules was discussed. Digestion enables these materials to enter the bloodstream as relatively small molecules. Once in the bloodstream, these small molecules undergo further metabolic reactions to enable their use for energy production and tissue synthesis. These metabolic processes are all rather complex and beyond the scope of this chapter. However, the main points are covered below.

4.3.1 An Overview of Catabolism

The overall process by which energy-yielding nutrients are broken down to provide the energy required for muscle movement, protein synthesis, nerve function, maintenance of body heat, and other energy-consuming functions is illustrated in Figure 4.4. The approximate empirical formula of the biomolecules from which energy is obtained in catabolism can be represented as $\{CH_2O\}$. The overall energy-yielding catabolic process is the following:

$$\{CH_2O\} + O_2 \rightarrow CO_2 + H_2O + energy \qquad (4.3.1)$$

Figure 4.4 as summarized in Reaction 4.3.1 represents **oxidative respiration**, in which glucose, other nutrients that can be converted to glucose, and the intermediates that glucose generates are oxidized completely to carbon dioxide and water, yielding large amounts of energy. Oxidative

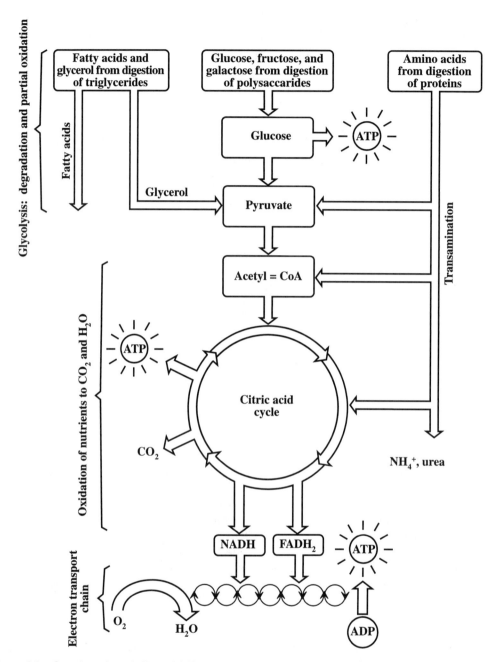

Figure 4.4 Overview of catabolic metabolism, the process by which nutrients are broken down to provide energy.

respiration is in fact a very complicated process involving many steps, numerous enzymes, and a variety of intermediate species. Discussed in more detail in Section 4.4, oxidative respiration in eukaryotic organisms begins with the conversion of glucose to pyruvic acid, a step that does not require oxygen. The second stage of oxidative respiration is the conversion of pyruvic acid to acetyl coenzyme A (acetyl-CoA). In the third stage, the acetyl-CoA goes through the citric acid cycle, in which chemical bond energy harvested in the oxidation of the biomolecules metabolized is converted primarily to a species designated as NADH. In the last stage of oxidative respiration, NADH transfers electrons to molecular O_2 and generates high-energy species (ATP) that are utilized for metabolic needs.

4.3.2 Carbohydrate Metabolism

As discussed in the preceding section, starch and the major disaccharides are broken down by digestive processes to glucose, fructose, and galactose monosaccharrides. Fructose and galactose are readily converted by enzyme action to glucose. Glucose is converted to the glucose 1-phosphate species:

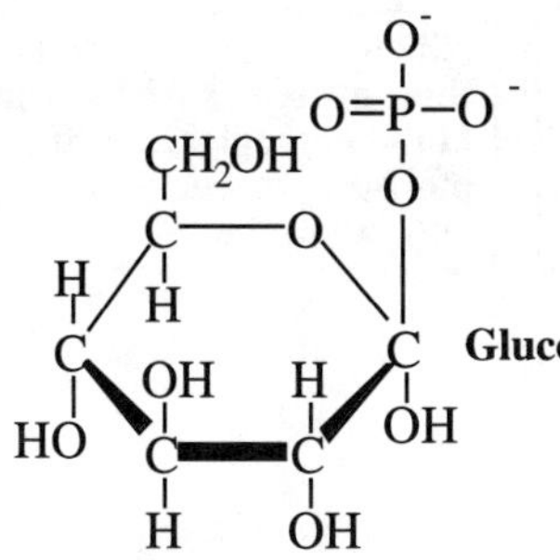

Glucose 1-phosphate (structural formula)

From the glucose 1-phosphate form, glucose may be incorporated into macromolecular (polymeric) glycogen for storage in the animal's body and to provide energy-producing glucose on demand. For the production of energy, the glucose 1-phosphate enters the catabolic process through glycolysis, discussed in Section 4.4.

4.3.3 Metabolism of Fats

Fats are stored and circulated through the body as triglycerides, which must undergo hydrolysis to glycerol and fatty acids before they are further metabolized. Glycerol is broken down via the glycolysis pathway discussed above for carbohydrate metabolism. The fatty acids are broken down in the **fatty acid cycle**, in which a long-chain fatty acid goes through a number of sequential steps to be shortened by two carbon fragments, producing CO_2, H_2O, and energy.

4.3.4 Metabolism of Proteins

A central feature of protein metabolism is the **amino acid pool**, consisting of amino acids in the bloodstream. Figure 4.5 illustrates the metabolic relationship of the amino acid pool to protein breakdown, synthesis, and storage.

Proteins are synthesized from amino acids in the amino acid pool as discussed in Section 3.3. This occurs through the joining of H_3N^+- and $-CO_2^-$ groups at peptide bonds, with the elimination of H_2O for each peptide bond formed. The body can make many of the amino acids it needs, but eight of them, the **essential amino acids**, cannot be synthesized in the human body and must be included in the diet.

The first step in the metabolic breakdown of amino acids is often the replacement of the $^-NH_2$ group with a C=O group by the action of α-ketoglutaric acid in a process called **transamination**. **Oxidative deamination** then regenerates the α-ketoglutaric acid from the glutamic acid product of transamination. These processes are illustrated in Figure 4.6. As a net result of transamination, N(-III) is removed from amino acids and eliminated from the body. For this to occur, nitrogen is first converted to urea:

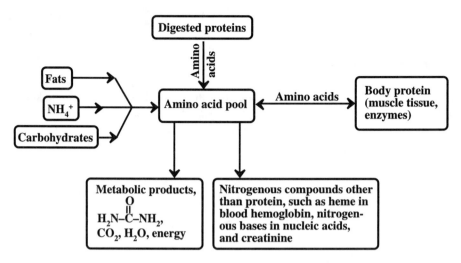

Figure 4.5 Main features of protein metabolism.

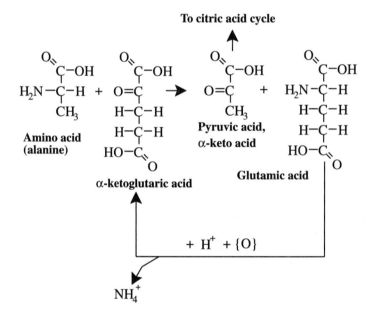

Figure 4.6 Transamination of an amino acid and regeneration of α-ketoglutaric acid by oxidative deamination.

Urea is a solute that is contained in urine, and it is eliminated from the body via the kidneys and bladder.

The α-keto acids formed by transamination of amino acids are further broken down in the citric acid (Krebs) cycle. This process yields energy, and the body's energy needs can be met with protein if sufficient carbohydrates or fats are not available.

4.4 ENERGY UTILIZATION BY METABOLIC PROCESSES

Energy in the form of **free energy** needed by organisms is provided by enzymatically mediated oxidation–reduction reactions. Oxidation in a biological system, as in any chemical system, is the loss of electrons, and reduction is the gain of electrons. A species that is oxidized by losing a negatively charged electron may maintain electrical neutrality by losing H^+ ion; the loss of both e^- and H^+ is equivalent to the loss of a hydrogen atom, H.

A large number of steps within several major cycles are involved in energy conversion, transport, and utilization in organisms. It is beyond the scope of this book to discuss all of these mechanisms in detail. However, it is useful to be aware of the main mechanisms involving energy in relation to biochemical processes in which chemical or photochemical energy is utilized by organisms. They are the following:

- **Glycolysis**, in which, through a series of enzymatic reactions, a six-carbon glucose molecule is converted to two three-carbon pyruvic acid (pyruvate) species with the release of a relatively small amount of the energy in the glucose
- **Cellular respiration**, which occurs in the presence of molecular oxygen, O_2, and involves the conversion of pyruvate to carbon dioxide, CO_2, with the release of relatively large amounts of energy by way of intermediate chemical species
- **Fermentation**, which occurs in the absence of molecular O_2 and produces energy-rich molecules, such as ethanol or lactic acid, with release of relatively little useable energy

4.4.1 High-Energy Chemical Species

Metabolic energy is provided by the breakdown and oxidation of energy-providing nutrients, especially glucose. Usually, however, the energy is needed in a different location and at a different time from the place and time where it is generated. This entails the synthesis of high-energy chemical species that require energy for their synthesis and release it when they break down. Of these, the most important is ATP:

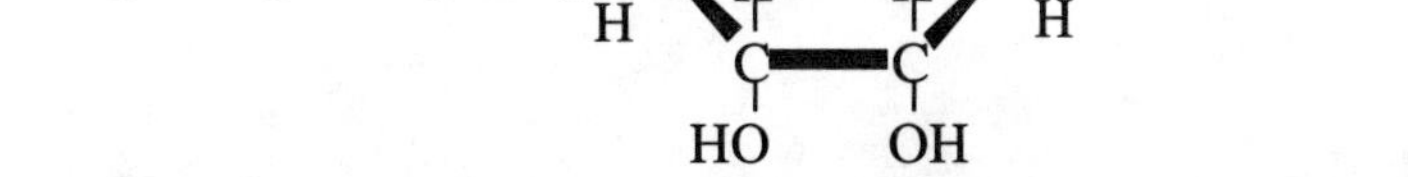

which is generated by the addition of inorganic phosphate, commonly represented as P_i, from **adenosine diphosphate** (ADP):

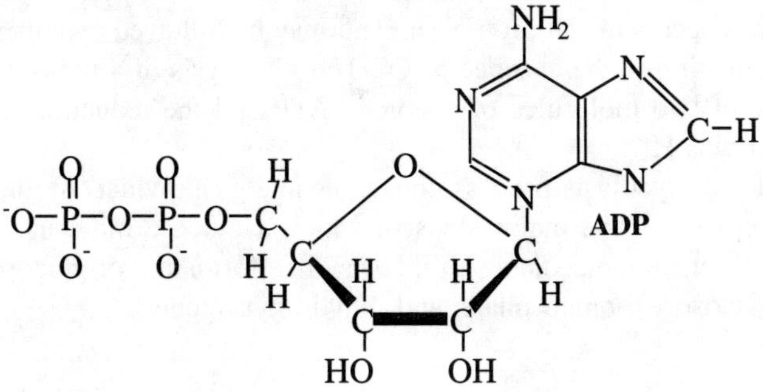

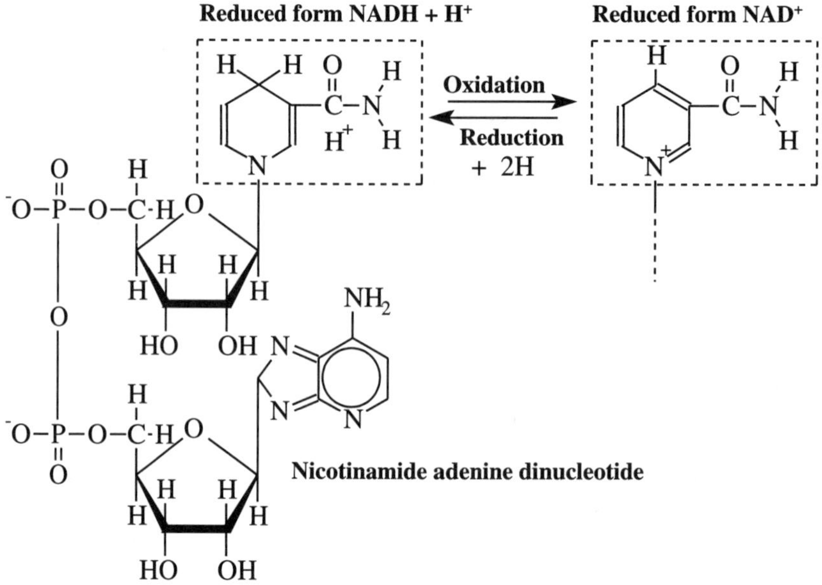

Figure 4.7 Structural formula of nicotinamide adenine dinucleotide in its reduced form of NADH + H⁺ and its oxidized form NAD⁺.

When ATP releases inorganic phosphate and reverts to ADP, a quantity of energy equivalent to 31 kJ of energy per mole of ATP is released that can be utilized metabolically. A pair of species that are similar in function to ATP and ADP are **guanine triphosphate** (GTP) and **guanine diphosphate** (GDP).

An important aspect of enzymatic oxidation–reduction reactions involves the transfer of hydrogen atoms. This transfer is mediated by coenzymes (substances that act together with enzymes) **nicotinamide adenine dinucleotide** (NAD) and **nicotinamide adenine dinucleotide phosphate** (NADP). These two species pick up H atoms to produce NADH and NADPH, respectively, both of which can function as hydrogen atom donors. Another pair of species involved in oxidation–reduction processes by hydrogen atom transfer consists of **flavin adenine triphosphate** (FAD) and its hydrogenated form **FADH$_2$**. The structural formulas of NAD and its cationic form, NAD⁺, are shown in Figure 4.7.

4.4.2 Glycolysis

Glycolysis is a multistepped, anaerobic (without oxygen) process in which a molecule of glucose is broken down in the absence of O_2 to produce two molecules of pyruvic acid (pyruvate anion) and energy. Glycolysis occurs in cell protoplasm and may be followed by either cellular respiration utilizing O_2 or fermentation in the absence of O_2. The glycolysis of a molecule of glucose results in the net formation of two molecules of energetic ATP and the reduction of two NAD⁺ to two molecules of NADH plus H⁺.

The first part of the glycolysis process consumes energy provided by the conversion of two ATPs to ADP. It consists of five major steps in which a glucose molecule is converted to two glylceraldehyde 3-phosphate molecules with intermediate formation of glucose 6-phosphate, fructose 6-phosphate, fructose 1,6-biphosphate, and dihydroxyacetone:

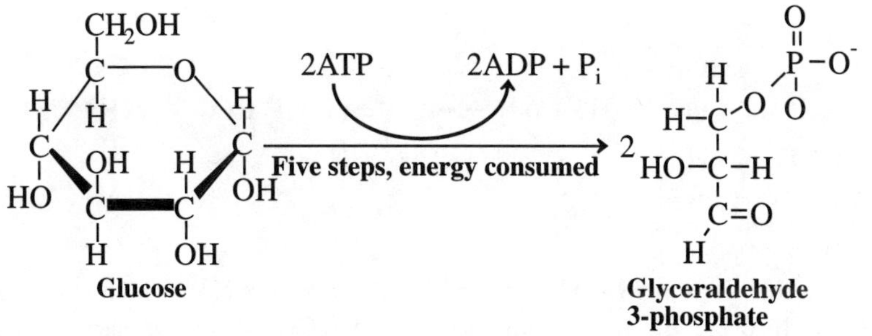

METABOLIC PROCESSES

$$\text{Glucose} \xrightarrow[\text{Five steps, energy consumed}]{2\text{ATP} \quad 2\text{ADP} + P_i} 2 \text{ Glyceraldehyde 3-phosphate} \quad (4.4.1)$$

The second part of the glycolyis process is the five-step conversion of glyceraldehyde to pyruvate accompanied by conversion of four ADPs to four ATPs:

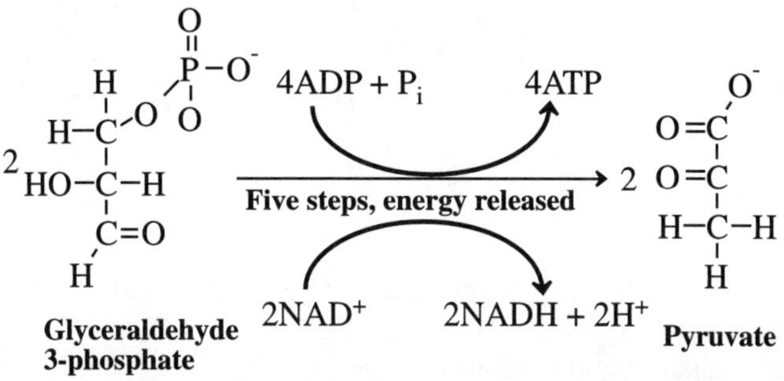

$$2 \text{ Glyceraldehyde 3-phosphate} \xrightarrow[\text{Five steps, energy released}]{4\text{ADP} + P_i \quad 4\text{ATP} \quad 2\text{NAD}^+ \quad 2\text{NADH} + 2\text{H}^+} 2 \text{ Pyruvate} \quad (4.4.2)$$

Since two molecules of ATP are converted to ADP in the first part of the glycolysis process, there is a net gain of two molecules of ATP. The second part of the glycolysis process also yields two molecules of NADH + H$^+$ per molecule of glucose. Subsequently, the energy-yielding conversion of the two molecules of ATP back to ADP and the oxidation of NADH,

$$2\text{NADH} + 2\text{H}^+ + \text{O}_2 \rightarrow \text{NAD}^+ + \text{H}_2\text{O} + \text{energy} \quad (4.4.3)$$

can provide energy for metabolic needs. The three conversions accomplished in glycolysis are (1) glucose to pyruvate, (2) ADP to ATP, and (3) NAD$^+$ to NADH. The net reaction for glycolysis may be summarized as

$$\text{Glucose} + 2\text{ADP} + 2\text{P}_i + 2\text{NAD}^+ \rightarrow 2\text{Pyruvate} + 2\text{ATP} + 2\text{NADH} + 2\text{H}^+ + 2\text{H}_2\text{O} \quad (4.4.4)$$

In addition to glucose, other monosaccharides and nutrients may be converted to intermediates in the glycolysis cycle and enter the cycle as these intermediates.

4.4.3 Citric Acid Cycle

Glycolysis yields a relatively small amount of energy. Much larger amounts of energy may be obtained by complete oxidation of bionutrients to CO$_2$ and 2H$_2$O, which occurs in heterotrophic organisms that utilize oxygen for respiration. The pyruvic acid product of glycolysis can be oxidized to the acetyl group, which becomes bound to coenzyme A in the highly energetic molecule acetyl-CoA, as shown by the following reaction:

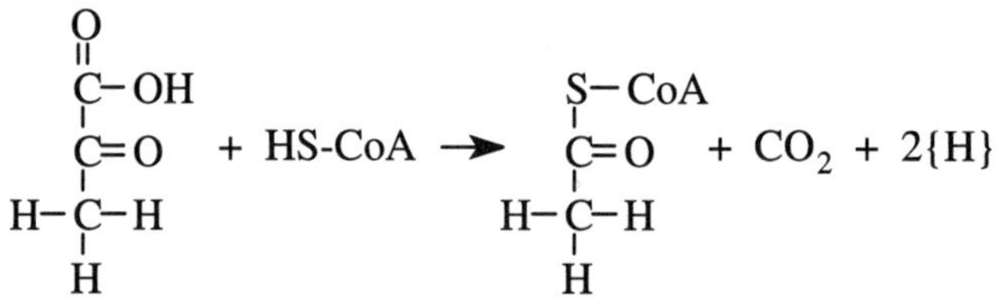

$$\begin{array}{c}\text{O}\\\|\\\text{C-OH}\\|\\\text{C=O}\\|\\\text{H-C-H}\\|\\\text{H}\end{array} + \text{HS-CoA} \rightarrow \begin{array}{c}\text{S-CoA}\\|\\\text{C=O}\\|\\\text{H-C-H}\\|\\\text{H}\end{array} + CO_2 + 2\{H\} \qquad (4.4.5)$$

Acetyl-CoA enters the **citric acid cycle** (also called the Krebs cycle), which occurs in cell mitochondria. In the Krebs cycle, the acetyl group is oxidized to CO_2 and water, harvesting a substantial amount of energy. This complex cycle starts with the reaction of oxaloacetate with acetyl-CoA,

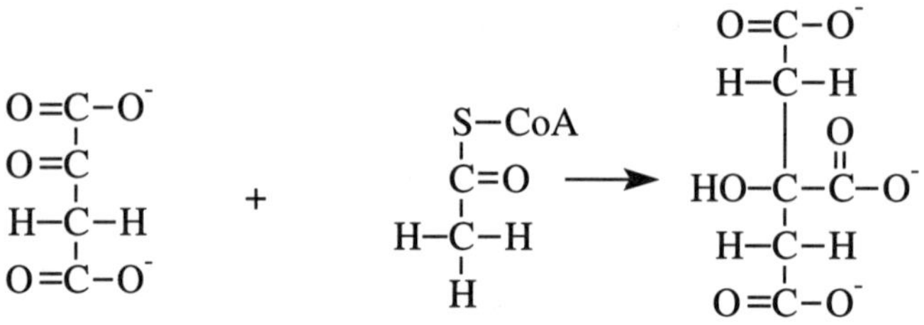

$$\begin{array}{c}\text{O=C-O}^-\\|\\\text{O=C}\\|\\\text{H-C-H}\\|\\\text{O=C-O}^-\end{array} + \begin{array}{c}\text{S-CoA}\\|\\\text{C=O}\\|\\\text{H-C-H}\\|\\\text{H}\end{array} \rightarrow \begin{array}{c}\text{O=C-O}^-\\|\\\text{H-C-H}\\|\quad\text{O}\\\text{HO-C-C-O}^-\\|\\\text{H-C-H}\\|\\\text{O=C-O}^-\end{array} \qquad (4.4.6)$$

Oxaloacetate **Acetyl = CoA** **Citrate**

to produce citrate. In a series of steps involving a number of intermediates, CO_2 is evolved and H is removed by NAD^+ to yield $NADH + H^+$ and by FAD to yield $FADH_2$. Guanosine triphosphate is also generated from guanosine diphosphate in the citric acid cycle and later generates ATP. The anions of several four-, five-, and six-carbon organic acids are generated as intermediates in the citric acid cycle, including citric, isocitric, ketoglutaric, succinic, fumaric, malic, and oxaloacetic acids; the last of these reacts with additional acetyl-CoA from glycolysis to initiate the cycle again. Structural formulas of the intermediate species generated in the citric acid cycle are shown in Figure 4.8. The reduced carrier molecules generated in the citric acid cycle, NADH and $FADH_2$, are later oxidized in the respiratory chain (see below) to produce ATP and are, therefore, the major conduits of energy from the citric acid cycle. The reaction for one complete cycle of the citric acid cycle can be summarized as follows:

$$\text{Acetyl-CoA} + 3NAD^+ + FAD + GDP + P_i + 2H_2O \rightarrow$$
$$3NADH + FADH_2 + 2CO_2 + 2H^+ + GTP + \text{HS-CoA} \qquad (4.4.7)$$

4.4.4 Electron Transfer in the Electron Transfer Chain

As indicated by Reaction 4.3.1, the driving force behind the high-energy yields of oxidative respiration is the reaction with molecular oxygen to produce H_2O. Electrons removed from glucose and its products during oxidative respiration are donated to O_2, which is the final **electron receptor**. To this point in the discussion of oxidative respiration, molecular oxygen has not entered any of the steps. It does so during **transfer of electrons** in the **electron transfer chain**, the step at which most of the energy is harvested from oxidative respiration. Electrons picked up from glycolysis and citric acid cycle intermediates are transferred to the electron transport chain via NAD^+/(NADH + H^+) and FAD/$FADH_2$. The overall process, which occurs in several small increments, is represented for NADH as

$$NADH + H^+ + \tfrac{1}{2}O_2 \rightarrow NAD^+ + H_2O \qquad (4.4.8)$$

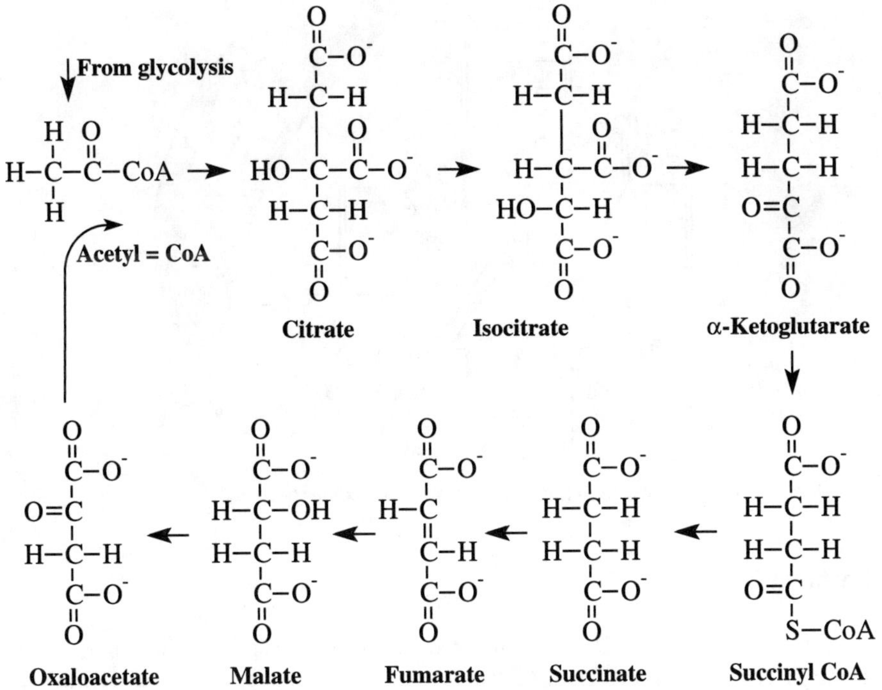

Figure 4.8 Intermediates in the citric acid cycle shown in the ionized forms in which they exist at physiological pH values. The final oxaloacetate product reacts with acetyl-CoA from glycolysis to start the cycle over again.

The electron transfer chain converts ADP to highly energetic ATP, which provides energy in cells. By occurring in several reactions, the oxidation of NADH + H$^+$ to NAD$^+$ releases energy in small increments enabling its efficient utilization.

4.4.5 Electron Carriers

Electron carriers are chemical species that exist in both oxidized and reduced forms capable of reversible exchange of electrons. Electron carriers consist of flavins, coenzyme Q, iron–sulfur proteins, and cytochromes. As shown in Figure 4.9, cytochromes contain iron bound with four N atoms attached to protein molecules, a group called the **heme group**. The iron ions in cytochromes are capable of gaining and losing electrons to produce Fe^{2+} and Fe^{3+}, respectively. Interference with the action of cytochromes is an important mode of the action of some toxicants. Cyanide ion, CN^-, has a strong affinity for Fe^{3+} in ferricytochrome, preventing it from reverting back to the Fe^{2+} form, thus stopping the transfer of electrons to O_2 and resulting in rapid death in the case of cyanide poisoning.

4.4.6 Overall Reaction for Aerobic Respiration

From the discussion above, it is obvious that aerobic respiration is a complex process involving a multitude of steps and a large number of intermediate species. These can be summarized by the following overall net reaction for the catabolic metabolism of glucose:

$$C_6H_{12}O_6 + 10NAD^+ + 2FAD^+ + 36ADP + 36P_i + 14H^+ + 6O_2 \rightarrow$$
$$6CO_2 + 36ATP + 6H_2O + 10NADH + 6FADH_2 \quad (4.4.9)$$

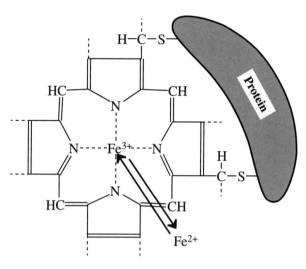

Figure 4.9 Heme group in a cytochrome involved in electron exchange.

4.4.7 Fermentation

Fermentation occurs when O_2 is not utilized for aerobic respiration through the citric acid cycle and the electron transport chain. Glycolysis still occurs as a prelude to fermentation with some production of ATP, but the utilization of energy from glucose is much less than when aerobic respiration occurs. Instead of complete oxidation of glucose to carbon dioxide and water, fermentation stops with an organic molecule. Fermentation is carried out by a variety of bacteria and by eukaryotic cells in the human body. Muscle cells carry out fermentation of pyruvate to lactate under conditions of insufficient oxygen, and the accumulation of lactate in muscle cells is responsible for the pain associated with extreme exertion. Nerve cells are incapable of carrying out fermentation, which is why brain tissue is rapidly destroyed when the brain is deprived of oxygen.

Lactic acid fermentation occurs when lactate is the end product of fermentation. Coupled with glycolysis, lactic acid fermentation can generate ATP from ADP and provide energy for cellular processes. The fermentation step in lactic acid fermentation generates NAD^+ from $NADH + H^+$, and the NAD^+ cycles back to the glycolysis process. The lactic acid fermentation cycle is illustrated in Figure 4.10.

Alcoholic fermentation occurs when the end product is ethanol, as shown in Figure 4.11. In this process the pyruvate is first converted enzymatically to acetaldehyde. The conversion of acetaldehyde to ethanol produces NAD^+ from $NADH + H^+$, and the NAD^+ is cycled through the glycolysis process. As with lactic acid fermentation, the glycolysis process produces usable energy contained in two molecules of ATP produced for each molecule of glucose metabolized.

4.5 USING ENERGY TO PUT MOLECULES TOGETHER: ANABOLIC REACTIONS

The preceding section has discussed in some detail the complicated processes by which complex molecules are disassembled to extract energy for metabolic needs. Much of this energy goes into *anabolic metabolic processes* to put small molecules together to produce large molecules needed for function and structure in organisms. Typical of the small molecules so put together are glucose monosaccharide molecules, assembled into starch macromolecules, and amino acids, assembled into proteins. In all cases of macromolecule synthesis, an H atom is removed from one molecule and an –OH group from the other to link the two together:

METABOLIC PROCESSES

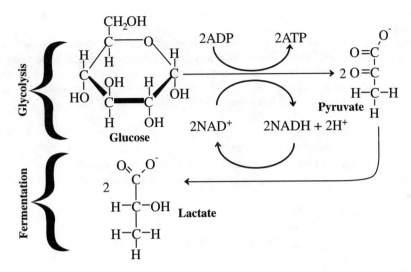

Figure 4.10 Lactic acid fermentation in which the conversion of pyruvate to lactate is coupled with glycolysis to produce energetic ATP.

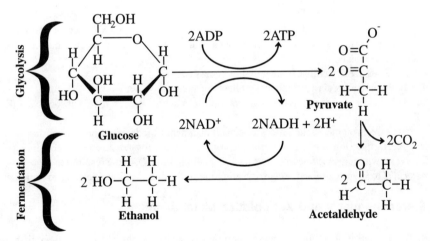

Figure 4.11 Alcoholic fermentation in which the conversion of pyruvate to ethanol through an acetaldehyde intermediate is coupled with glycolysis to produce energetic ATP.

$$\{\text{Molecule A}\}-\text{H} + \text{HO}-\{\text{Molecule B}\} \rightarrow \{\text{Molecule A}\}-\{\text{Molecule B}\} + H_2O \quad (4.5.1)$$

Since a molecule of water is eliminated for each linkage formed, the anabolic reactions leading to macromolecule formation are **dehydration** reactions. Recall from Section 4.2 that when macromolecular nutrient molecules are digested prior to their entering the body's system, a molecule of water is added for each linkage broken — a **hydrolysis** reaction. The energy required for the anabolic synthesis of macromolecules is provided by catabolic processes of glycolysis, the citric acid cycle, and electron transport.

A variety of macromolecules are produced anabolically. The more important of these are listed below:

- Polysaccharide glycogen (animals) and starch (plants) produced for energy storage from glucose

- Polysaccharide chitin composing shells of crabs and similar creatures produced from modified glucose
- Lipid triglyceride fats and oils used for energy storage produced from glycerol and three fatty acids
- Phospholipids present in cell membranes produced from glycerol, two fatty acids, and phosphate
- Globular proteins (in enzymes) and structural proteins (in muscle) produced from amino acids
- Nucleic acids that provide the genetic code and directions for protein synthesis composed of phosphate, nitrogenous bases, deoxyribose (DNA), and oxyribose (RNA)

The anabolic processes by which macromolecules are produced are obviously important in life processes. Remarkably, these processes generally occur properly, making the needed materials when and where needed. However, in some cases things go wrong with potentially catastrophic results. This can occur through the action of toxicants and is a major mode of the action of toxic substances.

4.6 METABOLISM AND TOXICITY

Metabolism is of utmost importance in toxicity. Details of the metabolism of toxic substances and their precursors are addressed in Chapter 7, "Toxicological Chemistry." At this point it should be noted that there are several major aspects of the relationship between toxic substances and metabolism, as listed below:

- Some substances that are not themselves toxic are metabolized to toxic species. Most substances regarded as causing cancer must be metabolically activated to produce species that are the ultimate carcinogenic agents.
- Toxic species are detoxified by metabolic processes.
- Metabolic processes act to counter the effects of toxic substances.
- Metabolic processes of fungi, bacteria, and protozoa act to degrade toxic substances in the water and soil environments.
- Adverse effects on metabolic processes constitute a major mode of action of toxic substances. For example, cyanide ion bonds with ferricytochrome oxidase, a form of an enzyme containing iron(III) that cycles with ferrouscytochrome oxidase, containing iron(II), in the respiration process by which molecular oxygen is utilized, thus preventing the utilization of O_2 and leading to rapid death.

4.6.1 Stereochemistry and Xenobiotics Metabolism

Recall from Section 1.9 that some molecules can exist as chiral *enantiomers* that are mirror images of each other. Although enantiomers may appear to be superficially identical, they may differ markedly in their metabolism and toxic effects. Much of what is known about this aspect of xenobiotics has been learned from studies of the metabolism and effects of pharmaceuticals. For example, one of the two enantiomers that comprise antiepileptic Mesantoin is much more rapidly hydroxylated in the body and eliminated than is the other enantiomer. The human cytochrome P-450 enzyme denoted CYP2D6 is strongly inhibited by quinidine, but is little affected by quinine, an optical isomer of quinidine. Cases are known in which a chiral secondary alcohol is oxidized to an achiral ketone, and then reduced back to the secondary alcohol in the opposite configuration of the initial alcohol.

SUPPLEMENTARY REFERENCES

Brody, T., *Nutritional Biochemistry,* 2nd. ed., Academic Press, San Diego, 1999.
Finley, J.W. and Schwass, D.E., Eds., *Xenobiotic Metabolism,* American Chemical Society, Washington, D.C., 1985.
Groff, J.L., Gropper, S.S., and Hunt, S.M., *Advanced Nutrition and Human Metabolism,* 2nd ed., West Publishers, Minneapolis/St. Paul, 1995.
Hutson, D.H., Caldwell, J., and Paulson, G.D., *Intermediary Xenobiotic Metabolism in Animals: Methodology, Mechanisms, and Significance,* Taylor & Francis, London, 1989.
Illing, H.P.A., Ed., *Xenobiotic Metabolism and Disposition,* CRC Press, Boca Raton, FL, 1989.
Salway, J.G., *Metabolism at a Glance,* 2nd ed., Blackwell Science, Malden, MA, 1999.
Stephanopoulos, G., Aristidou, A.A., and Nielsen, J., *Metabolic Engineering Principles and Methodologies,* Academic Press, San Diego, 1998.

QUESTIONS AND PROBLEMS

1. Define metabolism and its relationship to toxic substances.
2. Distinguish between digestion and metabolism. Why is digestion relatively unimportant in regard to toxic substances, most of which are relatively small molecules?
3. What is the fundamental difference between the digestion of fats and that of complex carbohydrates (starches) and proteins? What role is played by bile salts in the digestion of fats?
4. What are the functions of ADP, ATP, NAD^+, and NADH in metabolism?
5. What is the overall reaction mediated by the Krebs cycle? What does it produce that the body needs?
6. What is the amino acid pool? What purposes does it serve?
7. What is meant by an essential amino acid?
8. What is transamination? What product of amino acid synthesis is eliminated from the body by the kidneys?
9. Give the definition and function of an energy carrier species in metabolism.

CHAPTER 5

Environmental Biological Processes and Ecotoxicology

5.1 INTRODUCTION

Ecology is the study of groups of organisms interacting with each other and with their surroundings in a generally steady-state and sustainable manner. Such relationships exist within **ecosystems** that include the organisms and their surrounding environment. An ecosystem may be viewed on a rather limited scale. For example, a small pond, the photosynthetic phytoplankton (microscopic plants) that provide the basis of the food chain in it, and the animals that inhabit the pond compose a small ecosystem. The scale may be expanded to a much larger scale; indeed, all of Earth may be viewed as a vast ecosystem.

As discussed in detail in Chapter 6, **toxicology** refers to the detrimental effects of substances on organisms. Substances with such effects are called **toxic substances, toxicants,** or **poisons**. Whether or not a substance is toxic depends on the amount to which an organism is exposed and the manner of exposure. Some substances that are harmless or even beneficial at low levels are toxic at higher levels of exposure.

It should be obvious that toxic substances have a strong influence on ecosystems and the organisms in ecosystems. In the example of the small pond ecosystem mentioned above, copper sulfate added to the water can kill the microscopic algae, thus destroying the photosynthetic food base of the ecosystem and causing the ecoystem's collapse. It is logical, therefore, to consider the interactions between ecology and toxicology. These interactions may be complex and involve a number of organisms. They may involve food chains and complex food webs. For example, persistent organohalide compounds may become more highly concentrated through food chains and exert their most adverse effects on organisms such as birds of prey at the top of the food chain. The combination of ecology and toxicology — the study of the effects of toxic substances on ecosystems — has come to be known as **ecotoxicology**, which has been developed into an important discipline in environmental science. This chapter addresses the concept of ecotoxicology as it applies to toxicological chemistry.

Ecotoxicological effects may be considered through several different organizational levels, as shown in Figure 5.1. The first step consists of the introduction of a toxicant or pollutant into the system. This may result in biochemical changes at the molecular level. As a result, physiological changes may occur in tissues and organs. These may result in detrimental alterations of organisms. As a result, the organisms affected may undergo population changes, such as that which occurred with hawks exposed to DDT. Such changes can alter communities; for example, decreased numbers of hawks may allow increased numbers of rodents, accompanied by greater destruction of grain crops. Finally, whole ecosystems may be altered significantly.

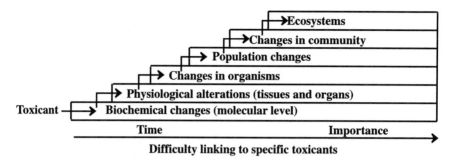

Figure 5.1 Responses to toxicants at different organizational levels in life systems.

5.2 TOXICANTS

There are numerous kinds of toxicants that may cause alterations in ecosystems. These are normally regarded as pollutants from anthropogenic sources, but they may come from natural sources as well. Examples of such naturally occurring toxicants are hydrogen sulfide from geothermal sources or heavy metals, such as lead, leached from minerals.

The metals of most concern are the **heavy metals**, especially cadmium, lead, and mercury. Although it is a metalloid with characteristics of both metals and nonmetals, arsenic is commonly classified as a heavy metal for a discussion of its toxicity. Though not particularly toxic, zinc is abundant and may reach toxic levels in some cases. For example, zinc accumulates in sewage sludge and crop productivity has been lowered on land fertilized with sludge because of zinc accumulation. Copper may be toxic to plants. Aluminum, a natural constituent of soil, may be leached from soil by polluted acidic rainwater and reach levels that are toxic to plants. Other metals that may be of concern because of their toxicities include chromium, cobalt, iron, nickel, and vanadium. Radium, a radioactive alpha particle-emitting metal, can be very toxic at even very low levels in water or food.

Unlike organic compounds, metals are not biodegradable. Most of the heavy metals of greatest concern form poorly soluble compounds with hydroxide (OH$^-$), carbonate (CO$_3^{2-}$), or sulfide (S^{2-}) and therefore tend to accumulate in sediments. This is generally beneficial in making the metals less available to organisms. However, creatures that live and feed in sediments and the organisms that feed on these creatures can acquire toxic levels of heavy metals. In the case of mercury, sediment-dwelling anaerobic bacteria can produce soluble methylated organometallic species in the form of monomethylmercury cation, HgCH$_3^+$, and dimethylmercury, Hg(CH$_3$)$_2$. These species can enter food chains, causing toxic accumulations of mercury in fish or birds at the top of the chain.

Several **nonmetallic inorganic species** may act as toxicants that affect ecosystems. Spills of salts of cyanide ion, CN$^-$, from mining operations have temporarily sterilized small streams. Nitrate ion, NO$_3^-$, in contaminated well water may be reduced to nitrite ion, NO$_2^-$, in the stomachs of ruminant animals and infants. The nitrite converts the iron(II) in blood hemoglobin to iron(III), thus producing methemoglobin, which is useless for transporting oxygen in blood. In extreme cases, fatalities have resulted in livestock and human infants. Excessive levels of carbon dioxide, CO$_2$, in water or air can be detrimental or even fatal.

Persistent organic toxicants are arguably the pollutants of most concern with respect to ecotoxicology because of their abilities to bioaccumulate in lipid (fat) tissue and to become increasingly concentrated in such tissue as they progress through food chains. The most common of such compounds are the organohalides discussed in Chapter 16. The classic persistent organohalide compound is DDT:

DDT

which was widely used as a very effective insecticide during the mid-1900s. DDT is a prime example of a chemical that is significant for its ecotoxicological effects. Its acute toxicological effects are low, and it is safe to use around humans and other mammals. Unfortunately, it persists through food chains and accumulates at high levels in birds of prey at the top of the food chain, causing them to produce thin-shelled eggs that break before hatching. Populations of eagles, ospreys, hawks, and similar birds were virtually wiped out before DDT was banned for general use. Several related organohalides used as insecticides, such as dieldrin and aldrin, were similarly implicated and have been banned.

Another notable class of persistent organohalide compounds is the polychlorinated biphenyl (PCB) class, discussed in Section 16.4. Widely used in industrial applications, these compounds have become widespread and persistent environmental pollutants. The compound 2,3,7,8-tetrachlorodibenzo-p-dioxin (TCDD), commonly known as dioxin, discussed in Section 16.6, is an extremely persistent manufacturing by-product that has caused significant environmental problems.

Another class of organic toxicants consists of relatively biodegradable organic compounds, many of which have been used as insecticides and herbicides. A common class of such compounds is composed of organophosphate insecticides such as those discussed in Section 18.7. Another common class of biodegradable organics contains the carbamates discussed in Section 15.9. Pyrethrins from plant sources and their synthetic analogs, pyrethroids, are discussed in Section 19.5. These substances are readily biodegraded compounds used as insecticides. Biodegradable organophosphates, carbamates, and pyrethroids can have some ecotoxicological effects because of their toxicities, but they are of relatively less concern because they biodegrade before having much of an opportunity to accumulate in organisms.

Gaseous toxicants that occur as air pollutants may have some significant ecotoxicological effects. Prominent among such pollutants are NO and NO_2 associated with photochemical smog. A particularly damaging smog component is ozone, O_3. Ozone can be very harmful to plants, greatly reducing the photosynthetic productivity of ecosystems.

5.3 PATHWAYS OF TOXICANTS INTO ECOSYSTEMS

The major pathways by which toxicants enter ecosystems are through (1) water, (2) food, (3) soil, and (4) air. Each of these will be considered briefly here. There is significant overlap among these pathways. In order to characterize the ecotoxicological behavior of toxic substances, it is necessary to understand their movement between and within the water, air, soil, and living compartments of the environment and to consider how toxic substances are metabolized, broken down, concentrated, and stored within these compartments. The science that deals with these aspects of chemical release and uptake, movement, degradation, and effects is called **chemodynamics**. An understanding of the chemodynamics of environmental chemicals must consider biotic factors as well as abiotic factors, such as temperature, wind, water flow, and sunlight.

For the most part, water that affects ecosystems is **surface water** in streams, lakes, ponds, and other bodies of water. Water held by sediments in such bodies may be important repositories of toxic substances. **Groundwater** in underground aquifers may be released to surface sources where it may contact organisms. Shallow groundwater and water held by soil may enter plant roots.

Toxicants may get into water from a number of different sources. Release from anthropogenic sources as water pollutants can be a significant source of toxicants in water. Sewage, both treated and untreated, may contaminate water. Sewage provides high concentrations of degradable organic substances that consume oxygen when they biodegrade in water. Sewage that has been treated, usually biologically, may still contain residual organic matter, salts, and other materials that can get into water. The sludge left over from biological treatment of sewage is often spread on land and may lose contaminants that get into water. Industrial wastewaters improperly released to natural waters may be sources of organic pollutants and heavy metals in water.

Toxicants may enter organisms by way of **food**. In some cases, residues of pesticides or other toxic organic compounds applied to food crops may be ingested by animals or even humans. In nature, such toxicants may enter food chains as organisms higher on the food chain consume organisms lower on the chain. This phenomenon was responsible for the toxic levels of DDT and other poorly biodegradable toxicants that became highly concentrated in birds of prey at the top of food chains, almost causing extinction of some kinds of raptors.

Exposure to toxicants from **terrestrial** sources is very common because of the intimate association between animals and plants that live on the land with soil. Modern agricultural practices call for the application of large quantities of herbicides and other pesticides to plants and soil; residues of these substances may readily enter living organisms. Soil is the repository of a variety of air pollutants, especially airborne particles that settle onto soil. Improper disposal of industrial wastes and toxic substances washed from or blown off of hazardous waste sites has contributed to exposure of soil.

Air-breathing animals and plants that grow on land surfaces are exposed to toxic substances that may be in the atmosphere. There are numerous sources of toxic substances in the atmosphere. Direct emissions from industrial sources and engines may contaminate the atmosphere with toxic organic materials, acidic substances, and heavy metals. Some harmful air pollutants are secondary pollutants formed by atmospheric chemical processes. Strong acids responsible for acid rain are formed by atmospheric oxidation of inorganic gases, especially SO_2 (forming H_2SO_4) and NO and NO_2 (forming HNO_3). Under appropriate conditions of stagnant air, low humidity, and intense sunlight, reactive hydrocarbons and nitrogen oxides emitted from automobile exhausts form a variety of noxious organic aldehydes and oxidants as well as ozone, all of which are toxic to plants and animals.

5.3.1 Transfers of Toxicants between Environmental Spheres

An important aspect of toxic substances related to ecotoxicology is their transfer between spheres of the environment, the atmosphere, hydrosphere, geosphere, biosphere, and anthrosphere. Important considerations in such transfers are illustrated in Figure 5.2. In general, toxicants enter the atmosphere by evaporation, the solids in the geosphere by adsorption, and the hydrosphere by dissolution. Movements of substances between the atmosphere and other spheres are very much influenced by vapor pressures of the substances; those with higher vapor pressures have a greater tendency to remain in the atmosphere. Temperature is also important. Because of the low viscosity of air and the constant circulation of air masses, toxic substances tend to be much more mobile in the atmosphere than in the other environmental spheres. One interesting aspect is the tendency of somewhat volatile organic compounds to evaporate into the atmosphere in warmer regions, and then condense in polar and cold mountainous areas, which often receive a high dose of organic toxicants as a consequence, even though they are far from industrial sources. Low water solubility is associated with the tendency of substances to accumulate in sediments and on soil, whereas highly soluble substances may be transported readily by water. High lipid solubility and low water solubility enable substances to accumulate in the lipid tissue of animals in the biosphere.

ENVIRONMENTAL BIOLOGICAL PROCESSES AND ECOTOXICOLOGY

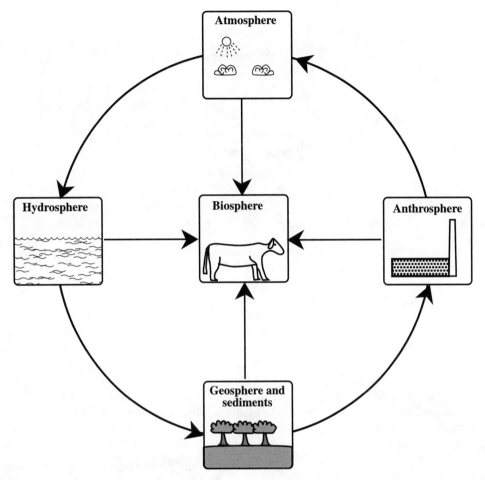

Figure 5.2 Transfers of toxic substances among the major environmental spheres are very important in determining their ecotoxicological effects.

5.3.2 Transfers of Toxicants to Organisms

A crucial aspect of the interchange of toxicants between environmental spheres is the transfer of toxicants to organisms in the biosphere. Sources of toxicants that organisms acquire from their surroundings may be divided into three general categories: those that are ingested with food or water into the alimentary tract, those that are absorbed from ambient surroundings, especially by fish that live in water, and those that are inhaled in air that is breathed into lungs. Persistent, lipophilic (fat-seeking), poorly metabolized toxicants tend to persist and to increase in concentration in higher trophic levels of organisms where a **trophic level** consists of a group of organisms that derive their energy from the same portion of the food web in a biological community. This can lead to **biomagnification** of toxicants in organisms higher in the food chain. Lipophilic toxicants may reach particularly high levels in some birds of prey, which are not only at the top of food chains, but also lack well-developed metabolic mechanisms for dealing with such toxicants. The mobility of organisms, especially birds, can cause toxic substances to spread out over great distances in food webs.

Suspended and dissolved toxic substances in water can be transferred directly into fish, aquatic invertebrates, and amphibians that live in water. In the case of fish, toxicants can be absorbed

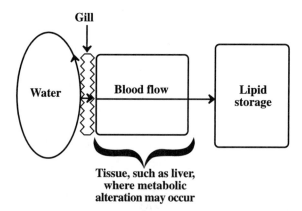

Figure 5.3 Overall pathway of bioconcentration.

primarily through the gills, and with aquatic invertebrates, through respiratory surfaces. There can also be some direct absorption from water through the skin of these organisms and amphibians.

An important consideration in the uptake of toxicants by organisms is the **bioavailability** of various substances, which is basically the tendency of a substance to enter an organism's system upon exposure. The bioavailability of lead has been studied to a large extent and varies significantly with the chemical form of the lead. Mercury also offers striking examples. Water-soluble inorganic mercury is very toxic when ingested. Mercury metal vapor is readily absorbed by the lungs. Fatal doses of lipid-soluble organometallic dimethylmercury can be absorbed through the skin.

The two main pathways for the uptake of toxic substances by plants are through their root systems and across their leaf cuticles. Stomata, the specialized openings in plant leaves that allow carbon dioxide required for photosynthesis to enter the leaves and oxygen and water vapor to exit, are also routes by which toxic substances may enter plants. The mechanisms by which plants take up systemic pesticides and herbicides, which become distributed within the plant, have been studied very intensively.

5.4 BIOCONCENTRATION

The tendency of a chemical to leave aqueous solution and enter a food chain is important in determining its environmental effects and is expressed through the concept of bioconcentration. **Bioconcentration** (Figure 5.3) may be viewed as a special case of bioaccumulation in which a *dissolved substance* is selectively taken up from water solution and concentrated in tissue by nondietary routes. Bioconcentration applies specifically to the concentration of materials from water into organisms that live in water, especially fish. As illustrated in Figure 5.3, the model of bioconcentration is based on a process by which contaminants in water traverse fish gill epithelium and are transported by the blood through highly vascularized tissues to lipid tissue, which serves as a storage sink for hydrophobic substances. Transport through the blood is affected by several factors, including rate of blood flow and degree and strength of binding to blood plasma protein. Prior to reaching the lipid tissue sink, some of the compound may be metabolized to different forms. The concept of bioconcentration is most applicable under the following conditions:

- The substance is taken up and eliminated via passive transport processes.
- The substance is metabolized slowly or not at all.
- The substance has a relatively low water solubility.
- The substance has a relatively high lipid solubility.

Substances that undergo bioconcentration are hydrophobic and lipophilic, and therefore tend to undergo transfer from water media to fish lipid tissue. The simplest model of bioconcentration views the phenomenon on the basis of the physical properties of the contaminant and does not account for physiologic variables (such as variable blood flow) or metabolism of the substance. Such a simple model forms the basis of the **hydrophobicity model** of bioconcentration, in which bioconcentration is regarded from the viewpoint of a dynamic equilibrium between the substance dissolved in aqueous solution and the same substance dissolved in lipid tissue.

5.4.1 Variables in Bioconcentration

There are several important variables in estimating bioconcentration. These are discussed briefly here.

A basic requirement for uptake of a chemical species from water is **bioavailability**. Normally uptake is viewed in terms of absorption from true water solution. Biouptake may be severely curtailed for substances with extremely low water solubilities or that are bound to particulate matter. Dissolved organic matter may also bind to substances and limit their biouptake.

Although the simple bioconcentration model assumes relatively unhindered movement of a contaminant across the barriers between water and lipid tissue, such is often not the case. The uptake of an organic species can be a relatively complex process in which the chemical must traverse membranes in the gills and skin to reach a final lipid sink. A **physiological component** of the process by which a chemical species moves across membranes tends to cause bioconcentration to deviate from predictions based on hydrophobicity alone.

Some evidence suggests that the **lipid content** of the subject organism affects bioconcentration. Higher lipid contents in an organism may to a degree be associated with relatively higher bioconcentration factors (BCFs) (see Section 5.5). The lipid levels at the site of entry (gill in fish) may be relatively more important than those in the whole organism.

Molecular shape and size seem to play a role in bioconcentration. There are steric hindrances to the movement of large molecules across membranes, compared to molecules of about the same mass but with smaller cross-sectional areas. For larger molecules, this results in slower transfer and a lower BCF.

Although a simple bioconcentration model assumes rapid movement of a hydrophobic contaminant through an organism, **distribution** may be relatively slow. The predominant limiting factor in this case is the blood flow. Slow transport to lipid tissue sinks can result in lower apparent BCF values than would be the case if true equilibrium were attained.

5.4.2 Biotransfer from Sediments

Because of the strong attraction of hydrophobic species for insoluble materials such as humic matter, many organic pollutants in the aquatic environment are held by sediments in bodies of water. Bioaccumulation of these materials must, therefore, consider transfer from sediment to water to organism, as illustrated in Figure 5.4.

5.5 BIOCONCENTRATION AND BIOTRANSFER FACTORS

5.5.1 Bioconcentration Factor

Quantitatively, the *hydrophobicity model* of bioconcentration is viewed in the classical thermodynamic sense as an equilibrium between the uptake and elimination of substance X:

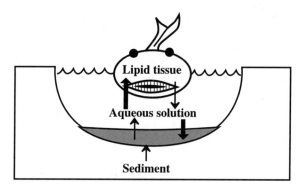

Figure 5.4 Partitioning of a hydrophobic chemical species among sediment, water, and lipid tissue. Thicker arrows denote the preference of the chemical for sediment and lipid tissue, compared to aqueous solution.

Table 5.1 Example Bioconcentration Factors

Chemical	Fish Species	Temperature (°C)	BCF
PCB	Sunfish	5	6.0×10^3
PCB	Sunfish	15	5.0×10^4
PCB	Trout	5	7.4×10^3
PCB	Trout	15	1.0×10^4
Hexachlorobenzene	Rainbow trout	15	5.5×10^3
Hexachlorobenzene	Fathead minnow	15	1.6×10^4

Source: From Barron, M.G., Bioconcentration, *Environ. Sci. Technol.*, 24, 1612–1618, 1990.

$$X(aq) \rightleftarrows X(lipid) \tag{5.5.1}$$

Using k_u as the rate constant for uptake and k_e as the rate constant for elimination leads to the following definition of **bioconcentration factor**:

$$\text{BCF} = \frac{k_u}{k_e} = \frac{[X(lipid)]}{[X(aq)]} \tag{5.5.2}$$

When $[X(lipid)]/[X(aq)] = \text{BCF}$, the rates of uptake and elimination are equal, the lipid concentration of the xenobiotic substance remains constant (at constant $[X(aq)]$), and the system is in a condition of **dynamic equilibrium** or **steady state**. Some typical values of BCF are given in Table 5.1.

Evidence for the validity of the hydrophobicity model of bioconcentration is provided by correlations of it with the **octanol–water partition coefficient**, K_{ow}, using *n*-octanol as a surrogate for fish lipid tissue. The measurement of K_{ow} consists of determining the concentration of a hydrophobic contaminant in water-immiscible *n*-octanol relative to water with which it is in equilibrium. Typical K_{ow} values range from 10 to 10^7, corresponding to BCF values of 1 to 10^6. Such K_{ow}/BCF correlations have proven to be reasonably accurate when narrowly defined for a specified class of compounds, most commonly poorly metabolized organohalides. Major inconsistencies appear when attempts are made to extrapolate from one class of contaminants to another.

Table 5.2 Example Biotransfer Factors for Beef and Milk[a]

Chemical	Log K_{ow}	Biotransfer Factor Beef	Milk
Arochlor 1254 (PCB)	6.47	−1.28	−1.95
Chordane	6.00	−2.13	−3.43
DDT	5.76	−1.55	−2.62
Endrin	5.16	−1.92	−2.76
Lindane	3.66	−1.78	−2.60
2,4,5-T	3.36	−4.82	−2.60
TCDD	6.15	−1.26	−1.99
Toxaphene	5.50	−2.79	−3.20

[a] For a more complete listing, see Travis, C.C. and Arms, A.D., Bioconcentration of organics in beef, milk, and vegetation, *Environ. Sci. Technol.*, 22, 271–274, 1988.

5.5.2 Biotransfer Factor

A useful measure of bioaccumulation from food and drinking water by land animals is the **biotransfer factor** (BTF), defined as

$$\text{BTF} = \frac{\text{Concentration in tissue}}{\text{Daily intake}} \quad (5.5.3)$$

where the concentration in tissue is usually expressed in mg/kg and daily intake in mg/day. This expression can be modified to express other parameters, such as concentration in milk. As is the case for bioconcentration factors for fish in water, BTF shows a positive correlation with K_{ow} values. Representative BTF values for biotransfer to beef are given in Table 5.2.

5.5.3 Bioconcentration by Vegetation

Like fish and mammals, vegetation can absorb organic contaminants. In the case of vegetation, the bioconcentration factor can be expressed relative to the mass of compound per unit mass of soil. The exact expression for vegetation is

$$\text{BCF} = \frac{\text{Concentration in plant tissue}}{\text{Concentration in soil}} \quad (5.5.4)$$

where the concentration in plant tissue is given in units of mg/kg dry plant tissue and the concentration in soil is in units of mg/kg dry soil. Table 5.3 gives some typical values of BCF for plants relative to log K_{ow}. It is seen that for uptake of hydrophobic substances by plants, BCF values are less than one and tend to decrease with increasing K_{ow}, the opposite of the trend observed in animals. This is explained by the transport of organic substances by water from soil to plant tissue, which increases with increasing water solubility of the compound and therefore with decreasing K_{ow}.

5.6 BIODEGRADATION

Biodegradation may involve relatively small changes in the parent molecule, such as substitution or modification of a functional group. In the most favorable cases, however, the compound is completely destroyed and the end result is conversion of relatively complex organic compounds to

Table 5.3 Example Bioconcentration Factors[a]

Chemical	Log K_{ow}	Log BCF
Arochlor 1254 (PCB)	6.47	−1.77
3,4-Dichloroaniline	2.69	−0.30
Diflubenzuron	3.82	−0.53
Heptachlor	5.44	−1.48
Lindane	3.66	−0.41
TCDD	6.15	−1.87

[a] For a more complete listing, see Travis, C.C. and Arms, A.D., Bioconcentration of organics in beef, milk, and vegetation, *Environ. Sci. Technol.*, 22, 271–274, 1988.

CO_2, H_2O, and inorganic salts, a process called **mineralization**. Usually the products of biodegradation are molecular forms that tend to occur in nature and that are in greater thermodynamic equilibrium with their surroundings.

5.6.1 Biochemical Aspects of Biodegradation

Several terms should be reviewed in considering the biochemical aspects of biodegradation. *Biotransformation* is what happens to any substance that is *metabolized* by the biochemical processes in an organism and is altered by these processes. *Metabolism* is divided into the two general categories of *catabolism*, which is the breaking down of more complex molecules, and *anabolism*, which is the building up of life molecules from simpler materials. The substances subjected to biotransformation may be naturally occurring or *anthropogenic* (made by human activities). They may consist of *xenobiotic* molecules that are foreign to living systems.

It should be emphasized that biodegradation of an organic compound occurs in a stepwise fashion and is usually not the result of the activity of a single specific organism. Usually several strains of microorganisms, often existing synergistically, are involved. These may utilize different metabolic pathways and a variety of enzyme systems.

Although biodegradation is normally regarded as degradation to simple inorganic species such as carbon dioxide, water, sulfates, and phosphates, the possibility must always be considered of forming more complex or more hazardous chemical species. An example of the latter is the production of volatile, soluble, toxic methylated forms of arsenic, and mercury from inorganic species of these elements by bacteria under anaerobic conditions. There is some evidence to suggest that toluene, an anthropogenic compound, can be produced metabolically from the amino acid phenylalanine by anaerobic bacteria. This would occur by successive deammination and decarboxylation as follows:

Phenylalanine → (Deammination) → (Two-step removal of carbon) → Toluene

It is well known that microbial communities exposed to xenobiotic compounds develop the ability to break these compounds down metabolically. This has become particularly obvious from studies of biocidal compounds in the environment. In general, such compounds are readily degraded by bacteria that have been exposed to the compounds for prolonged periods, but not by bacteria from unexposed sites. The development of microbial cultures with the ability to degrade materials to which they are exposed is described as **metabolic adaptation**. In rapidly multiplying microbial cultures, metabolic adaptation can include genetic changes that favor microorganisms that can degrade a specific kind of pollutant. Metabolic adaptation may also include increased numbers of microorganisms capable of degrading the substrate in question and enzyme induction.

5.6.2 Cometabolism

Xenobiotic compounds are usually attacked by enzymes whose primary function is to react with other compounds, a process that provides neither carbon nor energy called **cometabolism**. Cometabolism usually involves relatively small modifications of the substance that is cometabolized (the secondary substrate), compared to the primary substrate. The enzymes that carry out cometabolism tend to be relatively nonspecific. As an environmentally significant example of cometabolism, at least one strain of bacteria degrades trichloroethylene with an enzyme system that acts predominantly on phenol. The enzyme activity can be induced by exposure to phenol, after which it acts on trichloroethylene.

In pure cultures of microorganisms, the products of cometabolism tend to accumulate and often do not undergo further degradation. However, in mixed cultures, which are the norm for environmental systems, they may serve as substrates for other organisms so that complete biodegradation results. Therefore, studies of biodegradation in pure cultures are usually of limited utility in predicting what happens in the environment.

An example of cometabolism of pollutants is provided by the white rot fungus *Phanerochaete chrysosporium*, which degrades a number of kinds of organochlorine compounds, including DDT, PCBs, and chlorodioxins, under the appropriate conditions. The enzyme system responsible for this degradation is one that the fungus uses to break down lignin in plant material under normal conditions.

5.6.3 General Factors in Biodegradation

The rates and efficacy of biodegradation of organic substances depend on several obvious factors. These include the concentration of the substrate compound, the nature and concentration of the final electron acceptor (most commonly O_2), the presence of phosphorus and nitrogen nutrients, the availability of trace element nutrients, the presence of a suitable organism, the absence of toxic substances, and the presence of appropriate physical conditions (temperature, growth matrix). In addition to their biochemical properties, the physical properties of compounds, including volatility, water solubility, organophilicity, tendency to be sorbed by solids, and charge, play a role in determining the biodegradability of organic compounds.

To a large extent, xenobiotic compounds in the aquatic environment are bound with sediments and suspended solid materials, such as humic acids. This binding plays a large role in biodegradation. Indeed, the structure of the bound form of the xenobiotic, such as a humic acid complex with a synthetic organic compound, may largely determine its rate of enzymatic degradation.

Some studies suggest that biodegradation rates of substances at relatively higher concentrations are not extrapolatable to very low concentrations (see Figure 5.5). That could explain the persistence of very low residual levels of some biodegradable substances in water or soil.

An example of the effects of the presence of toxic materials is provided by the biodegradation of polycyclic aromatic hydrocarbons (PAHs). PAH compounds from fuel and petroleum sludges spilled on soil undergo biodegradation with relative ease, whereas PAHs from creosote contamination are poorly biodegradable. This observation may be explained by the bactericidal properties of creosote components that inhibit the growth of organisms responsible for degrading PAH compounds.

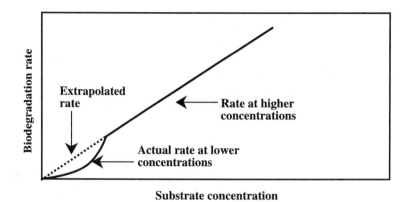

Figure 5.5 Biodegradation vs. substrate concentration showing that rates may not extrapolate well to lower concentrations.

Competition from other organisms may be a factor in biodegradation of pollutants. "Grazing" by protozoa may result in consumption of bacterial cells responsible for the biodegradation of particular compounds.

Trace amounts of micronutrients are needed to support biological processes and as constituents of enzymes. Important micronutrients are calcium, magnesium, potassium, sodium, chlorine, cobalt, iron, vanadium, and zinc. Sometimes sulfur, phosphorus, and micronutrients must be added to media in which microorganisms are used to degrade hazardous wastes in order for optimum growth to occur.

5.6.4 Biodegradability

The amenability of a compound to chemical attack by microorganisms is expressed as its **biodegradability**. The biodegradability of a compound is influenced by its physical characteristics, such as solubility in water and vapor pressure, and by its chemical properties, including molecular mass, molecular structure, and presence of various kinds of functional groups, some of which provide a "biochemical handle" for the initiation of biodegradation. With the appropriate organisms and under the right conditions, even substances that are biocidal to most microorganisms can undergo biodegradation. For example, normally bactericidal phenol is readily metabolized by the appropriate bacteria acclimated to its use as a carbon and energy source.

Recalcitrant or **biorefractory** substances are those that resist biodegradation and tend to persist and accumulate in the environment. Such materials are not necessarily toxic to organisms, but simply resist their metabolic attack. Even some compounds regarded as biorefractory may be degraded by microorganisms adapted to their biodegradation. Examples of such compounds and the types of microorganisms that can degrade them include endrin (*Arthrobacter*), DDT (*Hydrogenomonas*), phenylmercuric acetate (*Pseudomonas*), and raw rubber (*Actinomycetes*).

5.7 BIOMARKERS

A crucial aspect of ecotoxicology is the measurement of the effects of toxic substances on organisms in ecosystems and on ecosystems as a whole. This has traditionally been done by determining levels of toxic substances in organisms and relating these levels to detrimental effects on organisms. Because of factors such as biochemical alterations of toxicants, correlations between toxicant levels and observed toxic effects are often impossible to achieve. A better approach is the use of **biomarkers** consisting of observations and measurements of alterations in biological components, structures, processes, or behaviors attributable to exposure to xenobiotic substances. Biomarkers can often use animals and other organisms to infer the chemical hazards to humans. Biomarkers can be broadly divided into three main categories: exposure, effects, or susceptibility.

A **biomarker of exposure** consists of the measurement of a xenobiotic substance, a metabolite of a xenobiotic substance, or an effect directly attributable to such a substance in an organism. For example, a biomarker of human exposure to aniline might consist of measurement of it or its *p*-aminophenol metabolite in blood or urine. Xenobiotics or their metabolites can be measured directly

$$\text{Aniline} \quad \text{C}_6\text{H}_5\text{-NH}_2 \qquad \text{HO-C}_6\text{H}_4\text{-NH}_2 \quad p\text{-Aminophenol}$$

in tissues obtained by biopsy of live organisms or necropsy of deceased organisms. Urine, blood, exhaled air, feces, and breast milk can also serve as samples and are advantageous in measurements to be made at intervals over a period of time.

A particularly useful kind of biomarker used with increasing frequency during recent years consists of adducts of xenobiotics or their metabolites to biomolecules. A particularly straightforward example of such an adduct measured for many years as evidence of exposure is carboxyhemoglobin, COHb, produced when inhaled carbon monoxide adds to blood hemoglobin, Hb:

$$O_2Hb + CO \rightarrow COHb + O_2 \tag{5.7.1}$$

The COHb has a distinctly different color from the oxygenated form, O_2Hb, and can be measured spectrophotometrically. Cancer-causing compounds and carcinogenic metabolites are generally electrophilic (electron-seeking) species that cause the biochemical changes leading to cancer by adding to nucleophilic groups (electron-rich bound oxygen and nitrogen atoms) on biomolecules, particularly those in DNA. These adducts and adducts to hemoglobin can be measured as biomarkers of exposure.

Biomarkers of effect are alterations of physiology, biochemistry, or behavior directly attributable to exposure to a xenobiotic substance. As noted above, exposure to aniline can be determined by measuring it or its *p*-aminophenol in blood, but it can also be measured by its production of blood methemoglobin (a product of hemoglobin useless for carrying oxygen in blood in which the iron(II) in hemoglobin has been oxidized to iron(III)). Exposure to substances such as nerve gas organophosphates, organophosphate insecticides, and carbamate insecticides that inhibit cholinesterase enzymes required for nerve function can be determined by measurement of cholinesterase enzyme activity. Exposure to the insecticidal DDT metabolite, *p,p'*-DDE can be measured in the laboratory rat (*Rattus rattus*) by induction of enzymatic cytochrome P-4502B, an enzyme used by some organisms to detoxify xenobiotics.

Closely related to biomarkers of effect are **biomarkers of susceptibility**, which indicate increased vulnerability of organisms to disease, physical attack (such as low temperatures), or chemical attack from other toxicants. The most obvious biomarkers of susceptibility are those associated with weakened immune systems, which may make organisms more vulnerable to cancer, infectious diseases, or parasites.

5.8 ENDOCRINE DISRUPTERS AND DEVELOPMENTAL TOXICANTS

Toxicants that disturb the action of hormones involved in reproduction and development are of particular ecotoxicological significance. **Endocrine disrupters** may bind to hormone receptor sites, thus interfering with the essential functions of the endocrine system. Endocrine disrupters may lower production of essential hormones or induce excess production. Perchlorate ion, ClO_4^-, can compete with iodide ion, I^-, for uptake by the thyroid gland and disrupt synthesis of the thyroid hormone.

The most obvious ecotoxicological concern with substances that disrupt endocrine systems and affect development has to do with reproduction of species and their development into healthy adults. Toxicants that interfere with these functions have the potential to severely reduce key populations in ecosystems, thereby altering the nature of the ecosystems.

5.9 EFFECTS OF TOXICANTS ON POPULATIONS

For the most part, toxicology considers the effects of toxicants on individuals. However, ecotoxicology is more concerned with toxic effects on large numbers of individuals composing populations of organisms. A **population** in ecological terms refers to total numbers or numbers per unit area of a particular species in an ecosystem. The average population of a species that an ecosystem can support is its **carrying capacity**. Exposure to toxicants is likely to lower populations by lowering reproductive rates, causing mortality, or weakening members of the species so that they become more vulnerable to predation, disease, or parasites.

In severe cases of exposure to toxicants, a population may go to zero and stay there so long as exposure continues or, if the toxicant is removed, until immigration from outside the ecosystem restores the population. In other cases, the population is reduced and remains at a lower level so long as exposure to the toxicant continues. Another possibility is for a population to decline, and then rise again as the population develops resistance to the effects of the toxic substance. This effect has been observed in the case of exposure to insecticides of insects whose short reproductive times enable natural selection to build populations resistant to the insecticides.

Decreased populations due to exposure to toxic substances may result from mortality to members of the population or from adverse effects on reproduction. Increased mortality obviously decreases numbers directly, but it also results in decreased numbers because of fewer adults capable of producing young.

Effects of toxicants on populations may be indirect. There have been cases reported in which herbicides destroyed weed species that served as sources of food for certain kinds of insects that, in turn, were the food source for particular species of birds. Spraying with these herbicides resulted in decreases in bird populations because of the destruction of their insect food sources.

Toxicants have been known to result in increases in some populations. This occurs when organisms that are predators to a particular species are poisoned, allowing the species to increase in numbers.

5.10 EFFECTS OF TOXICANTS ON ECOSYSTEMS

An ecosystem consists of a variety of communities of organisms and their surrounding environment existing in a generally steady state. An important aspect of ecosystems is the flow of energy and materials. In an ecosystem intricate relationships exist among the organisms — with each other and with their surroundings — and these relationships may be perturbed by the effects of toxicants. The flow of materials inherent to ecosystems may be involved in the dispersal of and exposure to toxicants. Exposure to xenobiotics may alter the homeostasis (same-state status of equilibrium with surroundings) of individual organisms, leading to effects on populations of specific kinds of organisms, communities of organisms, and whole ecosystems.

Ecosystems can be divided generally into two categories: **terrestrial ecosystems**, consisting of those that exist primarily on land, and **aquatic ecosystems**, composed of those that exist in water. There are many interconnections between terrestrial and aquatic ecosystems, and many ecosystems have both terrestrial and aquatic components.

Aquatic ecotoxicology is relatively more simple than terrestrial ecotoxicology. Exposure of organisms to toxicants in aquatic systems is normally through direct contact of the organisms with contaminated water, which can be analyzed chemically for the presence of toxicants. The organisms in aquatic ecosystems are largely confined to relatively small areas, although there are exceptions, such as salmon, which migrate over great distances to reproduce.

Terrestrial ecotoxicology is complicated by a number of factors. Large numbers of organisms exist at various trophic levels, fitting into a variety of parts of food webs and occupying a variety of ecological niches. Terrestrial animals may move significant distances as they forage, migrate, defend their territories, and seek locations more amenable to their existence. Whereas a body of water exhibits a high degree of inertia, particularly toward changes in temperature because of water's high heat capacity, terrestrial environments may undergo sudden and dramatic changes in temperature, moisture, and other factors that affect the environment. It is against such a variable background that the terrestrial ecotoxicologist must attempt to determine the effects, often subtle and small, of xenobiotic chemicals, which are often present at low and variable concentrations. It is not surprising, therefore, that the earliest evidence of terrestrial ecotoxicological effects dealt with particularly high levels of obvious poisons, such as arsenic compounds that were applied in huge quantities to orchards. One of the defining works of ecotoxicology was performed on raptor bird populations in the United Kingdom in the 1960s. Studies of peregrine falcon (*Falco peregrinus*) and sparrow hawk (*Accipiter nisus*) revealed that declining numbers of these birds were due to low hatches, a consequence of eggshell thinning from exposure to DDT metabolites and other organochlorine compounds.

Populations of organisms in ecosystems may reflect exposures to toxicants. Observations of effects on populations of organisms or on entire ecosystems are considered to be higher order end points in toxicological studies than are observations of effects on individual organisms. A severe decline in populations of a particular kind of organism may be the result of the direct effects of a toxic substance on the population. This was the case of drastically reduced populations of eagles exposed to insecticidal DDT, which prevented production of eggs capable of hatching. Populations may reflect indirect effects of toxicants, such as an overpopulation of rodents resulting from reduced numbers of birds of prey killed by toxicants. Population distributions between young and older organisms and between sexes may be altered by toxic substance exposure. During times when DDT and other organochlorine insecticides were widely used, declines in some populations of female ducks were observed. These ducks died because they accumulated large amounts of fat tissue contaminated with organochlorine compounds prior to the season in which they laid eggs. These stores of fat were rapidly used during egg production and incubation, leading to a release of high levels of toxic organics into the blood of the female ducks and causing fatal poisoning to result.

Effects on population distributions may serve as evidence of kinds of toxicants. A marked decrease in numbers of juveniles may indicate the presence of reproductive poisons. A low population of older individuals can result from exposure to carcinogens, which take a relatively long time to act, so that older members develop cancer and die.

Many of the factors that can affect organisms adversely are inherent stressors, including availability of nutrition, water quality, temperature and other climatic extremes, disease, and predation. It is important to be able to separate effects of inherent stressors from those of toxic chemicals. There are often important synergistic relationships between inherent stressors and the effects of toxic chemicals. Organisms that are under stress from inherent stressors are likely to be more susceptible to the effects of xenobiotic toxicants.

Observation of deaths of organisms and resultant declines in populations, though straightforward and unequivocal, are often insufficient to fully explain the effects of toxic substances on ecosystems. It is also important to consider **sublethal effects**, which may cause ill effects in populations without directly killing individuals. Exposure to toxicants can lower reproductive rates of organisms and affect rates of survival of juveniles to adulthood. Ability to avoid predators may be curtailed, foraging capabilities may be reduced, and homing and migration instincts may be adversely affected. A number of tests are used to monitor sublethal effects. One such test examines the induction of mixed-function oxygenase enzymes, which are produced by organisms to metabolize toxicants. The presence of stress proteins that serve to repair proteins denatured by toxicants can be evidence of an organism attempting to cope with sublethal effects of toxicants. The activity of the acetylcholinesterase enzyme can be measured as evidence of the harmful effects of organophosphate and carbamate insecticides. Genotoxicity (such as damage to chromosomes) and immune system effects can also be monitored for evidence of sublethal effects.

Pollution-induced community tolerance has been observed in ecosystems in which communities that are resistant to the effects of some kinds of pollutants have survived. Additional exposure to the same pollutants may have relatively little effect because the organisms that have survived are those that tolerate the particular kinds of pollutants in question.

Exposure of an ecosystem to toxic substances may result in the disappearance of entire populations, which in turn may enable larger populations of competing organisms to develop or may result in loss of populations dependent upon those directly affected by the toxicants. Such an occurrence obviously involves loss of species diversity, which is an important characteristic of healthy ecosystems. More commonly, populations are reduced, resulting in a shift in the distribution of populations within the ecosystem.

Some of the most well studied effects of toxic substances on ecosystems have been those dealing with the effects of acid on freshwater ecosystems. Declines of pH from a nearly neutral range of 6.5 to 7.5 to values around 5.5 to 6.0 have been observed in some Scandinavian lakes during the approximate 30-year period following 1940. These declines have been accompanied by dramatic losses of commercially valuable salmon and trout populations. Lakes with pH values less than about 5.0 have become free of fish. Experimental studies of lake acidification have also shown changes in the phytoplankton and zooplankton communities, with a general loss of population diversity. In some cases, mats composed of just a few species of filamentous green algae changed the entire character of the lakes.

The influence of toxic substances on nutrient cycles can also be important. One of the more important nutrient cycles is the carbon cycle: carbon from atmospheric carbon dioxide is fixed as organic carbon in plant biomass, biomass is consumed by organisms, and carbon in decaying biomass is released back to the atmosphere as carbon dioxide, with a concurrent release of phosphate, nitrogen, and other nutrients. Toxic substances may cause perturbations in such a cycle, as has been observed when toxic heavy metals in soil have killed populations of earthworms, which are important in biomass recycling.

SUPPLEMENTARY REFERENCES

Connell, D.W., *Bioaccumulation of Xenobiotic Comounds,* CRC Press, Boca Raton, 1990.
Connell, D.W. et al., *Introduction to Ecotoxicology,* Blackwell Science, Malden, MA., 1999.
Forbes, V.E., Ed., *Genetics and Ecotoxicology,* Taylor & Francis, Philadelphia, 1999.
Haskell, P.T. and McEwen, P., Eds., *Ecotoxicology: Pesticides and Beneficial Organisms,* Chapman & Hall, New York, 1998.
Hoffman, D.J. et al., *Handbook of Ecotoxicology,* 2nd ed., Lewis Publishers, Boca Raton, 2002.
Lipnick, R.L. et al., Eds., *Persistent, Bioaccumulative, and Toxic Chemicals,* American Chemical Society, Washington, D.C., 2001.
Maltby, L. and Calow, P., *Methods in Ecotoxicology,* Blackwell Science, Oxford, 1995.
Rose, J., Ed., *Environmental Toxicology: Current Developments,* Australia: Gordon and Breach Science Publishers, Australia, 1998.
Sparks, T., Ed., *Statistics in Ecotoxicology,* Wiley, Chichester, 2000.
Walker, C.H. et al. *Principles of Ecotoxicology,* 2nd ed., Taylor & Francis, London, 2000.

QUESTIONS AND PROBLEMS

1. The combination of ecology and toxicology is known as ecotoxicology. Define ecotoxicology in a way that includes environmental chemistry.
2. Match the following pertaining to responses to toxicants at different organizational levels in life systems:
   1. Population changes
   2. Physiological alterations
   3. Biochemical changes
   4. Ecosystem changes

   (a) Parathion from insecticide spray binds with acetylcholinesterase enzyme
   (b) Animals with inhibited acetylcholinesterase enzyme cannot breathe properly
   (c) Numbers of animals of species most susceptible to acetylcholinesterase enzyme inhibition decrease
   (d) The decrease in animals susceptible to acetylcholinesterase enzyme inhibition significantly affects the distribution of various species in a defined area
3. Although lead and cadmium sulfate are both soluble, a body of water contaminated with these toxicants in the presence of sulfate and biodegradable organic matter shows very low concentrations of dissolved lead and cadmium, although levels are relatively high in the sediments of the body of water. Explain.
4. Explain why persistent organic toxicants such as DDT and PCBs are of particular concern in ecotoxicology, even though they are not notably acutely toxic.
5. Relate chemodynamics to both ecotoxicology and environmental chemistry.
6. Suggest how each of the following potential toxicants might transfer among environmental spheres based upon Figure 5.2:
   (a) Lead from a lead smelter deposited upon agricultural soil
   (b) Lipid-soluble, poorly biodegradable PCBs leached from a hazardous waste site into a farm pond used as a water source by animals
   (c) Sulfur dioxide released to the atmosphere as an air pollutant
   (d) Toxic radioactive radium present naturally in mineral formations
7. Suggest how the bioavailability of pollutants affects their ecotoxicology.
8. What are the two main pathways for the uptake of toxic substances by plants?
9. Explain how bioconcentration relates to bioaccumulation. How does the hydrophobicity model pertain to bioconcentration? What are the conditions under which bioconcentration is most applicable as a model?
10. Suggest how the bioconcentration factor may be used in discussing the ecotoxicology of toxic substances. How does the bioconcentration factor relate to the octanol-water partition coefficient.
11. Although the concept of bioconcentration applies well to fish in water, it is not useful for humans and land animals. What is used instead? Suggest how the concept of bioconcentration might pertain to humans with respect to their diets.
12. Complete the following: _____ is what happens to any substance that is _____ by the biochemical processes in an organism and is altered by these processes. If it proceeds all the way to simple organic species such as carbon dioxide and ammonia, the process is called _____.
13. Explain cometabolism as it relates to xenobiotic substances.
14. An herbicide contaminating a lake was measured on several consecutive days. Its concentration was found to decrease linearly with time and was at a level of 24 mg/L when first measured and 10 mg/L on the 5th day after measurement was initiated. Predict when the level would reach a value deemed safe by regulations governing this pollutant in water of 0.5 mg/L. Suggest some cautions that should be observed in relying on such a calculation.
15. The amenability of a compound to chemical attack by microorganisms is expressed as its _____ and substances that strongly resist biodegradation are called _____.

16. Match the following pertaining to biomarkers:
    1. Biomarker of exposure
    2. Biomarker of effect
    3. Biomarker of susceptibility
    (a) Fish fingerlings exposed to acidic runoff from acid rain are afflicted by fungi because of their weakened condition.
    (b) A child playing on a playground contaminated by lead shows elevated levels of carboxyhemoglobin.
    (c) The blood of an individual exposed to carbon monoxide shows elevated levels of carboxyhemoglobin.
    (d) An infant fed a formula made with nitrate-contaminated water develops methemglobinemia because of reduction of the nitrate to nitrite in the infant's stomach followed by conversion of blood hemoglobin to methemoglobin.
17. There have been cases in which male alligators in parts of Florida have developed feminine characteristics (perhaps roaring in a high-pitched voice), low sperm counts, and reduced size of essential sexual organs. Suggest the class of environmental pollutant that might cause such effects. Suggest how this effect, if it goes too far, might effect the population of alligators and the whole ecology of the Florida Everglades that are the natural habitat of alligators.
18. Justify the statement that aquatic ecotoxicology is relatively more simple than terrestrial ecotoxicology.
19. Explain why observation of deaths of organisms and resultant declines in populations, though straightforward and unequivocal, are often insufficient to fully explain the effects of toxic substances on ecosystems. Why is it important to have the capability to study sublethal effects?
20. Certain sheep that have been raised for centuries in coastal areas of Scotland exist on a diet of seaweed that is high enough in arsenic to kill other kinds of sheep. Does this observation illustrate pollution-induced community tolerance? Explain.

CHAPTER 6

Toxicology

6.1 INTRODUCTION

6.1.1 Poisons and Toxicology

A **poison**, or **toxicant**, is a substance that is harmful to living organisms because of its detrimental effects on tissues, organs, or biological processes. **Toxicology** is the science of poisons. A **toxicologist** deals with toxic substances, their effects, and the probabilities of these effects. These definitions are subject to a number of qualifications. Whether a substance is poisonous depends on the type of organism exposed, the amount of the substance, and the route of exposure. In the case of human exposure, the degree of harm done by a poison can depend strongly on whether the exposure is to the skin, by inhalation, or through ingestion. For example, a few parts per million of copper in drinking water can be tolerated by humans. However, at that level it is deadly to algae in their aquatic environment, whereas at a concentration of a few parts per billion copper is a required nutrient for the growth of algae. Subtle differences like this occur with a number of different kinds of substances.

6.1.2 History of Toxicology

The origins of modern toxicology can be traced to M.J.B. Orfila (1787–1853), a Spaniard born on the island of Minorca. In 1815 Orfila published a classic book,[1] the first ever devoted to the harmful effects of chemicals on organisms. This work discussed many aspects of toxicology recognized as valid today. Included are the relationships between the demonstrated presence of a chemical in the body and observed symptoms of poisoning, mechanisms by which chemicals are eliminated from the body, and treatment of poisoning with antidotes.

Since Orfila's time, the science of toxicology has developed at an increasing pace, with advances in the basic biological, chemical, and biochemical sciences. Prominent among these advances are modern instruments and techniques for chemical analysis that provide the means for measuring chemical poisons and their metabolites at very low levels and with remarkable sensitivity, thereby greatly extending the capabilities of modern toxicology.

This chapter deals with toxicology in general, including the routes of exposure and clinically observable effects of toxic substances. The information is presented primarily from the viewpoint of human exposure and readily observed detrimental effects of toxic substances on humans. To a somewhat lesser extent, this material applies to other mammals, especially those used as test organisms. It should be kept in mind that many of the same general principles discussed apply also to other living organisms. Although LD$_{50}$ (as discussed in Section 6.5, the lethal dose required to kill half of test subjects) is often the first parameter to come to mind in discussing degrees of toxicity, mortality is usually not a good parameter for toxicity measurement. Much more widespread than fatal poisoning, and certainly more subtle, are various manifestations of morbidity (unhealthiness). As discussed in this chapter, there are many ways in which morbidity is manifested. Some of these, such as effects on vital signs, are obvious. Others, such as some kinds of immune system impairment, can be observed only with sophisticated tests. Various factors must be considered, such as minimum dose or the latency period (often measured in years for humans) for an observable response to be observed. Furthermore, it is important to distinguish **acute toxicity**, which has an effect soon after exposure, and **chronic toxicity**, which has a long latency period.

6.1.3 Future of Toxicology

As with all other areas of the life sciences, toxicology is strongly affected by the remarkable ongoing advances in the area of mapping and understanding the deoxyribonucleic acid (DNA) that directs the reproduction and metabolism of all living things. This includes the human genome, as well as those of other organisms. It is known that certain genetic characteristics result in a predisposition for certain kinds of diseases and cancers. The action of toxic substances and the susceptibility of organisms to their effects have to be strongly influenced by the genetic makeup of organisms. The term **chemical idiosyncrasy** has been applied to the abnormal reaction of individuals to chemical exposure. An example of chemical idiosyncrasy occurs with some individuals who are affected very strongly by exposure to nitrite ion, which oxidizes the iron(II) in hemoglobin to iron(III), producing methemoglobin, which does not carry oxygen to tissues. These individuals have a low activity of the NADH–methemoglobin reductase enzyme that converts methemoglobin back to hemoglobin. An understanding of the reactions of organisms to toxic substances based on their genetic makeup promises tremendous advances in toxicological science.

6.1.4 Specialized Areas of Toxicology

Given the huge variety of toxic substances and their toxic effects, it is obvious that toxicology is a large and diverse area. Three specialized areas of toxicology should be pointed out. **Clinical toxicology** is practiced primarily by physicians who look at the connection between toxic substances and the illnesses associated with them. For example, a clinical toxicologist would be involved in diagnosing and treating cases of poisoning. **Forensic toxicology** deals largely with the interface between the medical and legal aspects of toxicology and seeks to establish the cause and responsibility for poisoning, especially where criminal activity is likely to be involved.[2] **Environmental toxicology** is concerned with toxic effects of environmental pollutants to humans and other organisms. Of particular importance are the sources, transport, effects, and interactions of toxic substances within ecosystems as they influence population dynamics within these systems. This area constitutes the branch of environmental toxicology called **ecotoxicology**.

6.1.5 Toxicological Chemistry

Toxicological chemistry relates chemistry to toxicology. It deals with the chemical nature of toxic substances, how they are changed biochemically, and how xenobiotic substances and their metabolites react biochemically in an organism to exert a toxic effect. Chapter 7 is devoted to defining and explaining toxicological chemistry, and Chapters 10–19 cover the toxicological chemistry of various kinds of toxic substances.

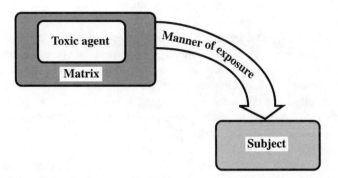

Figure 6.1 Toxicity is influenced by the nature of the toxic substance and its matrix, the subject exposed, and the conditions of exposure.

6.2 KINDS OF TOXIC SUBSTANCES

Toxic substances come in a variety of forms from a number of different sources. Those that come from natural sources are commonly called **toxins**, whereas those produced by human activities are called **toxicants**. They may be classified according to several criteria, including the following:

- Chemically, such as heavy metals or polycyclic aromatic hydrocarbons, some of which may cause cancer
- Physical form, such as dusts, vapors, or lipid-soluble liquids
- Source, such as plant toxins, combustion by-products, or hazardous wastes produced by the petrochemical industry
- Use, such as pesticides, pharmaceuticals, or solvents
- Target organs or tissue, such as neurotoxins that harm nerve tissue
- Biochemical effects, such as binding to and inhibiting enzymes or converting oxygen-carrying hemoglobin in blood to useless methemoglobin
- Effects on organisms, such as carcinogenicity or inhibition of the immune system

Usually several categories of classification are appropriate. For example, parathion is an insecticide that is produced industrially, to which exposure may occur as a mist from spray, and that binds to the acetylcholinesterase enzyme, affecting function of the nervous system.

Since toxicological chemistry emphasizes the chemical nature of toxic substances, classification is predominantly on the basis of chemical class. Therefore, there are separate chapters on elemental toxic substances, hydrocarbons, organonitrogen compounds, and other chemical classifications of substances.

6.3 TOXICITY-INFLUENCING FACTORS

6.3.1 Classification of Factors

It is useful to categorize the factors that influence toxicity within the following three classifications: (1) the toxic substance and its matrix, (2) circumstances of exposure, and (3) the subject and its environment (see Figure 6.1). These are considered in the following sections.

6.3.2 Form of the Toxic Substance and Its Matrix

Toxicants to which subjects are exposed in the environment or occupationally, particularly through inhalation, may be in several different physical forms. Gases are substances such as carbon monoxide in air that are normally in the gaseous state under ambient conditions of temperature and pressure. Vapors are gas-phase materials that can evaporate or sublime from liquids or solids. Benzene or naphthalene can exist in the vapor form. Dusts are respirable solid particles produced by grinding bulk solids, whereas fumes are solid particles from the condensation of vapors, often metals or metal oxides. Mists are liquid droplets.

Generally a toxic substance is in solution or mixed with other substances. A substance with which the toxicant is associated (the solvent in which it is dissolved or the solid medium in which it is dispersed) is called the matrix. The matrix may have a strong effect on the toxicity of the toxicant.

Numerous factors may be involved with the toxic substance itself. If the substance is a toxic heavy metal cation, the nature of the anion with which it is associated can be crucial. For example, barium ion, Ba^{2+}, in the form of insoluble barium sulfate, $BaSO_4$, is routinely used as an x-ray opaque agent in the gastrointestinal tract for diagnostic purposes (barium enema x-ray). This is a safe procedure; however, *soluble* barium salts such as $BaCl_2$ are deadly poisons when introduced into the gastrointestinal tract.

The pH of the toxic substance can greatly influence its absorption and therefore its toxicity. An example of this phenomenon is provided by aspirin, one of the most common causes of poisoning in humans. The chemical name of aspirin is sodium acetylsalicylate, the acidic form of which is acetylsalicylic acid (HAsc), a weak acid that ionizes as follows:

$$HAsc \rightleftarrows H^+ + Asc^- \qquad K_a = \frac{[H^+][Asc^-]}{[HAsc]} = 6 \times 10^{-4} \qquad (6.3.1)$$

The K_a expression is expressed in molar concentrations (denoted by brackets) of the neutral and ionized species involved in the ionization of the acetylsalicylic acid. The pK_a (negative log of K_a) of HAsc is 3.2, and at a pH substantially below 3.2, most of this acid is in the neutral HAsc form. This neutral form is easily absorbed by the body, especially in the stomach, where the contents have a low pH of about 1. Many other toxic substances exhibit acid–base behavior and pH is a factor in their uptake.

Solubility is an obvious factor in determining the toxicity of systemic poisons. These must be soluble in body fluids or converted to a soluble form in the organ or system through which they are introduced into the body. Some insoluble substances that are ingested pass through the gastrointestinal tract without doing harm, whereas they would be quite toxic if they could dissolve in body fluids (see the example of barium sulfate cited above).

As noted at the beginning of this section, the degree of toxicity of a substance may depend on its matrix. The solvent or suspending medium is called the **vehicle**. For laboratory studies of toxicity, several vehicles are commonly used. Among the most common of these are water and aqueous saline solution. Lipid-soluble substances may be dissolved in vegetable oils. Various organic liquids are used as vehicles. Dimethylsulfoxide is a solvent that has some remarkable abilities to carry a solute dissolved in it into the body. The two major classes of vehicles for insoluble substances are the natural gums and synthetic colloidal materials. Examples of the former are tragacanth and acacia, whereas methyl cellulose and carboxymethylcellulose are examples of the latter.

Some drug formulations contain **excipients** that have been added to give a desired consistency or form. In some combinations excipients have a marked influence upon toxicity. **Adjuvants** are excipients that may increase the effect of a toxic substance or enhance the pharmacologic action of a drug. For example, dithiocarbamate fungicides may have their activities increased by the addition of 2-mercaptothiazole.

A variety of materials other than those discussed above may be present in formulations of toxic substances. **Dilutents** increase bulk and mass. Common examples of these are salts, such as calcium carbonate and dicalcium phosphate; carbohydrates, including sucrose and starch; the clay, kaolin; and milk solids. Among the **preservatives** used are sodium benzoate, phenylmercuric nitrate, and butylated hydroxyanisole (an antioxidant). "Slick" substances such as cornstarch, calcium stearate, and talc act as **lubricants**. Various gums and waxes, starch, gelatin, and sucrose are used as **binders**. Gelatin, carnauba wax, and shellac are applied as **coating agents**. Cellulose derivatives and starch may be present as **disintegrators** in formulations containing toxicants.

Decomposition may affect the action of a toxic substance. Therefore, the stability and storage characteristics of formulations containing toxicants should be considered. A toxic substance may be contaminated with other materials that affect toxicity. Some contaminants may result from decomposition.

6.3.3 Circumstances of Exposure

There are numerous variables related to the ways in which organisms are exposed to toxic substances. One of the most crucial of these, **dose,** is discussed in Section 6.5. Another important factor is the **toxicant concentration**, which may range from the pure substance (100%) down to a very dilute solution of a highly potent poison. Both the **duration** of exposure per exposure incident and the **frequency** of exposure are important. The **rate** of exposure, inversely related to the duration per exposure, and the total time period over which the organism is exposed are both important situational variables. The exposure **site** and **route** strongly affect toxicity. Toxic effects are largely the result of metabolic processes on substances that occur after exposure, and much of the remainder of this book deals with these kinds of processes.

It is possible to classify exposures on the basis of acute vs. chronic and local vs. systemic exposure, giving four general categories. **Acute local** exposure occurs at a specific location over a time period of a few seconds to a few hours and may affect the exposure site, particularly the skin, eyes, or mucous membranes. The same parts of the body can be affected by **chronic local** exposure, but the time span may be as long as several years. **Acute systemic** exposure is a brief exposure or exposure to a single dose and occurs with toxicants that can enter the body, such as by inhalation or ingestion, and affect organs such as the liver that are remote from the entry site. **Chronic systemic** exposure differs in that the exposure occurs over a prolonged time period.

6.3.4 The Subject

The first of two major classes of factors in toxicity pertaining to the subject and its environment consists of **factors inherent to the subject**. The most obvious of these is the **taxonomic classification** of the subject, that is, the species and strain. With test animals it is important to consider the **genetic status** of the subjects, including whether they are littermates, half-siblings (different fathers), or the products of inbreeding. Body mass, sex, age, and degree of maturity are all factors in toxicity. **Immunological status** is important. Another area involves the general well-being of the subject. It includes disease and injury, diet, state of hydration, and the subject's psychological state as affected by the presence of other species and/or members of the opposite sex, crowding, handling, rest, and activity.

The other major class consists of **environmental factors**. Among these are ambient atmosphere conditions of temperature, pressure, and humidity, as well as composition of the atmosphere, including the presence of atmospheric pollutants, such as ozone or carbon monoxide. Light and noise and the patterns in which they occur are important. Social and housing (caging) conditions may also influence response of subjects to a toxicant.

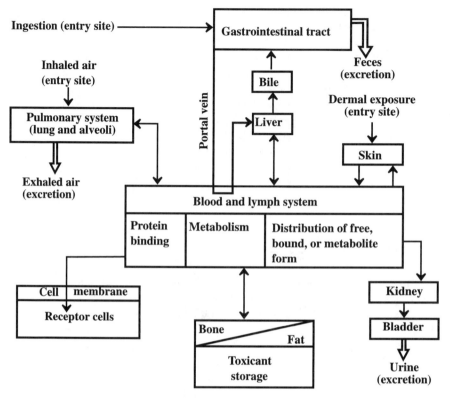

Figure 6.2 Major sites of exposure, metabolism, and storage, and routes of distribution and elimination of toxic substances in the body.

6.4 EXPOSURE TO TOXIC SUBSTANCES

Perhaps the first consideration in toxicology is **exposure** of an organism to a toxic substance. In discussing exposure sites for toxicants, it is useful to consider the major routes and sites of exposure, distribution, and elimination of toxicants in the body, as shown in Figure 6.2. The major routes of accidental or intentional exposure to toxicants by humans and other animals are the skin (percutaneous route), the lungs (inhalation, respiration, pulmonary route), and the mouth (oral route); minor means of exposure are the rectal, vaginal, and parenteral routes (intravenous or intramuscular, a common means for the administration of drugs or toxic substances in test subjects). The way that a toxic substance is introduced into the complex system of an organism is strongly dependent upon the physical and chemical properties of the substance. The pulmonary system is most likely to take in toxic gases or very fine, respirable solid or liquid particles. In other than a respirable form, a solid usually enters the body orally. Absorption through the skin is most likely for liquids, solutes in solution, and semisolids, such as sludges.

The defensive barriers that a toxicant may encounter vary with the route of exposure. For example, elemental mercury is more readily absorbed, often with devastating effects, through the alveoli in the lungs than through the skin or gastrointestinal tract. Most test exposures to animals are through ingestion or gavage (introduction into the stomach through a tube). Pulmonary exposure is often favored with subjects that may exhibit refractory behavior when noxious chemicals are administered by means requiring a degree of cooperation from the subject. Intravenous injection may be chosen for deliberate exposure when it is necessary to know the concentration and effect of a xenobiotic substance in the blood. However, pathways used experimentally that are almost

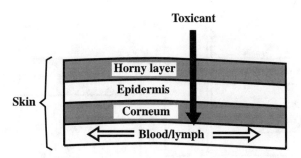

Figure 6.3 Absorption of a toxic substance through the skin.

certain not to be significant in accidental exposures can give misleading results when they avoid the body's natural defense mechanisms.

6.4.1 Percutaneous Exposure

Toxicants can enter the skin through epidermal cells, sebaceous gland cells, or hair follicles. By far the greatest area of the skin is composed of the epidermal cell layer, and most toxicants absorbed through the skin do so through epidermal cells. Despite their much smaller total areas, however, the cells in the follicular walls and in sebaceous glands are much more permeable than epidermal cells.

6.4.1.1 Skin Permeability

Figure 6.3 illustrates the absorption of a toxic substance through the skin and its entry into the circulatory system, where it may be distributed through the body. Often the skin suffers little or no harm at the site of entry of systemic poisons, which may act with devastating effects on receptors far from the location of absorption.

The permeability of the skin to a toxic substance is a function of both the substance and the skin. The permeability of the skin varies with both the location and the species that penetrates it. In order to penetrate the skin significantly, a substance must be a liquid or gas or significantly soluble in water or organic solvents. In general, nonpolar, lipid-soluble substances traverse skin more readily than do ionic species. Substances that penetrate skin easily include lipid-soluble endogenous substances (hormones, vitamins D and K) and a number of xenobiotic compounds. Common examples of these are phenol, nicotine, and strychnine. Some military poisons, such as the nerve gas sarin (see Section 18.8), permeate the skin very readily, which greatly adds to their hazards. In addition to the rate of transport through the skin, an additional factor that influences toxicity via the percutaneous route is the blood flow at the site of exposure.

6.4.2 Barriers to Skin Absorption

The major barrier to dermal absorption of toxicants is the **stratum corneum**, or horny layer (see Figure 6.3). The permeability of skin is inversely proportional to the thickness of this layer, which varies by location on the body in the following order: soles and palms > abdomen, back, legs, arms > genital (perineal) area. Evidence of the susceptibility of the genital area to absorption of toxic substances is to be found in accounts of the high incidence of cancer of the scrotum among chimney sweeps in London described by Sir Percival Pott, Surgeon General of Britain during the reign of King George III. The cancer-causing agent was coal tar condensed in chimneys. This material was more readily absorbed through the skin in the genital areas than elsewhere, leading to a high incidence of scrotal cancer. (The chimney sweeps' conditions were aggravated by their

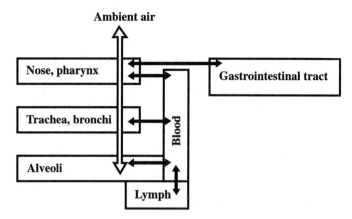

Figure 6.4 Pathways of toxicants in the respiratory system.

lack of appreciation of basic hygienic practices, such as bathing and regular changes of underclothing.) Breaks in epidermis due to laceration, abrasion, or irritation increase the permeability, as do inflammation and higher degrees of skin hydration.

6.4.2.1 Measurement of Dermal Toxicant Uptake

There are two principal methods for determining the susceptibility of skin to penetration by toxicants. The first of these is measurement of the dose of the substance received by the organism using chemical analysis, radiochemical analysis of radioisotope-labeled substances, or observation of clinical symptoms. Secondly, the amount of substance remaining at the site of administration may be measured. This latter approach requires control of nonabsorptive losses of the substance, such as those that occur by evaporation.

6.4.2.2 Pulmonary Exposure

The pulmonary system is the site of entry for numerous toxicants. Examples of toxic substances inhaled by human lungs include fly ash and ozone from polluted atmospheres, vapors of volatile chemicals used in the workplace, tobacco smoke, radioactive radon gas, and vapors from paints, varnishes, and synthetic materials used for building construction.

The major function of the lungs is to exchange gases between the bloodstream and the air in the lungs. This especially includes the absorption of oxygen by the blood and the loss of carbon dioxide. Gas exchange occurs in a vast number of alveoli in the lungs, where a tissue the thickness of only one cell separates blood from air. The thin, fragile nature of this tissue makes the lungs especially susceptible to absorption of toxicants and to direct damage from toxic substances. Furthermore, the respiratory route enables toxicants entering the body to bypass organs that have a screening effect (the liver is the major "screening organ" in the body and it acts to detoxify numerous toxic substances). These toxicants can enter the bloodstream directly and be transported quickly to receptor sites with minimum intervention by the body's defense mechanisms.

As illustrated in Figure 6.4, there are several parts of the pulmonary system that can be affected by toxic substances. The upper respiratory tract, consisting of the nose, throat, trachea, and bronchi, retains larger particles that are inhaled. The retained particles may cause upper respiratory tract irritation. Cilia, which are small hair-like appendages in the upper respiratory tract, move with a sweeping motion to remove captured particles. These substances are transported to the throat from which they may enter the gastrointestinal tract and be absorbed by the body. Gases such as ammonia (NH_3) and hydrogen chloride (HCl) that are very soluble in water are also removed from air predominantly in the upper respiratory tract and may be very irritating to tissue in that region.

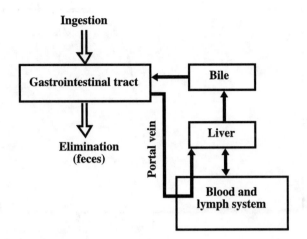

Figure 6.5 Representation of enterohepatic circulation.

6.4.3 Gastrointestinal Tract

The gastrointestinal tract may be regarded as a tube through the body from the mouth to the anus, the contents of which are external to the rest of the organism system. Therefore, any systemic effect of a toxicant requires its absorption through the mucosal cells that line the inside of the gastrointestinal tract. Caustic chemicals can destroy or damage the internal surface of the tract and are viewed as nonkinetic poisons that act mainly at the site of exposure.

6.4.4 Mouth, Esophagus, and Stomach

Most substances are not readily absorbed in the mouth or esophagus; one of several exceptions is nitroglycerin, which is administered for certain heart disfunctions and absorbed if left in contact with oral tissue. The stomach is the first part of the gastrointestinal tract where substantial absorption and translocation to other parts of the body may take place. The stomach is unique because of its high content of HCl and consequent low pH (about 1.0). Therefore, some substances that are ionic at pH values near 7 and above are neutral in the stomach and readily traverse the stomach walls. In some cases, absorption is affected by stomach contents other than HCl. These include food particles, gastric mucin, gastric lipase, and pepsin.

6.4.5 Intestines

The small intestine is effective in the absorption and translocation of toxicants. The pH of the contents of the small intestine is close to neutral, so that weak bases that are charged (HB^+) in the acidic environment of the stomach are uncharged (B) and absorbable in the intestine. The small intestine has a large surface area favoring absorption. Intestinal contents are moved through the intestinal tract by peristalsis. This has a mixing action on the contents and enables absorption to occur the length of the intestine. Some toxicants slow down or stop peristalsis (paralytic ileus), thereby slowing the absorption of the toxicant itself.

6.4.6 The Intestinal Tract and the Liver

The intestine–blood–liver–bile loop constitutes the **enterohepatic circulation** system (see Figure 6.5). A substance absorbed through the intestines goes either directly to the lymphatic system or to the **portal circulatory system**. The latter carries blood to the portal vein that goes directly

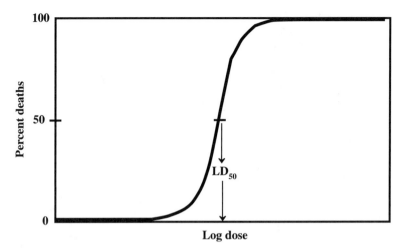

Figure 6.6 Illustration of a dose–response curve in which the response is the death of the organism. The cumulative percentage of deaths of organisms is plotted on the y axis. Although plotting log dose usually gives a better curve, with some toxic substances it is better to plot dose.

to the liver. The liver serves as a screening organ for xenobiotics, subjecting them to metabolic processes that usually reduce their toxicities, and secretes these substances or a metabolic product of them back to the intestines. For some substances, there are mechanisms of active excretion into the bile in which the substances are concentrated by one to three orders of magnitude over levels in the blood. Other substances enter the bile from blood simply by diffusion.

6.5 DOSE–RESPONSE RELATIONSHIPS

Toxicants have widely varying effects on organisms. Quantitatively, these variations include minimum levels at which the onset of an effect is observed, the sensitivity of the organism to small increments of toxicant, and levels at which the ultimate effect (particularly death) occurs in most exposed organisms. Some essential substances, such as nutrient minerals, have optimum ranges above and below which detrimental effects are observed.

Factors such as those just outlined are taken into account by the **dose–response** relationship, which is one of the key concepts of toxicology.[3] **Dose** is the amount, usually per unit body mass, of a toxicant to which an organism is exposed. **Response** is the effect on an organism resulting from exposure to a toxicant. In order to define a dose–response relationship, it is necessary to specify a particular response, such as death of the organism, as well as the conditions under which the response is obtained, such as the length of time from administration of the dose. Consider a specific response for a population of the same kinds of organisms. At relatively low doses, none of the organisms exhibit the response (for example, all live), whereas at higher doses, all of the organisms exhibit the response (for example, all die). In between, there is a range of doses over which some of the organisms respond in the specified manner and others do not, thereby defining a dose–response curve. Dose–response relationships differ among different kinds and strains of organisms, types of tissues, and populations of cells.

Figure 6.6 shows a generalized dose–response curve. Such a plot may be obtained, for example, by administering different doses of a poison in a uniform manner to a homogeneous population of test animals and plotting the cumulative percentage of deaths as a function of the log of the dose. The result is normally an S-shaped curve, as shown in Figure 6.6. The dose corresponding to the midpoint (inflection point) of such a curve is the statistical estimate of the dose that would cause death in 50% of the subjects and is designated as LD_{50}. The estimated doses at which 5% (LD_5)

and 95% (LD_{95}) of the test subjects die are obtained from the graph by reading the dose levels for 5 and 95% fatalities, respectively. A relatively small difference between LD_5 and LD_{95} is reflected by a steeper S-shaped curve and vice versa. Statistically, 68% of all values on a dose–response curve fall within ±1 standard deviation of the mean at LD_{50} and encompass the range from LD_{16} to LD_{84}.

The midrange of a dose–response curve is virtually a straight line. The slope of the curve in this range may vary. A very steep slope reflects a substance and organisms that have an abrupt onset of toxic effects, and only a small increase in dose causes a marked increase in response. A more gradual slope reflects a relatively large range, from a small percentage to a large percentage of responses. The LD_{50} might be the same in both cases, but with a sharp dose–response curve there is a small difference in dose between LD_5 and LD_{95}, whereas with a gradual curve the difference between these values is larger.

When exposure to toxic substances is in the air that animals breathe or in the water in which aquatic animals swim, exposure is commonly expressed as concentration. In such cases, LC_{50} values are obtained, where C stands for concentration, rather than dose.

6.5.1 Thresholds

An important concept pertinent to the dose–response relationship is that of **threshold** dose, below which there is no response. Threshold doses apply especially to acute effects and are very hard to determine, despite their crucial importance in determining safe levels of exposures to chemicals. In an individual, the response observed as the threshold level is exceeded may be very slight and subtle, making the threshold level very hard to determine. In a population, the number of subjects exhibiting the particular response at the threshold limit is very small and may be hard to detect above background effects (such as normal mortality rates of test organisms). For chronic effects, the determination of a threshold value is very difficult. This is especially true of cancer-causing substances that act by altering cellular DNA. For some of these substances, it is argued that there is no threshold and that the slightest exposure entails a risk.

6.6 RELATIVE TOXICITIES

Table 6.1 illustrates standard **toxicity ratings** that are used to describe estimated toxicities of various substances to humans. Reference is made to them in this book to denote toxicities of substances. Their values range from one (practically nontoxic) to six (supertoxic). In terms of fatal doses to an adult human of average size, a "taste" of a supertoxic substances (just a few drops or less) is fatal. A teaspoonful of a very toxic substance could have the same effect. However, as much as a quart of a slightly toxic substance might be required to kill an adult human.

When there is a substantial difference between LD_{50} values of two different substances, the one with the lower value is said to be the more **potent**. Such a comparison must assume that the dose–response curves for the two substances being compared have similar slopes (see Figure 6.6). If this is not the case, the substance for which the dose–response curve has the lesser slope may be toxic at a low dose, where the other substance is not toxic at all. Put another way, the relative LD_5 values of the substances may be reversed from the relative LD_{50} values.

6.6.1 Nonlethal Effects

It must be kept in mind that the acute toxicities of substances as expressed by LD_{50} values such as those in Table 6.1 have limited value in expressing hazards to humans. This is because death from exposure to a toxic substance is a relatively rare effect that is irreversible. Of much more concern are **sublethal effects** that are often **reversible**, such as allergies, and birth defects. Of

Table 6.1 Toxicity Scale with Example Substances[a]

Toxic Substance	Approximate LD_{50}	Toxicity Rating
	— 10^5	1. Practically nontoxic, $> 1.5 \times 10^4$ mg/kg
DEHP[b] ⟶ —		
Ethanol ⟶ —	10^4	2. Slightly toxic 5×10^3–1.5×10^4 mg/kg
Sodium chloride ⟶ —		
Malathion ⟶ —	10^3	3. Moderately toxic 500–5000 mg/kg
Chlorane ⟶ —		
Heptachlor ⟶ —	10^2	4. Very toxic 50–500 mg/kg
	—	
Parathion ⟶ —	10	5. Extremely toxic 5–50 mg/kg
	—	
TEPP[c] ⟶		
Nicotine ⟶ —	1	
	—	
Tetrodotoxin[d] ⟶ —	10^{-1}	
	— 10^{-2}	6. Supertoxic <5 mg/kg
	—	
TCDD[e] ⟶ —	10^{-3}	
	—	
	— 10^{-4}	
	—	
Botulinus toxin ⟶ —	10^{-5}	

[a] Doses are in units of mg of toxicant per kg of body mass. Toxicity ratings on the right are given as numbers ranging from 1 (practically nontoxic) to 6 (supertoxic), along with estimated lethal oral doses for humans in mg/kg. Estimated LD_{50} values for substances on the left have been measured in test animals, usually rats, and apply to oral doses.
[b] Bis(2-ethylhexyl)phthalate.
[c] Tetraethylpyrophosphate.
[d] Toxin from pufferfish.
[e] TCDD represents 2,3,7,8-tetrachlorodibenzodioxin, commonly called "dioxin."

particular concern is the development of cancer from exposure to toxic substances (carcinogenicity) that, although often fatal, is not an acute effect and does not register on tables of LD_{50} values.

Sublethal reversible effects are obviously important with drugs, where death from exposure to a registered therapeutic agent is rare, but other effects, both detrimental and beneficial, are usually observed. By their very nature, drugs alter biologic processes; therefore, the potential for harm is almost always present. The major consideration in establishing drug dose is to find a dose that has an adequate therapeutic effect without undesirable side effects. The difference between the effective dose and harmful dose reflects the **margin of safety** (see Figure 6.7).

When substances are used as pharmaceuticals to destroy disease-causing microorganisms or cancer tissue, or as pesticides to kill insects, weeds, or other pests, there is obviously an organism or tissue that is to be destroyed, commonly called the **uneconomic form**, and an organism or tissue that should remain unharmed, commonly called the **economic form**. Much of the ongoing research in pharmaceuticals and pesticides is designed to maximize the ratio of toxicities to uneconomic forms to those of economic forms. Several approaches are used. Some agents are about as toxic to both forms, but are accumulated more readily by the uneconomic form. Other agents take advantage of the higher susceptibility of receptors in tissues of the uneconomic form, which is therefore harmed more by the toxic agent. The selective toxicity of antibiotic penicillin to bacteria is due to the fact that it inhibits formation of cell walls, which bacteria have and need, whereas animals do not have cell walls. Another example is provided by genetically engineered soybeans

TOXICOLOGY

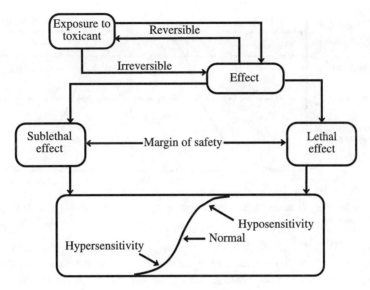

Figure 6.7 Effects of and responses to toxic substances.

that are not harmed by the hugely popular Roundup herbicide, whereas competing plants are destroyed by it.

In some cases there are striking differences between species in susceptibilities to toxic substances. One notable example is the 1000-fold greater susceptibility of guinea pigs over hamsters — both rodent species — to the effects of 2,3,7,8-tetrachlorodibenzo-*p*-dioxin (TCDD), commonly called "dioxin." Such differences may arise from the absence in the less susceptible species of receptors affected by the toxic substance. More commonly, they are due to the presence in the less susceptible organism of more effective mechanisms, usually detoxifying enzyme systems, for counteracting the effects of the toxic substance.

Individuals of the same species may differ significantly in their susceptibilities to various toxic agents. These differences are often genetic in nature. For example, some individuals lack tumor suppressor genes that other individuals possess and are thus more likely to develop some kinds of cancers, some of which are initiated by carcinogens. With increased knowledge of the human genome, these kinds of susceptibilities may become more apparent and appropriate preventive measures may be applied in some cases.

6.7 REVERSIBILITY AND SENSITIVITY

Sublethal doses of most toxic substances are eventually eliminated from an organism's system. If there is no lasting effect from the exposure, it is said to be **reversible**. However, if the effect is permanent, it is termed **irreversible**. Irreversible effects of exposure remain after the toxic substance is eliminated from the organism. Figure 6.7 illustrates these two kinds of effects. For various chemicals and different subjects, toxic effects may range from the totally reversible to the totally irreversible.

6.7.1 Hypersensitivity and Hyposensitivity

Examination of the dose–response curve, shown in Figure 6.6, reveals that some subjects are very sensitive to a particular poison (for example, those killed at a dose corresponding to LD_5), whereas others are very resistant to the same substance (for example, those surviving a dose

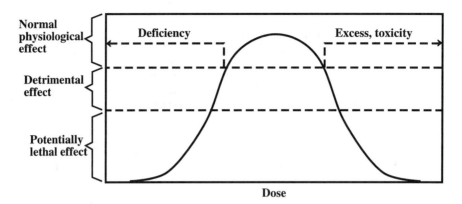

Figure 6.8 Biologic effect of an endogenous substance in an organism showing optimum level, deficiency, and excess.

corresponding to LD_{95}). These two kinds of responses illustrate **hypersensitivity** and **hyposensitivity**, respectively; subjects in the midrange of the dose–response curve are termed **normals**. These variations in response tend to complicate toxicology in that there is not a specific dose guaranteed to yield a particular response, even in a homogeneous population.

In some cases hypersensitivity is induced. After one or more doses of a chemical, a subject may develop an extreme reaction to it. This occurs with penicillin, for example, in cases where people develop such a severe allergic response to the antibiotic that exposure results in death if countermeasures are not taken.

A kind of hyposensitivity is that induced by repeated exposures to a toxic substance leading to **tolerance** and reduced toxicities from later exposures. Tolerance can be due to a less toxic substance reaching a receptor or to tissue building up a resistance to the effects of the toxic substance. An example of the former occurs with repeated doses of toxic heavy metal cadmium. Animals respond by generating larger quantities of polypeptide **metallothionein**, which is rich in –SH groups that bind with Cd^{2+} ion, making it less available to receptors.

6.8 XENOBIOTIC AND ENDOGENOUS SUBSTANCES

Xenobiotic substances are those that are foreign to a living system. It should be kept in mind that xenobiotic substances may come from natural sources, such as deadly botulinus toxin produced by bacteria. Substances that occur naturally in a biologic system are termed **endogenous**. Endogenous substances are usually required within a particular concentration range in order for metabolic processes to occur normally. Levels below a normal range may result in a toxic response or even death, and the same effects may occur above the normal range. This kind of response is illustrated in Figure 6.8.

6.8.1 Examples of Endogenous Substances

Examples of endogenous substances in organisms include various hormones, glucose (blood sugar), some vitamins, and some essential metal ions, including Ca^{2+}, K^+, and Na^+. Calcium in human blood serum exhibits the kind of behavior shown in Figure 6.8, with an optimum level that occurs over a rather narrow range of 9 to 9.5 mg/dL. Below these values, a toxic response known as hypocalcemia occurs, manifested by muscle cramping. At serum levels above about 10.5 mg/dL, hypercalcemia occurs, the major effect of which is kidney malfunction. Vitamin A is required for proper nutrition, but excessive levels damage the liver and may cause birth defects. Selenium, essential at low levels, can be toxic to the brain at higher levels.

6.9 KINETIC AND NONKINETIC TOXICOLOGY

Nonkinetic toxicology deals with generalized harmful effects of chemicals that occur at an exposure site; a typical example is the destruction of skin tissue by contact with concentrated nitric acid, HNO_3. Nonkinetic toxicology applies to those poisons that are not metabolized or transported in the body or subject to elimination processes that remove them from the body. The severity of a nonkinetic insult depends on both the characteristic of the chemical and the exposure site. Injury increases with increasing area and duration of the exposure, with the concentration of the toxicant in its matrix (for example, the concentration of HNO_3 in solution), and with the susceptibility of the exposure site to damage. The toxic action of the substance ceases when its chemical reaction with tissue is complete or when it is removed from the exposure site. Nonkinetic toxicology is also called **nonmetabolic** or **nonpharmacologic** toxicology.

6.9.1 Kinetic Toxicology

Kinetic toxicology, also known as **metabolic** or **pharmacologic** toxicology, involves toxicants that are transported and metabolized in the body. Such substances are called **systemic poisons** and they are studied under the discipline of **systemic toxicology**. Systemic poisons may cross cell membranes (see Chapter 3) and act on **receptors** such as cell membranes, bodies in the cells, and specific enzyme systems. The effect is dose responsive, and it is terminated by processes that may include metabolic conversion of the toxicant to a metabolic product, chemical binding, storage, and excretion from the organism.

In an animal, a xenobiotic substance may be bound reversibly to a plasma protein in an inactivated form. A polar xenobiotic substance, or a polar metabolic product, may be excreted from the body in solution in urine. Nonpolar substances delivered to the intestinal tract in bile are eliminated with feces. Volatile nonpolar substances such as carbon monoxide tend to leave the body via the pulmonary system. The ingestion, biotransformation, action on receptor sites, and excretion of a toxic substance may involve complex interactions of biochemical and physiological parameters. The study of these parameters within a framework of metabolism and kinetics is called **toxicometrics**.

6.10 RECEPTORS AND TOXIC SUBSTANCES

A toxic substance that enters the body through any of the possible entry sites (ingestion, inhalation, skin) undergoes biochemical transformations that can increase or decrease its toxicity, affect its ability to traverse cell membranes, or enable its elimination from the body. A substance involved in a kinetic toxicological process (see Section 6.7) generally enters the blood and lymph system before it has any effect. Plasma proteins may inactivate the toxic substance by binding reversibly to it. The substance often undergoes biotransformation, most commonly in the liver, but in other types of tissue as well. These reactions are catalyzed by enzymes, most frequently mixed-function oxidases. Toxicants can either stimulate or inhibit enzyme action. It is obvious that biochemical actions and transformations of toxicants are varied and complex. They are discussed in greater detail in Chapter 9.

6.10.1 Receptors

As noted in the preceding section, there are various *receptors* upon which xenobiotic substances or their metabolites act. In order to bind to a receptor, the substance has to have the proper structure or, more precisely, the right **stereochemical molecular configuration** (see Chapters 1 and 3). Receptors are almost always proteinaceous materials, normally enzymes. Nonenzyme receptors include opiate (nerve) receptors, gonads, or the uterus.

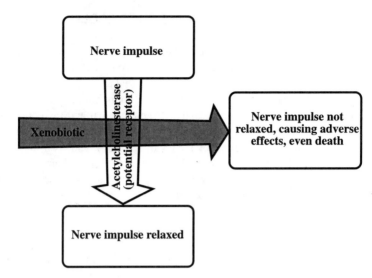

Figure 6.9 Example of a toxicant acting on a receptor to cause an adverse effect. When acetylcholinesterase is bound by a xenobiotic substance, the enzyme does not act to stop nerve impulse. This can result in paralysis of the respiratory system and death.

An example of a toxicant acting on a receptor will be cited here; the topic is discussed in greater detail in Chapter 9. One of the most commonly cited examples of an enzyme receptor that is adversely affected by toxicants is that of **acetylcholinesterase**. It acts on **acetylcholine** as shown by the reaction

$$(CH_3)_3 \overset{+}{N}-\underset{H}{\overset{H}{C}}-\underset{H}{\overset{H}{C}}-O-\underset{}{\overset{O}{\underset{\|}{C}}}-\underset{H}{\overset{H}{C}}-H + H_2O \xrightarrow{\text{Acetylcholinesterase}} (CH_3)_3 \overset{+}{N}-\underset{H}{\overset{H}{C}}-\underset{H}{\overset{H}{C}}-OH + H-\underset{H}{\overset{H}{C}}-\overset{O}{\underset{\|}{C}}-OH \quad (6.10.1)$$

Acetylcholine

Acetylcholine is a neurotransmitter, a key substance involved with transmission of nerve impulses in the brain, skeletal muscles, and other areas where nerve impulses occur. An essential step in the proper function of any nerve impulse is its cessation (see Figure 6.9), which requires hydrolysis of acetylcholine as shown by Reaction 6.10.1. Some xenobiotics, such as organophosphate compounds (see Chapter 18) and carbamates (see Chapter 15) inhibit acetylcholinesterase, with the result that acetylcholine accumulates and nerves are overstimulated. Adverse effects may occur in the central nervous system, in the autonomic nervous system, and at neuromuscular junctions. Convulsions, paralysis, and finally death may result.

6.11 PHASES OF TOXICITY

Having examined the routes by which toxicants enter the body, it is now appropriate to consider what happens to them in the body and what their effects are. The action of a toxic substance can be divided into two major phases, as illustrated in Figure 6.10. The **kinetic phase** involves absorption, metabolism, temporary storage, distribution, and, to a certain extent, excretion of the toxicant or its precursor compound, called the **protoxicant**. In the most favorable scenario for an organism, a toxicant is absorbed, detoxified by metabolic processes, and excreted with no harm resulting. In

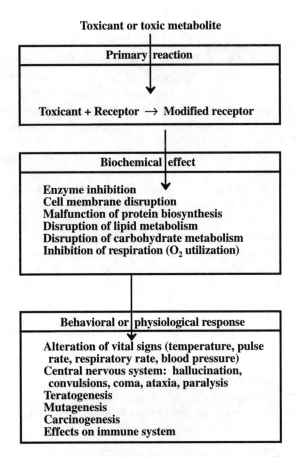

Figure 6.10 Different major steps in the overall process leading to a toxic response.

the least favorable case, a protoxicant that is not itself toxic is absorbed and converted to a toxic metabolic product that is transported to a location where it has a detrimental effect. The **dynamic phase** is divided as follows: (1) the toxicant reacts with a receptor or target organ in the **primary reaction** step, (2) there is a biochemical response, and (3) physiological or behavioral manifestations of the effect of the toxicant occur.

6.12 TOXIFICATION AND DETOXIFICATION

As shown for the kinetic phase in Figure 6.10, a xenobiotic substance may be (1) detoxified by metabolic processes and eliminated from the body, (2) made more toxic (toxified) by metabolic processes and distributed to receptors, or (3) passed on to receptors as a metabolically unmodified toxicant. A metabolically unmodified toxicant is called an **active parent compound**, and a substance modified by metabolic processes is an **active metabolite**. Both types of species may be involved in the dynamic phase.

During the kinetic phase an active parent compound can be present in blood, liver, or nonliver (extrahepatic) tissue; in the latter two, it may be converted to inactive metabolites. An inactive parent metabolite may produce a toxic metabolite or metabolites in the liver or in extrahepatic tissue; in both these locations a toxic metabolite may be changed to an inactive form. Therefore, the kinetic phase involves a number of pathways by which a xenobiotic substance is converted to a toxicant that can act on a receptor or to a substance that is eliminated from the organism.

6.12.1 Synergism, Potentiation, and Antagonism

The biological effects of two or more toxic substances can be different in kind and degree from those of one of the substances alone. One of the ways in which this can occur is when one substance affects the way in which another undergoes any of the steps in either the kinetic phase or the dynamic phase. Chemical interaction between substances may affect their toxicities. Both substances may act on the same physiologic function, or two substances may compete for binding to the same receptor. When both substances have the same physiologic function, their effects may be simply **additive** or they may be **synergistic** (the total effect is greater than the sum of the effects of each separately). **Potentiation** occurs when an inactive substance enhances the action of an active one, and **antagonism** when an active substance decreases the effect of another active one. Antagonism falls into several different categories. **Functional antagonism** occurs when two different substances have opposite functions and tend to balance each other. When a toxic substance reacts chemically with another toxic substance and is neutralized, the phenomenon is called **chemical antagonism**. The degree to which a toxic substance reaches a target organ can be reduced by the presence of another substance, a phenomenon called **dispositional antagonism**. The effects of a toxic substance can be reduced by the action of another substance (called a **blocker**) that competes with it for binding to a receptor, a phenomenon called **receptor antagonism**.

6.13 BEHAVIORAL AND PHYSIOLOGICAL RESPONSES

The final part of the overall toxicological process outlined in Figure 6.10 consists of **behavioral** and **physiological responses**, which are observable symptoms of poisoning. These are discussed here, primarily in terms of responses seen in humans and other animals. Nonanimal species exhibit other kinds of symptoms from poisoning; for example, plants suffer from leaf mottling, pine needle loss, and stunted growth as a result of exposure to some toxicants.

6.13.1 Vital Signs

Human subjects suffering from acute poisoning usually show alterations in the **vital signs**, which consist of **temperature**, **pulse rate**, **respiratory rate**, and **blood pressure**. These are discussed here in connection with their uses as indicators of toxicant exposure.

Some toxicants that affect body temperature are shown in Figure 6.11. Among those that increase body temperature are benzadrine, cocaine, sodium fluoroacetate, tricyclic antidepressants, hexachlorobenzene, and salicylates (aspirin). In addition to phenobarbital and ethanol, toxicants that decrease body temperature include phenothiazine, clonidine, glutethimide, and haloperidol.

Toxicants may have three effects on pulse rate: **bradycardia** (decreased rate), **tachycardia** (increased rate), and **arrhythmia** (irregular pulse). Alcohols may cause either bradycardia or tachycardia. Amphetamines, belladonna alkaloids, cocaine, and tricyclic antidepressants (see imiprimine hydrochloride in Figure 6.12) may cause either tachycardia or arrhythmia. Toxic doses of digitalis may result in bradycardia or arrhythmia. The pulse rate is decreased by toxic exposure to carbamates, organophosphates, local anesthetics, barbiturates, clonidine, muscaric mushroom toxins, and opiates. In addition to the substances mentioned above, those that cause arrhythmia are arsenic, caffeine, belladonna alkaloids, phenothizine, theophylline, and some kinds of solvents.

Among the toxicants that increase respiratory rate are cocaine, amphetamines, and fluoroacetate (all of which are shown in Figure 6.11), nitrites (compounds containing the NO_2^- ion), methanol (CH_3OH), salicylates, and hexachlorobenzene. Cyanide and carbon monoxide may either increase or decrease respiratory rate. Alcohols other than methanol, analgesics, narcotics, sedatives, phenothiazines, and opiates in toxic doses decrease respiratory rate. The structural formulas of some compounds that affect respiratory rate are shown in Figure 6.13.

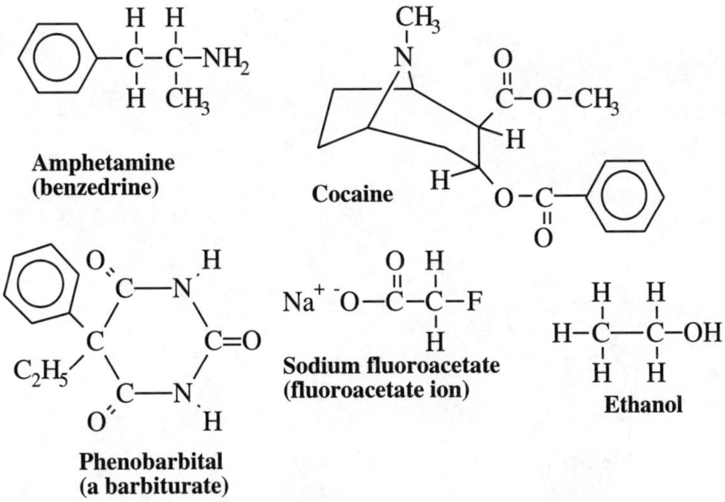

Figure 6.11 Examples of toxicants that affect body temperature. Amphetamine, cocaine, and fluoroacetate increase body temperature; phenobarbital and ethanol decrease it.

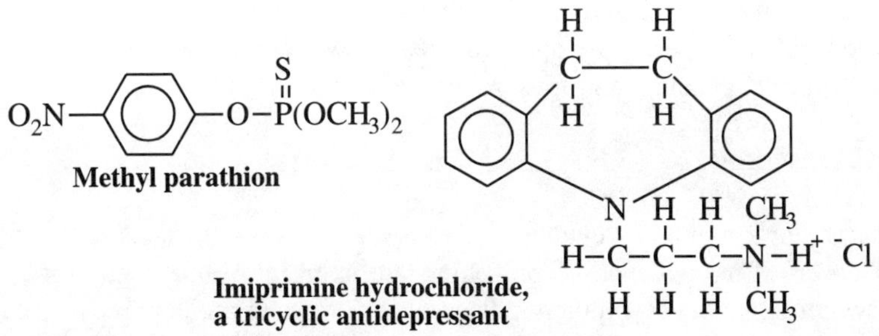

Figure 6.12 Structures of toxicants that can affect pulse rate. Methyl parathion, a commonly used plant insecticide, can cause bradycardia. Imiprimine hydrochloride, a tricyclic antidepressant, can cause either tachycardia or arrhythmia.

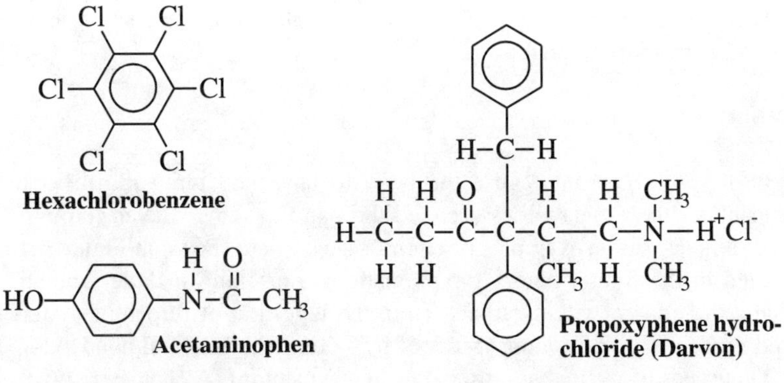

Figure 6.13 Some compounds that affect respiratory rate. Acetaminophen is one of the simple analgesics, which in therapeutic doses relieves pain without any effect on an individual's consciousness. Propoxyphene hydrochloride is a narcotic analgesic, a class of substances that can cause biochemical changes in the body leading to chemical dependency.

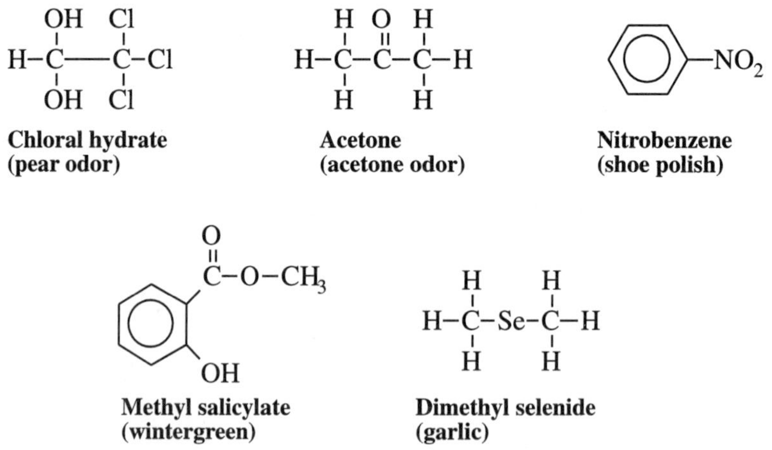

Chloral hydrate
(pear odor)
$(HO)_2CH-CCl_3$

Acetone
(acetone odor)
$CH_3-CO-CH_3$

Nitrobenzene
(shoe polish)
$C_6H_5-NO_2$

Methyl salicylate
(wintergreen)

Dimethyl selenide
(garlic)
$CH_3-Se-CH_3$

Figure 6.14 Some toxicants and the odors they produce in exposed subjects.

Amphetamines and cocaine (Figure 6.11), tricyclic antidepressants (see imiprimine hydrochloride in Figure 6.12), phenylcyclidines, and belladonna alkaloids at toxic levels increase blood pressure. Overdoses of antihypertensive agents decrease blood pressure, as do toxic doses of opiates, barbiturates, iron, nitrite, cyanide, and mushroom toxins.

6.13.2 Skin Symptoms

In many cases the skin exhibits evidence of exposure to toxic substances. The two main skin characteristics observed as evidence of poisoning are skin color and degree of skin moisture. Excessively dry skin tends to accompany poisoning by tricyclic antidepressants, antihistamines, and belladonna alkaloids. Among the toxic substances for which moist skin is a symptom of poisoning are mercury, arsenic, thallium, carbamates, and organophosphates. The skin appears flushed when the subject has been exposed to toxic doses of carbon monoxide, nitrites, amphetamines, monsodium glutamate, and tricyclic antidepressants. Higher doses of cyanide, carbon monoxide, and nitrites give the skin a **cyanotic** appearance (blue color due to oxygen deficiency in the blood). Skin may appear **jaundiced** (yellow because of the presence of bile pigments in the blood) when the subject is poisoned by a number of toxicants, including arsenic, arsine gas (AsH_3), iron, aniline dyes, and carbon tetrachloride.

6.13.3 Odors

Toxic levels of some materials cause the body to have unnatural **odors** because of parent compound toxicants or their metabolites secreted through the skin, exhaled through the lungs, or present in tissue samples. Some examples of odorous species are shown in Figure 6.14. In addition to the odors noted in the figure, others symptomatic of poisoning include aromatic odors from hydrocarbons and the odor of violets arising from the ingestion of turpentine. Alert pathologists have uncovered evidence of poisoning murders by noting the bitter almond odor of hydrogen cyanide (HCN) in tissues of victims of criminal cyanide poisoning. A characteristic rotten-egg odor is evidence of hydrogen sulfide (H_2S) poisoning. The same odor has been reported at autopsies of carbon disulfide poisoning victims. Even very slight exposures to some selenium compounds cause an extremely potent garlic breath odor.

6.13.4 Eyes

Careful examination of the eyes can reveal evidence of poisoning. The response, both in size and reactivity, of the pupils to light may indicate response to toxicants. Both voluntary and involuntary movement of the eyes can be significant. The appearance of eye structures, including optic disc, conjunctiva, and blood vessels, can be significant. Eye **miosis**, defined as excessive or prolonged contraction of the eye pupil, is a toxic response to a number of substances, including alcohols, carbamates, organophosphates, and phenycyclidine. The opposite response of excessive pupil dilation, **mydriasis**, is caused by amphetamines, belladonna alkaloids, glutethimide, and tricyclic antidepressants, among others. **Conjunctivitis** is a condition marked by inflammation of the conjunctiva, the mucus membrane that covers the front part of the eyeball and the inner lining of the eyelids. Corrosive acids and bases (alkalies) cause conjunctivitis, as do exposures to nitrogen dioxide, hydrogen sulfide, methanol, and formaldehyde. **Nystagmus**, the involuntary movement of the eyeballs, usually in a side-to-side motion, occurs in poisonings by some toxicants, including barbiturates, phenycyclidine, phentoin, and ethychlorovynol.

6.13.5 Mouth

Examination of the mouth provides evidence of exposure to some toxicants. Caustic acids and bases cause a moist condition of the mouth. Other toxicants that cause the mouth to be moister than normal include mercury, arsenic, thallium, carbamates, and organophosphates. A dry mouth is symptomatic of poisoning by tricyclic antidepressants, amphetamines, antihistamines, and glutethimide.

6.13.6 Gastrointestinal Tract

The gastrointestinal tract responds to a number of toxic substances, usually by pain, vomiting, or paralytic ileus (see "Intestines," Section 6.4.5). Severe gastrointestinal pain is symptomatic of poisoning by arsenic or iron. Both of these substances can cause vomiting, as can acids, bases, fluorides, salicylates, and theophyllin. Paralytic ileus can result from ingestion of narcotic analgesics, tricyclic antidepressants, and clonidine.

6.13.7 Central Nervous System

The central nervous system responds to poisoning by exhibiting symptoms such as **convulsions**, **paralysis**, **hallucinations**, and **ataxia** (lack of coordination of voluntary movements of the body). Other behavioral symptoms of poisoning include agitation, hyperactivity, disorientation, and delirium.

Among the many toxicants that cause convulsions are chlorinated hydrocarbons, amphetamines, lead, organophosphates, and strychnine. There are several levels of **coma**, the term used to describe a lowered level of consciousness. At level 0, the subject may be awakened and will respond to questions. At level 1, withdrawal from painful stimuli is observed and all reflexes function. A subject at level 2 does not withdraw from painful stimuli, although most reflexes still function. Levels 3 and 4 are characterized by the absence of reflexes; at level 4, respiratory action is depressed and the cardiovascular system fails. Among the many toxicants that cause coma are narcotic analgesics, alcohols, organophosphates, carbamates, lead, hydrocarbons, hydrogen sulfide, benzodiazepines, tricyclic antidepressants, isoniazid, phenothiazines, and opiates.

6.14 REPRODUCTIVE AND DEVELOPMENTAL EFFECTS

Some of the more serious effects of toxic substances are those that affect the reproduction of organisms and their development to adulthood.[4] Because of the serious nature of these effects, they are commonly examined in animal studies of pharmaceuticals, pesticides, and other chemicals.

Developmental toxic effects are those that adversely influence the growth and development of an organism to adulthood. These may occur from exposure of either parent of the organism to toxic substances even before conception. They may be the result of exposure of the embryo or fetus before birth. And they include effects resulting from exposure during the growth of the juvenile organism from birth to adulthood.

Teratology refers specifically to adverse effects of substances on an organism after conception up until birth. The teratogenic effects of thalidomide that resulted when women took this tranquilizing drug during early stages of pregnancy stand as one of the most distressing examples of teratogenic substances. Teratogens are most likely to cause harmful effects during the first trimester of pregnancy, when organs are becoming differentiated.

The reproductive systems of both males and females are suceptible to adverse effects of toxic substances. The study of these effects is called **reproductive toxicology**.

The chemical alteration of cell DNA that results in effects passed on through cell division is known as **mutagenesis**. Mutagenesis may occur in germ cells (female egg cells, male sperm cells) and cause mutations that appear in offspring. Mutagenesis may also occur in somatic cells, which are any body cells that are not sexual reproductive cells. Somatic cell mutagens are of particular concern because of the possibility that they will result in uncontrolled cell reproduction leading to cancer. Somatic cell mutations are easier to detect than germ cell mutations through observation of chromsomal aberrations and other effects.

REFERENCES

1. Orfila, M.J.B., *Traité des Poisons Tiré s des Règnes Minéral, Végétal, et Animal, ou, Toxicologie Générale Considérée sous les Rapports de la Physiologie, de la Pathologie, et de la Médicine Légale*, Crochard, Paris, 1815.
2. Leume, B. and Levine, B., *Principles of Forensic Toxicology*, AOAC Press, Baltimore, 1999.
3. Eaton, D.L. and Klaassen, C.D., Principles of toxicology, in *Casarett and Doull's Toxicology*, 6th ed., Amdur, M.O., Doull, J., and Klaassen, C.D., Eds., McGraw-Hill, New York, 2001, chap. 2, pp. 13–34.
4. Korach, K.S., Ed., *Reproductive and Developmental Toxicology*, Marcel Dekker, New York, 1998.

SUPPLEMENTARY REFERENCES

Bingham, E., Ed., *Patty's Toxicology*, 5th ed., John Wiley & Sons, New York, 2000.
Cheremisinoff, N.P., *Handbook of Industrial Toxicology and Hazardous Materials*, Marcel Dekker, New York, 1999.
Crosby, D.G. and Crosby, D.F, *Environmental Toxicology and Chemistry*, Oxford University Press, New York, 1998.
Derelanko, M.J. and Hollinger, M.A., *Handbook of Toxicology*, 2nd ed., CRC Press, Boca Raton, FL, 2001.
Ford, M.D. et al., *Clinical Toxicology*, W.B. Saunders Company, Philadelphia, 2000.
Harbison, R.D. and Hardy, H.L., Eds., *Hamilton and Hardy's Industrial Toxicology*, Mosby-Year Book, St. Louis, MO, 1998.
Harvey, P.W., Rush, K.C., and Cockburn, A., *Endocrine and Hormonal Toxicology*, John Wiley & Sons, New York, 1999.
Hayes, A.W., Ed., *Principles and Methods of Toxicology*, Taylor & Francis, London, 2000.
Hodgson, E., *Introduction to Toxicology*, Elsevier Science Ltd., Amsterdam, 2000.
Hodgson, E., Chambers, J.E., and Mailman, R.B., Eds., *Dictionary of Toxicology*, 2nd ed., Grove's Dictionaries, Inc., 1998.
Klaassen, C.D., *Casarett and Doull's Toxicology: The Basic Science of Poisons*, 6th ed., McGraw-Hill Professional Publishing, New York, 2001.
Krieger, R., Ed., *Handbook of Pesticide Toxicology*, 2nd ed., Academic Press, San Diego, CA, 2001.

Landis, W.G. and Yu, M.-H., *Introduction to Environmental Toxicology: Impacts of Chemicals upon Ecological Systems*, Lewis Publishers/CRC Press, Boca Raton, FL, 1998.
Lewis, R.A., Ed., *Lewis' Dictionary of Toxicology*, Lewis Publishers/CRC Press, Boca Raton, FL, 1998.
Marquardt, H., Schaefer, S.G., and McClellan, R.O., Eds., *Toxicology*, Academic Press, San Diego, CA, 1999.
Marrs, T., Syversen, T., and Ballantyne, B., Eds., *General and Applied Toxicology*, Grove's Dictionaries, Inc., New York, 1999.
Massaro, E.J., Ed., *Human Toxicology Handbook*, CRC Press, Boca Raton, FL, 1999.
Reiss, C., Parvez, S., and Labbe, G., *Advances in Molecular Toxicology*, V.S.P. International Science, Amsterdam, 1998.
Rose, J., Ed., *Environmental Toxicology: Current Developments*, G and B Science Publishers, London, 1998.
Ryan, R.P. and Terry, C.E., Eds., *Toxicology Desk Reference: The Toxic Exposure and Medical Monitoring Index*, 5th CD-ROM ed., Hemisphere Publishing Co., Washington, D.C., 1999.
Shaw, I. and Chadwick, J., *Principles of Environmental Toxicology*, Taylor & Francis, London, 1998.
Stelljes, M.E., *Toxicology for Non-Toxicologists*, Government Institutes, Rockville, MD, 1999.
Ware, G.W., Ed., *Reviews of Environmental Contamination and Toxicology*, Springer-Verlag, Heidelberg, 2001.
Wexler, P. and Gad, S.C., Eds., *Encyclopedia of Toxicology*, Academic Press, San Diego, CA, 1998.
Williams, P.L., James, R.C., and Roberts, S.M., Eds., *The Principles of Toxicology: Environmental and Industrial Applications*, John Wiley & Sons, New York, 2000.
Wright, D.A. and Welbourn, P., *Environmental Toxicology*, Cambridge University Press, London, 2001.
Yu, M.-H., Ed., *Environmental Toxicology: Impacts of Environmental Toxicants on Living Systems*, Lewis Publishers/CRC Press, Boca Raton, FL, 2000.
Zakrzewski, S.F., *Environmental Toxicology*, Oxford University Press, New York, 2001.

QUESTIONS AND PROBLEMS

1. Distinguish between acute toxicity and chronic toxicity.
2. Distinguish among acute local exposure, chronic local exposure, acute systemic exposure, and chronic systemic exposure to toxicants.
3. List and discuss the major routes and sites of exposure, distribution, and elimination of toxicants in the body.
4. What function is served by the stratum corneum in exposure of the body to toxic substances?
5. Explain why the lungs are regarded as the place where substances external to the body have the most intimate contact with body fluids. In what sense does pulmonary intake of a toxicant evade important "screening organs"?
6. Why are ammonia (NH_3) and hydrogen chloride (HCl) removed from air predominantly in the upper respiratory tract?
7. In what sense is the gastrointestinal tract "external" to the body?
8. How do the different regions of the gastrointestinal tract influence the uptake of toxicants, such as weak acids, that have different acid–base behaviors?
9. What are the major components of the enterohepatic circulation system? What is the portal circulatory system?
10. Describe the nature and significance of the dose–response curve. What is the significance of its inflection point (midpoint)? Define dose and response.
11. How do toxicity ratings relate to the potency of a toxicant?
12. Define sublethal effects, reversible effects, and margin of safety. What is an irreversible toxic effect?
13. What are hypersensitivity and hyposensitivity? Can these phenomena be related in any respect to the immune system?
14. What is the distinction between a xenobiotic substance and an endogenous substance? What are some examples of endogenous substances?
15. Define nonkinetic toxicology and how it relates to corrosive substances. What is kinetic toxicology and how does it relate to systemic poisons?
16. What is a receptor? In what way may acetylcholinesterase act as a receptor? What happens when this enzyme becomes bound to a toxic substance?

17. What is a protoxicant? What may happen to a protoxicant in the kinetic phase?
18. What are the three major divisions of the dynamic phase? In which of these is a receptor acted upon by a toxicant?
19. Distinguish between an active parent compound and an active metabolite in toxicology.
20. Differentiate among synergism, potentiation, and antagonism. What is an additive effect?
21. Define bradycardia, tachycardia, and arrhythmia. What are some of the toxicants that may cause each?
22. Distinguish between a cyanotic appearance of skin and a jaundiced appearance. Which kinds of toxicants may cause each?
23. List the major biological agents against which the body's immune system defends. How are leukocytes involved in this defense?
24. Define and give the significance of immunosuppression, hypersensitivity, uncontrolled proliferation, and autoimmunity.

CHAPTER 7

Toxicological Chemistry

7.1 INTRODUCTION

As defined in Section 1.1, **toxicological chemistry** is the chemistry of toxic substances, with emphasis on their interactions with biologic tissue and living systems. This chapter expands on this definition to define toxicological chemistry in more detail. Earlier chapters of the book have outlined the essential background required to understand toxicological chemistry. In order to comprehend this topic, it is first necessary to have an appreciation of the chemical nature of inorganic and organic chemicals, the topic of Chapter 1. An understanding of biochemistry, covered in Chapter 3, is required to comprehend the ways in which xenobiotic substances in the body undergo biochemical processes and, in turn, affect these processes. Additional perspective is provided by the discussion of metabolic processes in Chapter 4. The actual toxicities and biologically manifested effects of toxicants are covered in Chapter 6. Finally, an understanding of the environmental biochemistry of toxicants requires an appreciation of environmental chemistry, which is outlined in Chapter 2.

7.1.1 Chemical Nature of Toxicants

It is not possible to exactly define a set of chemical characteristics that make a chemical species toxic. This is because of the large variety of ways in which a substance can interact with substances, tissues, and organs to cause a toxic response. Because of subtle differences in their chemistry and biochemistry, similar substances may vary enormously in the degrees to which they cause a toxic response. For example, consider the toxic effects of carbon tetrachloride, CCl_4, and a chemically closely related chlorofluorocarbon, dichlorodifluoromethane, CCl_2F_2. Both of these compounds are completely halogenated derivatives of methane possessing very strong carbon–halogen bonds. As discussed in Section 16.2, carbon tetrachloride is considered to be dangerous enough to have been banned from consumer products in 1970. It causes a large variety of toxic effects in humans, with chronic liver injury being the most prominent. Dichlorodifluoromethane, a Freon compound, is regarded as nontoxic, except for its action as a simple asphyxiant and lung irritant at high concentrations.

An increasingly useful branch of toxicological chemistry is the one dealing with **quantitative structure-activity relationships** (QSARs). By relating the chemical structure and physical characteristics of various compounds to their toxic effects, it is possible to predict the toxicological effects of other compounds and classes of compounds.

With the qualification that there are exceptions to the scheme, it is possible to place toxic substances into several main categories. These are listed below:

- Substances that exhibit **extremes of acidity, basicity, dehydrating ability,** or **oxidizing power**. Examples include concentrated sulfuric acid (a strong acid with a tendency to dehydrate tissue), strongly basic sodium hydroxide, and oxidant elemental fluorine, F_2. Such species tend to be nonkinetic poisons (see Section 6.9) and corrosive substances that destroy tissue by massively damaging it at the site of exposure.
- **Reactive substances** that contain bonds or functional groups that are particularly prone to react with biomolecules in a damaging way. One reason that diethyl ether, $(C_2H_5)-O-(C_2H_5)$, is relatively nontoxic is because of its lack of reactivity resulting from the very strong C–H bonds in the ethyl groups and the very stable C–O–C ether linkage. A comparison of allyl alcohol with 1-propanol (structural formulas below)

Allyl alcohol   **Propyl alcohol**

shows that the former is a relatively toxic irritant to the skin, eyes, and respiratory tract that also damages liver and kidneys, whereas 1-propanol is one of the less toxic organic chemicals with an LD_{50} (see Section 6.5) about 100 times that of allyl alcohol. As shown by the structures, allyl alcohol differs from 1-propanol in having the relatively reactive alkenyl group C=C.
- **Heavy metals**, broadly defined, contain a number of members that are toxic by virtue of their interaction with enzymes, tendency to bond strongly with sulfhydryl (–SH) groups on proteins, and other effects.
- **Binding species** are those that bond to biomolecules, altering their function in a detrimental way. This binding may be reversible, as is the case with the binding of carbon monoxide with hemoglobin (see Chapter 11), which deprives hemoglobin of its ability to attach molecular O_2 and carry it from the lungs to body tissues. The binding may be irreversible. An example is that which occurs when an electron-deficient carbonium ion, such as H_3C^+ (an electrophile), binds to a nucleophile, such as an N atom on guanine attached to deoxyribonucleic acid (DNA).
- **Lipid-soluble compounds** are frequently toxic because of their ability to traverse cell membranes and similar barriers in the body. Lipid-soluble species frequently accumulate to toxic levels through biouptake and biomagnification processes (see Chapter 5).
- Chemical species that induce a toxic response based largely on their **chemical structures**. Such toxicants often produce an allergic reaction as the body's immune system recognizes the foreign agent, causing an immune system response. Lower-molecular-mass substances that act in this way usually must become bound to endogenous proteins to form a large enough species to induce an allergic response.

7.1.2 Biochemical Transformations

The toxicological chemistry of toxicants is strongly tied to their metabolic reactions and fates in the body.[1] Systemic poisons in the body undergo (1) biochemical reactions through which they have a toxic effect, and (2) biochemical processes that increase or reduce their toxicities, or change toxicants to forms that are readily eliminated from the body. In dealing with xenobiotic compounds, the body metabolizes them in ways that usually reduce toxicity and facilitate removal of the substance from the body, a process generally called **detoxication**. The opposite process by which nontoxic substances are metabolized to toxic ones or by which toxicities are increased by biochemical reactions is called **toxication** or **activation**. Most of the processes by which xenobiotic substances are handled in living organisms are phase I and phase II reactions discussed in the remainder of this chapter.

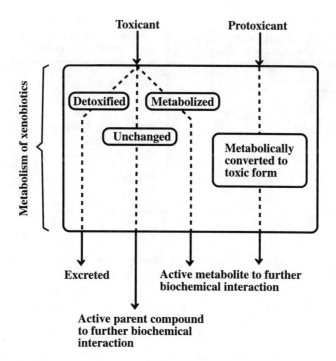

Figure 7.1 Pathways of xenobiotic species prior to their undergoing any biochemical interactions that could lead to toxic effects.

7.2 METABOLIC REACTIONS OF XENOBIOTIC COMPOUNDS

Toxicants or their metabolic precursors (**protoxicants**) may undergo absorption, metabolism, temporary storage, distribution, or excretion, as illustrated in Figure 7.1.[2] The modeling and mathematical description of these aspects as a function of time is called **toxicokinetics**.[3] Here are discussed the metabolic processes that toxicants undergo. Emphasis is placed on xenobiotic compounds, on chemical aspects, and on processes that lead to products that can be eliminated from the organism. Of particular importance is **intermediary xenobiotic metabolism**, which results in the formation of somewhat transient species that are different from both those ingested and the ultimate product that is excreted. These species may have significant toxicological effects. Xenobiotic compounds in general are acted on by enzymes that function on an **endogenous substrate** that is in the body naturally. For example, flavin-containing monooxygenase enzyme acts on endogenous cysteamine to convert it to cystamine, but also functions to oxidize xenobiotic nitrogen and sulfur compounds.

Biotransformation refers to changes in xenobiotic compounds as a result of enzyme action. Reactions not mediated by enzymes may also be important. As examples of nonenzymatic transformations, some xenobiotic compounds bond with endogenous biochemical species without an enzyme catalyst, undergo hydrolysis in body fluid media, or undergo oxidation–reduction processes. However, the metabolic phase I and phase II reactions of xenobiotics discussed here are enzymatic.

The likelihood that a xenobiotic species will undergo enzymatic metabolism in the body depends on the chemical nature of the species. Compounds with a high degree of polarity, such as relatively ionizable carboxylic acids, are less likely to enter the body system and, when they do, tend to be quickly excreted. Therefore, such compounds are unavailable, or available for only a short time, for enzymatic metabolism. Volatile compounds, such as dichloromethane or diethylether, are

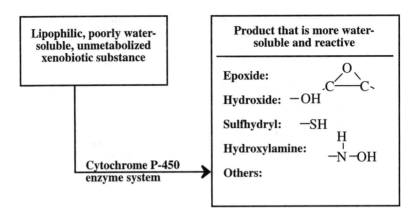

Figure 7.2 Overall process of phase I reactions.

expelled so quickly from the lungs that enzymatic metabolism is less likely. This leaves as the most likely candidates for enzymatic metabolic reactions **nonpolar lipophilic compounds**, those that are relatively less soluble in aqueous biological fluids and more attracted to lipid species. Of these, the ones that are resistant to enzymatic attack (polychlorinated biphenyls (PCBs), for example) tend to bioaccumlate in lipid tissue.

Xenobiotic species may be metabolized in a wide variety of body tissues and organs. As part of the body's defense against the entry of xenobiotic species, the most prominent sites of xenobiotic metabolism are those associated with entry into the body (see Figure 6.2). The skin is one such organ, as is the lung. The gut wall through which xenobiotic species enter the body from the gastrointestinal tract is also a site of significant xenobiotic compound metabolism. The liver is of particular significance because materials entering systemic circulation from the gastrointestinal tract must first traverse the liver.

7.2.1 Phase I and Phase II Reactions

The processes that most xenobiotics undergo in the body can be divided into two categories: phase I reactions and phase II reactions. A **phase I reaction** introduces reactive, polar functional groups (see Table 1.3) onto lipophilic (fat-seeking) toxicant molecules. In their unmodified forms, such toxicant molecules tend to pass through lipid-containing cell membranes and may be bound to lipoproteins, in which form they are transported through the body. Because of the functional group attached, the product of a phase I reaction is usually more water soluble than the parent xenobiotic species, and more importantly, it possesses a "chemical handle" to which a substrate material in the body may become attached so that the toxicant can be eliminated from the body. The binding of such a substrate is a **phase II reaction**, and it produces a **conjugation product** that normally (but not always) is less toxic than the parent xenobiotic compound or its phase I metabolite and more readily excreted from the body.

In general, the changes in structure and properties of a compound that result from a phase I reaction are relatively mild. Phase II processes, however, usually produce species that are much different from the parent compounds. It should be emphasized that not all xenobiotic compounds undergo both phase I and phase II reactions. Such a compound may undergo only a phase I reaction and be excreted directly from the body. Or a compound that already possesses an appropriate functional group capable of conjugation may undergo a phase II reaction without a preceding phase I reaction.

Phase I and phase II reactions are obviously important in mitigating the effects of toxic sustances. Some toxic substances act by inhibiting the enzymes that carry out phase I and phase II reactions, leading to toxic effects of other substances that normally would be detoxified.

7.3 PHASE I REACTIONS

Figure 7.2 shows the overall processes involved in a phase I reaction. Normally a phase I reaction adds a functional group to a hydrocarbon chain or ring or modifies one that is already present.[4] The product is a chemical species that readily undergoes conjugation with some other species naturally present in the body to form a substance that can be readily excreted. Phase I reactions are of several types, of which oxidation of C, N, S, and P is most important. Reduction may occur on reducible functionalities by addition of H or removal of O. Phase I reactions may also consist of hydrolysis processes, which require that the xenobiotic compound have a hydrolyzable group.

7.3.1 Oxidation Reactions

The most important phase I reactions are oxidation reactions, particularly those classified as microsomal monooxygenation reactions, formerly called mixed-function oxidations. Microsomes refer to a fraction collected from the centrifugation at about $100,000 \times g$ of cell homogenates and consisting of pellets. These pellets contain rough and smooth **endoplasmic reticulum** (extensive networks of membranes in cells) and Golgi bodies, which store newly synthesized molecules. **Monooxidations** occur with O_2 as the oxidizing agent, one atom of which is incorporated into the substrate, and the other going to form water:

$$\text{Substrate} + O_2 \xrightarrow{\text{Monooxidation}} \begin{array}{l} \text{Product-OH} \\ H_2O \end{array} \qquad (7.3.1)$$

The key enzymes of the system are the cytochrome P-450 enzymes, which have active sites that contain an iron atom that cycles between the +2 and +3 oxidation states. These enzymes bind to the substrate and molecular O_2 as part of the substrate oxidation process. Cytochrome P-450 is found most abundantly in the livers of vertebrates, reflecting the liver's role as the body's primary defender against systemic poisons. Cytochrome P-450 occurs in many other parts of the body, such as the kidney, ovaries, testes, and blood. The presence of this enzyme in the lungs, skin, and gastrointestinal tract may reflect their defensive roles against toxicants.

Epoxidation consists of adding an oxygen atom between two C atoms in an unsaturated system, as shown in Reactions 7.3.2 and 7.3.3. It is a particularly important means of metabolic attack on aromatic rings that abound in many xenobiotic compounds. Cytochrome P-450 is involved in epoxidation reactions. Both of the epoxidation reactions shown below have the effect of increasing the toxicities of the parent compounds, a process called **intoxication**. Some epoxides are unstable,

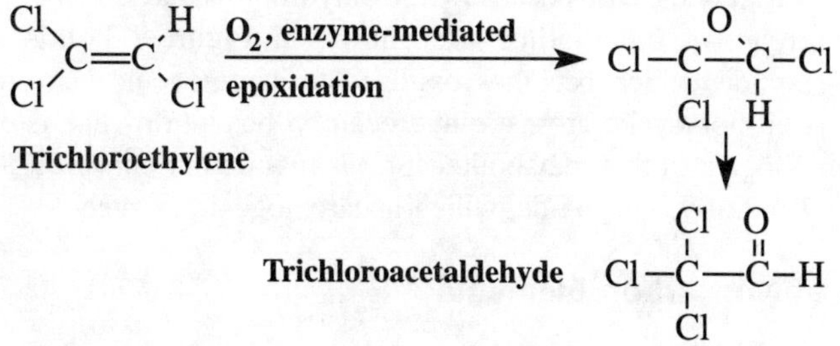

(7.3.2)

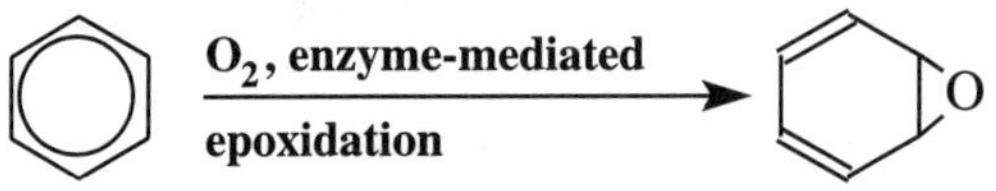

$$\text{benzene} \xrightarrow[\text{epoxidation}]{O_2, \text{ enzyme-mediated}} \text{benzene epoxide} \qquad (7.3.3)$$

tending to undergo further reactions, usually hydroxylation (see below). A well-known example of the formation of a stable epoxide is the conversion to aldrin of the insecticide dieldrin (Chapter 16).

7.3.2 Hydroxylation

Hydroxylation is the attachment of –OH groups to hydrocarbon chains or rings. **Aliphatic hydroxylation** of alkane chains can occur on the terminal carbon atom (–CH$_3$ group or ω-carbon) or on the C atom next to the last one (ω-1-carbon) by the insertion of an O atom between C and H, as shown below for the hydroxylation of the side chain on a substituted aromatic compound:

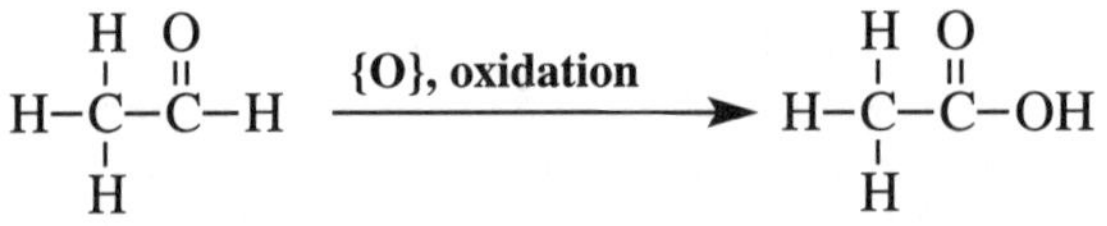

$$\underset{\text{Aldehyde}}{\text{H–CH(OH)–CHO}} \xrightarrow{\{O\}, \text{ oxidation}} \underset{\text{Carboxylic acid}}{\text{H–CH(OH)–COOH}} \qquad (7.3.4)$$

Hydroxylation can follow epoxidation, as shown by the following rearrangement reaction for benzene epoxide:

$$\underset{\text{Benzene epoxide}}{\text{benzene epoxide}} \longrightarrow \underset{\text{Phenol}}{\text{phenol}} \qquad (7.3.5)$$

7.3.3 Epoxide Hydration

The addition of H$_2$O to epoxide rings, a process called **epoxide hydration**, is important in the metabolism of some xenobiotic materials. This reaction can occur, for example, with benzo(a)pyrene 7,8-epoxide, formed by the metabolic oxidation of benzo(a)pyrene, as shown in Figure 7.3. Hydration of an epoxide group on a ring leads to the *trans* dihydrodiols in which the –OH groups are on opposite sides of the ring.

Formation of a dihydrodiol by hydration of epoxide groups can be an important detoxication process in that the product is often much less reactive to potential receptors than is the epoxide. However, this is not invariably the case because some dihydrodiols may undergo further epoxidation to form even more reactive metabolites. As shown in Figure 7.3, this can happen with benzo(a)pyrene 7,8-epoxide, which becomes oxidized to carcinogenic benzo(a)pyrene 7,8-diol-9,10-epoxide. The parent polycyclic aromatic hydrocarbon benzo(a)pyrene is classified as a procarcinogen, or precarcinogen, in that metabolic action is required to convert it to a species, in this case benzo(a)pyrene 7,8-diol-9,10-epoxide, which is carcinogenic as such.

7.3.4 Oxidation of Noncarbon Elements

As summarized in Figure 7.4, the oxidation of nitrogen, sulfur, and phosphorus is an important type of metabolic reaction in xenobiotic compounds. It can be an important intoxication mechanism

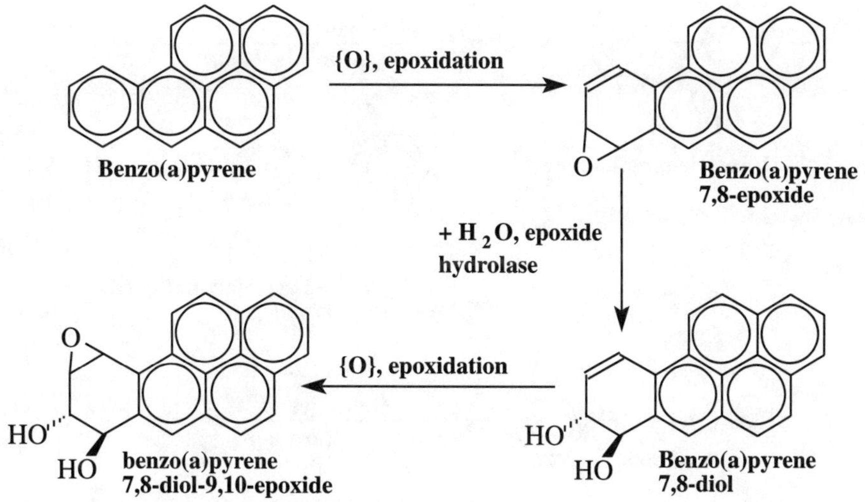

Figure 7.3 Epoxidation and hydroxylation of benzo(a)pyrene (left) to form carcinogenic benzo(a)pyrene 7,8-diol-9,10-epoxide.

by which compounds are made more toxic. For example, the oxidation of nitrogen in 2-acetylaminofluorene yields potently carcinogenic N-hydroxy-2-acetylaminofluorene. Two major steps in the metabolism of the plant systemic insecticide aldicarb (Figure 7.5) are oxidation to the sulfoxide and oxidation to the sulfone (see sulfur compounds in Chapter 17). The oxidation of phosphorus in parathion (replacement of S by O, oxidative desulfurization) yields insecticidal paraoxon, which is much more effective than the parent compound in inhibiting acetylcholinesterase enzyme (see Section 6.10).

In addition to cytochrome P-450 enzymes, another enzyme that mediates phase I oxidations is **flavin-containing monooxygenase** (FMO), likewise contained in the endoplasmic reticulum. It is especially effective in oxidizing primary, secondary, and tertiary amines. Additionally, it catalyzes oxidation of other nitrogen-containing xenobiotic compounds, as well as those that contain sulfur and phosphorus, but does not bring about hydroxylation of carbon atoms.

7.3.5 Alcohol Dehydrogenation

A common step in the metabolism of alcohols is carried out by **alcohol dehydrogenase** enzymes that produce aldehydes from primary alcohols that have the –OH group on an end carbon and produce ketones from secondary alcohols that have the –OH group on a middle carbon, as shown by the examples in Reactions 7.3.6 and 7.3.7. As indicated by the double arrows in these reactions, the reactions are reversible and the aldehydes and ketones can be converted back to alcohols. The oxidation of aldehydes to carboxylic acids occurs readily (Reaction 7.3.8). This is an important detoxication process because aldehydes are lipid soluble and relatively toxic, whereas carboxylic acids are more water soluble and undergo phase II reactions leading to their elimination.

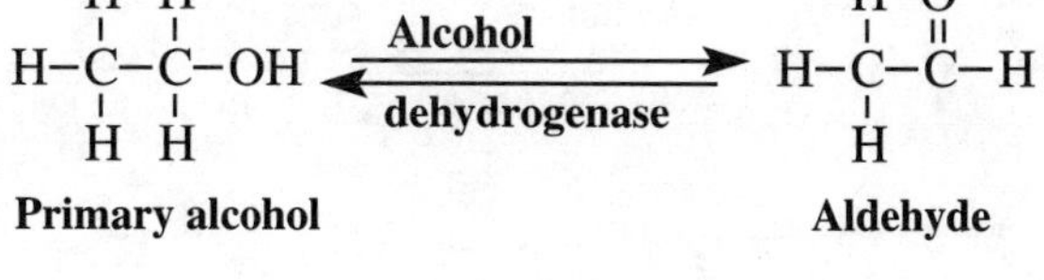

$$\underset{\text{Primary alcohol}}{\text{H-}\overset{\overset{\text{H}}{|}}{\underset{\underset{\text{H}}{|}}{\text{C}}}\text{-}\overset{\overset{\text{H}}{|}}{\underset{\underset{\text{H}}{|}}{\text{C}}}\text{-OH}} \quad \underset{\text{dehydrogenase}}{\overset{\text{Alcohol}}{\rightleftharpoons}} \quad \underset{\text{Aldehyde}}{\text{H-}\overset{\overset{\text{H}}{|}}{\underset{\underset{\text{H}}{|}}{\text{C}}}\text{-}\overset{\overset{\text{O}}{\|}}{\text{C}}\text{-H}} \qquad (7.3.6)$$

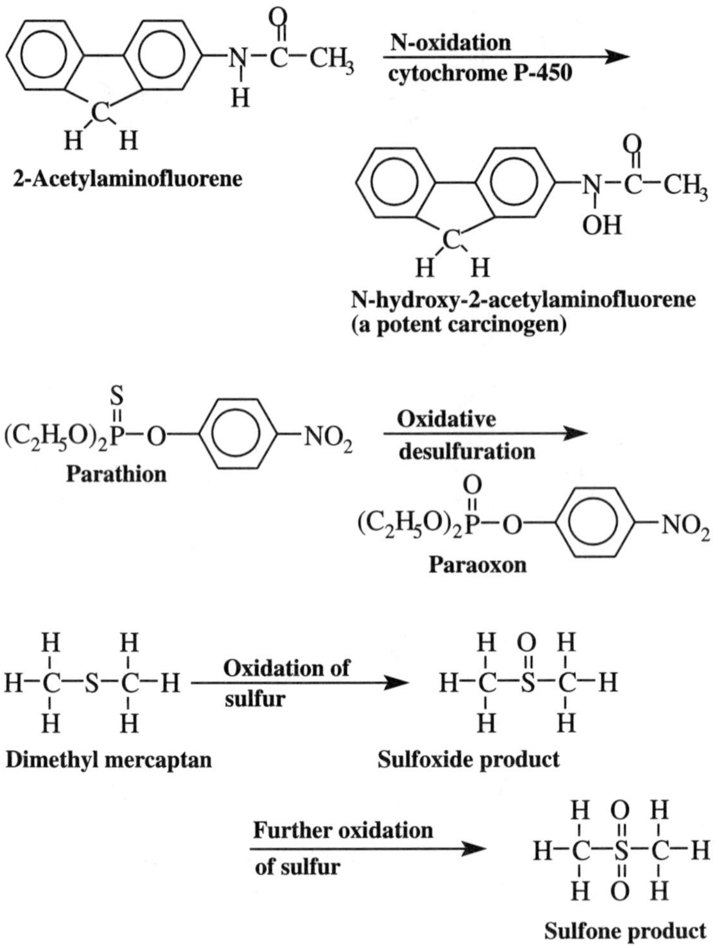

Figure 7.4 Metabolic oxidation of nitrogen, phosphorus, and sulfur in xenobiotic compounds.

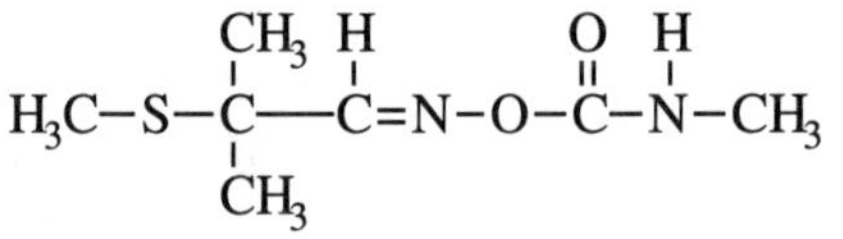

Figure 7.5 Structure of the plant systemic insecticide temik (aldicarb). The sulfur is metabolically oxidizable.

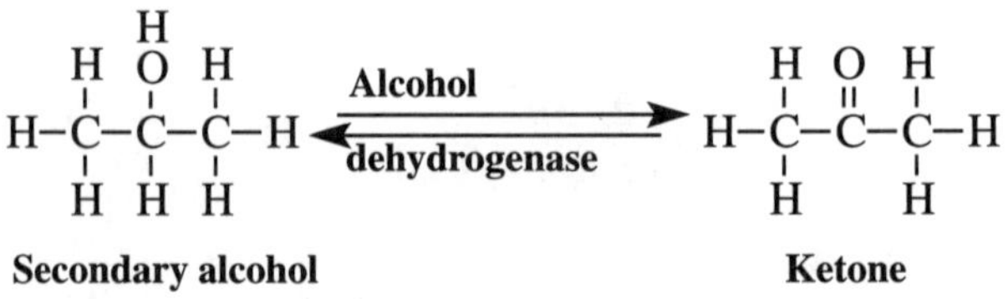

$$\text{Secondary alcohol} \underset{\text{dehydrogenase}}{\overset{\text{Alcohol}}{\rightleftharpoons}} \text{Ketone} \qquad (7.3.7)$$

Table 7.1 Functional Groups That Undergo Metabolic Reduction

Functional Group	Process	Product
$R-\underset{\underset{}{\parallel}}{C}(=O)-H$	Aldehyde reduction	$R-CH(H)-OH$
$R-\underset{}{C}(=O)-R'$	Ketone reduction	$R-CH(OH)-R'$
$R-S(=O)-R'$	Sulfoxide reduction	$R-S-R'$
$R-SS-R'$	Disulfide reduction	$R-SS-H$
$>C=C<$	Alkene reduction	$-CH-CH-OH$
$R-N=N-R'$	Azo reduction	$R-NH-NH-R' \longrightarrow R-NH_2 + H_2N-R'$
$R-NO_2$	Nitro reduction	$R-NO$, $R-NH_2$, $R-NH-OH$
$As(V)$	Arsenic reduction	$As(III)$

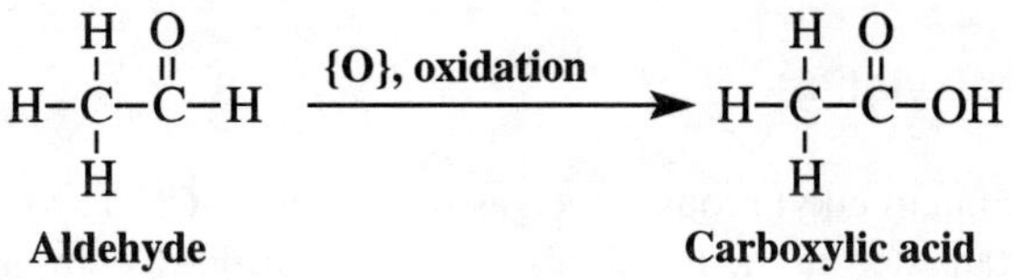

$$H-\underset{\underset{H}{|}}{\overset{H}{|}}{C}-\overset{O}{\underset{}{\parallel}}{C}-H \xrightarrow{\{O\},\ \text{oxidation}} H-\underset{\underset{H}{|}}{\overset{H}{|}}{C}-\overset{O}{\underset{}{\parallel}}{C}-OH \tag{7.3.8}$$

Aldehyde → Carboxylic acid

7.3.6 Metabolic Reductions

Table 7.1 summarizes the functional groups in xenobiotics that are most likely to be reduced metabolically. Reductions are carried out by **reductase enzymes**; for example, nitroreductase enzyme catalyzes the reduction of the nitro group. Reductase enzymes are found largely in the liver and to a certain extent in other organs, such as the kidneys and lungs. Most reductions of xenobiotic compounds are mediated by bacteria in the intestines, the **gut flora**. The contents of the lower bowel may contain a huge concentration of anaerobic bacteria. The compounds reduced by these bacteria may enter the lower bowel by either oral ingestion (without having been absorbed through the intestinal wall) or secretion with bile. In the latter case, the compounds may be parent materials or metabolic products of substances absorbed in upper regions of the gastrointestinal tract. Intestinal flora are known to mediate the reduction of organic xenobiotic sulfones and sulfoxides to sulfides:

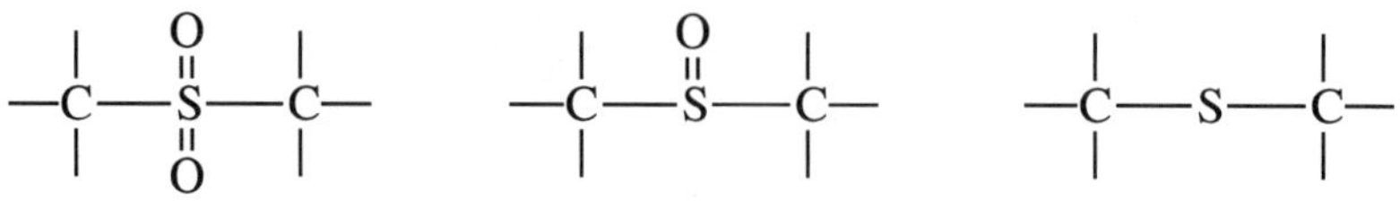

$$-\overset{|}{\underset{|}{C}}-\overset{O}{\underset{\underset{O}{\|}}{S}}-\overset{|}{\underset{|}{C}}- \qquad -\overset{|}{\underset{|}{C}}-\overset{\overset{O}{\|}}{S}-\overset{|}{\underset{|}{C}}- \qquad -\overset{|}{\underset{|}{C}}-S-\overset{|}{\underset{|}{C}}- \qquad (7.3.9)$$

        Sulfone   **Sulfoxide**   **Sulfide**

7.3.7 Metabolic Hydrolysis Reactions

Many xenobiotic compounds, such as pesticides, are esters, amides, or organophosphate esters, and hydrolysis is a very important aspect of their metabolic fates. **Hydrolysis** involves the addition of H_2O to a molecule accompanied by cleavage of the molecule into two species. The two most common types of compounds that undergo hydrolysis are esters

$$R-\overset{\overset{O}{\|}}{C}-OR' + H_2O \rightarrow R-\overset{\overset{O}{\|}}{C}-OH + HOR' \qquad (7.3.10)$$

and amides

$$R-\overset{\overset{O}{\|}}{C}-N\overset{R'}{\underset{R''}{}} + H_2O \rightarrow R-\overset{\overset{O}{\|}}{C}-OH + H-N\overset{R'}{\underset{R''}{}} \qquad (7.3.11)$$

The types of enzymes that bring about hydrolysis are **hydrolase enzymes**. Like most enzymes involved in the metabolism of xenobiotic compounds, hydrolase enzymes occur prominently in the liver. They also occur in tissue lining the intestines, nervous tissue, blood plasma, the kidney, and muscle tissue. Enzymes that enable the hydrolysis of esters are called **esterases**, and those that hydrolyze amides are **amidases**. Aromatic esters are hydrolyzed by the action of aryl esterases and alkyl esters by aliphatic esterases. Hydrolysis products of xenobiotic compounds may be either more or less toxic than the parent compounds.

7.3.8 Metabolic Dealkylation

Many xenobiotics contain alkyl groups, such as the methyl (–CH_3) group, attached to atoms of O, N, and S. An important step in the metabolism of many of these compounds is replacement of alkyl groups by H, as shown in Figure 7.6. These reactions are carried out by mixed-function oxidase enzyme systems. Examples of these kinds of reactions with xenobiotics include O-dealkylation of methoxychlor insecticides, N-dealkylation of carbaryl insecticide, and S-dealkylation of dimethyl mercaptan. Organophosphate esters (see Chapter 18) also undergo hydrolysis, as shown in Reaction 7.3.12 for the plant systemic insecticide demeton:

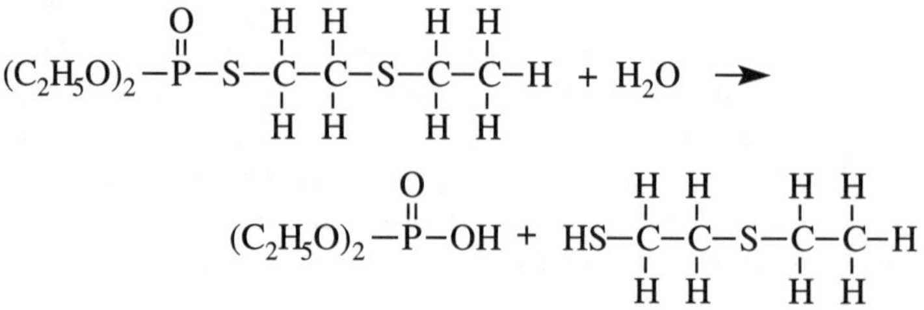

$$(C_2H_5O)_2-\overset{\overset{O}{\|}}{P}-S-\overset{H}{\underset{H}{C}}-\overset{H}{\underset{H}{C}}-S-\overset{H}{\underset{H}{C}}-\overset{H}{\underset{H}{C}}-H + H_2O \rightarrow$$

$$(C_2H_5O)_2-\overset{\overset{O}{\|}}{P}-OH + HS-\overset{H}{\underset{H}{C}}-\overset{H}{\underset{H}{C}}-S-\overset{H}{\underset{H}{C}}-\overset{H}{\underset{H}{C}}-H \qquad (7.3.12)$$

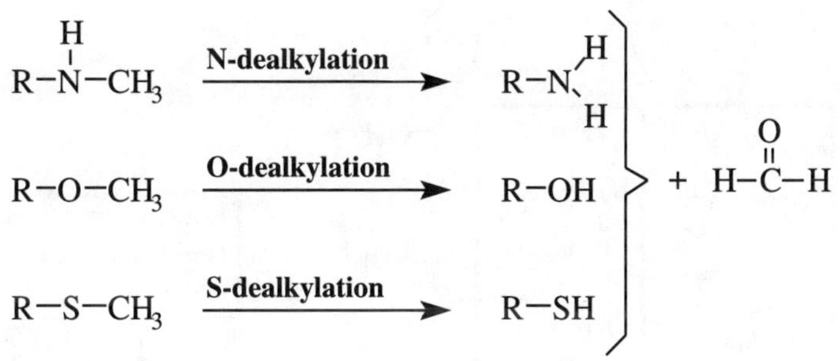

Figure 7.6 Metabolic dealkylation reactions shown for the removal of CH_3 from N, O, and S atoms in organic compounds.

7.3.9 Removal of Halogen

An important step in the metabolism of the many xenobiotic compounds that contain covalently bound halogens (F, Cl, Br, I) is the removal of halogen atoms, a process called **dehalogenation**. This may occur by **reductive dehalogenation**, in which the halogen atom is replaced by hydrogen, or two atoms are lost from adjacent carbon atoms, leaving a carbon–carbon double bond. These processes are illustrated by the following:

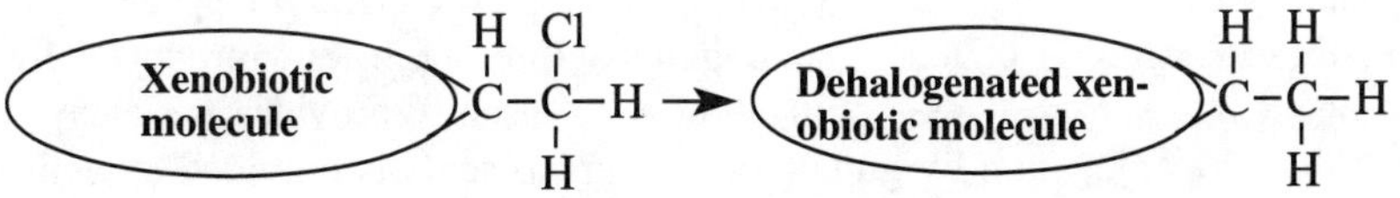

(7.3.13)

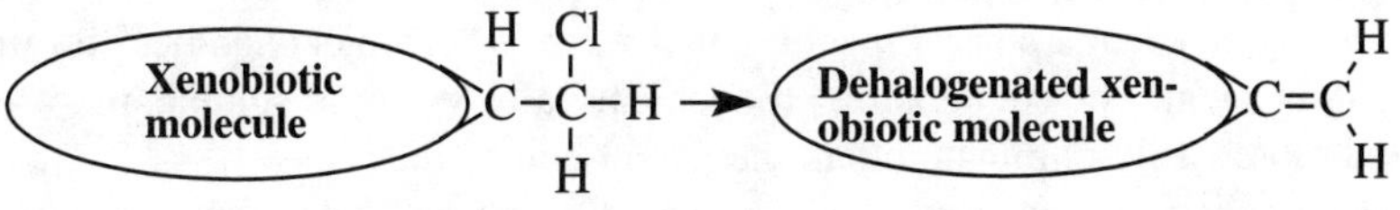

(7.3.14)

Oxidative dehalogenation occurs when oxygen is added in place of a halogen atom, as shown by the following reaction:

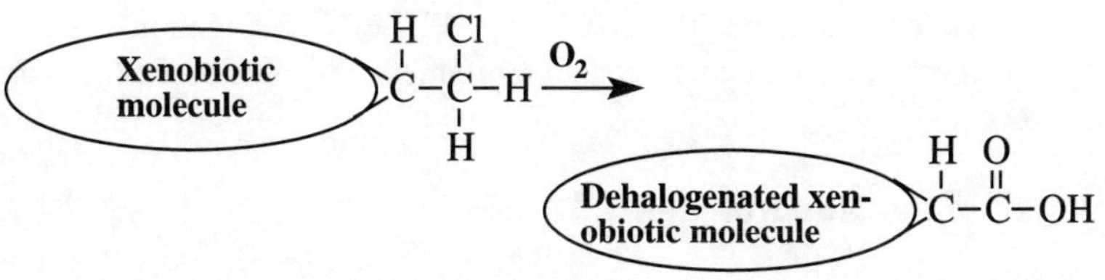

(7.3.15)

7.4 PHASE II REACTIONS OF TOXICANTS

Phase II reactions are also known as **conjugation reactions** because they involve the joining together of a substrate compound with another species that occurs normally in (is endogenous to) the organism.[5] This can occur with unmodified xenobiotic compounds, xenobiotic compounds that

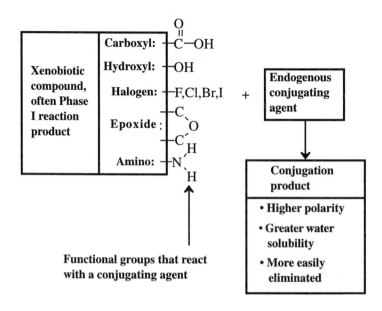

Figure 7.7 Overall process of conjugation that occurs in phase II reactions.

have undergone phase I reactions, and compounds that are not xenobiotic species. The substance that binds to these species is called an **endogenous** (present in and produced by the body) **conjugating agent**. Activation of the conjugating agent usually provides the energy needed for conjugation, although conjugation by glutathione or amino acids is provided by activation of the species undergoing conjugation preceding the reaction. The overall process for the conjugation of a xenobiotic compound is shown in Figure 7.7. Such a compound contains functional groups, often added as the consequence of a phase I reaction, that serve as "chemical handles" for the attachment of the conjugating agent. The conjugation product is usually less lipid soluble, more water soluble, less toxic, and more easily eliminated than the parent compound.

The conjugating agents that are attached as part of phase II reactions include glucuronide, sulfate, acetyl group, methyl group, glutathione, and some amino acids. Conjugation with glutathione is also a step in mercapturic acid synthesis. Glycine, glutamic acid, and taurine are common amino acids that act as conjugating agents. Most of the conjugates formed by these agents are more hydrophilic than the compounds conjugated, so the conjugates are more readily excreted. The exceptions are methylated and acetylated conjugates. Phase II conjugation reactions are usually rapid, and if they are performed on phase I reaction products, the rates of the latter are rate limiting for the overall process.

7.4.1 Conjugation by Glucuronides

Glucuronides are the most common endogenous conjugating agents in the body. They react with xenobiotics through the action of uridine diphosphate glucuronic acid (UDPGA). This transfer is mediated by glucuronyl transferase enzymes. These enzymes occur in the endoplasmic reticulum, where hydroxylated phase I metabolites of lipophilic xenobiotic compounds are produced. As a result, the lifetime of the phase I metabolites is often quite brief because the conjugating agent is present where they are produed. A generalized conjugation reaction of UDPGA with a xenobiotic compound can be represented as the following:

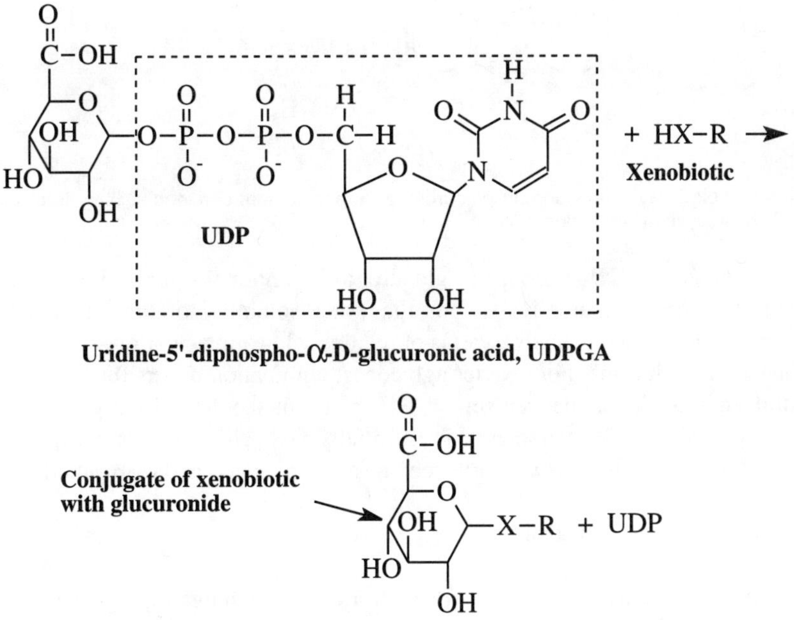

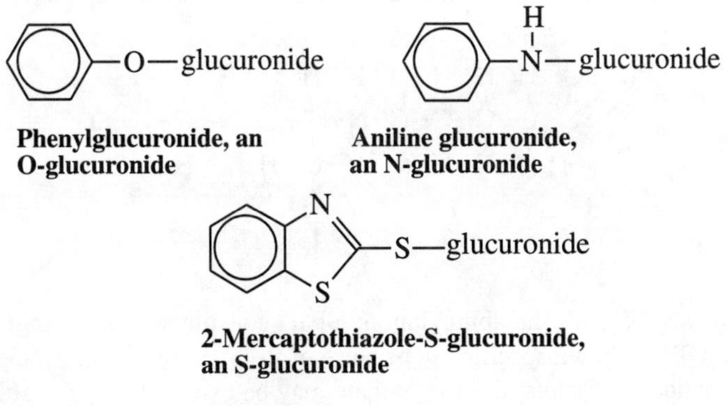

Figure 7.8 Examples of O-, N-, and S-glucuronides.

In this reaction HX–R represents the xenobiotic species in which HX is a functional group (such as –OH) and R is an organic moiety, such as the phenyl group (benzene ring less a hydrogen atom). The kind of enzyme that mediates this type of reaction is UDP glucuronyltransferase.

Glucuronide conjugation products may be classified according to the element to which the glucuronide is bound. The atoms to which the glucuronide most readily attaches are electron rich, usually O, N, or S (nucleophilic heteroatoms in the parlance of organic chemistry). Example glucuronides involving O, N, and S atoms are shown in Figure 7.8. When the functional group through which conjugation occurs is a hydroxyl group, –OH (HX in Reaction 7.4.1), an ether glucuronide is formed. A carboxylic acid group for HX gives an ester glucuronide. Glucuronides may be attached directly to N as the linking atom, as is the case with aniline glucuronide in Figure 7.8, or through an intermediate O atom. An example of the latter is N-hydroxyacetylaminoglucuronide, for which the structure is shown in Figure 7.9. This species is of interest because it is a stronger carcinogen than its parent xenobiotic compound, N-hydroxyacetylaminofluorene, contrary to the decrease in toxicity that usually results from glucuronide conjugation.

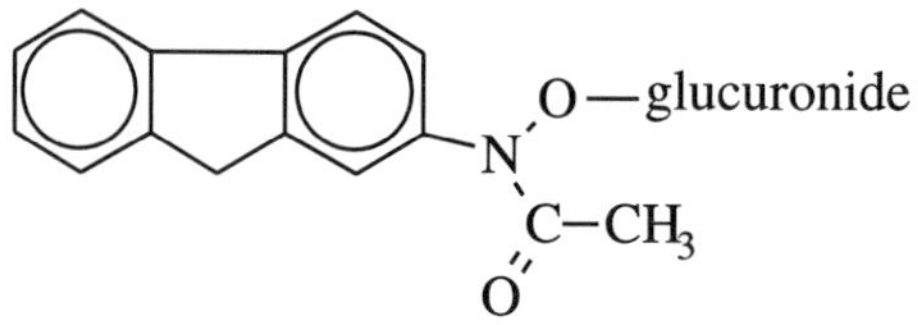

Figure 7.9 N-hydroxyacetylaminofluorene glucuronide, a more potent carcinogen than its parent compound, N-hydroxyacetylaminofluorene.

The carboxylic acid ($-CO_2H$ group) in glucuronides is normally ionized at the pH of physiological media, which is a major reason for the water solubility of the conjugates. When the compound conjugated (called the aglycone) is of relatively low molecular mass, the conjugate tends to be eliminated through urine. For heavier aglycones, elimination occurs through bile. **Enterohepatic circulation** provides a mechanism by which the metabolic effects of some glucuronide conjugates are amplified. This phenomenon is essentially a recycling process in which a glucuronide conjugate released to the intestine with bile becomes deconjugated and reabsorbed in the intestine.

7.4.2 Conjugation by Glutathione

Glutathione (commonly abbreviated GSH) is a crucial conjugating agent in the body. This compound is a tripeptide, meaning that it is composed of three amino acids linked together. These amino acids and their abbreviations are glutamic acid (Glu), cysteine (Cys), and glycine (Gly) (see structures in Figure 3.2). The structural formula of the glutathione tripeptide is the following:

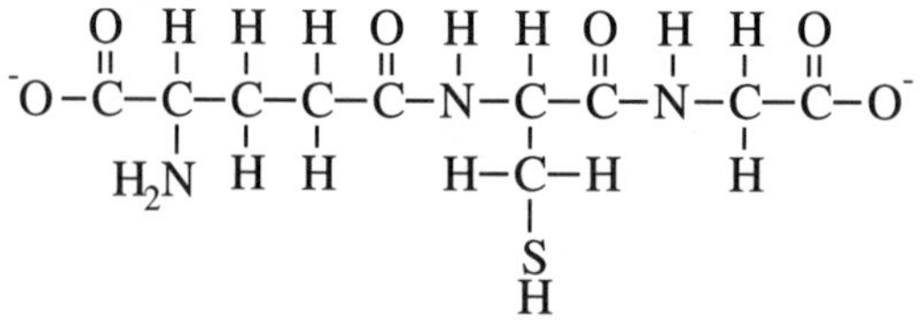

It may be represented with the abbreviations of its constituent amino acids, as illustrated in Figure 7.10, where SH is shown specifically because of its crucial role in forming the covalent link to a xenobiotic compound. A glutathione conjugate may be excreted directly, although this is rare. More commonly, the GSH conjugate undergoes further biochemical reactions that produce mercapturic acids (compounds with N-acetylcysteine attached) or other species. The overall process for the production of mercapturic acids as applied to a generic xenobiotic species, HX–R (see previous discussion), is illustrated in Figure 7.10.

There are numerous variations on the general mechanism outlined in Figure 7.10. Glutathione forms conjugates with a wide variety of xenobiotic species, including alkenes, alkyl epoxides (1,2-epoxyethylbenzene), arylepoxides (1,2-epoxynaphthalene), aromatic hydrocarbons, aromatic halides, alkyl halides (methyl iodide), and aromatic nitro compounds. The glutathione transferase enzymes required for the initial conjugation are widespread in the body.

The importance of glutathione in reducing levels of toxic substances can be understood by considering that loss of H^+ from –SH on glutathione leaves an electron-rich $-S^-$ group (nucleophile) that is highly attractive to electrophiles. Electrophiles are important toxic substances because of their tendencies to bind to nucleophilic biomolecules, including nucleic acids and proteins. Such binding can cause mutations (potentially cancer) and result in cell damage. Included among the toxic substances bound by glutathione are reactive intermediates produced in the metabolism of xenobiotic substances, including epoxides and free radicals (species with unpaired electrons).

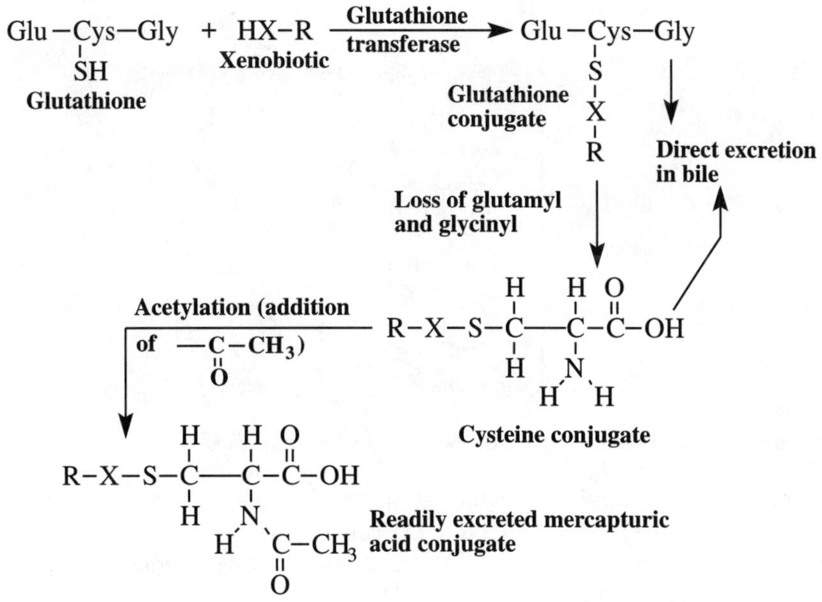

TOXICOLOGICAL CHEMISTRY

Figure 7.10 Glutathione conjugate of a xenobiotic species (HX–R), followed by formation of glutathione and cysteine conjugate intermediates (both of which may be excreted in bile) and acetylation to form readily excreted mercapturic acid conjugate.

7.4.3 Conjugation by Sulfate

Although conjugation by sulfate requires the input of substantial amounts of energy, it is very efficient in eliminating xenobiotic species through urine because the sulfate conjugates are completely ionized and therefore highly water soluble. The enzymes that enable sulfate conjugation are sulfotransferases, which act with the 3'-phosphoadenosine-5'-phosphosulfate (PAPS) cofactor:

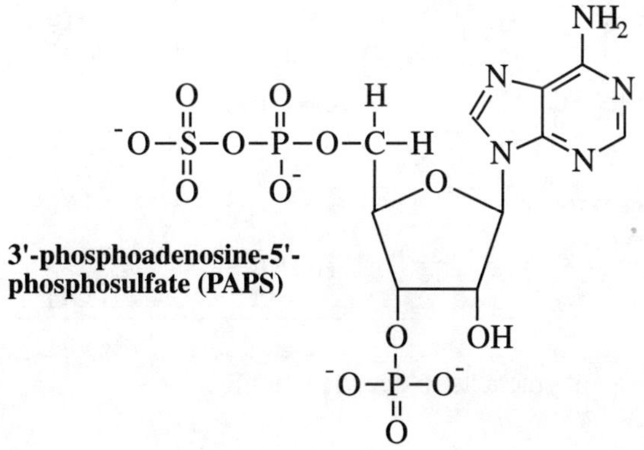

The types of species that form sulfate conjugates are alcohols, phenols, and aryl amines, as shown by the examples in Figure 7.11.

Although sulfation is normally an effective means of reducing toxicities of xenobiotic substances, there are cases in which the sulfate conjugate is reactive and toxic. An interesting example of such a substance is produced by the sulfate conjugation of 1'-hydroxysafrole, which is a phase

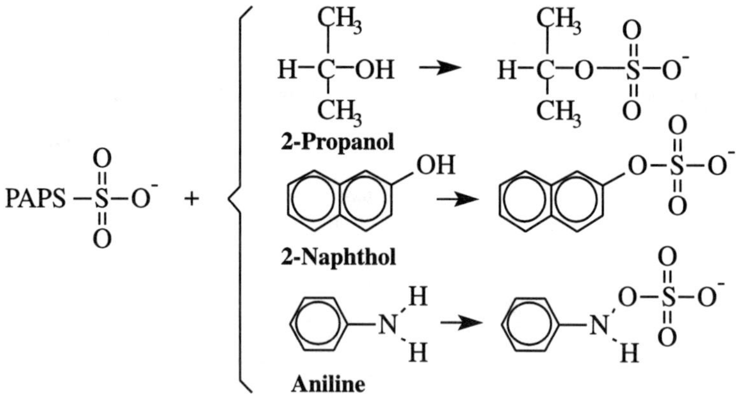

Figure 7.11 Formation of sulfate conjugates of some xenobiotic compounds.

I hydroxylation product of safrole, an ingredient of sassafras, used as a flavoring ingredient until its carcinogenic nature was revealed. Figure 7.12 shows the transformation of safrole through a sulfate conjugate intermediate to a positively charged electrophilic carbonium ion species that can bind with DNA and lead to tumor formation.

7.4.4 Acetylation

Acetylation reactions catalyzed by acetyltransferase enzymes involve the attachment of the acetyl moiety, shown as the final step in glutathione conjugation and the production of a mercapturic acid conjugate in Figure 7.10. The cofactor upon which the acetyltransferase enzyme acts in acetylation is acetyl coenzyme A:

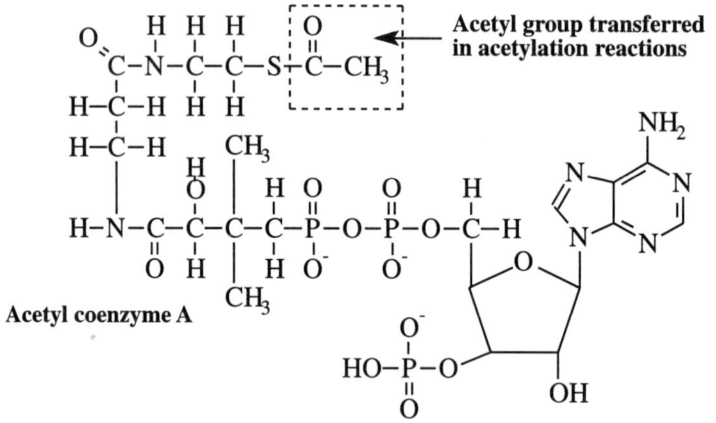

The acetyl transferase enzyme acts to acetylate aniline:

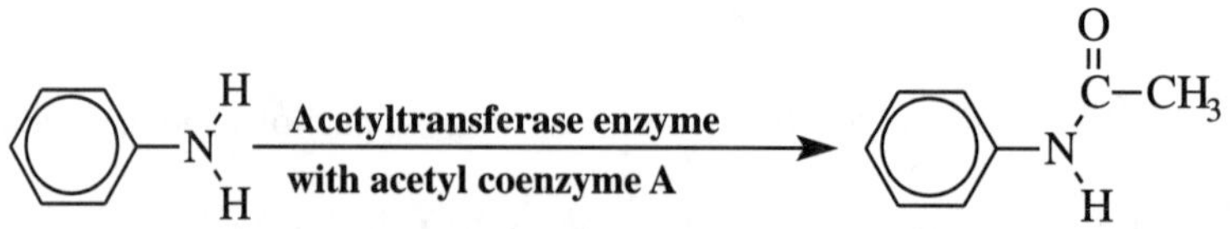

$$\text{(7.4.2)}$$

The most important kind of acetylation reaction is the acetylation of aromatic amines. This converts the ionizable amine group to a nonionizable group, to which the acetyl group is attached.

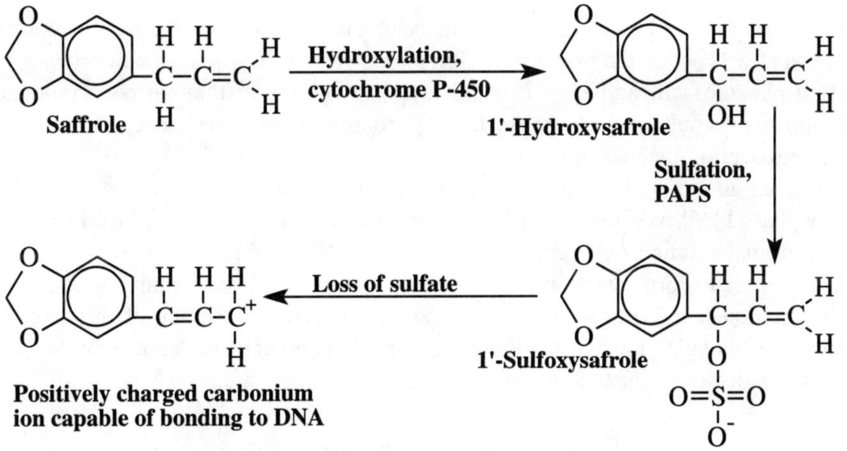

TOXICOLOGICAL CHEMISTRY

Figure 7.12 Formation of a positively charged carbonium ion capable of binding to DNA and causing cancer formed by the phase I hydroxylation of saffrole, followed by sulfation and loss of sulfate. In this case, sulfate conjugation forms a more toxic species.

As a consequence, some acetylated products are not as soluble in water as the parent compounds. In some cases, acetylation of aromatic amines makes them less active as toxicants, particularly in binding with DNA, whereas in other cases, they are made more active. In the latter case, activity can be due to a cytochrome P-450 catalyzed attachment of an –OH group to the acetylated nitrogen, leading to a positively charged electrophilic species capable of binding with DNA.

7.4.5 Conjugation by Amino Acids

Common amino acids that conjugate xenobiotics are glycine, glutamine, taurine, and serine, the anionic forms of which are shown below:

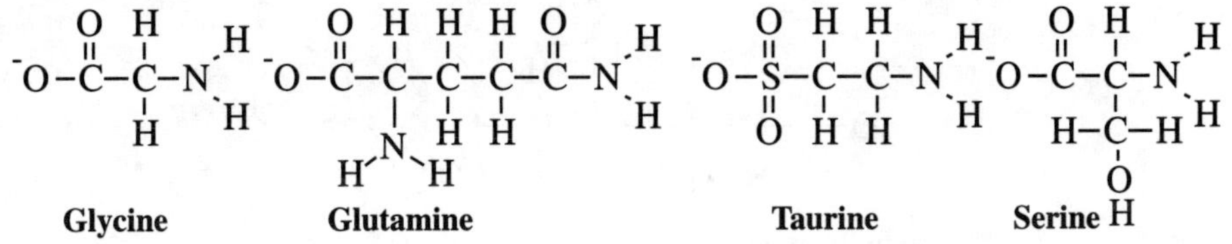

Glycine, **Glutamine**, **Taurine**, **Serine**

In addition to single amino acids, dipeptides consisting of two amino acids connected by a peptide linkage, such as glycylglycine and glycyltaurine, may conjugate xenobiotics to produce **peptide conjugates**. Amino acids have both an acid group and an amino (–NH$_2$) group at which conjugation to a xenobiotic may occur. Both types of binding are involved in amino acid conjugation.

The classic example of amino acid conjugation to a carboxylic acid group is the production of hippuric acid from benzoic acid and glycine:

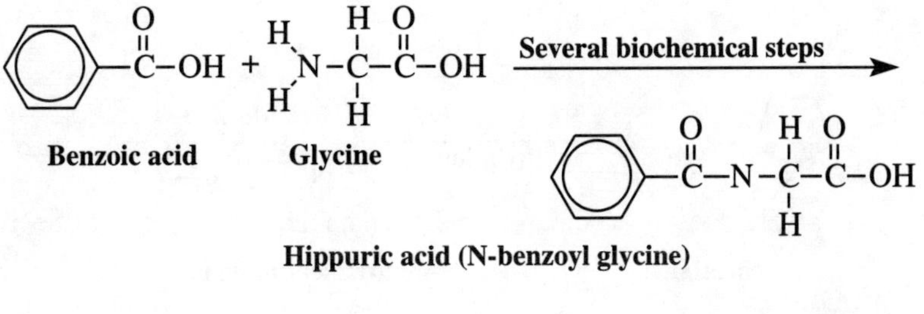

Benzoic acid + Glycine → Hippuric acid (N-benzoyl glycine) (7.4.3)

This is the oldest known biosynthesis, having been discovered in 1842. Before concerns over possible health effects ended the practice, it used to be performed by students of organic chemistry, who ingested benzoic acid and then isolated hippuric acid from their urine. Conjugation of a xenobiotic substance containing a carboxylic acid group with the –NH$_2$ group of an amino acid is generally a detoxication mechanism.

Binding of an amino acid through its carboxylic acid group can occur with hydroxylamines generated by phase I hydroxylation of aromatic amino compounds. This is shown in Reaction 7.4.4 for a generic aromatic amine represented as Ar–NH. The N-esters formed by reactions such as the one above can react to form electrophilic cations (carbonium and nitrenium) that can bind with nucleophilic biomolecules to produce toxic responses. Therefore, binding of the carboxylic acid group of an amino acid with the hydroxylamino group of a xenobiotic material should be considered an intoxication pathway rather than detoxication.

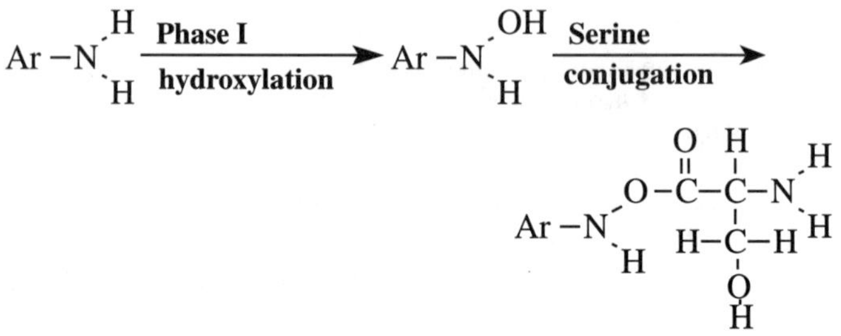

$$\text{Ar}-\text{NH}_2 \xrightarrow{\text{Phase I hydroxylation}} \text{Ar}-\text{NHOH} \xrightarrow{\text{Serine conjugation}} \text{Ar}-\text{N(H)}-\text{O}-\text{C(O)}-\text{CH(NH}_2\text{)}-\text{CH}_2\text{OH} \quad (7.4.4)$$

7.4.6 Methylation

Phase II **methylation** occurs with the S-adenosylmethionine (SAM) cofactor acting as a methylating agent:

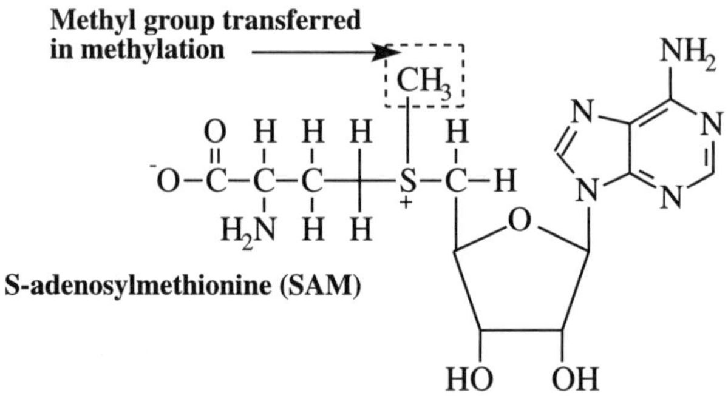

S-adenosylmethionine (SAM)

The methyl group on SAM behaves as an electrophilic $^+\text{CH}_3$ positively charged carbocation that is attracted to electron-rich nucleophilic O, N, and S atoms on a xenobiotic compound; methylation of carbon is rare. Therefore, the kinds of compounds commonly methylated include amines, heterocyclic nitrogen compounds, phenols, and compounds containing the –SH group. A typical methylation reaction is that of nicotine:

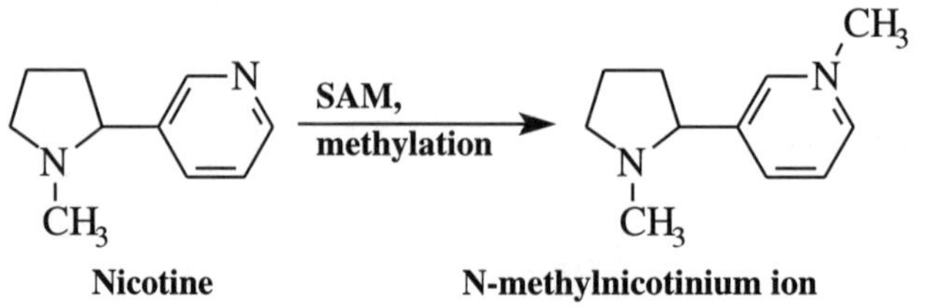

$$\text{Nicotine} \xrightarrow{\text{SAM, methylation}} \text{N-methylnicotinium ion} \quad (7.4.5)$$

Because of the hydrocarbon nature of the methyl group, it generally makes xenobiotic substrates less hydrophilic, which is the opposite of most other conjugation processes.

7.5 BIOCHEMICAL MECHANISMS OF TOXICITY

A critical aspect of toxicological chemistry is that which deals with the biochemical mechanisms and reactions by which xenobiotic compounds and their metabolites interact with biomolecules to cause an adverse toxicological effect.[6,7] The remainder of this chapter addresses the major aspects of biochemical mechanisms and processes of toxicity.

As discussed earlier in this chapter, metabolic processes make toxic agents from nontoxic ones or make toxic substances more toxic. In order to cause a toxic response, substances are often quite reactive and, if introduced into an organism directly, would react before reaching a target at which they could cause a toxic response. However, when reactive substances are produced metabolically,[8] it may be in a location where they can rapidly interact with a biomolecule, membrane, or tissue to cause a toxic response. Such agents generally fall into the following four categories:

- **Electrophilic species** that are positively charged or have a partial positive charge and therefore a tendency to bond to electron-rich atoms and functional groups, particularly N, O, and S, that abound on nucleic acids and proteins (including proteinaceous enzymes), which are commonly affected by toxic substances.
- **Nucleophilic species** that are negatively charged or partially so and have a tendency to bind with electron-deficient targets. These are much less common toxicants than electrophilic species, but include agents such as CO, formed metabolically by loss of halogen and oxidation of dihalomethane compounds or cyanide, CN^-, produced by the metabolic breakdown of acrylonitrile, a biochemically reactive organic compound containing both a –CN group and a reactive C=C bond. Carbon monoxide bonds with Fe^{2+} in hemoglobin, depriving it of its ability to carry oxygen to tissues, and nucleophilic CN^- ion bonds with Fe^{3+} in *ferricytochrome oxidase* enzyme, preventing the utilization of oxygen in respiration.
- **Free radicals** that consist of neutral or ionic species that have unpaired electrons. Free radicals include the superoxide anion radical, O_2^-, produced by adding an electron to O_2, and the hydroxyl radical, $HO\cdot$, produced by splitting (homolytic cleavage) of the H_2O_2 molecule. These species can react with larger molecules to generate other free radical species. Electron transfer from cytochrome P-450 enzyme to xenobiotic carbon tetrachloride, CCl_4, can produce the reactive, damaging $Cl_3C\cdot$ radical.
- **Redox-reactive** reagents that bring about harmful oxidation–reduction reactions. An example is the generation from nitrite esters of nitrite ion, NO_2^-, which causes oxidation of Fe^{2+} in hemoglobin to Fe^{3+}, producing methemoglobin, which does not transport oxygen in blood.

In understanding the kinds of processes by which toxic substances harm an organism, it is important to understand the concept of receptors.[9] Here a **receptor** is taken to mean a biochemical entity that interacts with a toxicant to produce some sort of toxic effect. Generally receptors are macromolecules, such as proteins, nucleic acids, or phospholipids of cell membranes, inside or on the surface of cells. In the context of toxicant–receptor interactions, the substance that interacts with a receptor is called a **ligand**. Ligands are normally relatively small molecules. They may be endogenous, such as hormone molecules, but in discussions of toxicity are normally regarded as xenobiotic materials.

The function of a receptor depends on its high specificity for particular ligands. This often involves the stereochemical fit between a ligand and a receptor, the idea of a "lock and key," similar to the interaction of enzymes with various substrates. It should be noted, however, that toxicant–receptor interactions are often around 100 times as strong as enzyme–substrate interactions. Furthermore, whereas an enzyme generally alters a substrate chemically (such as by hydrolysis), a toxicant does not usually change the chemical nature of a receptor other than binding to it. In

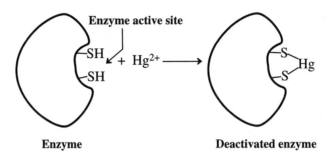

Figure 7.13 Binding of a heavy metal to an enzyme active site.

many cases, the identity of a receptor is not known, as is the case, for example, with pyrethroid insecticides. In such a case, for toxicant X, reference may be made to the X receptor.

Several major categories of toxicant–receptor interactions occur. What is known about these kinds of interactions is largely based on studies of pharmaceuticals, which act by binding with various receptors. This information is now being applied to reactions of toxicants with receptors. In considering such interactions, it may be assumed that the receptor normally binds to some endogenous substance, causing a normal effect, such as a nerve impulse. In some cases, the toxicant may activate the receptor, causing an effect similar to that of the endogenous ligand, but different enough in degree that some adverse effect results. Another possibility is that the toxicant binds to a receptor site, preventing an endogenous ligand from binding; this is known as an **antagonist action**. Yet another possibility is for the toxicant to bind to a site different from, but close enough to, the normal binding site to interfere with the binding of an endogenous substance. As a final possibility, the receptor may not have any endogenous ligands, but being bound by a toxicant nevertheless has some sort of effect.

Advantage is taken of antagonist action to treat poisoning. A simple example is provided by treatment for carbon monoxide poisoning, in which blood hemoglobin, which normally carries molecular O_2 to tissues, is the receptor that is bound strongly by CO. By treating the subject with pure oxygen or even pressurized oxygen, the oxygen competes with the receptor sites, driving off carbon monoxide and reversing the effects of this toxic substance.

7.6 INTERFERENCE WITH ENZYME ACTION

Enzymes are extremely important because they must function properly to enable essential metabolic processes to occur in cells. Substances that interfere with the proper action of enzymes obviously have the potential to be toxic. Many xenobiotics that adversely affect enzymes are **enzyme inhibitors**, which slow down or stop enzymes from performing their normal functions as biochemical catalysts. Stimulation of the body to make enzymes that serve particular purposes, a process called **enzyme induction**, is also important in toxicology.

The body contains numerous endogenous enzyme inhibitors that serve to control enzyme-catalyzed processes. When a toxicant acts as an enzyme inhibitor, however, an adverse effect usually results. An important example of this is the action of ions of heavy metals, such as mercury (Hg^{2+}), lead (Pb^{2+}), and cadmium (Cd^{2+}), which have strong tendencies to bind to sulfur-containing functional groups, especially –SS–, –SH, and –S–CH_3. These functional groups are often present on the active sites of enzymes, which, because of their specific three-dimensional structures, bind with high selectivity to the substrate species upon which the enzymes act. Toxic metal ions may bind strongly to sulfur-containing functional groups in enzyme active sites, thereby inhibiting the action of the enzyme. Such a reaction is illustrated in Figure 7.13 for Hg^{2+} ion binding to sulfhydryl groups on an enzyme active site.

7.6.1 Inhibition of Metalloenzymes

Substitution of foreign metals for the metals in metalloenzymes (those that contain metals as part of their structures) is an important mode of toxic action by metals. A common mechanism for cadmium toxicity is the substitution of this metal for zinc, a metal that is present in many metalloenzymes. This substitution occurs readily because of the chemical similarities between the two metals (for example, Cd^{2+} and Zn^{2+} behave alike in solution). Despite their chemical similarities, however, cadmium does not fulfill the biochemical function of zinc and a toxic effect results. Some enzymes that are affected adversely by the substitution of cadmium for zinc are adenosine triphosphate, alcohol dehydrogenase, and carbonic anhydrase.

7.6.2 Inhibition by Organic Compounds

The covalent bonding of organic xenobiotic compounds to enzymes, as shown in Reaction 7.6.1, can cause enzyme inhibition. Such bonding occurs most commonly through hydroxyl (–OH) groups on enzyme active sites. Covalent bonding of xenobiotic compounds is one of the major ways in which acetylcholinesterase (an

$$(C_3H_7O)_2\overset{\overset{O}{\|}}{P}-F \; + \; HO-(\text{Acetylcholinesterase}) \; \rightarrow$$
$$\text{Ligand} \quad\quad\quad \text{Receptor}$$
$$HF \; + \; (C_3H_7O)_2\overset{\overset{O}{\|}}{P}-O-(\text{Acetylcholinesterase})$$
$$\text{Modified receptor (inhibited acetylcholinesterase)}$$

(7.6.1)

enzyme crucial to the function of nerve impulses) can be inhibited. An organophosphate compound, such as the nerve gas compound diisopropylphosphorfluoridate (a reactant in Reaction 7.6.1), may bind to acetylcholinesterase, thereby inhibiting the enzyme.

7.7 BIOCHEMISTRY OF MUTAGENESIS

Mutagenesis is the phenomenon in which inheritable traits result from alterations of DNA. Although mutation is a normally occurring process that gives rise to diversity in species, most mutations are harmful. The toxicants that cause mutations are known as **mutagens**. These toxicants, often the same as those that cause cancer or birth defects, are a major toxicological concern.

To understand the biochemistry of mutagenesis, it is important to recall from Chapter 3 that DNA contains the nitrogenous bases adenine, guanine, cytosine, and thymine. The order in which these bases occur in DNA determines the nature and structure of newly produced ribonucleic acid (RNA), a substance produced as a step in the synthesis of new proteins and enzymes in cells. Exchange, addition, or deletion of any of the nitrogenous bases in DNA alters the nature of RNA produced and can change vital life processes, such as the synthesis of an important enzyme. This phenomenon, which can be caused by xenobiotic compounds, is a mutation that can be passed on to progeny, usually with detrimental results.

There are several ways in which xenobiotic species may cause mutations. It is beyond the scope of this work to discuss these mechanisms in detail. For the most part, however, mutations due to xenobiotic substances are the result of chemical alterations of DNA, such as those discussed in the following two examples.

Nitrous acid, HNO_2, is an example of a chemical mutagen that is often used to cause mutations in bacteria. To understand the mutagenic activity of nitrous acid, it should be noted that three of

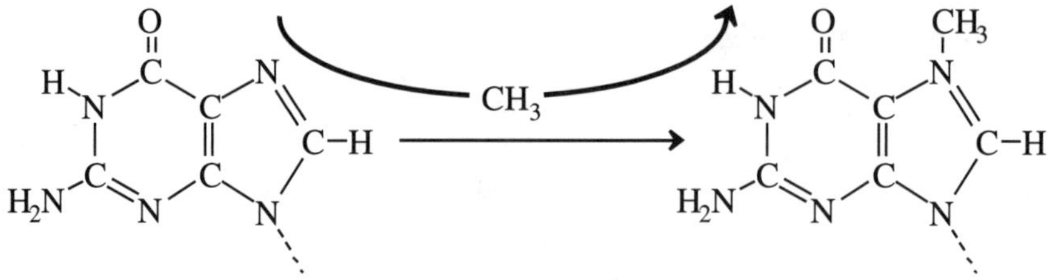

Guanine bound to DNA **Methylated guanine in DNA**

Figure 7.14 Alkylation of guanine in DNA.

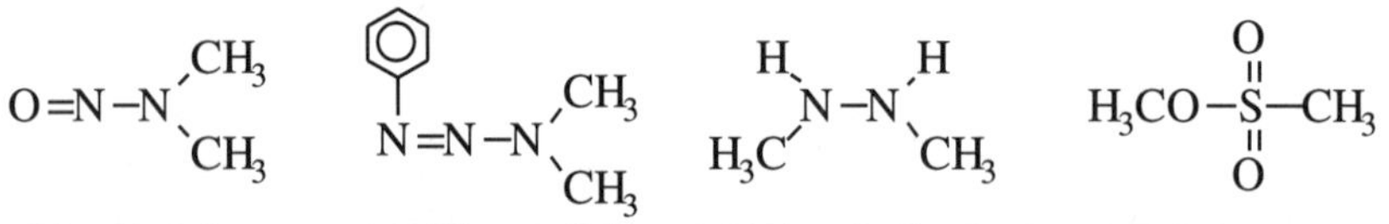

Dimethylnitrosamine	3,3-Dimethyl-1-phenyltriazine	1,2-Dimethylhydrazine	Methylmethanesulfonate

Figure 7.15 Examples of simple alkylating agents capable of causing mutations.

the nitrogenous bases — adenine, guanine, and cytosine — contain the amino group $-NH_2$. Nitrous acid acts to replace amino groups with doubly bonded oxygen atoms, thereby placing keto groups (C=O) in the rings of the nitrogenous bases and converting them to other compounds. When this occurs, the DNA may not function in the intended manner, and a mutation may occur.

Alkylation consisting of the attachment of a small alkyl group, such as $^-CH_3$ or $^-C_2H_5$, to an N atom on one of the nitrogenous bases in DNA is one of the most common mechanisms leading to mutation. The methylation of 7 nitrogen in guanine in DNA to form N^7 guanine is shown in Figure 7.14. O-alkylation may also occur by attachment of a methyl or other alkyl group to the oxygen atom in guanine. A number of mutagenic substances act as alkylating agents. Prominent among these are the compounds shown in Figure 7.15.

Alkylation occurs by way of generation of positively charged electrophilic species that bond to electron-rich nitrogen or oxygen atoms on the nitrogenous bases in DNA. The generation of such species usually occurs by way of biochemical and chemical processes. For example, dimethylnitrosamine (structural formula in Figure 7.15) is activated by oxidation through cellular NADPH (see Section 4.3) to produce the following highly reactive intermediate:

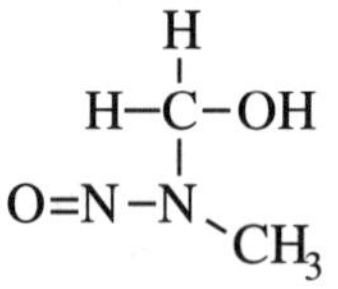

This product undergoes several nonenzymatic transitions, losing formaldehyde and generating a carbonium ion, $^+CH_3$, that can methylate nitrogenous bases on DNA:

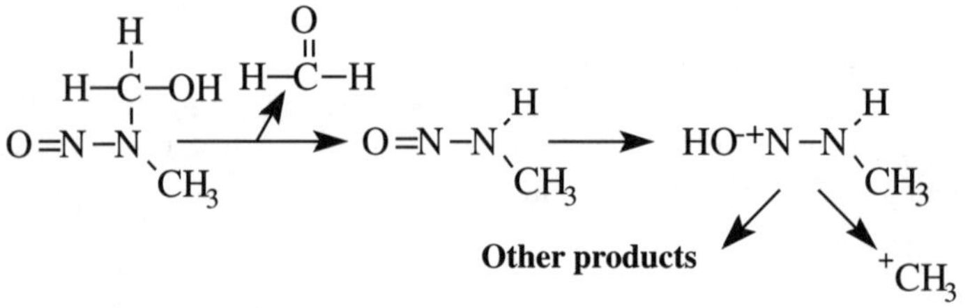

(7.7.1)

One of the more notable mutagens is tris(2,3-dibromopropyl)phosphate, commonly called tris, which was used as a flame retardant in children's sleepwear. Tris was found to be mutagenic in experimental animals, and metabolites of it were found in children wearing the treated sleepwear. This strongly suggested that tris is absorbed through the skin, and its use was discontinued.

7.8 BIOCHEMISTRY OF CARCINOGENESIS

Cancer is a condition characterized by the uncontrolled replication and growth of the body's own cells (somatic cells). **Carcinogenic agents** may be categorized as follows:

- Chemical agents, such as nitrosamines and polycyclic aromatic hydrocarbons
- Biological agents, such as hepadna viruses or retroviruses
- Ionizing radiation, such as x-rays
- Genetic factors, such as selective breeding

Clearly, in some cases, cancer is the result of the action of synthetic and naturally occurring chemicals. The role of xenobiotic chemicals in causing cancer is called **chemical carcinogenesis**.[10] It is often regarded as the single most important facet of toxicology and clearly the one that receives the most publicity.

Chemical carcinogenesis has a long history. In 1775, Sir Percivall Pott, surgeon general serving under King George III of England, observed that chimney sweeps in London had a very high incidence of cancer of the scrotum, which he related to their exposure to soot and tar from the burning of bituminous coal. (This occupational health hazard was exacerbated by their aversion to bathing and changing underwear.) A German surgeon, Ludwig Rehn, reported elevated incidences of bladder cancer in dye workers exposed to chemicals extracted from coal tar; 2-naphthylamine

2-Naphthylamine

was shown to be largely responsible. Other historical examples of carcinogenesis include observations of cancer from tobacco juice (1915), oral exposure to radium from painting luminescent watch dials (1929), tobacco smoke (1939), and asbestos (1960).

Large expenditures of time and money on the subject in recent years have yielded a much better understanding of the biochemical bases of chemical carcinogenesis. The overall processes for the induction of cancer may be quite complex, involving numerous steps. However, it is generally recognized that there are two major steps in carcinogenesis: an initiation stage followed by a promotional stage. These steps are further subdivided, as shown in Figure 7.16.

Initiation of carcinogenesis may occur by reaction of a **DNA-reactive species** with DNA or by the action of an **epigenetic carcinogen** that does not react with DNA and is carcinogenic by some other mechanism.[11] Most DNA-reactive species are **genotoxic carcinogens** because they are also mutagens. These substances react irreversibly with DNA. They are either electrophilic or, more commonly, metabolically activated to form electrophilic species, as is the case with electrophilic $^{+}CH_3$ generated from dimethylnitrosamine, as discussed under mutagenesis above. Cancer-causing substances that require metabolic activation are called **precarcinogens** or **procarcinogens**. The metabolic species actually responsible for carcinogenesis is termed an **ultimate carcinogen**. Some species that are intermediate metabolites between precarcinogens and ultimate carcinogens are called **proximate carcinogens**. These definitions can be illustrated by the species shown in Figure 7.3, in which benzo(a)pyrene is a procarcinogen, benzo(a)pyrene 7,8-epoxide is the proximate carcinogen,

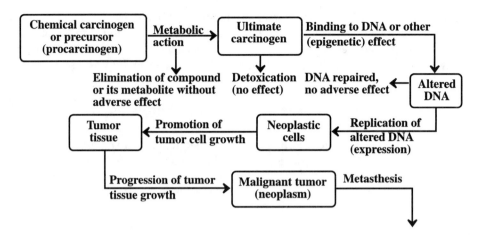

Figure 7.16 Outline of the carcinogenic process.

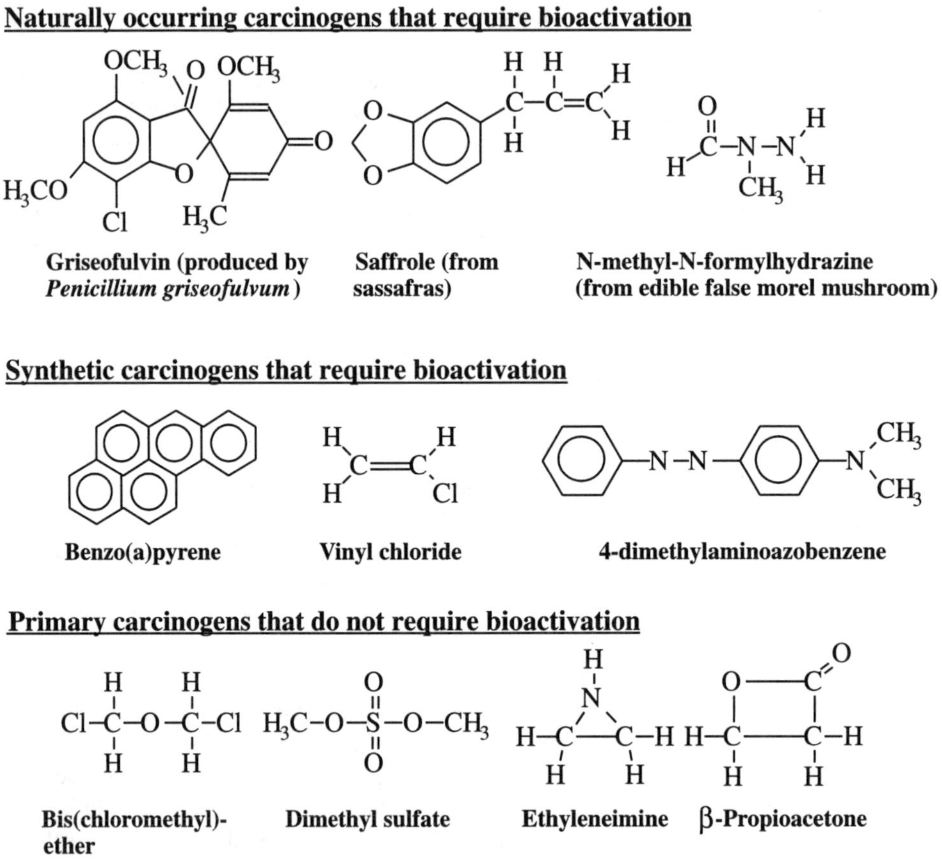

Naturally occurring carcinogens that require bioactivation

Griseofulvin (produced by *Penicillium griseofulvum*) Saffrole (from sassafras) N-methyl-N-formylhydrazine (from edible false morel mushroom)

Synthetic carcinogens that require bioactivation

Benzo(a)pyrene Vinyl chloride 4-dimethylaminoazobenzene

Primary carcinogens that do not require bioactivation

Bis(chloromethyl)-ether Dimethyl sulfate Ethyleneimine β-Propioacetone

Figure 7.17 Examples of the major classes of naturally occurring and synthetic carcinogens, some of which require bioactivation and others of which act directly.

and benzo(a)pyrene 7,8-diol-9,10-epoxide is the ultimate carcinogen. Carcinogens that do not require biochemical activation are categorized as **primary** or **direct-acting carcinogens**. Some example procarcinogens and primary carcinogens are shown in Figure 7.17.

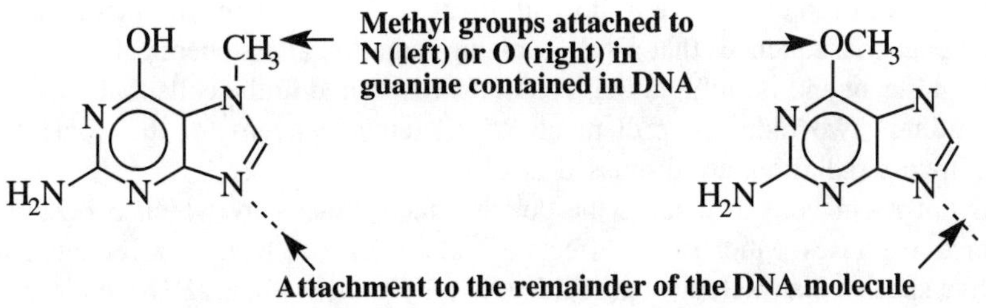

Figure 7.18 Alkylated (methylated) forms of the nitrogenous base guanine.

Most substances classified as epigenetic carcinogens are **promoters** that act after initiation. Manifestations of promotion include increased numbers of tumor cells and decreased length of time for tumors to develop (shortened latency period). Promoters do not initiate cancer, are not electrophilic, and do not bind with DNA. The classic example of a promotor is a substance known chemically as decanoyl phorbol acetate or phorbol myristate acetate, a substance extracted from croton oil.

7.8.1 Alkylating Agents in Carcinogenesis

Chemical carcinogens usually have the ability to form covalent bonds with macromolecular life molecules. Such covalent bonds can form with proteins, peptides, RNA, and DNA. Although most binding is with other kinds of molecules, which are more abundant, the DNA adducts are the significant ones in initiating cancer. Prominent among the species that bond to DNA in carcinogenesis are the alkylating agents that attach alkyl groups — such as methyl (CH_3) or ethyl (C_2H_5) — to DNA. A similar type of compound, **arylating agents**, act to attach aryl moieties, such as the phenyl group,

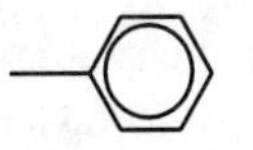

Phenyl group

to DNA. As shown by the examples in Figure 7.18, the alkyl and aryl groups become attached to N and O atoms in the nitrogenous bases that compose DNA. This alteration in the DNA can initiate the sequence of events that results in the growth and replication of neoplastic (cancerous) cells. The reactive species that donate alkyl groups in alkylation are usually formed by metabolic activation, as shown for dimethylnitrosamine in the discussion of mutagenesis above.

7.8.2 Testing for Carcinogens

In some cases, chemicals are known to be carcinogens from epidemiological studies of exposed humans. Animals are used to test for carcinogenicity, and the results can be extrapolated with some uncertainty to humans. The most broadly applicable test for potential carcinogens is the **Bruce Ames** procedure, which actually reveals mutagenicity. The principle of this method is the reversion of mutant histidine-requiring *Salmonella* bacteria back to a form that can synthesize their own histidine.[12] The test is discussed in more detail in Section 8.4.

7.9 IONIZING RADIATION

Although not a chemical agent as such, ionizing radiation, such as x-rays or alpha particles from ingested alpha emitters, causes chemical reactions that have toxic, even fatal, effects. The toxicologic effects of radiation have to do with its physical and chemical interactions with matter and the biological consequences that result. Ionizing radiation alters chemical species in tissue and can lead to significant and harmful alterations in the tissue and in the cells that make up the tissue. Radon and radium, two radioactive elements of particular concern for their potential to expose humans to ionizing radiation, are discussed in Chapter 10.

There is not room here to discuss the detailed mechanisms by which exposure to radiation causes adverse responses. Much of the effects of radiation result from its interaction with water to produce active species that include superoxide (O_2^-), hydroxyl radical (HO·), hydroperoxyl radical (HOO·), and hydrogen peroxide (H_2O_2). These species oxidize cellular macromolecules. When DNA is so affected, mutagenesis and carcinogenesis may result. Ionizing radiation can also interact with organic substances to produce a carbonium ion, such as $^+CH_3$, that can alkylate nitrogenous bases on DNA.

REFERENCES

1. Parkinson, A., Biotransformation of xenobiotics, in *Casarett and Doull's Toxicology: The Basic Science of Poisons*, 5th ed., Klaassen, C.D., Ed., McGraw-Hill, New York, 1996, chap. 6, pp. 113–186.
2. Rozman, K.K. and Klaassen, C.D., Principles of toxicology, in *Casarett and Doull's Toxicology: The Basic Science of Poisons*, 6th ed., Klaassen, C.D., Ed., McGraw-Hill, New York, 2001, chap. 5, pp. 107–132.
3. Welling, P.G. and de la Iglesia, F.A., Eds., *Drug Toxicokinetics*, Marcel Dekker, New York, 1993.
4. Hodgson, E. and Goldstein, J.A., Metabolism of toxicants: phase I reactions and pharmacogenetics, in *Introduction to Biochemical Toxicology*, 3rd ed., Hodgson, E. and Smart, R.C., Eds., Wiley-Interscience, New York, 2001, chap. 5, pp. 67–113.
5. LeBlanc, G.A. and Dauterman, W.C., Conjugation and elimination of toxicants, in *Introduction to Biochemical Toxicology*, 3rd ed., Hodgson, E. and Smart, R.C., Eds., Wiley-Interscience, New York, 2001, chap. 3, pp. 115–136.
6. Zoltán, G. and Klaassen, C.D., Mechanisms of toxicity, in *Casarett and Doull's Toxicology: The Basic Science of Poisons*, 6th ed., Klaassen, C.D., Ed., McGraw-Hill, New York, 2001, chap. 3, pp. 35–82.
7. Timbrell, J.A., *Principles of Biochemical Toxicology*, Taylor and Francis, London, 1982.
8. Hutson, D.H., Caldwell, J., and Paulson, G.D., Eds., *Intermediary Xenobiotic Metabolism in Animals: Methodology, Mechanisms, and Significance*, Taylor and Francis, London, 1989.
9. Mailman, R.B. and Lawler, C.P., Toxicant–receptor interactions: fundamental principles, in *Introduction to Biochemical Toxicology*, 3rd ed., Hodgson, E. and Smart, R.C., Eds., Wiley-Interscience, New York, 2001, chap. 12, pp. 277–308.
10. Pitot, H.C., III and Dragan, Y.P., Chemical carcinogenesis, in *Casarett and Doull's Toxicology: The Basic Science of Poisons*, 6th ed., Klaassen, C.D., Ed., McGraw-Hill, New York, 2001, chap. 3, pp. 241–320.
11. Smart, R.C. and Akunda, J.K., Carcinogenesis, in *Introduction to Biochemical Toxicology*, 3rd ed., Hodgson, E. and Smart, R.C., Eds., Wiley-Interscience, New York, 2001, chap. 15, pp. 343–396.
12. Ames, B.N., The detection of environmental mutagens and potential carcinogens, *Cancer*, 1984, chap. 53, pp. 2034–2040.

QUESTIONS AND PROBLEMS

1. Define toxicological chemistry. What is the significance of structure-activity relationships in toxicological chemistry?
2. One of the main categories of toxic substances consists of those that exhibit extremes of acidity, basicity, dehydrating ability, or oxidizing power. Give an example of a substance in each of these categories.
3. Define and distinguish the two main phases of biochemical transformations that toxicants can undergo in the body.

4. Which of the following processes is **least likely** to be a phase I reaction? epoxide formation, reduction, dealkylation, bonding with sulfate, hydrolysis, or removal of halogen
5. Explain what the following process shows in terms of a biochemical mechanism of toxicity:

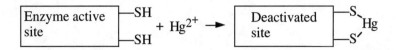

6. What are the roles of nitrogenous bases in mutagenesis? What often happens to these nitrogenous bases that results in a mutation?
7. What does the following species have to do with the mutagenicity of dimethylnitrosamine compound?

$$\begin{array}{c} H \\ | \\ H-C-OH \\ | \\ O=N-N \\ \diagdown CH_3 \end{array}$$

8. Match the following:
   (a) Procarcinogen   1. An irreversible process that does not necessarily lead to cancer
   (b) Genotoxic carcinogens 2. Do not require bioactivation
   (c) Initiation   3. The metabolic species actually responsible for carcinogenesis
   (d) Ultimate carcinogen   4. Are also mutagens
   (e) Direct-acting carcinogens 5. Require metabolic activation
9. Explain what alkylating agents have to do with carcinogenesis. Which kind of biomolecule is a receptor for alkylation?
10. What are the two main types of groups to which alkylating agents may attach in nitrogenous bases?
11. What is one of the main distinctions between vinyl chloride and dimethyl sulfate as carcinogens?
12. In what sense is the Bruce Ames test not strictly a test for carcinogenicity? Why is homogenized liver tissue used in this test?
13. List some of the main ways in which teratogens may act biochemically to cause birth defects.
14. What are two therapeutic drugs known to be potent teratogens?
15. What are some of the major clinical symptoms associated with allergy?
16. List the active species produced by ionizing radiation. How are these species related to the effects of oxidants?

CHAPTER 8

Genetic Aspects of Toxicology

8.1 INTRODUCTION

Recall from Chapter 3 that the directions for reproduction and metabolic processes in organisms are contained in *nucleic acids*, which are huge biopolymeric molecules consisting of *nucleotide* units each composed of a sugar, a nitrogenous base, and a phosphate group. There are two kinds of nucleic acids. The first of these is deoxyribonucleic acid (DNA), in which the sugar is 2-deoxyribose and the bases may be thymine, adenine, guanine, and cytosine. The second kind of nucleic acid is ribonucleic acid (RNA), in which the sugar is ribose and the bases may be adenine, guanine, cytosine, and uracil. The monomeric units of nucleic acids are summarized in Figure 8.1, and an example nucleotide is shown. A nucleic acid molecule, which typically has a molecular mass of billions, consists of many nucleotides joined together. Alternate sugar and phosphate groups compose the chain skeleton, and the nitrogenous base in each nucleotide gives it its unique identity. Since there are four possible bases for each kind of nucleic acid, the nucleic acid chain functions like a four-letter alphabet that carries a message for cell metabolism and reproduction.

As discussed in Chapter 3, the structure of DNA is that of a double helix, in which there are two complementary strands of DNA counterwound around each other. In this structure, guanine (G) is opposite cytosine (C), and adenine (A) is opposite thymine (T) in the opposing strand. The structures of these nitrogenous bases are such that hydrogen bonds form between them on the two strands, bonding the strands together. During cell division, the strands of DNA unwind and each generates a complementary copy of itself, so that each new cell has an exact duplicate of the DNA in the parent cell.

8.1.1 Chromosomes

The nuclei of eukaryotic cells contain multiply coiled DNA bound with proteins in bodies called **chromosomes**. The number of chromosomes varies with the organism. Humans have 46 chromosomes in their body cells (**somatic cells**) and 23 chromosomes in each **germ cell**, the eggs and sperm that fuse to initiate sexual reproduction. During cell division, each chromosome is duplicated and the DNA in it is said to be **replicated**. The production of duplicates of a molecule as complicated as DNA has the potential to go wrong and is a common mode of action of toxic substances. Uncontrolled cell duplication is another problem that can be caused by toxic substances and can result in the growth of cancerous tissue. This condition can be caused by exposure to some kinds of toxicants.

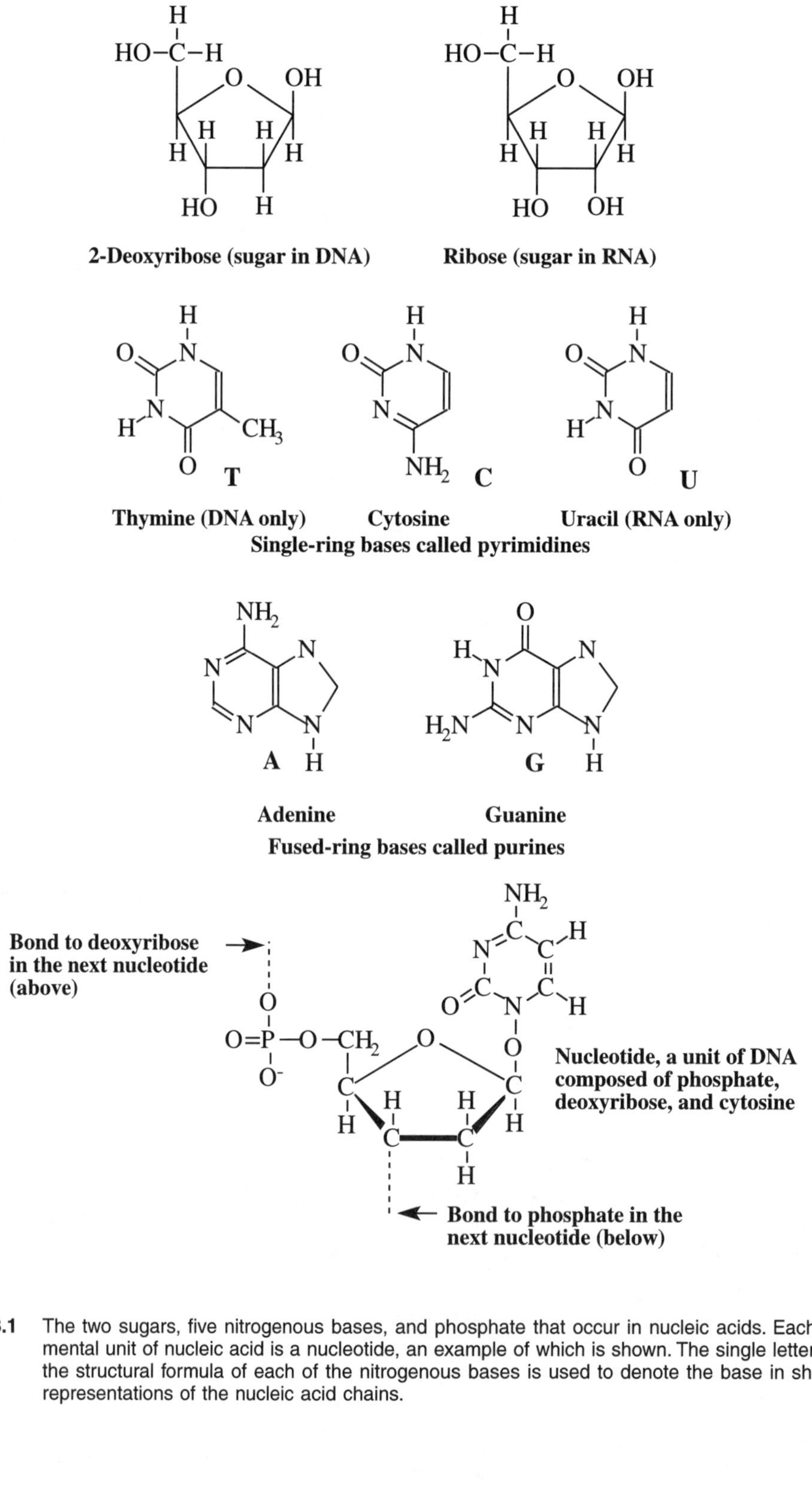

Figure 8.1 The two sugars, five nitrogenous bases, and phosphate that occur in nucleic acids. Each fundamental unit of nucleic acid is a nucleotide, an example of which is shown. The single letter beside the structural formula of each of the nitrogenous bases is used to denote the base in shorthand representations of the nucleic acid chains.

GENETIC ASPECTS OF TOXICOLOGY

8.1.2 Genes and Protein Synthesis

The basic units of heredity consist of segments of the DNA molecule composed of varying numbers of nucleotides called **genes**. Each gene gives directions for the synthesis of a particular protein, such as an essential enzyme. Cellular DNA remains in the cell nucleus, from which it sends out directions to synthesize various proteins. The first step in this process is **transcription**, in which a segment of the DNA molecule generates an RNA molecule called **messenger RNA** (mRNA). The nucleotides in a gene are arranged in active groups called **exons**, separated by inactive groups called **introns**, of which only the exons are translated during protein synthesis. In producing mRNA, adenine, thymine, cytosine, and guanine in DNA cause formation of uracil, adenine, guanine, and cytosine, respectively, in the mRNA chain. The mRNA generated by transcription travels from the nucleus to cell **ribosomes**. The mRNA attached to a ribosome operates with **transfer RNA** (tRNA) to cause the synthesis of a specific protein in a process called **translation**. Sequences of three bases on a chain of mRNA, a base triplet called a **codon**, specify a particular amino acid to be assembled on a protein. Each codon matches with a complementary sequence of amino acids, called an **anticodon**, on a tRNA molecule, each of which carries a specific amino acid to be assembled in the protein being synthesized. For example, a codon of GUA on mRNA pairs with tRNA having the anticodon CAU. The tRNA with this anticodon always carries the amino acid valine, which becomes bound in the protein chain through peptide linkages. So by matching successive codons on mRNA with the complementary anticodons on tRNA carrying specific amino acids, a protein chain with the appropriate order of amino acids is assembled.

There are 20 naturally occurring amino acids that are assembled into proteins. If codons consisted of only two base pairs, each of which could be one of four nitrogenous bases, directions could be given for only $4 \times 4 = 16$ amino acids. Using three bases per codon gives a total of $4 \times 4 \times 4 = 64$ possibilities, which is more than sufficient. This provides for some redundancies; for example, six different codons specify arginine. Codons also signal initiation and termination of a protein chain.

8.1.3 Toxicological Importance of Nucleic Acids

In discussing the toxicological importance of nucleic acids, it is useful to define two terms relating to the genetic makeup of organisms and their manifestations in organisms. The **genotype** of an individual describes the genetic constitution of that individual. It may refer to a single trait or to a set of interrelated traits. The **phenotype** of an individual consists of all of the individual's observable properties, as determined by both genetic makeup and environmental factors to which the individual has been exposed. Until relatively recently, genetic effects were largely inferred from observations of genotype, such as by observations of strange mutant offspring of fruit flies irradiated with x-rays. With the ability to perform DNA sequencing, it has become possible to determine genotypes exactly through the science of **genomics**, which gives an accurate description of the complete set of genes, called the **genome**. This capability makes possible accurate observations of the effects of toxicants on genotype.

Nucleic acids are very important in toxicology for two reasons. The first of these is that heredity as directed by DNA determines susceptibility to the effects of certain kinds of toxicants. This phenomenon makes different species respond differently to the same toxicant; for example, the LD_{50} for dioxin in hamsters is 10,000 times that in guinea pigs. In addition, differences in genotype cause substantial differences in the susceptibilities of individuals within a species to effects of toxicants.

The second reason that nucleic acids are so important in toxicology is that the intricate processes of reproduction and protein synthesis in organisms as carried out by nucleic acids can be altered in destructive ways by the effects of toxic substances. This can result in effects such as harmful mutations, uncontrolled replication of somatic cells (cancer), and the synthesis of altered proteins that do not perform a needed function in an organism.

8.2 DESTRUCTIVE GENETIC ALTERATIONS

Toxic substances and radiation can damage genetic material in three major ways: gene mutations, chromosome aberrations, and changes in the number of chromosomes.[1] Each of these has the potential to be quite damaging. They are discussed separately here.

It should be kept in mind that cellular DNA is susceptible to damage from spontaneous processes that are not caused by xenobiotic toxicants. These include hydrolysis reactions, oxidation, nonenzymatic methylation, and effects from background ionizing radiation. To cope with these insults, organisms have developed a variety of mechanisms to repair DNA. These fall into two broad categories, the first of which is **reversal**, consisting of direct repair of a damaged site (such as removal of a methyl group from a methylated DNA base, see below). The second category of coping with damage to DNA is **excision**, in which a faulty sequence of DNA bases is removed and replaced with a new segment, a process called **nucleotide excision**, or **base excision**, in which the damaged base molecule is removed and replaced with the correct one. In both cases, the remaining strand of DNA is used as a template to replace the correct complementary bases on the damaged strand.

8.2.1 Gene Mutations

When the sequence of bases in DNA is altered, a **gene mutation** (also called **point mutation**) may result. One way in which this may occur is through a **base-pair substitution**, where a base pair refers to two nitrogenous bases, one a purine and the other a pyrimidine, bonded together between two strands of DNA. If the purine–pyrimidine orientation remains the same, the alteration is called a **transition**. For example, using the abbreviations of bases given in Figure 8.1 and keeping in mind that guanine (G) always pairs with cytosine (C), whereas adenine (A) always pairs with thymine (T), switching an A:T pair on DNA with a G:C pair results in a transition. A **transversion** occurs when a purine on one strand is replaced by a pyrimidine, and on the corresponding location of the opposite strand, a pyrimidine is replaced by a purine. For example, the switch of A:T $\rightarrow$ C:G means that the purine adenine on one strand is switched with the pyrimidine cytosine on the second strand, whereas the pyrimidine thymine on the first chain is switched with the purine guanine on the second chain.

The two possible consequences of base-pair substitution are that the gene encodes for either no amino acid or the wrong amino acid. Effects can range from minor results to termination of protein synthesis.

The loss or gain of one or two base pairs in a gene causes an incorrect reading of the DNA and is known as a **frameshift mutation**. This is illustrated in Figure 8.2, which shows the insertion of a single base pair into a gene. It is seen that subsequent codons are changed, which almost always means that there are "nonsense" codons that specify no amino acid. So either no protein or a useless protein is likely to result.

8.2.2 Chromosome Structural Alterations, Aneuploidy, and Polyploidy

Chromosome structural alterations occur when genetic material is changed to such an extent that visible alterations in chromosomes are apparent under examination by light microscopy. These changes may include both breakage of chromosomes and rearrangements. In some cases, chromosome alterations can be passed on to progeny cells. Chromosomes may break during replication and then rejoin incorrectly.

Not only can there be changes in structures of chromosomes, but it is also possible to have altered numbers of them. **Aneuploidy** refers to a circumstance in which a cell has a number of

GENETIC ASPECTS OF TOXICOLOGY

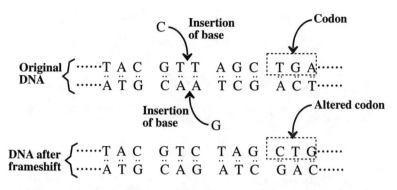

Figure 8.2 Illustration of a frameshift mutation in which a base pair is inserted into a DNA sequence, altering the codons that code for kinds of amino acids in a protein.

chromosomes differing by one to several from the normal number of chromosomes; for example, a human cell with 44 chromosomes rather than the normal 46. **Polyploidy** occurs when there is a large excess of numbers of chromosomes (such as half again as many as normal).

8.2.3 Genetic Alteration of Germ Cells and Somatic Cells

Genetic alterations or abnormalities of germ cells, some of which can be caused by toxicant exposure, can be manifested by adverse effects on progeny. The important health effects of these kinds of alterations may be appreciated by considering the kinds of human maladies that are caused by inherited recessive mutations. One such disease is cystic fibrosis, in which the clinical phenotype has thick, dry mucus in the tubes of the respiratory system such that inhaled bacterial and fungal spores cannot be cleared from the system. This results in frequent, severe infections. It is the consequence of a faulty chloride transporter membrane protein that does not properly transport Cl⁻ ion from inside cells to the outside, where they normally retain water characteristic of healthy mucus. The faulty transporter protein is the result of a change of a *single amino acid* in the protein.

Genetic alteration of somatic cells, which may also occur by the action of toxicants, is most commonly associated with cancer, the uncontrolled replication of somatic cells. Replication and growth of cells is a normal and essential biological process. However, there is a fine balance between a required rate of cell proliferation and the uncontrolled replication characteristic of cancer, that is, between the promotion and restriction of cell growth. The transformation of normal cells to cancer cells results from the excessive growth-stimulating activity of **oncogenes**, which are produced from genes called **proto-oncogenes** that promote normal cell growth.

The body has defensive mechanisms against the development of cancer in the form of **tumor suppressor genes**. Whereas the activation of oncogenes can cause cancer to develop, the inactivation of tumor suppressor genes disables the normal mechanisms that prevent cancerous cells from developing. Both the activation of oncogenes and the inactivation of tumor suppressor genes contribute to the development of many kinds of cancer.

Gene mutations, chromosome structural alterations, and aneuploidy may all be involved in the development of cancer. These effects are involved in the initiation of cancer (altered DNA, see Figure 7.16). However, they may also be involved in the progression of cancer through genetic effects such as damage to tumor suppressor genes.

8.3 TOXICANT DAMAGE TO DNA

Toxicants can cause destructive alteration of DNA, specifically the nitrogenous bases on the DNA nucleotides. There are three ways in which this may occur. One of these is **oxidative**

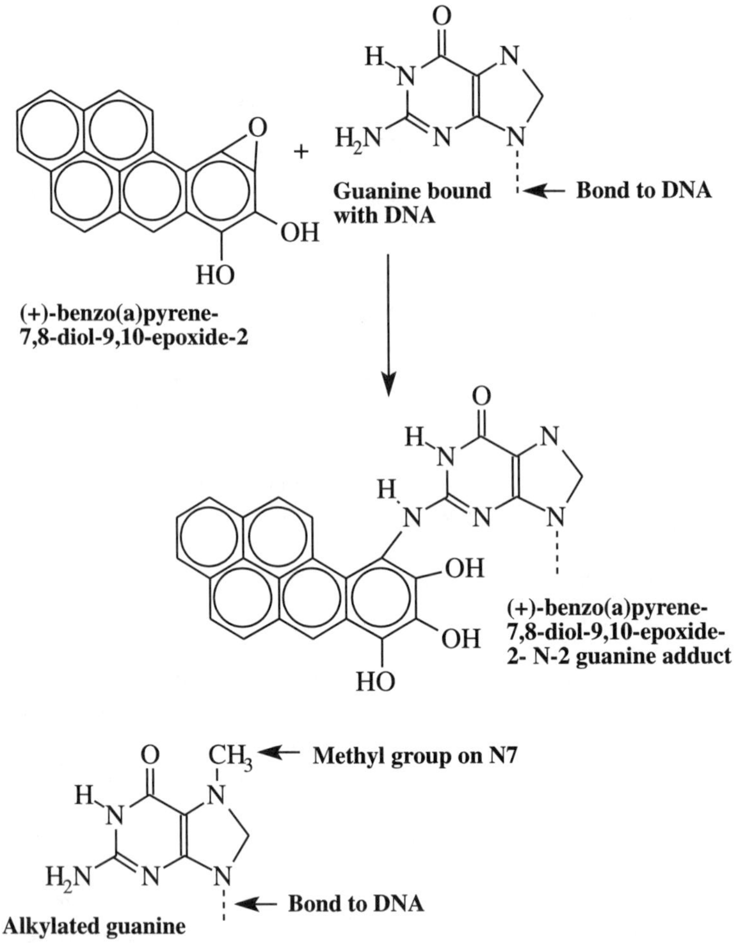

Figure 8.3 Formation of the bulky guanine adduct of (+)-benzo(a)pyrene-7,8-diol-9,10-epoxide-2.

alteration, in which a functional group on a base is oxidized. The other two modes of damage are by binding of electrophilic molecules or molecular fragments to the electron-rich N and O atoms on the bases to form **DNA adducts**. There are two major kinds of such adducts. One kind is produced by **alkylating agents** that add methyl (–CH$_3$) groups or other alkyl groups to bases. The other kind of adduct is that in which a **large bulky group** is attached.

The attachment of a methyl group to guanine in DNA is shown in Figure 7.14. This is an alkylation reaction in which the small methyl group is attached. The attachment of a large bulky group is illustrated by the binding to guanine of benzo(a)pyrene-7,8-diol-9,10-epoxide, a substance formed by the epoxidation of the polycyclic aromatic hydrocarbon benzo(a)pyrene, followed by hydroxylation and a second epoxidation (see Figure 7.3). There are actually four stereoisomers of this compound, depending on the orientations of the epoxide group and the two hydroxide groups above or below the plane of the molecule. Only one of these stereoisomers, designated (+)-benzo(a)pyrene-7,8-diol-9,10-epoxide-2, is active in binding to guanine to initiate cancer. The binding of this substance to guanine is shown in Figure 8.3.

A major effect of binding of a base on DNA can be altered pairing as the DNA replicates. For example, the normal pairing of guanine is with cytosine, a G:C pair. Guanine to which an alkyl group has been attached to oxygen may pair with thymine, which subsequently pairs with adenine during cell replication. This leads to a G:C → A:T transition, hence to altered DNA, which may initiate cancer.

Another effect on DNA can result when alkylated bases are lost from the DNA polymer. For example, guanine alkylated in the N^7 position has a much weakened bond to DNA and may split off from the DNA molecule:

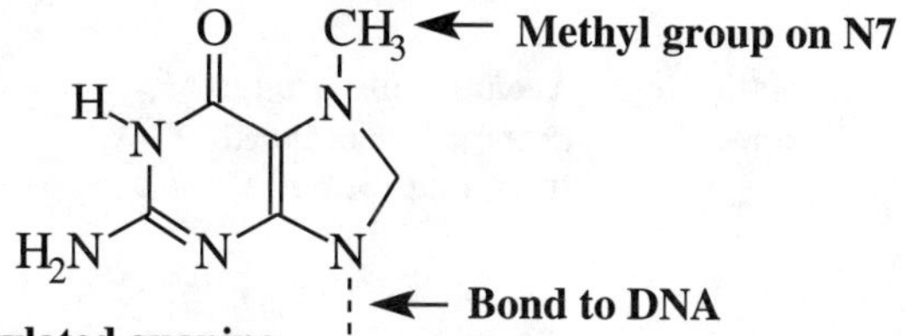

Alkylated guanine — Methyl group on N7 — Bond to DNA

This leaves an AP site (where AP stands for apurinic or apyrimidinic). This site may become occupied by a different base, leading again to alteration of DNA.

The DNA alterations described above have involved covalent bonding of groups to nitrogenous bases. Another type of interaction is possible with highly planar (flat) molecules that are able to fit between base pairs (somewhat like slipping a sheet of paper between pages of a book), a phenomenon called **intercalation**. This can cause deletion or addition of base pairs, leading to mutation and cancer. A compound known to cause this phenomenon is 9-aminoacridine:

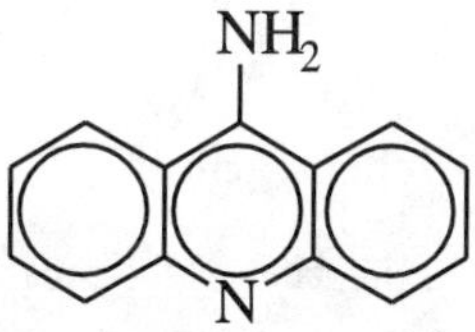

9-Aminoacridine

8.4 PREDICTING AND TESTING FOR GENOTOXIC SUBSTANCES

The ability to predict and test for genotoxic substances is important in preventing exposure to these substances. One way in which this is done is by the use of *structure-activity relationships* (see Section 7.1). Several classes of chemicals are now recognized as being potentially genotoxic (mutagenic) based on their structural features.[2] These are summarized in Figure 8.4. The single most important indicator of potential mutagenicity of a compound is electrophilic functionality showing a tendency to react with nucleophilic sites on DNA bases. Steric hindrance of the electrophilic functionalities may reduce the likelihood of reacting with DNA bases. Some substances do not react with DNA directly, but generate species that may do so. Compounds that generate reactive free radicals fall into this category.

8.4.1 Tests for Mutagenic Effects

In addition to structure-activity relationships, dozens of useful tests have been developed for mutagenicity to germ cells and somatic cells and inferred carcinogenicity. The most straightforward means of testing for effects on DNA is an examination of DNA itself. This is normally difficult to do, so indirect tests are used. One useful test measures the activity of DNA repair mechanisms (unscheduled DNA synthesis); a higher activity is indicative of prior damage to DNA.

Commonly used tests for mutagenic effects are most effective in revealing gene mutations and chromosome aberrations. Mammals, especially laboratory mice and rats, have long been used for these tests. As sophistication in cell culture has developed, mammalian cells have come into widespread use for genotoxicity testing. Insects and plants have been used, as well as bacteria, fungi, and viruses. Tests on insects favor *Drosophila* (fruit flies), on which much of the pioneering

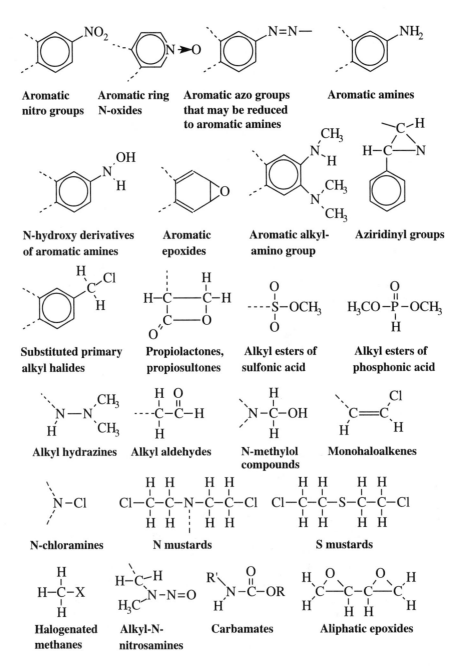

Figure 8.4 Functionalities commonly associated with genotoxicity and mutagenicity. These groups are used in structure-activity relationships to alert for possible carcinogenic substances.

studies of basic genetics were performed. For reasons of speed, simplicity, and low cost, tests on microorganisms and cell cultures are favored.

Microorganisms used in genetic testing may consist of wild-type microorganisms that have not been preselected for a particular mutation and mutant microorganisms that have a readily identifiable characteristic, such as an inability to make a particular amino acid. These classes of microorganisms give rise to two general categories of mutagenicity tests based on observation of phenotypes (offspring after exposure to the potential mutagen). The first of these involves **forward mutations**, in which the organism loses a gene function that can be observed in the phenotype. The second type of test entails **back mutation (reversion)**, in which the function of a gene is restored to a mutant. Testing of cultured mammalian cells usually involves forward mutations that confer resistance of the cells to a toxicant, that is, some of the cells exposed to the test compound reproduce in the presence of another substance that is normally toxic to the cells. Testing with microorganisms favors reversion with restoration of a gene function that has been lost in a previous mutation through which the test microorganisms were developed. Microbial tests are particularly useful for changes that occur at low frequencies because of the large number of test organisms that can be exposed to a potential mutagen.

8.4.2 The Bruce Ames Test and Related Tests

The most widely used test for mutagenicity is the Bruce Ames test, named after the biochemist who developed it. A number of variations and improvements of this test have evolved since it was first published. The Bruce Ames test and related ones make use of **auxotrophs**, mutant microorganisms that require a particular kind of nutrient and will not grow on a medium missing the nutrient, unless they have mutated back to the wild type. The Bruce Ames test uses bacterial *Salmonella typhimurium* that cannot synthesize the essential amino acid histidine and do not normally grow on histidine-free media. The bacteria are inoculated onto a medium that does not contain histidine, and those that mutate back to a form that can synthesize histidine establish colonies, which are assayed on the growth medium, thereby providing both a qualitative and quantitative indication of mutagenicity. The test chemicals are mixed with homogenized liver tissue to simulate the body's alteration of chemicals (conversion of procarcinogens to ultimate carcinogens). Up to 90% correlation has been found between mutagenesis on this test and known carcinogenicity of test chemicals.

8.4.3 Cytogenetic Assays

Cytogenetic assays use microscopic examination of cells for the observation of damage to chromosomes by genotoxic substances. These tests are based on the cellular **karyotype**, that is, the number of chromosomes, their sizes, and their types. The standard test cell for cytogenetic testing is the Chinese hamster ovary cell. In addition to a well-defined karyotype, these cells have the desired characteristics of a low number of large chromosomes and a short generation time. In order to test a substance, the cells have to be exposed to it at a suitable part of the cell cycle and examined after the first mitotic division. (Mitosis refers to the process by which the nucleus of a eukaryotic cell divides to form two daughter nuclei.) This means that the examination is performed on cells in the metaphase of nuclear division, in which the chromosomes are conducive to microscopic examination and abnormalities are most apparent. Abnormalities in the chromosomes are then scored systematically as a measure of the effects of the test subsance. A complication in these assays can be the requirement to use such high doses of a test substance that it is toxic to the cell in general, resulting in chromosomal aberrations that may not be due to specific genotoxicity.

In addition to performing cytogenetic assays on cell cultures, it is often desirable to perform **in vivo cytogenetic assays** consisting of microscopic examination of cells of whole animals — most commonly mice, rats, and Chinese hamsters — that have been exposed to toxicants. Bone marrow cells are commonly used because they are abundant and replicate rapidly. A disadvantage to in vivo cytogenetic assays is that the system is much less controlled than in assays on cell cultures. The major advantage is that the test substance has had the opportunity to be metabolized (which can produce a more genotoxic metabolite), and normal processes such as DNA repair can occur.

8.4.4 Transgenic Test Organisms

As discussed above, in vivo assays reproduce the metabolic and other processes that a xenobiotic substance undergoes in an organism. However, microbial systems are much simpler and more straightforward to detect mutations. A clever approach to combining these two techniques makes use of transgenic recombinant DNA techniques to introduce bacterial genes into test animals for chemical testing, and then transfers the genes back to bacteria for assay of mutagenic effects. Genes most commonly used for this purpose are the *lac* genes from *Escherichia coli* bacteria.[3] These genes are involved with the expression of the *β-galactosidase* lactose-metabolizing enzymes, which consist of three proteins. Either the *lacI* genes, which suppress formation of the enzymes, or the *lacZ* genes, which allow formation of the enzymes, may be used. When *lacI* genes are used that are inserted transgenically into the test mouse (known by the rather picturesque brand name of Big Blue Mouse), the mouse is treated with potential mutagen for a sufficient time to allow for mutant expression. Samples are then collected from various tissues of the mouse. The segment of DNA involved with the *lacI* genes is then extracted from these samples and put back into *Escherichia coli* bacteria, which are grown in an appropriate medium containing lactose. The bacteria with unaltered *lacI* genes (*lacI*$^+$) do not produce *β-galactosidase*, whereas the mutants (*lacI*$^-$) do produce *β-galactosidase*. Another kind of mouse (brand name MutaMouse) has been used that contains *lacZ* genes that encode for expression of *β-galactosidase*. In this case, the procedure is exactly the same, except that the nonmutants (*lacZ*$^+$) produce *β-galactosidase* and the mutants (*lacZ*$^-$) do not produce it.

One reason for the popularity of this test is the facile detection of *β-galactosidase* activity. This is accomplished with the chromogenic substrate 5-bromo-4-chloro-3-indoyl-*β*-D-galactopyranoside, which is metabolized by *β-galactosidase* to form a blue product. Therefore, when colonies of the *Escherichia coli* bacteria are grown in an assay, the *lac*$^+$ colonies are blue and the *lac*$^-$ colonies are white.

Despite the rather involved nature of the *lac* test described above, it has several very important advantages. The simplicity of assaying microorganisms is one advantage. The fact that the potential mutagens act within a complex organism (the mouse) where they are subject to a full array of absorption, distribution, metabolism, and excretion processes is another advantage. Finally, the procedure allows sampling from specific tissues, such as liver or kidney tissue.

8.5 GENETIC SUSCEPTIBILITIES AND RESISTANCE TO TOXICANTS

The discussion in this chapter so far has focused on the toxicological implications of damage to DNA by toxic agents. However, the genetic implications of toxicology are much broader than damage to DNA because of the strong influence of genetic makeup on susceptibility and resistance to toxicants. It is known that susceptibility to certain kinds of cancers is influenced by genetic makeup. In Section 8.2, mention was made of **oncogenes**, associated with the development of cancer, and **tumor suppressor genes**, which confer resistance to cancer. Susceptibility to certain kinds of cancers, some of which are potentially initiated by toxicants, clearly have a genetic component. Breast cancer is a prime example in that women whose close relatives (mother, sisters) have developed breast cancer have a much higher susceptibility to this disease, to the extent that some women have had prophylactic removal of breast tissue based on the occurrence of this disease in close relatives. It is now possible to run genetic tests for two common gene mutations, BRCA1 and BRCA2, that indicate a much increased susceptibility to breast cancer.

Another obvious genetic aspect of toxicology has to do with the level in skin of **melanin**, a pigment that makes skin dark. Melanin levels vary widely with genotype. Melanin confers resistance to the effects of solar ultraviolet radiation, which is absorbed by DNA in skin cells, causing damage that in the worst-case results in deadly melanoma skin cancer. Skin melanin is a chromophore (a substance that selectively absorbs light and ultraviolet radiation) that absorbs visible light and, more importantly, ultraviolet radiation in the UVB wavelength region of 290 to 320 nm. Melanin's presence confers resistance to sunburn and other toxic effects of ultraviolet radiation.

Genetic susceptibilities exist to the chemically induced adverse effects of ultraviolet radiation and visible light, a condition known as **photosensitivity**. **Porphyria**, an abnormal extreme sensitivity to sunlight, can result from chemical exposure in genetically susceptible individuals. Lupus erythematosus, a heritable disease manifested by red, scaly skin patches, is characterized by abnormal sensitivity to ultraviolet radiation. Porphyrias in genetically susceptible individuals, which can be induced by chemicals such as hexachlorobenzene and dioxin, occur through the malfunction of enzymes involved in producing the porphyrin heme used in hemoglobin. This results in the accumulation of porphyrin precursors in the skin. Exposed to ultraviolet light at 400 to 410 nm, these precursors reach excited states (see Chapter 2), which may generate damaging free radicals through interaction with cellular macromolecules and O_2. Phototoxicity can also be caused by xenobiotic substances either applied to skin or distributed systemically. Photoallergy is a condition in which exposure to a xenobiotic substance, either through application to skin or systemically, results in sensitization to ultraviolet radiation.

Although many smokers develop lung emphysema (see Section 9.2) with age, some do so extremely early, suggesting a genetic susceptibility to this malady. It is now believed that early onset of emphysema occurs with a rare mutation that prevents production of the protein $alpha_1$-antiprotease in the lungs. In normal individuals this substance retards the protein-digesting activity of elastase enzyme. The elastin protein that constitutes elastic tissue in the lung is readily destroyed by the action of elastin enzyme in the lung in individuals with the mutation that does not allow for generation of $alpha_1$-antiprotease. This allows for the loss of lung elasticity characteristic of emphysema at a very early stage of smoking.

Since the early 1940s, it has been known that there is a genetic predisposition to allergic contact dermatitis, a skin condition that is one of the most common maladies caused by workplace exposure to xenobiotics and to cosmetics (see Section 9.3). A study published in 1993 revealed that some individuals have a genetic predisposition to produce human leukocyte (white blood cell) antigen, resulting in allergy to nickel, chromium, and cobalt.[4]

8.6 TOXICOGENOMICS

Genomics was mentioned in Section 8.1 as the science dealing with a description of all the genes in an organism, its genome. Because of the known relationship of gene characteristics to disease, the decision was made in the mid-1980s to map all the genes in the human body. This collective body of genes is called the human genome, and the project to map it is called the Human Genome Project. The original impetus for this project in the U.S. arose because of interest in the damage to human DNA by radiation, such as that from nuclear weapons. But from the beginning it was recognized that the project had enormous commercial potential, especially in the pharmaceutical industry, and could be very valuable in human health.

The sequencing of the human genome has been done on individual chromosomes. Each chromosome consists of about 50 million base pairs. However, it is possible to sequence only about 500 to 800 base pairs at one time, so the DNA has to be broken into segments for sequencing. There are two approaches to doing this. The publicly funded consortium working on the Human Genome Project identified short marker sequences on the DNA that could be recognized in reassembling the information from the sequencing. The private concern involved in the effort used a process in which the DNA was broken randomly into fragments, each of which was sequenced. The data from the sequencing were then analyzed using powerful computer programs to show overlap, and the complete gene sequence was then assembled.

In 2001, a joint announcement from the parties involved in the Human Genome Project revealed that the genome had been sequenced. Details remain to be worked out, but the feasibility of the project has been demonstrated. This accomplishment is leading to a vast effort to understand genetically based diseases in humans, to develop pharmaceutical agents based on genetic information, and to develop other areas that can use information about the genome. The benefits and consequences of mapping the human genome will be felt for many decades to come.

The mapping of the human genone, as well as those of other organisms, has enormous potential consequences for toxicology. This has given rise to the science of toxicogenomics, which relates toxicity and the toxicological chemistry of toxicants to genomes at the molecular level.[5] More broadly, toxicogenetics relates genetic variations of subjects in their response to toxicants. Toxicogenomics has the potential to revolutionize understanding of toxic substances, how they act, and how to develop effective antidotes to them. Techniques are being developed to examine at the molecular level the interaction of specific toxicants and their metabolites with genes, even including genetic material such as DNA arrays printed on plates.

It may be anticipated that much of what will eventually be learned about toxicogenomics will be based on knowledge acquired through the science of pharmacogenomics, in which genetic variabilities to pharmaceuticals are determined in an effort to develop much more effective, sharply focused drugs.[6] Such variations arise from differences in targets or receptors (see Section 6.10 for a discussion of receptors and toxic substances) and differences in drug-metabolizing enzymes. Pharmacogenomics applies to both pharmacokinetics, which is how an organism processes a pharmaceutical agent, and pharmacodynamics, which is how the agent affects a target in an organism or a disease against which the agent acts. By analogy, toxicogenomics can be applied to toxicokinetics, the metabolism of a toxic agent, and toxicodynamics, its effect on a target.

8.6.1 Genetic Susceptibility to Toxic Effects of Pharmaceuticals

Pharmaceuticals have provided numerous examples of genetic susceptibilities to toxicants. This is because major pharmaceutical drugs are given to hundreds of thousands, or even millions, of people so that genetic defects that result in toxic effects will show up even if only very small fractions of the population (estimated to be 1 in 10,000 or less for cases of toxic effects to the liver) are genetically predisposed to adverse effects. Unfortunately, at such low levels of occurrence, there is as of yet no good way to predict such rare adverse effects in advance.

An example of genetic susceptibility to toxic effects of a drug is provided by mercaptopurine drugs, such as 6-mercaptopurine, used as antitumor agents. The active forms of these drugs are the methylated metabolites, as shown for the methylation of 6-mercaptopurine in Reaction 8.6.1:

$$\text{6-mercaptopurine} \xrightarrow[\text{SAM cofactor}]{\text{Thiopurine S-methyltransferase}} \text{6-methylmercaptopurine} \tag{8.6.1}$$

The methylation reaction occurs by the action of thiopurine S-methyltransferase enzyme with the S-adenosylmethionine (SAM) cofactor, discussed as a methylating agent in Section 7.4. In some children, the gene responsible for making the enzyme is mutated, the enzyme is not synthesized, and toxic effects occur due to the accumulation of 6-mercaptopurine. A knowledge of this genetic condition prior to treatment could prevent this toxic effect. In general, genetic screening for adverse drug reactions could be very helpful in increasing the safety of medical treatment.

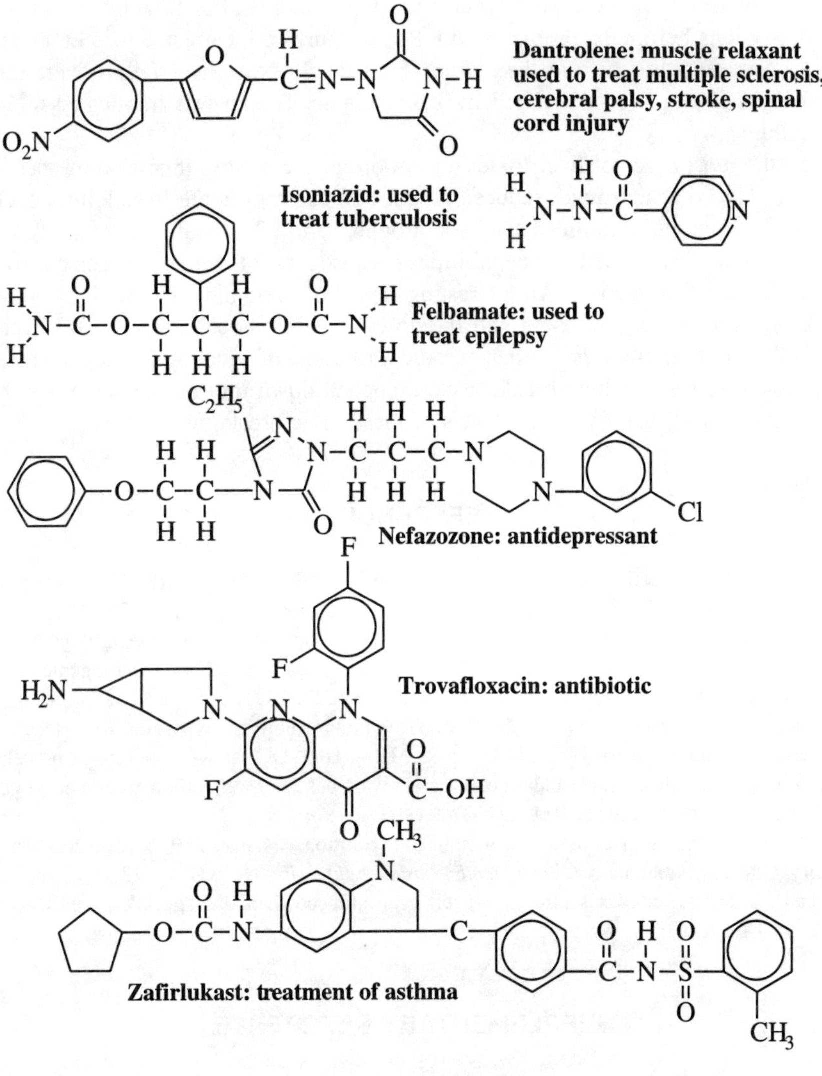

GENETIC ASPECTS OF TOXICOLOGY

Figure 8.5 Examples of pharmaceuticals that have caused liver damage. Instances of hepatotoxicity have been rare, suggesting a genetic susceptibility in some cases.

The most common adverse effect of drugs in genetically susceptible individuals is hepatotoxicity (toxic effects to the liver). The reason that the liver is damaged in these cases is that it is the first major organ to process substances taken orally and has an abundance of a wide variety of active enzymes that metabolize drugs (or fail to do so, as is the case with 6-mercaptopurine, discussed above). Although total liver failure from toxic side effects of pharmaceuticals is rare (only slightly more than 2000 cases per year in the U.S.), somewhat more than half the cases seen at liver transplant centers are caused by drugs.[7] However, far more people are afflicted with liver disease as the result of taking prescribed drugs.

Many drugs have been implicated in liver damage to relatively few individuals, many of whom probably have a genetic susceptibility to adverse effects from the drugs. Several examples of a number of drugs implicated in hepatotoxicity in rare cases are shown in Figure 8.5.

One of the most prominent examples of a drug that caused liver failure in a small percentage of genetically susceptible people is Rezulin. This oral diabetes drug was approved for use in 1997 and rapidly became very popular. However, within three years it had been implicated in 90 cases of liver failure, of which 63 were fatal. This led to the withdrawal of Rezulin from the market in March 2000. Not long before the problems with Rezulin surfaced, Duract, a pain killer, and Trovan, an antibiotic, were withdrawn from the market. Four deaths from liver failure were attributed to Duract, and eight other patients required liver transplants. Trovan was implicated in 14 cases of acute liver failure.

Arguably, the most cases of liver toxicity from drugs occur from ingestion of acetaminophen, a widely used pain killer and fever reducer. About 800 cases of acute liver failure each year are attributed to this drug, and fatalities have been around 80 to 90 annually. Most of these cases are probably not attributable to genetic susceptibilities because most of them have been due to accidental or intentional (suicidal) overdoses. An interesting aspect of acetaminophen toxicity is that chronic heavy drinkers can tolerate only about half the dose of acetaminophen that causes liver failure in nondrinkers. The reason for this is that chronic ingestion of alcohol increases the activity of cytochrome P-450 enzyme, which breaks acetaminophen down into products that can be toxic to the liver, leading to significantly higher levels of these toxic breakdown products.

REFERENCES

1. Hoffmann, G.B., Genetic toxicology, in *Casarett and Doull's Toxicology: The Basic Science of Poisons*, 5th ed., Klaassen, C.D., Ed., McGraw-Hill, New York, 1996, chap. 9, pp. 269–300.
2. Tennant, R.W. and Ashby, J., Classification according to chemical structure, mutagenicity to *Salmonella* and level of carcinogenicity of a further 39 chemicals tested for carcinogenicity by the U.S. National Toxicology Program, *Mutat. Res.*, 257, 209–227, 1991.
3. Josephy, P.D., The *Escherichia coli LacZ* reversion mutagenicity assay, *Mutat. Res.*, 455, 71–80, 2000.
4. Emtestam, L., Zetterqulist, H., and Olerup, O., HLA-DR, -DQ, and -DP alleles in nickel, chromium and/or cobalt-sensitive individuals: genomic analysis based on restriction fragment length polymorphisms, *J. Invest. Dermatol.*, 100, 271–274, 1993.
5. Lovett, R.L., Toxicologists brace for genomics revolution, *Science*, 289, 53–57, 2000.
6. Henry, C.M., Pharmacogenomics, *Chemical and Engineering News*, Aug. 13, 2001, pp. 37–42.
7. Tarkan, L., F.D.A. increases efforts to avert drug-induced liver damage, *New York Times*, Aug. 14, 2001, p. D5.

SUPPLEMENTARY REFERENCE

Choy, W.N., *Genetic Toxicology and Cancer Risk Assessment*, Marcel Dekker, New York, 2001.

QUESTIONS AND PROBLEMS

1. What is the basic structure of chromosomes?
2. Match the following:
   1. Genome   (a) Manifested by a large excess of numbers of chromosomes
   2. Phenotype   (b) Description of the genetic constitution of that individual
   3. Genotype   (c) All observable properties of an individual as determined by both its genetic makeup and environmental factors to which it has been exposed.
   4. Polyploidy (d) An accurate description of the complete set of genes
3. What are the meanings of reversal and excision as applied to damage to DNA?
4. What is the toxicological significance of chromosome structural alterations? How are such alterations observed?
5. What may happen when oncogenes are activated? What kind of gene may be involved in the opposite phenomenon?

6. What is the significance of oxidative alteration, alkylating agents, large bulky groups, and intercalation in respect to damage to DNA?
7. How are structure-activity relationships utilized in testing for genotoxic substances?
8. What is the difference between observation of forward mutations and reversions in evaluating genotoxic substances?
9. What is the significance of the Bruce Ames test and how does it operate?
10. What are the genetic aspects of melanin as related to DNA damage?
11. What is toxicogenomics and how does this science related to toxicological chemistry? How does this science relate to toxicokinetics and toxicodynamics.
12. How do studies of the toxic effects of pharmaceuticals relate to genetic aspects of toxic substances?

CHAPTER 9

Toxic Responses

9.1 INTRODUCTION

Toxicants can affect any type of tissue in any organ. Different classes of toxicants affect various tissues to varying degrees and in a variety of ways, depending on the nature of the toxicant, the kind of receptor that it attacks, the nature of the binding of the toxicant to the receptor, and the routes, transport, and metabolism of the toxicant in the organism.

Before considering chemical classes of toxicants, it is useful to survey the major toxic responses and effects. It is obviously beyond the scope of a single chapter to discuss the whole broad array of toxic responses in all kinds of organisms. Instead, this chapter divides these responses on the basis of responses in the major tissues and organs of mammals, particularly humans. The systems considered are the respiratory system, skin, liver, blood and cardiovascular system, immune system, endocrine system, nervous system, reproductive system, and kidney and bladder. This order is somewhat parallel to the major pathways of exposure, transport, and elimination of toxicants in the human body. As shown in Figure 6.2, toxicants can be inhaled through the respiratory system or absorbed through the skin. Those that are ingested through the digestive system normally pass through the liver. Systemic toxicants are carried by blood and through the lymph system to various organs and can affect the endocrine system, nervous system, and reproductive system. Finally, the kidney and urinary tract constitute the main route of elimination of the metabolites of systemic toxicants from the body.[1]

A common result of exposure of tissue to toxicants is cell death. In **cell necrosis**, a cell dies, swells, and disintegrates. The cell membrane and membranes of cell organelles break up, and a large amount of cell debris is released at the site of cell death. These fragments induce inflammatory action of other cells, making the injury worse. In addition to a toxic effect that may have caused the cell death in the first place, secondary processes caused by the presence of soil debris make the injury much worse.

In contrast to cell necrosis, **apoptosis** is a process by which cells are systematically destroyed and removed. The nuclear body and organelles within the cell become enclosed in a membrane to form residues called **apoptotic bodies**. These are then said to be phagocytosed as they are enclosed within membranes and actively removed from the site of the injury. Apoptosis is an active process for eliminating dead cellular material.

Many of the detrimental effects of toxicants have been observed with pharmaceutical agents. These are often administered to millions of people so that even a miniscule fraction of subjects who have adverse reactions will be observed, medical supervision of subjects is usually rather thorough, and cause and effect is relatively easy to establish. On August 8, 2001, the U.S. Food and Drug Administration announced a voluntary recall of Bayer Pharmaceutical's cholesterol-lowering drug Baycol (also called Lipobay). Only recently approved and used by about 700,000 people in the U.S., Baycol was implicated in 52 deaths due to rhabdomyolysis. This acute and sometimes fatal disease is first manifested by pain in the muscles of the calves and lower back. The cells in muscles can break down, leading to symptoms of fever, fatigue, and nausea. As part of the muscle cell breakdown, phosphokinase and myoglobin are released into the bloodstream, eventually causing kidney failure. Virtually all of the fatalities occurred in patients who were simultaneously taking another cholesterol-lowering drug, gemfibrozil.[2]

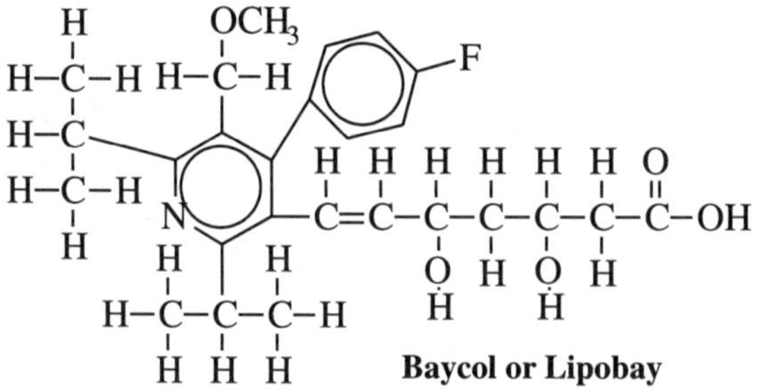

Baycol or Lipobay

Another even more widely used pharmaceutical that has been linked indirectly to numerous deaths is sildenafil citrate, marketed as the impotence drug Viagra. Used by millions during the few years since it was introduced, numerous deaths have been reported among men taking the drug along with nitroglycerin or nitrates administered for the treatment of heart disease. It is believed that the combination of these drugs with Viagra reduce blood flow through damaged coronary arteries, causing fatal heart attacks in some cases.

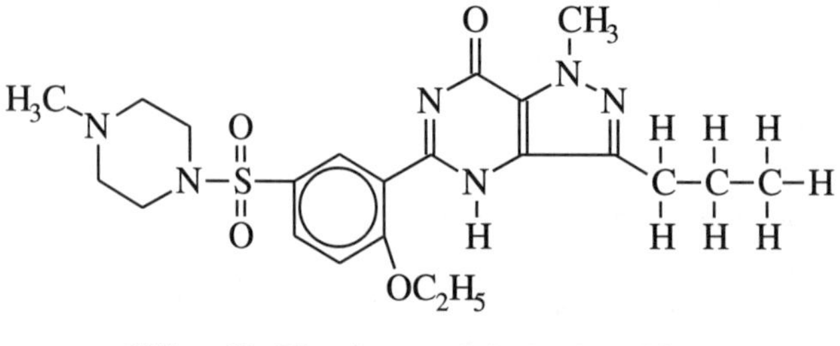

Sildenafil. The citrate salt is the drug Viagra, widely used to treat male sexual disfunction.

9.2 RESPIRATORY SYSTEM

In discussing toxic responses of the respiratory system, it is important to make a distinction between the respiratory system as an entryway for toxic substances, referred to as **inhalation toxicology**, and the respiratory system adversely affected by toxicants, referred to as **respiratory tract toxicology**. The term **pulmonary** refers to lungs. Generally, toxicants that adversely affect the respiratory system are those that have been inhaled, including such well-known substances as asbestos, chromate, and silica. However, it is possible for systemic poisons transported from elsewhere in the body to act as respiratory system toxicants. One interesting possibility that has been suggested is lung cancer caused by the diol epoxide of benzo(a)pyrene (see Figure 7.3), which is formed by inhalation of benzo(a)pyrene, converted to the ultimate carcinogen in the liver, and transported back to the lung, where cancer develops.

A highly simplified outline of the respiratory system is shown in Figure 6.4. The function of the respiratory system is to exchange gases with ambient air, taking oxygen from inhaled air into blood and releasing carbon dioxide from respiration back to air that is exhaled. As shown in Figure 6.4, air enters through the nose and travels through the pharynx, trachea, and bronchi, reaching the small sacs in the lung called alveoli, where gas exchange with blood occurs. The alveoli have walls that are as thin as a single cell and highly susceptible to damage; they constitute the alveolar–capillary barrier across which gases are exchanged. A large variety of potentially toxic substances enter with incoming air, including air pollutant particles and gases, disease-causing bacteria and viruses, and airborne allergens, such as pollen. Volatile substances are expelled with exhaled air, and the respiratory tract has mechanisms to eliminate solid and liquid particles and residues from respiratory tract infections.

The respiratory tract may suffer from a variety of ailments that can result from exposure to toxicants. A common one of these is acute or chronic **bronchitis**, manifested by inflammation of the membrane lining of the bronchial tubes, which can be caused by toxicants or by infections. **Emphysema**, the bane of aging heavy smokers, is the result of abnormal enlargement and loss of elasticity of pulmonary air spaces, resulting in difficulty in breathing. Interstitial disorders, predominantly **pulmonary fibrosis**, in which excess fibrous connective tissues develop in the lungs can result from exposure to toxicants. Often indicative of acute lung injury, **pulmonary edema** is the accumulation of fluid in the lungs; in severe cases, the subject literally drowns from those fluids. And, of course, **lung cancer** is a major concern with exposure to some kinds of toxicants.

A common toxic effect to the lung is the result of **oxidative burden**.[3] Oxidative burden occurs as the result of active oxidants, especially free radicals that are generated by a variety of toxic agents and the action of lung defense cells. Ozone, O_3, the air pollutant most commonly associated with photochemical smog, is a particularly active oxidant in polluted air, and smog contains other oxidants as well. NO_2, also associated with photochemical smog and polluted air, contributes to the oxidative burden. Much of the oxidative damage to lungs is probably done by free radicals, such as hydroxyl radical, HO·, and superoxide ion, O^-, which initiate and mediate oxidative chain reactions. Lungs of animals exposed to oxidants have shown elevated levels of enzymes that scavenge free radicals, providing evidence for their role in oxidative damage. There is evidence to suggest that lung cells damaged by toxicants release species that convert lung O_2 to reactive superoxide anion, O_2^-.

Lungs are subject to both acute and chronic injury from toxicants. A common manifestation of acute injury is pulmonary edema, in which liquid exudes into lung alveoli and other lung cavities, increasing the alveolar–capillary barrier and making breathing more difficult. Among the toxicants that cause pulmonary edema are ozone, phosgene ($COCl_2$), and perchloroethylene (C_2Cl_4).

There are several major types of chronic lung disorders that can be caused by exposure to toxicants. A common symptom of chronic lung damage is chronic bronchitis. Among the toxicants that cause this condition are ammonia, arsenic, cotton dust (brown lung disease), and iron oxide from exposure to welding fumes.

Lung fibrosis occurs with a buildup of fibrous material inside lung cavities. The fibers are rich in collagen, the tough, fibrous protein that gives strength to bone and connective tissue. Chronic fibrosis can result from pulmonary exposure to aluminum dust, aluminum abrasives, chromium(VI), coal dust, kaolin clay dust, ozone, phosgene, silica, and finely divided mineral talc.

Snider et al. have defined emphysema as "a condition of the lung characterized by abnormal enlargement of the air spaces distal to the terminal bronchiole, accompanied by destruction of the walls without obvious fibrosis."[4] Emphysema is characterized by enlarged lungs that do not expel air adequately and do not exchange gases well. Cigarette smoke is the overwhelming cause of emphysema. Inhalation of aluminum abrasives and cadmium oxide fumes can also cause emphysema.

Lung cancer is the best-known example of cancer caused by exposure to a toxicant, in this case cigarette smoke. As much as 90% of lung cancers are the result of exposure to tobacco smoke. The latency period for the development of lung cancer from this source is usually at least 20 years and may range up to 40 years or longer. Inhalation of other agents can cause lung cancer, although they are usually associated with synergistic effects from cigarette smoke. The most well established of these are asbestos and radon gas, a radioactive alpha particle emitter.

9.3 SKIN

The pulmonary system and skin constitute the major routes of entry for xenobiotic materials into the body. The skin has a large surface area of up to two m^2 for adults. This large area, along with skin's external exposure, means that it is a common site of contact with toxic substances, especially in the workplace. It has been estimated that about one third of all reported occupational exposures to toxic substances is through skin, and much larger numbers that produce relatively minor symptoms remain unreported.[5] Skin maladies constitute a large fraction of occupational and consumer problems with industrial chemicals and consumer products.

Figure 6.3 shows that skin has a layered structure. As with most organs, the structure of skin is relatively complicated. In addition to the layers shown in Figure 6.3, skin has a number of structures, including blood vessels, hair follicles, sweat glands and ducts, and sebacious glands that secrete oils, fat, and connective tissue. The two major layers of skin are the inner **dermis** and the outer **epidermis**. Skin cells are continually being generated and eventually end up in a much modified form on the surface layer, called the **stratum corneum**. The cells in this layer are not living and are continuously being shed. They consist mostly of **keratin**, a sulfur-containing protein that also constitutes nails and animal horns.

Although the stratum corneum acts as a simple physical barrier to outside influences, skin tissue as a whole is very active. It is crucial in maintaining the body's homeostasis, its essential steady-state environment. Skin maintains temperature and balance of electrolytes, the dissolved salts in internal body fluids. It is metabolically active and participates in hormonal and immune regulatory processes. More than serving as a passive barrier, it is proactive in response to xenobiotic insults and can be damaged in the defensive process by developing rashes and other symptoms.

Absorption through the skin, **percutaneous absorption**, is an important mechanism by which xenobiotic substances can enter the body. The surface stratum corneum layer is the main barrier to such absorption, and when it is shed or compromised, skin is much more susceptible to penetration by xenobiotic substances. A test for the ability of a hydrophobic xenobiotic substance to penetrate skin is to measure the partitioning of such a substance between water and powdered stratum corneum cells.

Modern medical science is taking advantage of skin permeability to chemicals by using it as a delivery system for some kinds of drugs. A single skin "patch" can deliver a drug at a uniform low rate for up to a week. The most long-standing transdermal drug delivery system is used to deliver nitroglycerin to blood for the relief of painful heart angina symptoms. Nitroglycerin is metabolized within minutes in the human body, so delivery through the gastrointestinal tract, where the substance first goes through the metabolically active liver, is relatively ineffective. Delivered transdermally, nitroglycerin enters the bloodstream directly and quickly reaches the heart, where its therapeutic effect is manifested. More recently, skin patches have been developed to deliver nicotine to relieve the cravings for this substance experienced by people trying to stop smoking. Estradiol, scopolamine, clonidine, and fetanyl have also been delivered by this means, and other drug delivery systems are under development.

In addition to serving as a physical barrier to the entry of xenobiotics, skin is active in metabolizing topically applied substances such as steroids and retinoids, primarily through the action of cytochrome P-450 enzymes. Active lipase, protease, glycosidase, and phosphatase enzymes have also been observed in skin. Skin contains some enzymes capable of catalyzing phase II conjugation reactions. In some cases, these metabolic processes detoxify xenobiotics. However, in other cases, they act in sensitization to make xenobiotics active in causing skin symptoms of poisoning.

9.3.1 Toxic Responses of Skin

The most common skin affliction resulting from exposure to toxic substances and the most common skin condition from occupational exposure is **contact dermatitis**, characterized by generally irritated, itching, and sometimes painful skin surface. Skin afflicted with contact dermatitis shows several symptoms. One of these is **erythema**, or redness. The skin surface may be subject to **scaling**, in which the surface flakes off. Thickening and hardening can occur, a condition clinically known as **induration**. Blistering, a condition called **vesiculation**, may also occur. Skin afflicted with contact dermatitis typically exhibits edema, with accumulation of fluid between skin cells. There are two general categories of contact dermatitis: irritant dermatitis and allergic contact dermatitis.

Irritant dermatitis does not involve an immune response and is typically caused by contact with corrosive substances that exhibit extremes of pH, oxidizing capability, dehydrating action, or tendency to dissolve skin lipids. In extreme cases of exposure, skin cells are destroyed and a permanent scar results. This condition is known as a **chemical burn**. Exposure to concentrated sulfuric acid, which exhibits extreme acidity, or to concentrated nitric acid, which denatures skin protein, can cause bad chemical burns. The strong oxidant action of 30% hydrogen peroxide likewise causes a chemical burn. Other chemicals causing chemical burns include ammonia, quicklime (CaO), chlorine, ethylene oxide, hydrogen halides, methyl bromide, nitrogen oxides, elemental white phosporous, phenol, alkali metal hydroxides (NaOH, KOH), and toluene diisocyanate.

Allergic contact dermatitis occurs when individuals become sensitized to a chemical by an initial exposure, after which subsequent exposures evoke a response characterized by skin dermatitis. Allergic contact dermatitis is a type IV hypersensitivity involving T cells and macrophages instead of antibodies. It is a delayed response, occurring one or two days after exposure, and often requires only very small quantities of allergen to cause it. Literally dozens of substances have been implicated as causative agents of contact dermatitis. Some of these are substances applied to skin directly as hygiene products. Included in this category are antibiotic bacitracin and neomycin, preservative benzalkonium chloride, therapeutic corticosteroids, and antiseptic dichlorophene. Among other substances causing allergic contact dermatitis are formaldehyde, abietic acid from plants, hydroquinone, acrylic monomers, triphenylmethane dyes, 2-mercaptobenzthiazole, *p*-phenylene diamine, tetramethylthiuram, 2,4-dinitrochlorobenzene, pentaerythritol triacrylate, epoxy resins, dichromate salts, mercury, and nickel.

The effects of poison ivy constitute a type of allergic contact dermatitis with which people who spend time camping and in other outdoor pursuits may have an unfortunate familiarity. Poison ivy, poison oak, and poison sumac contain toxicodendron, in which the active antigen is pentadecylcatechol:

Pentadecylcatechol $C_{15}H_{27}$

Urticaria, commonly known as hives, is a type I allergic reaction that results very rapidly from exposure to a toxicant to which the subject has become sensitized. It is characterized by the release of histamine from a type of white blood cell. Histamine causes many of the symptoms of allergic reaction, including tissue edema. In addition to edema, erythema, and accompanying raised welts on skin, urticaria is accompanied by severe itching. In severe cases, such as happen in some people as the result of bee or wasp stings, urticaria can result in systemic anaphylaxis, a potentially fatal allergic reaction.

9.3.2 Phototoxic Responses of Skin

Phototoxic responses of skin occur as the result of absorption of radiation, primarily sunlight and ultraviolet radiation in the UVB region of 290 to 320 nm. Because UVB radiation is vastly more effective in causing phototoxic symptoms than either UVA radiation (320 to 400 nm) or visible light (400 to 700 nm), reference will be made to it in discussions of phototoxicity. Photons of radiation are absorbed by functional groups called chromophores on biomolecules. The most significant chromophores in skin are molecules of DNA, which can be modified by absorbing photon energy. Also serving as chromophores in skin are amino acids and materials released by protein breakdown, including tryptophan and urocanic acid. The skin contains a protective pigment, **melanin**, synthesized from the amino acid tyrosine, that effectively absorbs UVB and protects people from the effects of sunlight. Levels of melanin differ widely in people, being high in darker-skinned individuals and very low in those with lighter skin. Production of melanin (suntan) can be promoted by exposure to natural or artificial sunlight.

The most common acute effect of exposure to toxic doses of UVB is erythema, commonly known as sunburn, the result of photooxidation processes in skin. Because of substances released from skin cells exposed to excessive UVB, systemic effects, including fever, chills, and a generally ill feeling, may result as well. Chronic symptoms of exposure to excessive UVB include changes in pigmentation, such as freckles, and general skin deterioration and wrinkling. Of greatest concern is the potential to form cancerous lesions. These include both basal and squamous cell carcinomas. The most serious such effect is the development of malignant melanoma, a particularly serious form of skin cancer.

Photosensitivity, or **porphyria**, is an abnormal sensitivity to ultraviolet radiation and visible light. A genetic predisposition to an inability to repair damage to skin from sunlight can cause photosensitivity, as can exposure to some chemicals, especially chlorinated aromatic compounds. These effects are tied with enzymatic malfunctions in the biosynthesis of heme, the protein molecule contained in blood hemoglobin. When this biosynthesis does not function properly, molecular fragments of heme (porphyrins) accumulate in skin, where they reach an excited state when exposed to light of 400 to 410 nm (the Soret band) and react with molecular O_2 to generate free radicals that are destructive to biomolecules in skin tissue.

Phototoxicity occurs when skin exposed to sunlight, especially in the UVA region of 320 to 400 nm, reddens and develops blisters as a consequence of the presence of certain chemical species. The phototoxic chemical species that result in such reactions are ones to which an individual is exposed either directly on the skin or systemically. These compounds absorb ultraviolet radiation and, like the porphyrins discussed above, enter excited states interacting with O_2 to generate destructive oxidant species and free radicals. Numerous chemical species, including furocoumarins, polycyclic aromatic hydrocarbons, tetracyclines, and sulfonamides, can be phototoxic.

Photoallergy is similar in symptoms and mechanism to allergic contact dermatitis discussed above, except that symptoms develop after exposure to sunlight. The subject develops an allergic response to light after sensitization with a chemical agent. This condition was observed in the mid-1900s in individuals who had used soaps containing antibacterial agents, including tetrachlorosalicylanilide and tribrom osalicylanilide, which had to be taken from the personal care product market.

9.3.3 Damage to Skin Structure and Pigmentation

Defects in skin pigments can result from chemical exposure. **Hyperpigmentation** occurs from increased production and deposition of melanin. **Hypopigmentation** occurs with loss of skin pigments, giving it a white, albino appearance. Among the chemicals that cause hyperpigmentation are volatile organics from coal tar, anthracene, mercury, lead, and hydroquinone. Hypopigmentation can result from exposure to hydroquinone and its derivatives, mercaptoamines, phenolic germicides, and butylated hydroxytoluene.

Acne, characterized by skin eruptions commonly known as blackheads or whiteheads plus a variety of pustules, cysts, and pits on the skin surface, can be caused by exposure to chemicals. The most notable kind of chemically induced acne is **chloracne**, resulting from exposure to chlorinated hydrocarbons. Of these, the most notorious is dioxin, 2,3,7,8-tetrachlorodibenzo-*p*-dioxin (TCDD):

2,3,7,8-tetrachlorodibenzo-*p*-dioxin

In addition to lesions on the face, in severe cases chloracne is characterized by cysts and other manifestations of acne on the shoulders, back, and even genitalia.

Granulomatous inflammation occurs in cases where skin tissue builds up around the site of exposure to an irritant. Introduction of foreign materials such as talc or silica into skin can cause this condition. In some cases, it occurs in response to exposure to some metals, including beryllium and chromium.

Toxic epidermal necrolysis occurs when the skin epidermis is destroyed by the action of toxicants and becomes detached from the dermis. This condition severely disrupts the ability of skin to regulate the release of heat, fluids, and electrolytes. Metabolites of the anticonvulsive drug carbamazepine have been implicated in toxic epidermal necrolysis.

9.3.4 Skin Cancer

Skin cancer is the most common type of cancer. Damage to skin DNA from sunlight is the most common cause of skin cancer. This causes mutations that result in formation of cancer cells and that suppress the immune responses that normally prevent replication of such cells. The class of chemicals most commonly associated with causing skin cancer are the polycyclic aromatic hydrocarbons from sources such as coal tar. These can be metabolized to electrophilic substances that bind with DNA to initiate cancer (see Figures 7.3 and 8.3). Arsenic in drinking water has been established as a cause of precancerous lesions, called arsenical keratoses, and squamous cell carcinoma of skin.

9.4 THE LIVER

The liver is often the first major metabolizing organ that an ingested toxicant encounters, and it has very high metabolic activity. The major function of the liver is to metabolize, store, and release nutrients, that is, to maintain nutrient homeostasis. When blood levels of nutrient molecules are high, the liver can convert them to glycogen and fats and store them. When blood levels of nutrients are low, the liver can convert some amino acids, pyruvate, and lactate to glucose, which is released to blood. It uses amino acids to synthesize proteins that are released to blood, including albumin, clotting factors, and transport proteins. The liver is the major site of fat metabolism and releases fat to blood as needed. It produces bile, which acts to emulsify fats in the small intestine.

Materials enter and leave the liver as blood in arteries and veins. Nutrients, drugs, and ingested xenobiotics absorbed from the small intestine go directly to the liver through the portal vein. The liver has another mechanism for excretion in the form of bile discharged back to the intestines. Bile discharge is a major route of elimination of xenobiotic compounds and their metabolites.

Insofar as toxicological chemistry is concerned, the major function of the liver is to metabolize xenobiotic substances through phase I and phase II reactions. Because of this function, the liver is a crucial organ in the study of toxicological chemistry. Since it processes xenobiotic chemicals, the liver is often the organ that is damaged by such chemicals and their metabolites. Recall from Chapter 8 that the livers of genetically susceptible individuals can be damaged by therapeutic doses of some drugs, such as those shown in Figure 8.5. Other xenobiotics ingested accidentally can damage the liver.

Toxic effects to the liver are studied under the topic of **hepatotoxicity**, and substances that are toxic to the liver are called **hepatotoxins**. Much is known about hepatotoxicity from the many cases of liver toxicity that are a manifestation of chronic alcoholism.[6] Liver injury from excessive alcohol ingestion initially hampers the ability of the organ to remove lipids, resulting in their accumulation in the liver (fatty liver). The liver eventually loses its ability to perform its metabolic functions and accumulates scar tissue, a condition known as cirrhosis. Inability to synthesize clotting factors can cause fatal hemorrhage in the liver.

A wide range of substances can cause hepatotoxicity. Even an essential vitamin, vitamin A,

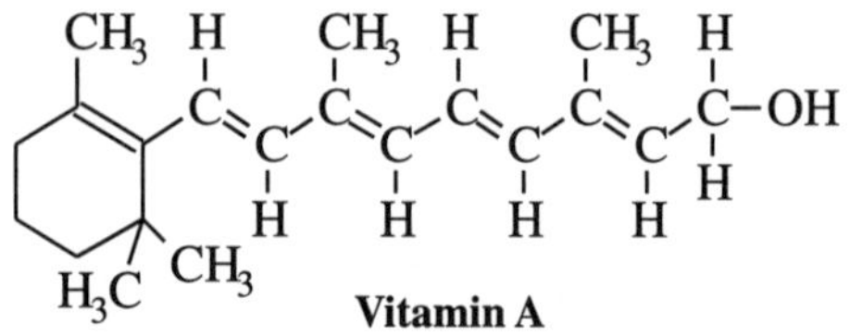

Vitamin A

is hepatotoxic in overdoses, a fact that should be kept in mind by health food fans who drink large amounts of carrot juice. Other hepatotoxins include toxins in hormones, tea (germander), and drinking water infested with the photosynthetic cyanobacteria *Microcystis aeruginosa*. Each year people are killed by eating toxic mushrooms, especially the appropriately named "death cap" mushroom, *Amanita phalloides*. This fungus produces a mycotoxin consisting of seven amino acid residues, a heptapeptide called phalloidin.

A great deal of information about hepatotoxicity has resulted from observed effects of pharmaceuticals, a number of which have been discontinued because of their damaging effects to the liver. An example of such a hepatotoxic compound tested as a pharmaceutical is fialuridine,

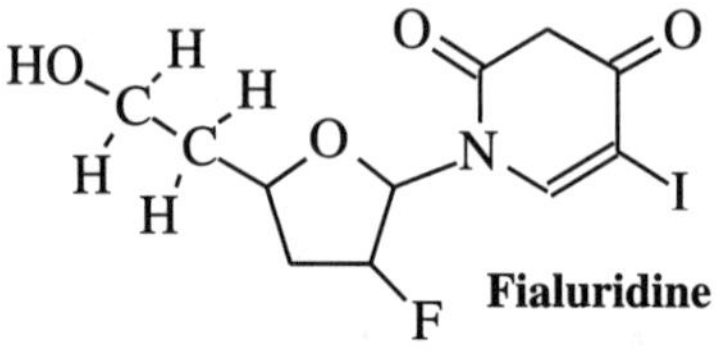

Fialuridine

which was tested during the mid-1990s as a treatment for viral chronic hepatitis B, a liver disease.[7] Seven of 13 patients in the test developed debilitating hepatotoxicity with severe jaundice, along with lactic acidosis due to accumulation of lactic acid, a metabolic intermediate. The seven patients were given liver transplants, but five of them died.

Steatosis, commonly known as fatty liver, is a condition in which lipids accumulate in the liver in excess of about 5%. It may result from toxicants that cause an increase in lipid synthesis, a decrease in lipid metabolism, or a decrease in the secretion of lipids as lipoproteins. An example of a substance that causes steatosis is valproic acid, once used as an anticonvulsant:

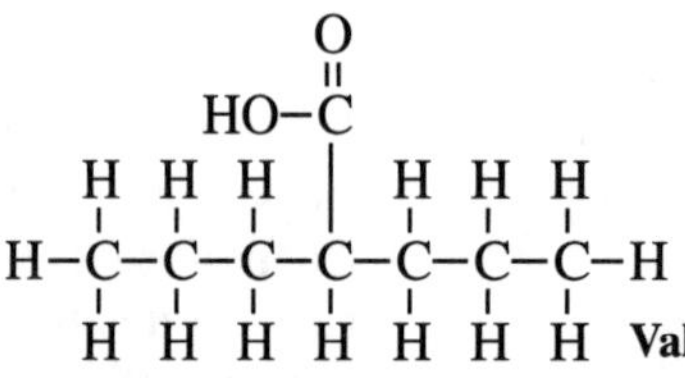

Valproic acid, 2-propylpentanoic acid

Other than ethanol, the xenobiotic chemical best known to cause steatosis is carbon tetrachloride, CCl_4. This compound was once widely used in industry as a solvent, and even in consumer items as a stain remover. As discussed in some detail in Chapter 16, it is converted by enzymatic action in the liver to $Cl_3C\cdot$ radical, then by reaction with O_2 to $Cl_3COO\cdot$ radical, which reacts with unsaturated lipids in the liver to cause fatty liver.

The general term hepatitis is used to describe conditions under which the liver becomes inflamed when liver cells that are damaged by a toxic substance, a substance that causes an immune response, or disease die, and their remnants are released to liver tissue. A number of toxicants can cause liver cell death. This is most damaging when it occurs through necrosis of liver cells, in which they rupture and leave remnants in the vicinity, which can lead to inflammation and other adverse effects. Dimethylformamide is a xenobiotic industrial chemical known to cause liver cell death:

$$H-\overset{\overset{O}{\|}}{C}-N\overset{CH_3}{\underset{CH_3}{}} \quad \text{N,N-dimethylformamide}$$

A more orderly type of cell death is apoptosis, in which the cells become encapsulated and are systematically removed from the organ. This essential housekeeping function is accomplished in the liver by special cells called **Kupffer** cells. These cells perform **phagocytosis**, in which a solid particle, such as a cell remnant or other foreign matter in the liver, becomes encapsulated in a plasma membrane and incorporated into the Kupffer cell, which is then eliminated.

Reduced bile output can result in an accumulation of bilirubin, a dark-colored pigment produced by the breakdown of blood heme. When this product is not discharged at a sufficient rate with bile, it accumulates in skin and eyes, giving the characteristic sickly color of jaundice. Impaired production and excretion of bile is known as **canalicular choleostasis**. It can be caused by a number of xenobiotic substances, such as chlorpromazine. Reduced bile output can also result from damage to bile ducts. Methylene dianiline used in epoxy resins is known to harm bile ducts.

Chlorpromazine   **Methylene dianiline**

Cirrhosis, which was mentioned in connection with chronic alcoholism above, is an often fatal end result of liver damage. It is often the result of repeated exposure to toxic agents, such as occurs with alcohol imbibed by heavy drinkers. Cirrhosis is characterized by deposition and buildup of fibrous collagen tissue, which replaces active liver cells and eventually forms barriers in the liver that prevent it from functioning.

Liver tumors have been directly attributed to exposure to some toxicants. Androgens (associated with male sex hormones), aflatoxins (from fungi, see Chapter 19), arsenic, and thorium dioxide (administered as a suspension to many patients between 1920 and 1950 as a radioactive contrast agent for diagnostic purposes) are known to cause liver cancer.

Arguably the most clearly documented human carcinogen is vinyl chloride, which has been shown to cause a type of liver tumor called **hemangiosarcoma**, which is virtually unobserved except in workers heavily exposed to vinyl chloride.[8] Up until about 1970, workers were exposed to high levels of up to several parts per thousand in air in the polyvinylchloride manufacturing industry. Poisoning was so common that reference was even made to "vinyl chloride disease," characterized by damage to skin, bones, and liver. It is believed that hemangiosarcoma resulted from the action of the metabolically produced reactive epoxide generated by enzymatic oxidation of vinyl chloride in the liver:

$$\underset{\text{Vinyl chloride}}{\overset{H}{\underset{H}{>}}C=C\overset{Cl}{\underset{H}{<}}} \xrightarrow{\text{Cytochrome P-450}} \underset{\text{Epoxide of vinyl chloride}}{\overset{H}{\underset{H}{>}}C\underset{O}{-}C\overset{Cl}{\underset{H}{<}}} \quad (9.4.1)$$

9.5 BLOOD AND THE CARDIOVASCULAR SYSTEM

Blood is a fluid tissue in which particles, the **formed elements**, are suspended in a circulating fluid medium. When blood is centrifuged, the formed elements, which constitute around 40% of the volume, settle out; the liquid that is left is the **plasma**. The study of blood is called **hematology**.

In considering toxic responses of the blood, it is useful to consider more than just the blood fluid. So consideration is given as well to the bone marrow, where red blood cells are produced; the **spleen**, which is a reservoir for blood cells and which breaks down old red blood cells; and reticuloendothelial tissue, which produces macrophages capable of engulfing and digesting small foreign particles, such as microorganisms. In addition, consideration should be given to **lymph**, a fluid largely derived from blood that accumulates in spaces outside of blood vessels.

The system that circulates blood is the **cardiovascular system**, shown in schematic form in Figure 9.1. It consists of two major components: the muscular part of the heart, called the **myocardium**, and the network of **blood vessels** composed of arteries, veins, and capillaries. Blood circulation is the body's transportation system that supplies tissues with the oxygen, nutrients and their metabolites, and hormones that they need for their function. Blood carries carbon dioxide, encapsulated dead cell matter, and other wastes away from tissues. Circulating blood is crucial to maintaining body homeostasis, with temperature, pH, and other crucial parameters kept within the narrow ranges required for good health. A number of toxicants have adverse effects on the cardiovascular system.

The toxicological chemistry aspects of blood and the cardiovascular system are very important. Systemic poisons and their metabolites are carried to receptors in the body through this system. It carries phase I and phase II reaction products to the kidney and bladder for elimination. Toxicants can bind to blood proteins. Analysis of blood for toxicants and their metabolites is a common way of determining exposure to toxic substances and generally the most accurate means for evaluating systemic exposure.

9.5.1 Blood

The formed elements of blood are red blood cells, platelets, and leukocytes. Red blood cells, or **erythrocytes** (Figure 9.2), are flexible biconcave disk-shaped bodies whose main function is to carry oxygen to tissue bound to the **hemoglobin** that they contain. They are generated in the marrow of various bones by the action of **stem cells**. The hormone **erythropoietin** stimulates erythrocyte production in response to tissue needs for oxygen. Marrow stem cells also produce **platelets**, tiny cell fragments that contain the biochemicals necessary for blood clotting. The third kind of formed elements consists of **leukocytes**, which are defensive white blood cells.

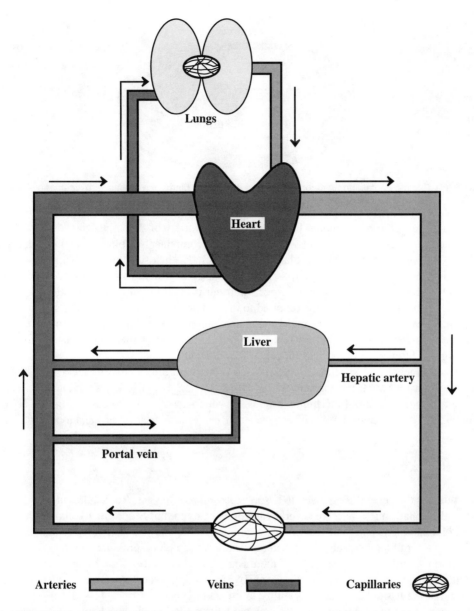

Figure 9.1 Important aspects of the cardiovascular system.

Figure 9.2 Erythrocytes are shaped as disks that are concave on both sides.

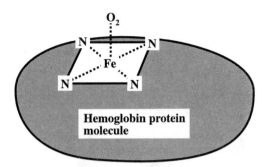

Figure 9.3 The iron-containing heme group on hemoglobin that binds with molecular oxygen and enables red blood cells to carry oxygen to tissues.

Blood plasma is a straw-colored liquid laden with a variety of salts, nutrients, dissolved gases, and biomolecules, especially proteins. The major ionic component of blood is sodium chloride as Na^+ and Cl^- ions. Nutrients that circulate in blood include glucose (blood sugar), various amino acids, lipids, lactic acid, and cholesterol. An important glycoprotein (protein bound with carbohydrate) in blood is transferrin, which transports nutrient iron absorbed by the intestine to the liver, spleen, and bone marrow, where it is required to make hemoglobin. Plasma is very similar to other tissue fluids, with which it readily undergoes interchange, except that it has much more protein. When blood is caused to clot before separating out the formed elements, the pale yellow liquid remaining is a fraction of blood plasma called **blood serum**.

The key biochemical species in erythrocytes is **hemoglobin**, a high-molecular-mass protein containing iron for which the formula is $C_{3032}H_{4816}O_{780}N_{780}S_8Fe_4$. The key functionalities on each hemoglobin molecule are four **heme** groups in which iron(II) is bound with four N atoms (Figure 9.3). These groups bind with oxygen and carry it to tissues as the blood erythrocytes circulate.

9.5.2 Hypoxia

Hypoxia is the general term given to tissue deprivation of oxygen. Toxicants can cause hypoxia by several mechanisms. There are several categories of hypoxia.[9] **Stagnant hypoxia** is a lowered flow of blood, which can result from reduced pumping efficiency of the heart or **vasodilation**, in which the walls of blood vessels are caused to relax, lowering blood pressure and flow. When blood flow is normal, hypoxia can also occur if there is a reduced capacity of the blood to carry oxygen, a condition called **anemic hypoxia**. **Histotoxic hypoxia** occurs when oxygen is delivered normally to tissue, but the tissue has a reduced ability to utilize oxygen.

A common cause of anemic hypoxia is competitive binding on the heme sites for oxygen, usually the result of exposure to carbon monoxide, CO. Carbon monoxide has a greater affinity for the iron(II) on heme sites than does molecular oxygen, forming a stable complex called **carboxyhemoglobin** in preference to the oxygen-bound **oxyhemoglobin**. Carboxyhemoglobin formation is considered in greater detail in Chapter 11.

The other major cause of anemic hypoxia from chemical exposure is **methemoglobinemia**, in which the iron(II) in hemoglobin is oxidized to iron(III). The methemoglobin product is a dark-colored substance in which the iron does not preferentially bind molecular oxygen, binding with OH^- or Cl^- ions instead, so methemoglobin does not carry oxygen and the poisoning victim may die of oxygen deprivation. Nitrite ion, NO_2^- (see Chapter 11), aniline, and nitrobenzene (see Chapter 15) are toxicants that can cause methemoglobinemia.

Hypoxia can be the long-term result of reduction of blood cell formation in bone marrow. Some toxicants reduce the production of both erythrocytes and leukocytes in marrow, resulting in a condition called **aplastic anemia**. Exposure to benzene (Chapter 13) can cause this condition. The major biochemical effect of toxic lead is interference with the process by which heme is synthesized.

Though not strictly a blood malady, histotoxic hypoxia deprives tissue of oxygen, even when it is delivered by blood, by preventing its utilization. The most common toxicant that causes histotoxic hypoxia is hydrogen cyanide, which binds strongly to the iron(III) form of the endogenous cytochrome species involved in molecular oxygen utilization, so that it cannot be reduced back to iron(II) in the electron transfer processes involved with O_2 utilization in tissue. Interestingly, an antidote to cyanide poisoning (if given rapidly in those cases where the victim survives long enough) is to administer nitrite compounds that form methemoglobin that has iron(III) capable of binding competitively for the cyanide. Hydrogen sulfide, H_2S, causes histotoxic hypoxia by a mechanism similar to that of hydrogen cyanide.

9.5.3 Leukocytes and Leukemia

Leukocytes are much more complex than erythrocytes and perform an entirely different function. Although they are present in blood and carried by blood flow, they perform their activities largely outside the bloodstream. Their main activity is in defending the body against foreign bodies and agents such as pathogenic microorganisms. In defense against foreign bodies such as bacterial cells, leukocytes perform phagocytosis, in which they envelop the object, resulting in its eventual elimination. To defend against foreign agents, such as toxicants attached to blood proteins, leukocytes generate antibodies.

Uncontrolled production of leukocytes is a form of cancer called **leukemia**. Although toxicants are suspected of causing some cases of leukemia, the evidence for such cases is not very strong. However, benzene exposure is now regarded as a cause of this kind of cancer.

9.5.4 Cardiotoxicants

Circulation of blood occurs by the action of a beating heart and is also influenced by conditions in the remainder of the vascular system. Heartbeat involves both electrical (nerve impulse) and mechanical (heart muscle contraction and relaxation) events. Some toxicants can adversely affect these finely coordinated actions. Adverse effects such as **bradycardia** (decreased pulse rate), **tachycardia** (increased rate), and **arrhythmia** (irregular pulse) can result.

A number of pharmaceutical agents have shown toxic side effects involving the heart.[10] These effects usually occur as the result of overdoses or in subjects with preexisting heart conditions. Some common antibiotics used to combat bacterial infections have shown a depressant effect on heart function. Antineoplastic agents used for cancer chemotherapy have caused cases of cardiotoxicity. Widely used 5-fluoruoracil has caused a variety of cardiac symptoms, including severe hypotension (low blood pressure). As might be expected from their potential to affect nerve impulses involved with the function of the cardiovascular system, drugs that act on the central nervous system may be cardiotoxic, causing symptoms such as reduced cardiac output and arrhythmia. Such drugs have included antidepressants such as imiprimine, antipsychotic agents, and general anesthetic. High systemic levels of local anesthetics such as lidocaine can cause cardiac irregularities because of their action in blocking nerve axon conduction. Synthetic catecholamines used to treat respiratory and cardiovascular disorders have caused toxic cardiac symptoms, including heart cell necrosis (cell death).

Some biochemical natural products have caused cardiotoxicity. Synthetic estrogens and progestins have been linked to cardiovascular disorders in women taking them for contraceptive purposes. Various animal and insect venoms and plant alkaloids may have adverse cardiovascular effects. There is some evidence to suggest that anabolic steroids, commonly linked to scandals involving athletes who take them to enhance performance, have caused cardiovascular disorders.

(Though not a natural product, another performance-enhancing substance, Viagra, which is used to treat erectile dysfunction in men, has been suspected of contributing to cardiovascular problems, including heart attacks.)

A number of industrial chemicals have been linked to cardiotoxicity. Aldehydes and primary alcohols that can be metabolically oxidized to aldehydes have exhibited cardiodepressant effects. Acute exposure to ethanol has caused arrhythmia. Isopropyl alcohol (2-propanol), a widely used industrial chemical and personal care product, may cause cardiovascular depression and excessively rapid heartbeat. Some halogenated hydrocarbons, including chloroform, ethyl bromide, and trichlorofluoromethane, have been implicated in cardiovascular disorders, including arrhythmia.

9.5.5 Vascular Toxicants

A number of toxicants have effects on the arteries, veins, and capillaries comprising the vascular system. An important factor in vascular toxicity is that toxic substances are transported by blood, which means that they contact the cells making up the structure of the vascular system, which may be adversely affected as a consequence. It is likely that a significant fraction of organ toxicities are actually the result of damage to blood vessels in the organs.

Adverse toxic effects on the vascular system can be manifested in a number of ways, generally falling into the classes of degenerative and inflammatory effects. Deteriorated blood vessel walls may hemorrhage or leak fluid, causing edema. Damage to blood vessels in the lungs by agents such as hydrogen fluoride, nitric oxide, and ozone can cause the fluid accumulation known as pulmonary edema. A common toxic effect is abnormal thickening of arterial walls accompanied by loss of elasticity, a condition called **arteriosclerosis**. Another common effect is **atherosclerosis**, a form of arteriosclerosis in which the inner lining of artery walls becomes covered with plaque produced by the deposition of fatty substances. Cholesterol, carbon monoxide, dinitrotoluenes, polycyclic aromatic hydrocarbons, and amino acid homocysteine have been implicated as causes of atherosclerosis.

The number of substances that have been implicated for potential vascular toxic effects is too great to discuss in any detail here. Of such substances, one of the most well documented is tobacco smoke, which has been shown to contribute to arterial degeneration leading to myocardial infarction (heart attack). Acrolein, an ingredient of tobacco smoke and engine exhausts, is biochemically very active by virtue of its aldehyde group in close proximity to a carbon–carbon double bond and is thought to be involved in damage to vascular cells. Allylamine is a reactive, unsaturated amine that has been implicated in arterial hypertrophy (enlargement), hemorrhage of lung alveoli, and pulmonary (lung) edema, probably associated with damage to blood vessels. These toxic effects are believed to be the result of metabolic conversion of allylamine to acrolein.

Acrolein

Allylamine

Arsenic has been implicated as a cause of arteriosclerosis. Blackfoot disease, a malady suffered in areas of Taiwan having high soil and water levels of arsenic, is a very severe form of arteriosclerosis. Dilation of arteries and capillaries is a symptom of acute arsenic poisoning.

9.6 IMMUNE SYSTEM

The **immune system** is the body's defense against biological systems that would harm it. The most obvious of these consist of **infectious agents**, such as viruses or bacteria. Also included are

neoplastic cells, which give rise to cancerous tissue. The immune system produces **immunoglobin**, a substance consisting of proteins bound to carbohydrates. This material functions as **antibodies** against **immunogen** or **antigen** macromolecules of polysaccharides, nucleic acids, or proteins characteristic of invasive foreign virus, bacteria, or other biological materials. The cells that the immune system uses to provide protection are called **leukocytes**.

The immune system response to toxicants is called **immunogenesis** and can occur in several ways. **Immunosuppression** occurs when the body's natural defense mechanisms are impaired by agents such as toxicants. Radiation and drugs such as chemotherapeutic agents, anticonvulsants, and corticosteroids can have immunosuppressive effects. Immunosuppressants are deliberately used to prevent rejection of transplanted organs. In some cases, toxicants adversely alter the mechanisms by which the immune system defends the body against pathogens and neoplastic cells. Another effect of toxicants on the immune system occurs through the loss of its ability to control proliferation of cells, resulting in leukemia or lymphoma.

Foreign agents can cause the immune system to overreact with an extreme, self-destructive response, called **allergy** or **hypersensitivity**, that can be quite severe or even fatal. **Chemical allergy** occurs after the subject has been exposed to a substance and developed a sensitivity to it. Allergic reactions develop to large molecules, much larger than those of common synthetic substances. Therefore, chemical allergy develops after the foreign agents or their metabolites, called **haptens**, become associated with large molecules endogenous to the body. Among the many substances that cause allergy are metals (beryllium, chromium, nickel), penicillin, formaldehyde, pesticides, food additives, resins, and plasticizers. Allergic reactions may range from minor skin irritation to rapidly fatal anaphylactic shock. Allergies are most commonly expressed in humans by skin conditions, such as dermatitis, and by conjunctivitis of the eye. A particular concern is potentially fatal chemically induced asthma manifested by severe bronchiolar constriction.

Uncontrolled proliferation is another immune system dysfunction that can occur. It may be manifested by lymphoma, leukemia, or related conditions.

In some cases, xenobiotic compounds adversely alter **host defense mechanisms**. This may reduce the body's ability to resist pathogenic bacteria or viruses or to combat neoplasia (tumor tissue).

Autoimmunity develops as a condition in which the body develops an allergic response to its own biomolecules. What essentially occurs is that to a degree it loses the ability to distinguish foreign antigens from its own antigens. Several important diseases are caused by autoimmune response, including rheumatoid arthritis and systemic lupus erythematosus. Binding of xenobiotic molecules to body proteins can induce autoimmunity. Chemicals that have been so implicated include heavy metals, hydrazine, epoxy resins, and chlorinated ethylene compounds.

A variety of chemical and therapeutic agents are known to cause allergic reactions in susceptible individuals. Several prominent examples of these are the following:

- Formaldehyde, used in a large number of consumer products, resins, and wood products. This agent causes type I hypersensitivity manifested by respiratory symptoms, including rhinitis, bronchial asthma, and asthmatic bronchitis.
- Trimellitic anhydride, used in chemical synthesis. Type II hypersensitivity manifested by adverse effects on blood, including hemolytic anemia and bone marrow depression, may be caused by exposure to trimellitic anhydride. This agent may also cause type III hypersensitivity, resulting from deposition of antigen–antibody complexes in tissue and causing symptoms such as rheumatoid disease or pneumonitis.
- A variety of agents, including antibiotic penicillin, beryllium, mercaptobenzothiazole, phthalic anhydride, and dichromate salts, cause type IV allergic reactions, one common symptom of which is contact dermatitis.
- Immunosuppression can result from exposure to a number of agents, including ozone, benzene, asbestos, silica, nitrogen mustards, and several metals.

Effects on the immune system are gaining increasing recognition as factors in toxicology and in evaluating the toxicity of various substances. There are numerous ways of evaluating potential effects on the immune system. The most modern methods are reproducible and sensitive and have been standardized and validated. One is a two-tier method that makes use of numerous tests made directly on the test organism and on samples taken from it. The first tier consists of relatively simple tests, such as measurements of body mass, blood count, examination of tissue (histology), and ability to form antibodies. The second tier consists of more sophisticated tests, such as the *Streptococcus challenge*, that measure host resistance. Bone marrow evaluations are also employed as second-tier tests.

9.7 ENDOCRINE SYSTEM

Whereas nerve signals represent an almost instantaneous response to stimuli and conditions affecting an organism, **hormones** are chemical messengers that take longer to deliver and act and are involved in long-term regulation of an organism's functions. They are crucial to the maintenance of homeostasis. Hormones are produced largely by special glands called **endocrine glands**, as well as by some tissues that are not part of such glands. Hormones are secreted by **endocrine cells**, enter the bloodstream, and are carried to **target cells**, where they have some sort of effect. Binding of hormone molecules to target cells can cause developmental, physiological, or behavioral responses. The major endocrine glands are shown in Figure 3.9. Table 9.1 is a list of important hormones, their sources, and their targets.

Table 9.1 Major Kinds of Hormones Produced and Used by Humans

Hormone	Target	Major Properties and Actions
Adrenal Medulla		
Epinephrine, norepinephrine (modified amino acids)	Cardiovascular, liver, fat cells	Enhance reactions; increase heart rate; raise blood sugar and send blood to muscles
Adrenal Cortex		
Glucocorticoids or cortisol (steroids)	Various tissues, including muscles, immune system	Response to stress; affect metabolism of glucose, protein, fat; reduce inflammation and immune response
Mineralocorticoids, aldosterone (steroids)	Kidneys	Stimulate reabsorption of Na$^+$, excretion of K$^+$
Anterior Pituitary		
Thyrotropin (glycoprotein)	Thyroid gland	Causes synthesis and secretion of thyroxine hormone
Adrenocorticotropin (polypeptide)	Adrenal cortex	Stimulation of hormone release
Luteinizing hormone (glycoprotein)	Ovary or testis	Stimulation of sex hormone release
Follicle-stimulating hormone (glycoprotein)	Ovary or testis	Stimulation of female egg production, male sperm production
Growth hormone	Muscle, bones, liver	Stimulation of protein synthesis (protein)
Prolactin (protein)	Mammary glands	Milk production
Melanocyte-stimulating hormone (peptide)	Melanocytes (skin)	Skin pigmentation
Endorphins and enkephalins (peptides)	Neurons on spinal cord	Reduce pain

TOXIC RESPONSES

Table 9.1 Major Kinds of Hormones Produced and Used by Humans (continued)

Hormone	Target	Major Properties and Actions
	Heart	
Atrial natriuretic hormone (peptide)	Kidneys	Increase Na^+ excretion
	Hypothalamus	
Release control hormones (peptides)	Anterior pituitary	Secretion of hormones from anterior pituitary
	Ovaries	
Estrogens (steroids)	Breast, uterus, other	Female sexual development and behavior
	Pancreas	
Insulin (protein)	Various tissues, fat, muscle	Glucose uptake, metabolism
	Parathyroid	
Parathormone (protein)	Bones	Transfer Ca^{2+} from bone to blood
	Pineal	
Melatonin (modified amino acid)	Hypothalmus	Control biological rhythms
	Posterior Pituitary	
Oxytocin (peptide)	Breasts, uterus	Stimulates labor, milk flow
Antidiuretic vasopressin (peptide)	Kidneys	Stimulates water reabsorption, increases blood pressure
	Various Tissue Cells	
Prostaglandins (modified fatty acids)	Variety of tissues	Variety of actions
	Small Intestine Lining	
Cholecystokinin (peptide)	Liver, gall bladder, pancreas	Stimulates digestive agents' release from pancreas and liver, contractions of gall bladder
	Stomach Lining	
Gastrin (peptide)	Stomach	Stimulates stomach movement and digestive juices
	Testes	
Androgens (steroids)	Several tissues	Stimulates male sexual development and behavior, sperm
	Thymus	
Thymosins (peptides)	Immune system	Stimulates T cell immune responses in lymphatic system
	Thyroid	
Thyroxine, triiodothyronine (iodine-containing amino acids)	Various tissues	Regulate metabolism required for development and growth
Calcitonin (peptide) bones	Uptake of Ca^{2+} from blood to bone	

The endocrine glands and the hormones that they produce are essential in determining development, growth, reproduction, and behavior of organisms. In humans they regulate many crucial metabolic and biochemical functions, including blood glucose levels, blood pressure, brain function, and nervous system response. These systems are obviously of crucial importance in maintaining human health and well-being, and any toxicological threats to their function are potentially quite serious.

Toxicants can affect the endocrine system in several ways. Those that impair the function of specific endocrine glands lower or stop the production of essential hormones. Carcinogens that harm endocrine glands would clearly have such an effect. Excessive stimulation of endocrine glands by exposure to toxicants has the potential to result in overproduction of hormones with detrimental effects.

Many chemicals are suspected of being **endocrine disruptors**. Such substances mimic the action of natural hormones, tricking the body into thinking that they are hormonal. In so doing, they may cause excessive action of the natural hormones or may act in an improper manner, thus having some sort of toxic effect. The greatest concern with endocrine disruptors is with reproductive effects and sexual development. But there is also concern that they may increase risk of some kinds of cancer, vascular disease, and diabetes.

Endocrine disruptors may act by binding to hormone receptors, such as steroid receptors, thereby preventing an endogenous hormone from binding with and activating the receptor. Or the disruptor may bind with a receptor and mimic the action of an endogenous hormone, excessively or inappropriately. Whereas natural hormones are released as needed by carefully choreographed action of the endocrine glands, a hormone-mimicking substance is introduced by exposure through food or other sources. Another potential mode of action of endocrine disruptors is that they bind to a receptor without blocking binding of the endogenous hormone, but prevent the activated receptor from properly triggering the signals by which it turns on hormone-responsive genes.

A number of compounds have been implicated as endocrine disruptors. They include 2,3,7,8-tetrachlorodibenzo-p-dioxin, polychlorinated biphenyls (PCBs), and some phenolic compounds, including nonylphenol and bisphenol-A:

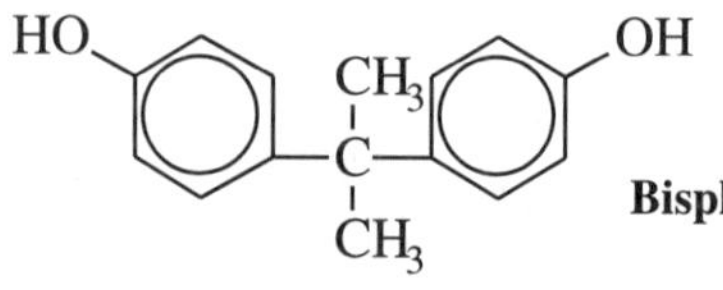

9.8 NERVOUS SYSTEM

The nervous system consists of the brain, spinal cord, and peripheral nerves. Whereas the hormones generated by the endocrine system discussed in the preceding section direct longer-term activities of the body, the nervous system sends impulses very rapidly to direct movement and response of the body. At intervals that are generally of slightly less than 1 sec, a sequence of nerve impulses directs the heart to beat from before birth to death, 24 h each day, 7 days of the week. If just a few of these impulses fail, life ends.

The nervous system is toxicologically important because of potential damage from **neurotoxins** that attack it. Beyond that, much of what is known about the nervous system has been the result of exposure to neurotoxins known to selectively attack certain kinds of cells or inhibit certain processes in the nervous system.[11]

In humans and other more developed animals, most nerve cells are located in the **brain** and in the **spinal cord**, which together make up the **central nervous system**. The brain acts to process and integrate information. It is composed of several parts, the "thinking" portion of which is composed of two hemispheres at the top and front of the brain, called the **cerebrum**. This part of the brain is covered with a thin layer of gray matter called the **cerebral cortex**.

Information is received to the brain from remote parts of the body, and impulses in turn are transmitted back through special cells called **neurons**. The system of neurons that links the central nervous system with other parts of the body constitutes the **peripheral nervous system**. A single neuron can encompass a very long distance, such as that from the spinal column to the toes. Neurons generally consist of four regions. The neuron cell nucleus and most cell organelles are contained in a compact **cell body**. Attached to the cell body are long, branched structures called **dendrites**, a name derived from the Greek *dendron* for tree. Dendrites carry information to the cell body. One of the dendrites that is normally the longest is the **axon**, which carries information away from the cell body. Axons are interfaced with a target cell, which may be a gland or muscle cell or another neuron. At this interface, the axon is divided into a number of nerve endings, which constitute an **axon terminal**. The interface of an axon terminal with a target cell constitutes a **synapse**, consisting of a specialized membrane of the axon next to a specialized plasma membrane of the target cell and separated by a distance of only about 25 nm, a gap called the **synaptic cleft**. When a nerve impulse is transmitted by the axon, it releases **neurotransmitters** that diffuse across the synaptic cleft, and then bind to receptors on the target cell plasma membrane. There are a number of different neurotransmitters, including acetylcholine, norepinephrine, dopamine, histamine, serotonin, γ-aminobutyric acid, and glutamate (a cause of adverse reactions in some people sensitive to monosodium glutamate added to food).

There is not space here to explain the process of nerve impulse transmission. It is an electrical process and involves pumping of Na^+ and K^+ ions across barriers. It should be noted that there are cells other than neurons in the nervous system, of which the most abundant are **glial cells**.

An important characteristic of the brain that largely determines its susceptibility to toxicants is the **blood–brain barrier**, which restricts transfer of substances between the blood and brain tissue. Brain cells have very tight junctions between each other, such that toxicants and their metaboliltes must move across cell membranes either by active transport processes or by virtue of their lipophilicity.

Brain function is dependent upon ready availability of energy by aerobic metabolism. This energy is provided by **aerobic glycolysis**, the breakdown of glucose blood sugar to pyruvic acid with O_2 as an electron acceptor. Therefore, brain cells and other nerve cells are highly susceptible to interruptions in the supply of either O_2 or blood glucose.

Neurotoxins may selectively attack neurons or even specific kinds of neurons. This can cause injury to the neurons. In severe cases, the neuron cells are killed, leading to irreversible loss of the neuron and associated dentrites, axons, and the insulating **myelin** sheathing around the axons.

The effects of neurotoxins may be manifested in a number of ways, divided broadly into two categories: encephelopathy and peripheral neuropathy. **Encephelopathy** refers to brain disorders, many of which may be caused by neurotoxins. It may entail cerebral edema (accumulation of fluid in the brain), degeneration and loss of brain neurons, and necrosis of the cerebral cortex. Symptoms of encephelopathy include loss of coordination (ataxia), convulsions, seizures, cerebral palsy (partial paralysis and tremors), and coma. Neurotoxins can cause symptoms of Parkinson's disease, which include rigidity, a shuffling mode of walking, and tremor of the hands and fingers. Psychological symptoms, such as shyness, uncontrolled anger, and extreme anxiety, may be symptomatic of damage by neurotoxins to brain tissue. Another effect of neurotoxins can be the development of dementia, characterized by loss of memory, impaired reasoning ability, and usually disturbed behavior.

As its name implies, **peripheral neuropathy** refers to damage to nerves outside the central nervous system. It is especially evident as damage to the motor nerves involved with voluntary muscle movement. Victims of peripheral neuropathy often have problems with movement and are afflicted with symptoms such as "foot drag" or "Jake leg," a malady that got its name from toxic effects of contaminated Jamaican ginger.

Maladies caused by the effects of substances that attack neurons are said to cause **neuronopathies** of various kinds. A number of toxicants cause neuronopathic symptoms. Metals that cause encephelopathy include aluminum, bismuth, lead, and arsenic (a metalloid). Arsenic causes peripheral neuropathy, bismuth causes emotional disturbances, lead causes learning deficits in children, manganese causes emotional disturbances and symptoms of Parkinson's disease, and thallium causes emotional disturbances, ataxia, and peripheral neuropathy. Elemental mercury inhaled as the vapor can result in a variety of psychological symptoms, including emotional disturbances, fatigue, and tremor. Methylated mercury compounds are highly neurotoxic, causing ataxia and paresthesia (abnormal tingling and pricking sensations, "pins and needles"). Carbon monoxide poisoning may result in loss of neurons in the cortex and symptoms of encephelopathy and parkinsonism. The second most common symptom of carbon tetrachloride poisoning after liver damage is encephelopathy. Victims who survive cyanide poisoning may suffer delayed parkinsonism. Antibiotic chloramphenicol has produced peripheral neuropathy, and pharmaceutical diphenylhydantoin has caused ataxia, dizziness, and nystagmus

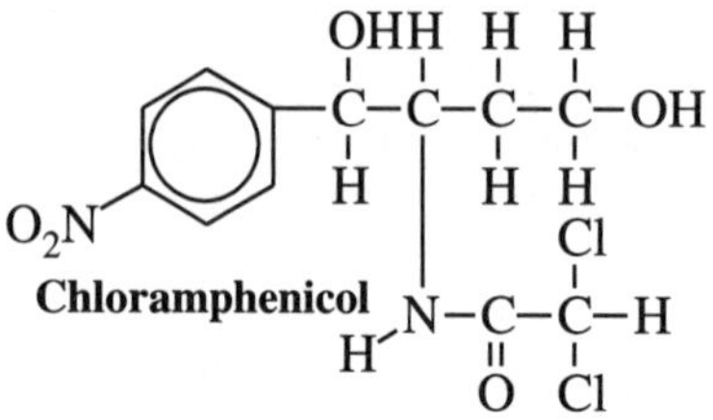

Chloramphenicol

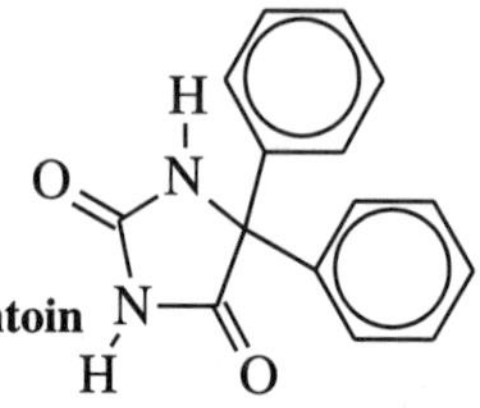

Diphenyldantoin

(involuntary, rapid, lateral eye movement) because of damage to cerebellum cells. Methyl bromide acts as a neurotoxin causing peripheral neuropathy and impairment of speech and vision.

The neuronopathic symptoms described above are caused by substances that attack and destroy the cell bodies of neurons. Another class of toxic effects occurs as the result of deterioration of nerve axons and its surrounding myelin. Symptoms resulting from this effect are called **axonopathies**. A classic toxicant cause of axonopathies is that of γ-diketones, most commonly 2,5-hexanedione:

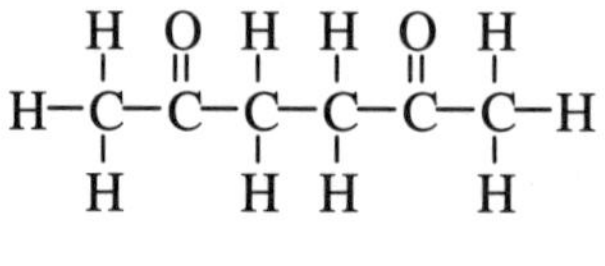

2,5-hexanedione

Substances that can be metabolized to γ-diketones, such as *n*-hexane, which is metabolized to 2,5-hexanedione, cause the same disorders. Examples of the many other substances known to cause axonopathies are colchicine, disulfiram, hydralazine, misonidazole, and insecticidal pyrethroids. Peripheral neuropathy is the most common kind of axonopathic disorder. However, other symptoms may be observed. Numerous cases of manic psychoses were produced in workers exposed to carbon disulfide, CS_2, in the viscose rayon and vulcan rubber industries.

Some neurotoxic effects are caused by attack on and disintegration of the myelin insulation around axons. A substance that was found to have such an effect is hexachlorophene,

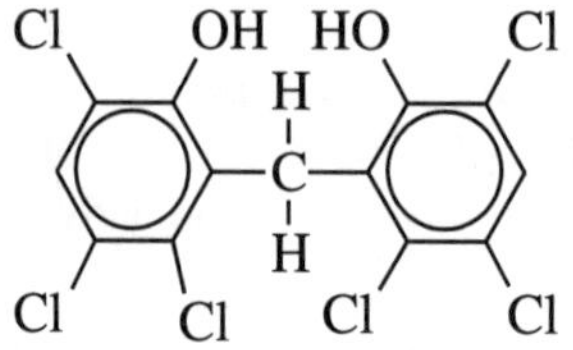

Hexachlorophene

used until the early 1970s as an antibacterial agent for bathing babies. Disorders caused by damage to myelin are called **myelinopathies**.

Some neurotoxins do not alter nerve cell structure, but interfere with **neurotransmission**, the transmission of nerve impulses. In some cases, pharmaceutical agents are administered to interfere with nerve impulses in beneficial ways in the practice of **neuropharmacology**.

One of the most common substances known to interfere with neurotransmission is **nicotine**. Neurotoxic effects from nicotine have occurred in children who have ingested nicotine, people who have accidentally ingested nicotine-based insecticides, and even workers who have absorbed nicotine through the skin from handling wet tobacco leaves. The first symptoms of nicotine intoxication include accelerated heart rate, perspiration, and nausea. Later, the heart may slow to such an extent that blood pressure becomes too low. The subject may become drowsy and confused and lapse into a coma. Death occurs from respiratory muscle paralysis.

A particularly devastating substance that affects neurotransmission is the illicit drug cocaine, which blocks catecholamine uptake at nerve terminals. Addictive cocaine is particularly dangerous because it can cross the blood–brain barrier readily.

Some amino acids are called **excitatory amino acids** because of their ability to excite neurotransmission. The most publicized disorder caused by excitatory amino acids is the "Chinese restaurant syndrome," manifested by a burning sensation of the skin, particularly on the face, neck, and chest. This can be caused by ingestion of monosodium glutamate, which is widely used to season some kinds of Oriental food:

$$\text{HO}-\overset{\overset{O}{\|}}{C}-\overset{\overset{H}{|}}{\underset{\underset{H}{|}}{C}}-\overset{\overset{H}{|}}{\underset{\underset{H}{|}}{C}}-\overset{\overset{H}{|}}{\underset{\underset{NH_2}{|}}{C}}-\overset{\overset{O}{\|}}{C}-O^-\ ^+Na \quad \text{Monosodium glutamate}$$

9.9 REPRODUCTIVE SYSTEM

Humans reproduce sexually to produce young that are born live. Sexual reproduction occurs when two **gametes**, **sperm** from males and **eggs** from females, each carrying a single set of chromosomes, unite in a process called **fertilization**. The fertilized egg, called a **zygote**, contains two sets of chromosomes, one from each of the gametes. The zygote begins the process of division and cell differentiation that results in an individual capable of living outside the womb at birth. The whole process is rather complicated and can be affected in many stages by toxic substances. Therefore, one of the primary concerns in toxicology is the influence of toxicants on the reproductive system.

The glands that produce sperm are the **testes**. Prior to copulation, the sperm are stored and undergo further development in the epididymis, located on the testicles. For delivery, sperm are incorporated into seminal fluid produced by seminal vesicles, the prostate gland, and the bulbourethral gland, and ejaculated through the urethra of the penis. The process of forming sperm and other male sexual functions and characteristics are promoted by **testosterone**, the male sex hormone.

Whereas healthy males can produce sperm at any time, the production of fertilizable eggs by females is rather complicated. It involves the **ovarian** cycle of normally around 28 days. During the first half of the cycle usually one egg is produced and expelled from the ovary in a process called **ovulation**. If the egg is not fertilized and implanted in the uterus, the endometrium lining the uterus breaks down and is expelled through the vagina, a process called menstruation that occurs during the second half of the ovarian cycle. The egg produced by the ovary moves slowly toward the uterus through **oviducts** (fallopian tubes). The uterus is connected to the vagina, through which sperm enters by an opening called the **cervix**. Fertilization occurs in the upper region of the oviducts.

In successful pregnancies, the zygote formed by the merging of the egg and sperm begins to divide in the oviduct, forming a **blastocyst** that becomes implanted in the endometrium wall of the uterus.

Several hormones are involved in the ovulation process. These include gonadotropin-releasing hormone from the hypothalamus, lutenizing hormone and follicle-stimulating hormone from the anterior pituitary, and estrogen from the ovaries (see endocrine glands in Section 9.7). If the blastocyst becomes implanted in the endometrium, a layer of cells covering it begin to secrete **human chorionic gonadotropin**, produced only by pregnant females and used as the basis for pregnancy testing. These tissues also produce high levels of estrogen and progesterone that prevent the pituitary from generating gonadotropins, thus stopping ovulation and menstruation during pregnancy. Synthetic analogs of estrogen and progesterone in oral contraceptives act to prevent ovulation but not the uterine cycle of menstruation.

Various toxicants are toxic to sperm or adversely affect semen quality. Common parameters for detecting damage to sperm include sperm production, numbers, transit time, and mobility. The ultimate measure of sperm quality is the ability to produce pregnancy resulting in normal offspring. Toxicants may interfere with the process of sperm development. In rodents, these include heavy metals (cadmium), hormones (estrogen), herbicides (linuron), industrial chemicals (dimethyl formamide), and pharmaceuticals, such as antihypertensive reserpine. Prominent among pharmaceuticals that are spermatotoxic are antimetabolites, such as cyclophosphamide, used in cancer chemotherapy. The cottonseed pigment gossypol adversely affects sperm, as does antifungal benomyl.

Cyclophosphamide

Some toxicants are known to affect the female reproductive system and processes. Exposure to the alkylating agents cyclophosphamide and vincristine can lead to loss of female sexual function. Cyclophosphamide may attack and damage the oocytes, cells that lead to egg formation. Pharmaceutical busulfan damages ovaries. The 7,8-diol-9,10-epoxide of benzo(a)pyrene, as well as some other metabolites of polycyclic aromatic hydrocarbons, can be toxic to oocytes.

Prominent among toxicants that adversely affect both male and female reproductive systems are endocrine disruptors (see Section 9.7). Toxicants that mimic the actions of sex hormones are *agonists*, and those that prevent hormonal action or bind competitively to hormone receptor sites are *antagonists*.[12] Male patients treated with cimetidine for peptic ulcers have exhibited low sperm counts and abnormal breast enlargement, a condition called **gynecomastia**. Gynecomastia has also been caused in men working in oral contraceptive production. Ketoconozole inhibits the enzymes required to produce hormones involved in sperm production and can immobilize sperm in seminal fluid.

Because of the complex hormonal cycle experienced by women, it is more difficult to study the effects of endocrine-disrupting toxicants in females. Estrogen-mimicking compounds have the potential to disrupt female hormonal cycles, causing adverse reproductive effects. The classic and tragic case of estrogen-disrupting toxicants is that of diethylstilbestrol, given to pregnant women in the 1950s to prevent miscarriages. Female children of these women developed vaginal cancer after reaching puberty. It is believed that the toxic agent responsible for the effect is a reactive epoxide intermediate in the metabolism of estradiol that binds with the receptor for the sex steroid estradiol (Reaction 9.9.1):

(9.9.1)

The possibility that high estrogen levels may be linked to breast cancer has led to concern that estrogen-mimicking chemicals might also increase the likelihood of breast cancer. Among the possibilities are polychlorinated biphenyls and DDT. Elevated blood serum levels of the DDT metabolite DDE have shown a positive correlation with breast cancer incidence.

9.10 DEVELOPMENTAL TOXICOLOGY AND TERATOLOGY

Developmental toxicology deals with the effects of toxic substances on development of an organism from conception to birth. Toxic substances may interfere with embryo growth, homeostatis, differentiation, and development of physical and behavioral characteristics.

Much of developmental toxicology deals with **teratology**, the branch of embryology pertaining to abnormal development and birth defects due to exposure to toxicants called **teratogens**. Teratogens can act in a number of ways. Many of their effects are due to harmful alterations of DNA, the expression of genes from DNA, and the processes by which DNA generates RNA in protein synthesis. Teratogens can alter chromosomes, leading to defects. Enzymatic processes can be altered. Other teratogenic effects include oxidative stress, interference with the normal processes of programmed cell death, and alterations in cell membranes.

Teratology is the science of birth defects caused by radiation, viruses, and chemicals, including drugs. Xenobiotic chemical species that cause birth defects are called teratogens. Teratogens affect developing embryos adversely, often with remarkable specificity in regard to effect and stage of embryo development when exposed. A teratogen may cause a specific effect when exposure occurs on a definite number of days after conception; if exposure occurs only a few days sooner or later, no effect, or an entirely different one, may be observed. Although mutations in germ cells (egg or sperm cells) may cause birth defects (e.g., Down's syndrome), teratology usually deals with defects arising from damage to embryonic or fetal cells.

The biochemical aspects of teratology are not particularly well understood. Several kinds of biochemical mechanisms are probably involved. One such mechanism is interference with DNA synthesis, which alters the function of nucleic acids in cell replications, resulting in effects that are expressed as birth defects. Exposure to teratogenic xenobiotic substances may result in either an absence or excess of chromosomes. Enzyme inhibition (see Section 7.6) by xenobiotics can result in birth defects. Xenobiotics that deprive the fetus of essential substrates (for example, vitamins), that interfere with energy supply, or that alter the permeability of the placental membrane may all cause birth defects.

9.10.1 Thalidomide

Perhaps the most notorious teratogen is thalidomide, a sedative–hypnotic drug used in Europe and Japan in 1960 and 1961. Some infants born to women who had taken thalidomide from days 35 through 50 of their pregnancies were born suffering from amelia or phocomelia, the absence or severe shortening, respectively, of the limbs. About 10,000 children were affected. The biochemical action of thalidomide leading to teratogenesis is not well understood. Possibilities include adverse modification of DNA and interference with the metabolism of folic acid or glutamic acid.

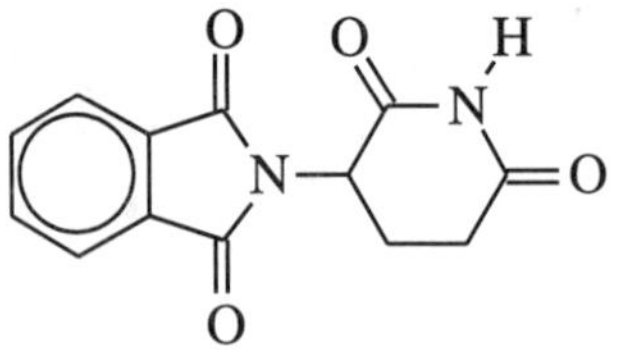

Thalidomide

9.10.2 Accutane

In 1988, the U.S. Food and Drug Administration estimated that **Accutane**, used as an antiacne medication, may have been reponsible for approximately 1000 birth defects in children born to women taking the drug during the period of 1982 to 1986. The chemical name for Accutane is isoretinoin, and it is chemically related to retinoic acid, vitamin A, which likewise is teratogenic at excessive levels. Exposure of the fetus to Accutane over a period of only several days can result in birth defects such as severe facial malformations, heart defects, thymus defects, and mental retardation.[13]

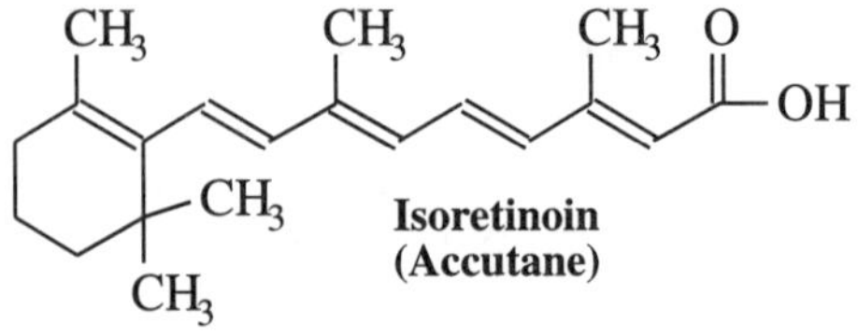

Isoretinoin (Accutane)

9.10.3 Fetal Alcohol Syndrome

One of the more common and devastating teratogenic effects is **fetal alcohol syndrome**, which can afflict offspring of women who have regularly and heavily ingested alcohol during pregnancy. This syndrome is most visibly manifested by abnormal facial features, including a very thin upper lip, an upturned nose, and a broadened nasal bridge. The brain is affected and may be smaller than normal, resulting in mental retardation.

9.11 KIDNEY AND BLADDER

The kidney and bladder are very important in toxicology because they are the main route of elimination of hydrophilic toxicant metabolites and because damage to them in the form of impaired kidney function or bladder cancer is one of the major adverse effects of toxicants. The kidney plays a key role in maintaining body homeostasis. The basic unit of the kidney, through which the organ performs its crucial blood filtration action, is the **nephron**. As the main organ through which fluid is lost from the body, it is vital in the maintenance of extracellular fluid volume. It acts to maintain the critical electrolyte balance in the blood and other parts of the body. It also maintains acid–base balance. It is in a sense the main "filtering" organ for the body's blood, removing wastes and toxicants from blood and discharging them through the bladder with urine while conserving essential ions, amino acids, and glucose. Another crucial function of the kidney is production of the active dihydroxy form from vitamin D_3, crucial in regulating the absorption of calcium from the intestines and deposition of calcium in bones. The kidney produces essential hormones, including renin and erythropoietin.

Though only about 1% of body mass, the kidney receives about one fourth of the heart output of blood. This high level of blood combined with the kidney's ability to concentrate substances in the kidney tubular fluid, from which most of the water and sodium removed by the kidney are returned to the blood, often means that the kidney is exposed to especially high levels of toxicants. Such concentrations have been known to cause deposition of substances such as sulfonamides and oxalates in the kidney, resulting in cell necrosis. Fortunately, the kidney has a good ability to compensate for damage.

Toxic effects to the kidney may be manifested by acute and chronic **renal failure**. Many substances are known to be **nephrotoxic**.[14] Included among such substances are therapeutic agents. Some of these include organic mercury compounds administered as diuretics to increase urine output, anti-infective agents such as sulfonamides and vancomycin, antineoplastic (cancer therapeutic) adriamycin and mitomycin C, immunosuppressive cyclosporin A, analgesic and anti-inflammatory acetaminophen, and enflurane and lithium used to treat disorders of the central nervous system. A number of substances to which environmental and occupational exposure may occur have also been implicated in kidney damage. Some metals, including cadmium, lead, mercury, nickel, and chromium, are nephrotoxic. Some substances derived from bacteria (mycotoxins) and plants (especially alkaloids) are nephrotoxic. These include aflatoxin B, citrinin, pyrrolizidine alkaloids, and rubratoxin B. Nephrotoxic halogenated hydrocarbons include bromobenzene, chloroform, carbon tetrachloride, and tetrafluoroethylene, which is transported to the kidney as the cysteine S-conjugate. Ethylene glycol and diethylene glycol harm kidneys because of their bioconversion to oxalates that clog kidney tubules. Herbicidal paraquat, diquat, and 2,4,5-trichlorophenoxyacetate also have toxic effects on the kidney.

REFERENCES

1. Zoltán, G. and Klaassen, C.D., Mechanisms of toxicity, in *Casarett and Doull's Toxicology: The Basic Science of Poisons*, 5th ed., Klaassen, C.D., Ed., McGraw-Hill, New York, 1996, chap. 3, pp. 35–74.
2. Andrews, E.L., Drug's removal exposes holes in Europe's net, *New York Times*, Aug. 22, 2001, p. C1.
3. Witschi, H. and Last, J.A., Toxic responses of the respiratory system, in *Casarett and Doull's Toxicology: The Basic Science of Poisons*, 5th ed., Klaassen, C.D., Ed., McGraw-Hill, New York, 1996, chap. 15, pp. 443–462.
4. Snider, G.L. et al., The definition of emphysema: report of a National Heart, Lung, and Blood Institute workshop, *Am. Rev. Respir. Dis.*, 132, 182–185, 1985.
5. Rice, R. and Cohen, D.E., Toxic responses of the skin, in *Casarett and Doull's Toxicology: The Basic Science of Poisons*, 5th ed., Klaassen, C.D., Ed., McGraw-Hill, New York, 1996, chap. 18, pp. 529–546.
6. Moslen, M.T., Toxic responses of the liver, in *Casarett and Doull's Toxicology: The Basic Science of Poisons*, 5th ed., Klaassen, C.D., Ed., McGraw-Hill, New York, 1996, chap. 13, pp. 403–416.
7. McKenzie, R. et al., Hepatic failure and lactic acidosis due to fialuridine (FIAU), an investigational nucleoside analog for chronic hepatitis B, *N. Engl. J. Med.*, 333, 1099–1105, 1995.
8. Biochemical mechanisms of toxicity: specific examples, in *Principles of Biochemical Toxicology*, 3rd ed., Timbrell, J., Ed., Taylor and Francis, London, 2000, chap. 7, pp. 259–353.
9. Smith, R.P, Toxic responses of the blood, in *Casarett and Doull's Toxicology: The Basic Science of Poisons*, 5th ed., Klaassen, C.D., Ed., McGraw-Hill, New York, 1996, chap. 11, pp. 335–354.

10. Ramos, K.S., Chacon, E., and Acosta, D., Toxic responses of the heart and vascular systems, in *Casarett and Doull's Toxicology: The Basic Science of Poisons*, 5th ed., Klaassen, C.D., Ed., McGraw-Hill, New York, 1996, chap. 17, pp. 487–527.
11. Anthony, D.C., Montine, T.J., and Graham, D.G., Toxic responses of the nervous system, in *Casarett and Doull's Toxicology: The Basic Science of Poisons*, 5th ed., Klaassen, C.D., Ed., McGraw-Hill, New York, 1996, chap. 16, pp. 463–486.
12. Branch, S., Reproductive and developmental toxicity, in *Introduction to Biochemical Toxicology*, Hodgson, E. and Smart, R.C., Eds., Wiley Interscience, New York, 2001, chap. 21, pp. 539–559.
13. Rogers, J.M. and Kavlock, R.J., Developmental toxicology, in *Casarett and Doull's Toxicology: The Basic Science of Poisons*, 5th ed., Klaassen, C.D., Ed., McGraw-Hill, New York, 1996, chap. 10, pp. 301–331.
14. Goldstein, R.S. and Schnellmann, R.C., Toxic responses of the kidney, in *Introduction to Biochemical Toxicology*, Hodgson, E. and Smart, R.C., Eds., Wiley Interscience, New York, 2001, chap. 24, pp. 417–442.

SUPPLEMENTARY REFERENCES

Bennett, E.O., *Dermatitis in Machinists: Causes and Solutions*, Biotech Publishing, Angleton, TX, 1993.
Cros, D., Ed., *Peripheral Neuropathy: a Practical Approach to Diagnosis and Management*, Lippincott Williams & Wilkins, Philadelphia, 2001.
Derkins, S., *The Immune System*, Rosen Publishing Group, 2001.
Harvey, P.W., Rush, K.C., and Cockburn, A., Eds., *Endocrine and Hormonal Toxicology*, Wiley, Chichester, 1999.
Iannucci, L., *Birth Defects*, Enslow Publishers, Berkley Heights, NJ, 2000.
Langston, J.W. and Young, A., Eds., *Neurotoxins and Neurodegenerative Disease*, New York Academy of Sciences, New York, 1992.
Naor, D., *Immunosuppression and Human Malignancy*, Humana Press, Clifton, NJ, 1989.
Phalen, R.F., Ed., *Methods in Inhalation Toxicology*, CRC Press, Boca Raton, 1997.
Porter, G.A., Ed., *Nephrotoxic Mechanisms of Drugs and Environmental Toxins*, Plenum Medical Book Co., New York, 1982.
Rietschel, R.L. and Fowler, Joseph F. Jr., Eds., *Fisher's Contact Dermatitis*, 5th ed., Lippincott Williams & Wilkins, Philadelphia, 2001.
Shinton, N.K., Ed., CRC Desk Reference for Hematology, CRC Press, Boca Raton, 1998.

QUESTIONS AND PROBLEMS

1. Why is apoptosis advantageous over cell necrosis in ridding the body of defective cells?
2. In what sense do the alveoli of lungs provide the closest contact of the internal parts of the body with the outside surroundings? What are the implications of this for toxicology?
3. What is meant by oxidative burden? What are its major toxicological implications?
4. The skin's stratum corneum is "dead," and yet it is a vital defense against toxic substances. Explain.
5. Match the following pertaining to toxic responses of the skin:
   1. Contact dermatitis
   2. Allergic contact dermatitis
   3. Porphyria
   4. Chloracne

   (a) Occurs with exposure to sunlight after sensitization by a toxic substance
   (b) Common result of exposure to TCDD
   (c) Characterized by generally irritated, itching, and sometimes painful skin surface
   (d) Requires initial sensitization to a substance
6. Match the following pertaining to toxic effects to the liver:
   1. Hepatotoxicity
   2. Steatosis
   3. Cirrhosis
   4. Haemangiosarcoma

   (a) Commonly known as fatty liver
   (b) General term applied to effects of toxic substances to the liver
   (c) Caused by exposure to vinyl chloride
   (d) Characterized by deposition and buildup of fibrous collagen tissue that replace active liver cells

7. Distinguish between anemic hypoxia and methemoglobinemia. In what respect are they similar? Name a toxicant responsible for each of these toxic effects on blood.
8. Distinguish among bradycardia, tachycardia, and arrhythmia as related to toxic substances.
9. Two general effects on the immune system occur when it (1) becomes too active and (2) becomes too inactive. Explain and give examples. What is autoimmunity?
10. Match the following pertaining to toxic effects on the endocrine system:
    1. Endocrine disruptors   (a) Chemical messengers produced by endocrine glands
    2. Insulin   (b) Steroids
    3. Hormones   (c) Greatest concern with reproductive effects and sexual development
    4. Estrogens   (d) A hormone produced and used by humans
11. What is the function of the blood-brain barrier? How does it relate to toxic substances?
12. Both encephelopathy and peripheral neuropathy are disorders of the nervous system. What is the distinction between them?
13. What is a major toxicological concern with estrogen-mimicking chemicals?
14. What do accutane, thalidomide, ethanol, radiation, and some viruses have in common toxicologically?
15. What are nephrotoxic substances? What are examples of such substances?

CHAPTER 10

Toxic Elements

10.1 INTRODUCTION

It is somewhat difficult to define what is meant by a toxic element. Some elements, such as white phosphorus, chlorine, and mercury, are quite toxic in the elemental state. Others, such as carbon, nitrogen, and oxygen, are harmless as usually encountered in their normal elemental forms. But, with the exception of those noble gases that do not combine chemically, all elements can form toxic compounds. A prime example is hydrogen cyanide. This extremely toxic compound is formed from three elements that are nontoxic in the uncombined form, and produce compounds that are essential constituents of living matter, but when bonded together in the simple HCN molecule constitute a deadly substance.

The following three categories of elements are considered here:

- Those that are notable for the toxicities of most of their compounds
- Those that form very toxic ions
- Those that are very toxic in their elemental forms

Elements in these three classes are discussed in this chapter as **toxic elements**, with the qualification that this category is somewhat arbitrary. With a few exceptions, elements known to be essential to life processes in humans have not been included as toxic elements.

10.2 TOXIC ELEMENTS AND THE PERIODIC TABLE

It is most convenient to consider elements from the perspective of the periodic table, which is shown in Figure 1.3 and discussed in Section 1.2. Recall that the three main types of elements, based on their chemical and physical properties as determined by the electron configurations of their atoms, are metals, nonmetals, and metalloids. Metalloids (B, Si, Ge, As, Sb, Te, At) show some characteristics of both metals and nonmetals. The nonmetals consist of those few elements in groups 4A to 7A above and to the right of the metalloids. The noble gases, only some of which form a limited number of very unstable chemical compounds of no toxicological significance, are in group 8A. All the remaining elements, including the lanthanide and actinide series, are metals. Elements in the periodic table are broadly distinguished between representative elements in the A groups of the periodic table and transition metals constituting the B groups, the lanthanide series, and the actinide series.

10.3 ESSENTIAL ELEMENTS

Some elements are essential to the composition or function of the body. Since the body is mostly water, hydrogen and oxygen are obviously essential elements. Carbon (C) is a component of all life molecules, including proteins, lipids, and carbohydrates. Nitrogen (N) is in all proteins. The other essential nonmetals are phosphorus (P), sulfur (S), chlorine (Cl), selenium (Se), fluorine (F), and iodine (I). The latter two are among the essential trace elements that are required in only small quantities, particularly as constituents of enzymes or as cofactors (nonprotein species essential for enzyme function). The metals present in macro amounts in the body are sodium (Na), potassium (K), and calcium (Ca). Essential trace elements are chromium (Cr), manganese (Mn), iron (Fe), cobalt (Co), copper (Cu), zinc (Zn), magnesium (Mg), molybdenum (Mo), nickel (Ni), and perhaps more elements that have not yet been established as essential.

10.4 METALS IN AN ORGANISM

Metals are mobilized and distributed through environmental chemical processes that are strongly influenced by human activities. A striking example of this phenomenon is illustrated by the lead content of the Greenland ice pack. Starting at very low levels before significant industrialization had occurred, the lead content of the ice increased in parallel with the industrial revolution, showing a strongly accelerated upward trend beginning in the 1920s, with the introduction of lead into gasoline. With the curtailment of the use of leaded gasoline, some countries are now showing decreased lead levels, a trend that hopefully will extend globally within the next several decades.

Metals in the body are almost always in an oxidized or chemically combined form; mercury is a notable exception in that elemental mercury vapor readily enters the body through the pulmonary route. The simplest form of a chemically bound metal in the body is the hydrated cation, of which $Na(H_2O)_6^+$ is the most abundant example. At pH values ranging upward from somewhat less than seven (neutrality), many metal ions tend to be bound to one or more hydroxide groups; an example is iron(II) in $Fe(OH)(H_2O)_5^+$. Some metal ions have such a strong tendency to lose H^+ that, except at very low pH values, they exist as the insoluble hydroxides. A common example of this phenomenon is iron(III), which is very stable as the insoluble hydrated iron(III) oxide, $Fe_2O_3 \cdot xH_2O$, or hydroxide, $Fe(OH)_3$. Metals can bond to some anions in body fluids. For example, in the strong hydrochloric acid medium of the stomach, some iron(III) may be present as $HFeCl_4$, where the acid in the stomach prevents formation of insoluble $Fe(OH)_3$ and a high concentration of chloride ion is available to bond to iron(III). Ion pairs may exist that consist of positively charged metal cations and negatively charged anions endogenous to body fluids. These do not involve covalent bonding between cations and anions, but rather an electrostatic attraction, such as in the ion pairs $Ca^{2+} HCO_3^-$ or $Ca^{2+}Cl^-$.

10.4.1 Complex Ions and Chelates

With the exception of group 1A metals and the somewhat lesser exception of group 2A metals, there is a tendency for metals to form **complexes** with **electron donor** functional groups on **ligands** consisting of anionic or neutral inorganic or organic species. In such cases, covalent bonds are formed between the **central metal ion** and the ligands. Usually the resulting complex has a net charge and is called a complex ion; $FeCl_4^-$ is such an ion. In many cases, an organic ligand has two or more electron donor functional groups that may simultaneously bond to a metal ion to form a complex with one or more rings in its structure. A ligand with this capability is called a **chelating agent**, and the complex is a **metal chelate**. Copper(II) ion forms such a chelate with the anion of the amino acid glycine, as shown in Figure 10.1. This chelate is very stable.

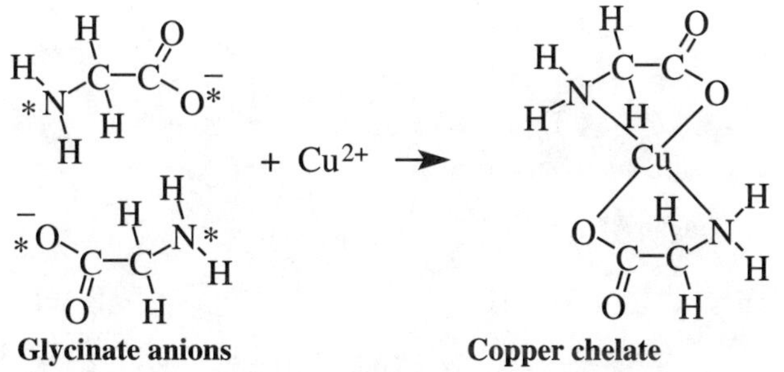

Figure 10.1 Chelation of Cu^{2+} by glycinate anion ligands to form the glycinate chelate. Each electron donor group on the glycinate anion chelating agents is designated with an asterisk. In the chelate, the central copper(II) metal ion is bonded in four places and the chelate has two rings composed of the five-atom sequence Cu–O–C–C–N.

Organometallic compounds constitute a large class of metal-containing species with properties quite different from those of the metal ions. These are compounds in which the metal is covalently bonded to carbon in an organic moiety, such as the methyl group, $-CH_3$. Unlike metal complexes, which can reversibly dissociate to the metal ions and ligands, the organic portions of organometallic compounds are not normally stable by themselves. The chemical and toxicological properties of organometallic compounds are discussed in detail in Chapter 12, so space will not be devoted to them here. However, it should be mentioned that neutral organometallic compounds tend to be lipid soluble, a property that enables their facile movement across biologic membranes. They often remain intact during movement through biological systems and so become distributed in these systems as lipid-soluble compounds.

A phenomenon not confined to metals, **methylation** is the attachment of a methyl group to an element and is a significant natural process responsible for much of the environmental mobility of some of the heavier elements. Among the elements for which methylated forms are found in the environment are cobalt, mercury, silicon, phosphorus, sulfur, the halogens, germanium, arsenic, selenium, tin, antimony, and lead.

10.4.2 Metal Toxicity

Inorganic forms of most metals tend to be strongly bound by protein and other biologic tissue. Such binding increases bioaccumulation and inhibits excretion. There is a significant amount of tissue selectivity in the binding of metals. For example, toxic lead and radioactive radium are accumulated in osseous (bone) tissue, whereas the kidneys accumulate cadmium and mercury. Metal ions most commonly bond with amino acids, which may be contained in proteins (including enzymes) or polypeptides. The electron-donor groups most available for binding to metal ions are amino and carboxyl groups (see Figure 10.2). Binding is especially strong for many metals to thiol (sulfhydryl) groups; this is particularly significant because the –SH groups are common components of the active sites of many crucial enzymes, including those that are involved in cellular energy output and oxygen transport. The amino acid that usually provides –SH groups in enzyme active sites is cysteine, as shown in Figure 10.2. The imidazole group of the amino acid histidine is a common feature of enzyme active sites with strong metal-binding capabilities.

The absorption of metals is to a large extent a function of their chemical form and properties. Pulmonary intake results in the most facile absorption and rapid distribution through the circulatory system. Absorption through this route is often very efficient when the metal is in the form of respirable particles less than 100 μm in size, as volatile organometallic compounds (see Chapter 12) or (in the case of mercury) as the elemental metal vapor. Absorption through the gastrointestinal tract is affected by pH, rate of movement through the tract, and presence of other materials. Particular combinations of these factors can make absorption very high or very low.

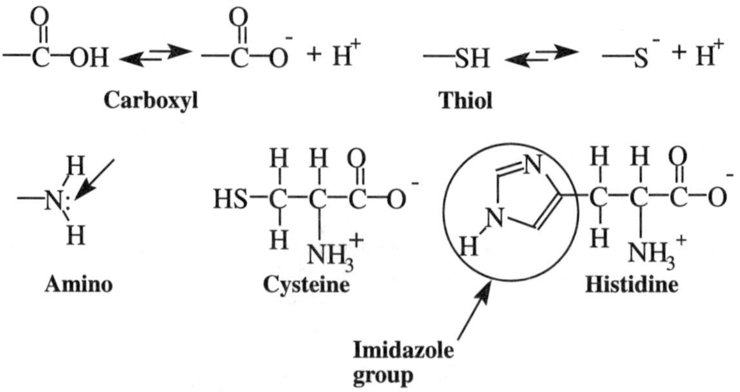

Figure 10.2 Major binding groups for metal ions in biologic tissue (carboxyl, thiol, amino) and amino acids with strong metal-binding groups in enzyme active sites (cysteine, histidine). The arrow pointing to the amino group designates an unshared pair of electrons available for binding metal ions. The thiol group is a weak acid that usually remains unionized until the hydrogen ion is displaced by a metal ion.

Metals tend to accumulate in target organs, and a toxic response is observed when the level of the metal in the organ reaches or exceeds a threshold level. Often the organs most affected are those involved with detoxication or elimination of the metal. Therefore, the liver and kidneys are often affected by metal poisoning. The form of the metal can determine which organ is adversely affected. For example, lipid-soluble elemental or organometallic mercury damages the brain and nervous system, whereas Hg^{2+} ion may attack the kidneys.

Because of the widespread opportunity for exposure, combined with especially high toxicity, some metals are particularly noted for their toxic effects. These are discussed separately in the following sections in the general order of their appearance in groups in the periodic table.

10.4.3 Lithium

Lithium, Li, atomic number 3, is the lightest group 1A metal that should be mentioned as a toxicant because of its widespread use as a therapeutic agent to treat manic-depressive disorders. It is also used in a number of industrial applications, where there is potential for exposure.

The greatest concern with lithium as a toxicant is its toxicity to kidneys, which has been observed in some cases in which lithium was ingested within therapeutic ranges of dose. Common symptoms of lithium toxicity include high levels of albumin and glucose in urine (albuminuria and glycosuria, respectively). Not surprisingly, given its uses to treat manic-depressive disorders, lithium can cause a variety of central nervous system symptoms. One symptom is psychosomatic retardation, that is, retardation of processes involving both mind and body. Slurred speech, blurred vision, and increased thirst may result. In severe cases, blackout spells, coma, epileptic seizures, and writhing, turning, and twisting choreoathetoid movements are observed. Neuromuscular changes may occur as irritable muscles, tremor, and ataxia (loss of coordination). Cardiovascular symptoms of lithium poisoning may include cardiac arrhythmia, hypertension, and, in severe cases, circulatory collapse. Victims of lithium poisoning may also experience an aversion to food (anorexia) accompanied by nausea and vomiting.

Lithium exists in the body as the Li^+ ion. Its toxic effects are likely due to its similarity to physiologically essential Na^+ and K^+ ions. Some effects may be due to the competition of Li^+ ion for receptor sites normally occupied by Na^+ or K^+ ions. Lithium toxicity may be involved in G protein expression and in modulating receptor–G protein coupling.[1]

10.4.4 Beryllium

Beryllium (Be) is in group 2A and is the first metal in the periodic table to be notably toxic. When fluorescent lamps and neon lights were first introduced, they contained beryllium phosphor; a number of cases of beryllium poisoning resulted from the manufacture of these light sources and the handling of broken lamps. Modern uses of beryllium in ceramics, electronics, and alloys require special handling procedures to avoid industrial exposure.

Beryllium has a number of toxic effects. Of these, the most common involve the skin. Skin ulceration and granulomas have resulted from exposure to beryllium. Hypersensitization to beryllium can result in skin dermatitis, acute conjunctivitis, and corneal laceration.

Inhalation of beryllium compounds can cause **acute chemical pneumonitis**, a very rapidly progressing condition in which the entire respiratory tract, including nasal passages, pharynx, tracheobronchial airways, and alveoli, develops an inflammatory reaction. Beryllium fluoride is particularly effective in causing this condition, which has proven fatal in some cases.

Chronic berylliosis may occur with a long latent period of 5 to 20 years. The most damaging effect of chronic berylliosis is lung fibrosis and pneumonitis. In addition to coughing and chest pain, the subject suffers from fatigue, weakness, loss of weight, and dyspnea (difficult, painful breathing). The impaired lungs do not transfer oxygen well. Other organs that can be adversely affected are the liver, kidneys, heart, spleen, and striated muscles.

The chemistry of beryllium is atypical compared to that of the other group 1A and group 2A metals. Atoms of Be are the smallest of all metals, having an atomic radius of 111 pm. The beryllium ion, Be^{2+}, has an ionic radius of only 35 pm, which gives it a high polarizing ability, a tendency to form molecular compounds rather than ionic compounds, and a much greater tendency to form complex compounds than other group 1A or 2A ions. The ability of beryllium to form chelates is used to treat beryllium poisoning with ethylenediaminetetraacetic acid (EDTA) and another chelating agent called Tiron[2]:

10.4.5 Vanadium

Vanadium (V) is a transition metal that in the combined form exists in the +3, +4, and +5 oxidation states, of which +5 is the most common. Vanadium is of concern as an environmental pollutant because of its high levels in residual fuel oils and subsequent emission as small particulate matter from the combustion of these oils in urban areas. Vanadium occurs as chelates of the porphyrin type in crude oil, and it concentrates in the higher boiling fractions during the refining process. A major industrial use of vanadium is in catalysts, particularly those in which sulfur dioxide is oxidized in the production of sulfuric acid. The other major industrial uses of vanadium are for hardening steel, as a pigment ingredient, in photography, and as an ingredient of some insecticides. In addition to environmental exposure from the combustion of vanadium-containing fuels, there is some potential for industrial exposure.

Probably the vanadium compound to which people are most likely to be exposed is vanadium pentoxide, V_2O_5. Exposure normally occurs via the respiratory route, and the pulmonary system is the most likely to suffer from vanadium toxicity. Bronchitis and bronchial pneumonia are the most common pathological effects of exposure; skin and eye irritation may also occur. Severe exposure can also adversely affect the gastrointestinal tract, kidneys, and nervous system.

Both V(IV) and V(V) have been found to have reproductive and developmental toxic effects in rodents. In addition to decreased fertility, lethal effects to embryos, toxicity to fetuses, and teratogenicity have been observed in mice, rats, and hamsters exposed to vanadium.[3]

It has been observed that vanadium has insulin-like effects on the main organs targeted by insulin — skeletal muscles, adipose, and liver — and vanadium has been shown to reduce blood glucose to normal levels in rats that have diabetic conditions. In considering the potential of vanadium to treat diabetes in humans, the toxicity of vanadium is a definite consideration. Several organically chelated forms of vanadium have been found to be more effective in treating diabetes symptoms and less toxic than inorganic vanadium.[4]

10.4.6 Chromium

Chromium (Cr) is a transition metal. In the chemically combined form, it exists in all oxidation states from +2 to +6, of which +3 and +6 are the more notable.

In strongly acidic aqueous solution, chromium(III) may be present as the hydrated cation $Cr(H_2O)_6^{3+}$. At pH values above approximately 4, this ion has a strong tendency to precipitate from solution the hydroxide:

$$Cr(H_2O)_6^{3+} \rightarrow Cr(OH)_3 + 3H^+ + 3H_2O \qquad (10.4.1)$$

The two major forms of chromium(VI) in solution are yellow chromate, CrO_4^{2-}, and orange dichromate, $Cr_2O_7^{2-}$. The latter predominates in acidic solution, as shown by the following reaction, the equilibrium of which is forced to the left by higher levels of H^+:

$$Cr_2O_7^{2-} + H_2O \rightleftharpoons 2HCrO_4^- \rightleftharpoons 2H^+ + 2CrO_4^{2-} \qquad (10.4.2)$$

Chromium in the +3 oxidation state is an essential trace element (see Section 10.3) required for glucose and lipid metabolism in mammals, and a deficiency of it gives symptoms of diabetes mellitus. However, chromium must also be discussed as a toxicant because of its toxicity in the +6 oxidation state, commonly called **chromate**. Exposure to chromium(VI) usually involves chromate salts, such as Na_2CrO_4. These salts tend to be water soluble and readily absorbed into the bloodstream through the lungs. The carcinogenicity of chromate has been demonstrated by studies of exposed workers. Exposure to atmospheric chromate may cause bronchogenic carcinoma with a latent period of 10 to 15 years. In the body, chromium(VI) is readily reduced to chromium(III), as shown in Reaction 10.4.3; however, the reverse reaction does not occur in the body.

$$CrO_4^{2-} + 8H^+ + 3e^- \rightarrow Cr^{3+} + 4H_2O \qquad (10.4.3)$$

An interesting finding regarding potentially toxic chromium (and cobalt) in the body is elevated blood and urine levels of these metals in patients who have undergone total hip replacement.[5] The conclusion of the study was that devices such as prosthetic hips that involve metal-to-metal contact may result in potentially toxic levels of metals in biological fluids.

10.4.7 Cobalt

Cobalt is an essential element that is part of vitamin B_{12}, or cobalamin, a coenzyme that is essential in the formation of proteins, nucleic acids, and red blood cells. Although cobalt poisoning is not common, excessive levels can be harmful. Most cases of human exposure to toxic levels of cobalt have occurred through inhalation in the workplace. Many exposures have been suffered by workers working with hard metal alloys of cobalt and tungsten carbide, where very fine particles of the alloy produced from grinding it were inhaled. The adverse effects of cobalt inhalation have been on the lungs, including wheezing and pneumonia as well as allergic asthmatic reactions and skin rashes. Lung fibrosis has resulted from prolonged exposures. Human epidemiology and animal studies suggest an array of systemic toxic effects of cobalt, including, in addition to respiratory effects, cardiovascular, hematological hepatic, renal, ocular, and body weight effects.

Exposure to cobalt is also possible through food and drinking water. An interesting series of cobalt poisonings occurred in the 1960s when cobalt was added to beer at levels of 1 to 1.5 ppm to stabilize foam. Consumers who drank excessive amounts of the beer (4 to 12 liters per day) suffered from nausea and vomiting, and in several cases, heart failure and death resulted.

10.4.8 Nickel

Nickel, atomic number 28, is a transition metal with a variety of essential uses in alloys, catalysts, and other applications. It is strongly suspected of being an essential trace element for human nutrition, although definitive evidence has not yet established its essentiality to humans. A nickel-containing urease metalloenzyme has been found in the jack bean.

Toxicologically, nickel is important because it has been established as a cause of respiratory tract cancer among workers involved with nickel refining. The first definitive evidence of this was an epidemiological study of British nickel refinery workers published in 1958. Compared to the general population, these workers suffered a 150-fold increase in nasal cancers and a 5-fold increase in lung cancer. Other studies from Norway, Canada, and the former Soviet Union have shown similar increased cancer risk from exposure to nickel. Nickel subsulfide, Ni_3S_2, has been shown to cause cancer in rats at sites of injection and in lungs from inhalation of nickel subsulfide.

The other major toxic effect of nickel is nickel dermatitis, an allergic contact dermatitis arising from contact with nickel metal. About 5 to 10% of people are susceptible to this disorder. It almost always occurs as the result of wearing nickel jewelry in contact with skin. Nickel carbonyl, $Ni(CO)_4$, is an extremely toxic nickel compound discussed further in Chapter 12.

10.4.9 Cadmium

Along with mercury and lead, cadmium (Cd) is one of the "big three" heavy metal poisons. Cadmium occurs as a constituent of lead and zinc ores, from which it can be extracted as a by-product. Cadmium is used to electroplate metals to prevent corrosion, as a pigment, as a constituent of alkali storage batteries, and in the manufacture of some plastics.

Cadmium is located at the end of the second row of transition elements. The +2 oxidation state of the element is the only one exhibited in its compounds. In its compounds, cadmium occurs as the Cd^{2+} ion. Cadmium is directly below zinc in the periodic table and behaves much like zinc. This may account in part for cadmium's toxicity; because zinc is an essential trace element, cadmium substituting for zinc could cause metabolic processes to go wrong.

The toxic nature of cadmium was revealed in the early 1900s as a result of workers inhaling cadmium fumes or dusts in ore processing and manufacturing operations. Welding or cutting metals plated with cadmium or containing cadmium in alloys, or the use of cadmium rods or wires for brazing or silver soldering, can be a particularly dangerous route to pulmonary exposure. In general, cadmium is poorly absorbed through the gastrointestinal tract. A mechanism exists for its active absorption in the small intestine through the action of the low-molecular-mass calcium-binding protein CaBP. The production of this protein is stimulated by a calcium-deficient diet, which may aggravate cadmium toxicity. Cadmium is transported in blood bound to red blood cells or to albumin or other high-molecular-mass proteins in blood plasma. Cadmium is excreted from the body in both urine and feces. The mechanisms of cadmium excretion are not well known.

Acute pulmonary symptoms of cadmium exposure are usually caused by the inhalation of cadmium oxide dusts and fumes, which results in cadmium pneumonitis, characterized by edema

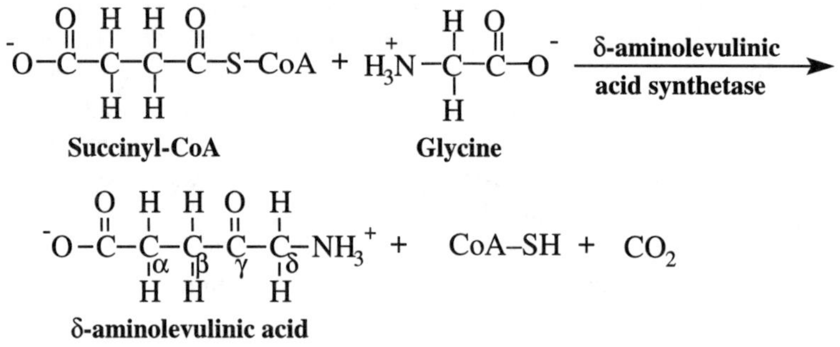

$$\overset{O}{\underset{}{\|}}\overset{H}{\underset{H}{C}}-\overset{H}{\underset{H}{C}}-\overset{O}{\underset{}{\|}}-S-CoA + H_3\overset{+}{N}-\overset{H}{\underset{H}{C}}-\overset{O}{\underset{}{\|}}-O^- \xrightarrow{\text{δ-aminolevulinic acid synthetase}}$$

Succinyl-CoA .... Glycine

$$^-O-\overset{O}{\underset{}{\|}}C-\underset{\alpha}{\overset{H}{\underset{H}{C}}}-\underset{\beta}{\overset{H}{\underset{H}{C}}}-\underset{\gamma}{\overset{O}{\underset{}{\|}}C}-\underset{\delta}{\overset{H}{\underset{H}{C}}}-NH_3^+ + CoA-SH + CO_2$$

δ-aminolevulinic acid

Figure 10.3 Path of synthesis of delta-aminolevulinic acid (coenzyme A abbreviated as CoA). Cadmium tends to inhibit the enzyme responsible for this process.

and pulmonary epithelium necrosis. Chronic exposure sometimes produces emphysema severe enough to be disabling. The kidney is generally regarded as the organ most sensitive to chronic cadmium poisoning. The function of renal tubules is impaired by cadmium, as manifested by excretion of both high-molecular-mass proteins (such as albumin) and low-molecular-mass proteins. Chronic toxic effects of cadmium exposure may also include damage to the skeletal system, hypertension (high blood pressure), and adverse cardiovascular effects. Based largely on studies of workers in the cadmium–nickel battery industry, cadmium is regarded as a human carcinogen, causing lung tumors and possibly cancer of the prostate.

Cadmium is a highly **cumulative** poison with a biologic half-life estimated at about 20 to 30 years in humans. About half of the body burden of cadmium is found in the liver and kidneys. The total body burden reaches a plateau in humans around age 50. Cigarette smoke is a source of cadmium, and the body burden of cadmium is about 1.5 to 2 times greater in smokers than in nonsmokers of the same age.

Cadmium in the body is known to affect several enzymes. It is believed that the renal damage that results in proteinuria is the result of cadmium adversely affecting enzymes responsible for reabsorption of proteins in kidney tubules. Cadmium also reduces the activity of delta-aminolevulinic acid synthetase (Figure 10.3), arylsulfatase, alcohol dehydrogenase, and lipoamide dehydrogenase, whereas it enhances the activity of delta-aminolevulinic acid dehydratase, pyruvate dehydrogenase, and pyruvate decarboxylase.

The most spectacular and publicized occurrence of cadmium poisoning resulted from dietary intake of cadmium by people in the Jintsu River Valley, near Fuchu, Japan. The victims were afflicted by *itai, itai* disease, which means "ouch, ouch" in Japanese. The symptoms are the result of painful osteomalacia (bone disease) combined with kidney malfunction. Cadmium poisoning in the Jintsu River Valley was attributed to irrigated rice contaminated from an upstream mine producing lead, zinc, and cadmium.

10.4.10 Mercury

Mercury is directly below cadmium in the periodic table, but has a considerably more varied and interesting chemistry than cadmium or zinc. Elemental mercury is the only metal that is a liquid at room temperature, and its relatively high vapor pressure contributes to its toxicological hazard. Mercury metal is used in electric discharge tubes (mercury lamps), gauges, pressure-sensing devices, vacuum pumps, valves, and seals. It was formerly widely used as a cathode in the chlor-alkali process for the manufacture of NaOH and Cl_2, a process that has been largely discontinued, in part because of the mercury pollution that resulted from it.

In addition to the uses of mercury metal, mercury compounds have a number of applications. Mercury(II) oxide, HgO, is commonly used as a raw material for the manufacture of other mercury compounds. Mixed with graphite, it is a constituent of the Ruben–Mallory dry cell, for which the cell reaction is

$$Zn + HgO \rightarrow ZnO + Hg \qquad (10.4.4)$$

Mercury(II) acetate, $Hg(C_2H_3O_2)_2$, is made by dissolving HgO in warm 20% acetic acid. This compound is soluble in a number of organic solvents. Mercury(II) chloride is quite toxic. The dangers of exposure to $HgCl_2$ are aggravated by its high water solubility and relatively high vapor pressure, compared to other salts. Mercury(II) fulminate, $Hg(ONC)_2$, has been used as a detonator for explosives. In addition to the +2 oxidation state, mercury can also exist in the +1 oxidation state as the dinuclear Hg_2^{2+} ion. The best-known mercury(I) compound is mercury(I) chloride, Hg_2Cl_2, commonly called calomel. It is a constituent of calomel reference electrodes, such as the well-known saturated calomel electrode (SCE).

A number of organomercury compounds are known. These compounds and their toxicities are discussed further in Chapter 12.

10.4.10.1 Absorption and Transport of Elemental and Inorganic Mercury

Monatomic elemental mercury in the vapor state, $Hg(g)$, is absorbed from inhaled air by the pulmonary route to the extent of about 80%. Inorganic mercury compounds are absorbed through the intestinal tract and in solution through the skin.

Although elemental mercury is rapidly oxidized to mercury(II) in erythrocytes (red blood cells), which have a strong affinity for mercury, a large fraction of elemental mercury absorbed through the pulmonary route reaches the brain prior to oxidation and enters that organ because of the lipid solubility of mercury(0). This mercury is subsequently oxidized in the brain and remains there. Inorganic mercury(II) tends to accumulate in the kidney.

10.4.10.2 Metabolism, Biologic Effects, and Excretion

Like cadmium, mercury(II) has a strong affinity for sulfhydryl groups in proteins, enzymes, hemoglobin, and serum albumin. Because of the abundance of sulfhydryl groups in active sites of many enzymes, it is difficult to establish exactly which enzymes are affected by mercury in biological systems.

The effect on the central nervous system following inhalation of elemental mercury is largely psychopathological. Among the most prominent symptoms are tremor (particularly of the hands) and emotional instability characterized by shyness, insomnia, depression, and irritability. These symptoms are probably the result of damage to the blood–brain barrier, which regulates the transfer of metabolites, such as amino acids, to and from the brain. Brain metabolic processes are probably disrupted by the effects of mercury. Historically, the three symptoms of increased excitability, tremors, and gum inflammation (gingivitis) have been recognized as symptoms of mercury poisoning from exposure to mercury vapor or mercury nitrate in the fur, hat, and felt trades.

The kidney is the primary target organ for Hg^{2+}. Chronic exposure to inorganic mercury(II) compounds causes proteinuria. In cases of mercury poisoning of any type, the kidney is the organ with the highest bioaccumulation of mercury.

Mercury(I) compounds are generally less toxic than mercury(II) compounds because of their lower solubilities. Calomel, a preparation containing Hg_2Cl_2, was once widely used in medicine. Its use as a teething powder for children has been known to cause a hypersensitivity response in children called "pink disease," manifested by a pink rash and swelling of the spleen and lymph nodes.

Excretion of inorganic mercury occurs through the urine and feces. The mechanisms by which excretion occurs are not well understood.

10.4.10.3 Minimata Bay

The most notorious incident of widespread mercury poisoning in modern times occurred in the Minimata Bay region of Japan during the period of 1953 to 1960. Mercury waste from a chemical plant draining into the bay contaminated seafood consumed regularly by people in the area. Overall, 111 cases of poisoning with 43 deaths and 19 congenital birth defects were documented. The seafood was found to contain 5 to 20 ppm of mercury.

10.4.11 Lead

Lead (Pb) ranks fifth behind iron, copper, aluminum, and zinc in industrial production of metals. About half of the lead used in the U.S. goes for the manufacture of lead storage batteries. Other uses include solders, bearings, cable covers, ammunition, plumbing, pigments, and caulking.

Metals commonly alloyed with lead are antimony (in storage batteries), calcium and tin (in maintenance-free storage batteries), silver (for solder and anodes), strontium and tin (as anodes in electrowinning processes), tellurium (pipe and sheet in chemical installations and nuclear shielding), tin (solders), and antimony and tin (sleeve bearings, printing, high-detail castings).

Lead(II) compounds are predominantly ionic (for example, $Pb^{2+}SO_4^{2-}$), whereas lead(IV) compounds tend to be covalent (for example, tetraethyllead, $Pb(C_2H_5)_4$). Some lead(IV) compounds, such as PbO_2, are strong oxidants. Lead forms several basic lead salts, such as $Pb(OH)_2 \cdot 2PbCO_3$, which was once the most widely used white paint pigment and the source of considerable chronic lead poisoning to children who ate peeling white paint. Many compounds of lead in the +2 oxidation state (lead(II)) and a few in the +4 oxidation state (lead(IV)) are useful. The two most common of these are lead dioxide and lead sulfate, which are participants in the following reversible reaction that occurs during the charge and discharge of a lead storage battery:

$$Pb + PbO_2 + 2H_2SO_4 \rightleftharpoons 2PbSO_4 + 2H_2O \qquad (10.4.5)$$

$$\text{Charge} \rightleftharpoons \text{Discharge}$$

In addition to the inorganic compounds of lead, there are a number of organolead compounds, such as tetraethyllead. These are discussed in Chapter 12.

10.4.11.1 Exposure and Absorption of Inorganic Lead Compounds

Although industrial lead poisoning used to be very common, it is relatively rare now because of previous experience with the toxic effects of lead and the protective actions that have been taken. Lead is a common atmospheric pollutant (though much less so now than when leaded gasoline was in general use), and absorption through the respiratory tract is the most common route of human exposure. The greatest danger of pulmonary exposure comes from inhalation of very small respirable particles of lead oxide (particularly from lead smelters and storage battery manufacturing) and lead carbonates, halides, phosphates, and sulfates. Lead that reaches the lung alveoli is readily absorbed into blood.

The other major route of lead absorption is the gastrointestinal tract. Dietary intake of lead reached average peak values of almost 0.5 mg per person per day in the U.S. around the 1940s. Much of this lead came from lead solder used in cans employed for canned goods and beverages. Currently, daily intake of dietary lead in the U.S. is probably only around 20 µg per person per

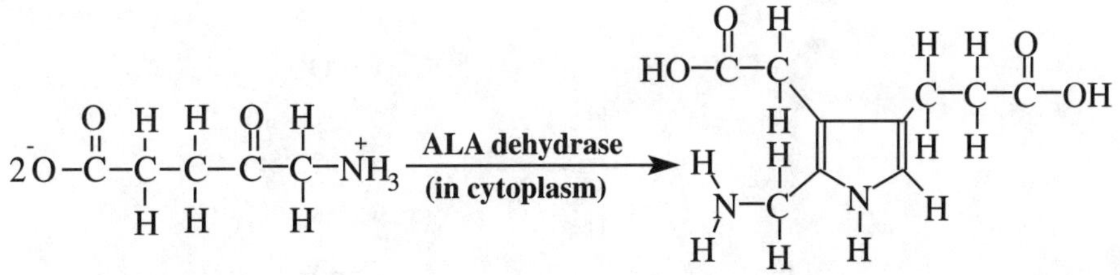

Figure 10.4 Synthesis of porphobilinogen from delta-aminolevulinic acid, a major step in the overall scheme of heme synthesis that is inhibited by lead in the body.

day. Lead(II) may have much the same transport mechanism as calcium in the gastrointestinal tract. It is known that lead absorption decreases with increased levels of calcium in the diet and vice versa.

10.4.11.2 Transport and Metabolism of Lead

A striking aspect of lead in the body is its very rapid transport to bone and storage there. Lead tends to undergo bioaccumulation in bone throughout life, and about 90% of the body burden of lead is in bone after long-term exposure. The half-life of lead in human bones is estimated to be around 20 years. Some workers exposed to lead in an industrial setting have as much as 500 mg of lead in their bones. Of the soft tissues, the liver and kidney tend to have somewhat elevated lead levels.

About 90% of blood lead is associated with red blood cells. Measurement of the concentration of lead in the blood is the standard test for recent or ongoing exposure to lead. This test is used routinely to monitor industrial exposure to lead and in screening children for lead exposure.

The most common biochemical effect of lead is inhibition of the synthesis of heme, a complex of a substituted porphyrin and Fe^{2+} in hemoglobin and cytochromes. Lead interferes with the conversion of delta-aminolevulinic acid to porphobilinogen, as shown in Figure 10.4, with a resulting accumulation of metabolic products. Hematological damage results. Lead inhibits enzymes that have sulfhydryl groups. However, the affinity of lead for the –SH group is not as great as that of cadmium or mercury.

10.4.11.3 Manifestations of Lead Poisoning

Lead adversely affects a number of systems in the body. The inhibition of the synthesis of hemoglobin by lead has just been noted. This effect, plus a shortening of the life span of erythrocytes, results in anemia, a major manifestation of lead poisoning.

The central nervous system is adversely affected by lead, leading to encephalopathy, including neuron degeneration, cerebral edema, and death of cerebral cortex cells. Lead may interfere with the function of neurotransmitters, including dopamine and γ-butyric acid, and it may slow the rate of neurotransmission. Psychopathological symptoms of restlessness, dullness, irritability, and memory loss, as well as ataxia, headaches, and muscular tremor, may occur with lead poisoning. In extreme cases, convulsions followed by coma and death may occur. Lead affects the peripheral nervous system, causing peripheral neuropathy. Lead palsy used to be a commonly observed symptom in lead industry workers and miners suffering from lead poisoning.

Lead causes reversible damage to the kidney through its adverse effect on proximal tubules. This impairs the processes by which the kidney absorbs glucose, phosphates, and amino acids prior to secretion of urine. A longer-term effect of lead ingestion on the kidney is general degradation of the organ (chronic nephritis), including glomular atrophy, interstitial fibrosis, and sclerosis of vessels.

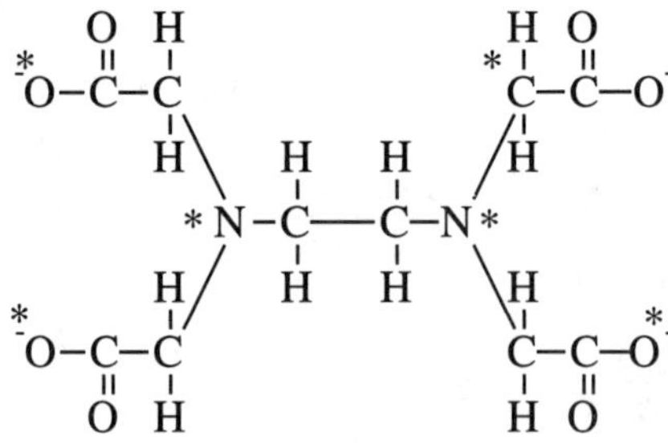

Anion of ethylenediaminetetraacetic acid, EDTA

Figure 10.5 The ionized form of EDTA. Asterisks denote binding sites.

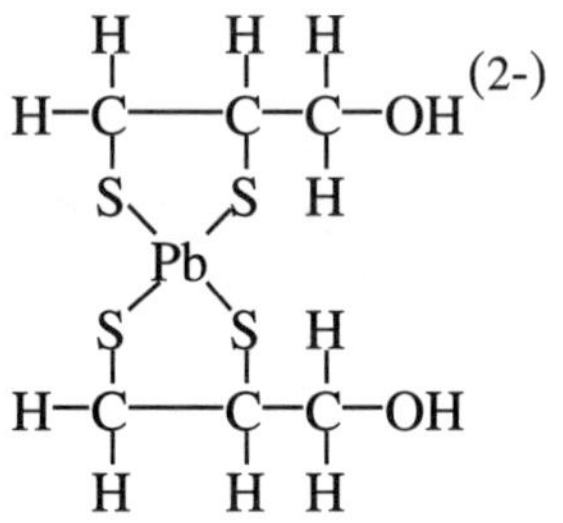

Figure 10.6 Lead chelated by the lead antidote BAL.

10.4.11.4 Reversal of Lead Poisoning and Therapy

Some effects of lead poisoning, such as those on proximal tubules of the kidney and inhibition of heme synthesis, are reversible upon removal of the source of lead exposure. Lead poisoning can be treated by chelation therapy, in which the lead is solubilized and removed by a chelating agent. One such chelating agent is ethylenediaminetetraacetic acid, which binds strongly to most +2 and +3 cations (Figure 10.5). It is administered for lead poisoning therapy in the form of the calcium chelate. The ionized Y^{4-} form chelates metal ions by bonding at one, two, three, or all four carboxylate groups ($-CO_3^{2-}$) and one or both of the two N atoms (see glycinate-chelated structure in Figure 10.1). EDTA is administered as the calcium chelate for the treatment of lead poisoning to avoid any net loss of calcium by solubilization and excretion.

Another compound used to treat lead poisoning is British anti-Lewisite (BAL), originally developed to treat arsenic-containing poison gas Lewisite. As shown in Figure 10.6, BAL chelates lead through its sulfhydryl groups, and the chelate is excreted through the kidney and bile.

10.4.12 Defenses Against Heavy Metal Poisoning

Organisms have some natural defenses against heavy metal poisoning. Several factors are involved in regulating the uptake and physiological concentrations of heavy metals. For example, higher levels of calcium in water tend to lower the bioavailability of metals such as cadmium, copper, lead, mercury, and zinc by fish, and the presence of chelating agents affects the uptake of such metals. Some evidence suggests that mechanisms developed to maintain optimum levels of essential metals, such as zinc and copper, are utilized to minimize the effects of chemically somewhat similar toxic heavy metals, of which cadmium, lead, and mercury are prime examples.

An interesting feature of heavy metal metabolism is the role of intracellular **metallothionein**, which consists of two similar proteins with a low molecular mass of about 6500. As a consequence of a high content of the amino acid cysteine,

$$\text{HS}-\underset{\underset{H}{|}}{\overset{\overset{H}{|}}{C}}-\underset{\underset{NH_3^+}{|}}{\overset{\overset{H}{|}}{C}}-\overset{\overset{O}{\|}}{C}-O^- \quad \text{Cysteine}$$

metallothionein contains a large number of thiol (sulfhydryl, –SH) groups. These groups bind very strongly to other heavy metals, particularly mercury, silver, zinc, and tin. The metal most investigated for its interaction with metallothionein is cadmium. The general reaction of metallothionein with cadmium ion is the following:

$$Cd^{2+} + \boxed{\overset{\overset{H}{|}}{\underset{S}{}}\ \overset{\overset{H}{|}}{\underset{S}{}}\ \text{Metallothionein}} \rightarrow \boxed{\overset{S-Cd-S}{\text{Metallothionein}}} + H^+ \quad (10.4.6)$$

By binding with metallothionein, the mobility of metals by diffusion is greatly reduced and the metals are prevented from binding to enzymes or other proteins essential to normal metabolic function.

Metallothionein has been isolated from virtually all of the major mammal organs, including liver, kidney, brain, heart, intestine, lung, skin, and spleen. Nonlethal doses of cadmium, mercury, and lead induce synthesis of metallothionein. In test animals, nonlethal doses of cadmium followed by an increased level of metallothionein in the body have allowed later administration of doses of cadmium at a level fatal to nonacclimated animals, but without fatalities in the test subjects.

Endogenous substances other than metallothionein may be involved in minimizing the effects of heavy metals and excreting them from the body. Hepatic (liver) glutathione, discussed as a phase II conjugating agent in Section 7.4, plays a role in the excretion of several metals in bile. These include the essential metals copper and zinc; toxic cadmium, mercury(II), and lead(II) ions; and organometallic methyl mercury.

Some plants have particularly high tolerances for cadmium and some other heavy metals by virtue of their content of cysteine-rich peptides, known as **phytochelatins**, sulfur-rich peptides that perform in plants much like metallothionein acts in animals. Plants that resist the effects of heavy metals through the action of phytochelatins require a high activity of cysteine synthase enzyme that makes the sulfur-containing cysteine amino acid from hydrogen sulfide and O-acetylserine. Cadmium-resistant transgenic tobacco plants have been bred that have a high activity of cysteine synthase from genes taken from rice.[6]

10.5 METALLOIDS: ARSENIC

10.5.1 Sources and Uses

Arsenopyrite and loellingite are both arsenic minerals that can be smelted to produce elemental arsenic. Both elemental arsenic and arsenic trioxide (As_2O_3) are produced commercially; the latter is the raw material for the production of numerous arsenic compounds. Elemental arsenic is used to make alloys with lead and copper. Arsenic compounds have a number of uses, including applications in catalysts, bactericides, herbicides, fungicides, animal feed additives, corrosion inhibitors, pharmaceuticals, veterinary medicines, tanning agents, and wood preservatives. Arsenicals were the first drugs to be effective against syphilis, and they are still used to treat amebic dysentery. Arsobal, or Mel B, an organoarsenical, is the most effective drug for the treatment of the neurological stage of African trypanosomiasis, for which the infectious agents are *Trypanosoma gambiense* or *T. rhodesiense*.

10.5.2 Exposure and Absorption of Arsenic

Arsenic can be absorbed through both the gastrointestinal and pulmonary routes. Although the major concern with arsenic is its effect as a systemic poison, arsenic trichloride ($AsCl_3$) and the organic arsenic compound Lewisite (used as a poison gas in World War I) can penetrate skin; both of these compounds are very damaging at the point of exposure and are strong vesicants (causes of blisters). The common arsenic compound As_2O_3 is absorbed through the lungs and intestines. The degree of coarseness of the solid is a major factor in how well it is absorbed. Coarse particles of this compound tend to pass through the gastrointestinal tract and to be eliminated with the feces.

The chemistry of arsenic is so varied that it is difficult to regard as a single element.[7] Arsenic occurs in the +3 and +5 oxidation states; inorganic compounds in the +3 oxidation state (arsenite) are generally more toxic. The conversion to arsenic(V) is normally favored in the environment, which somewhat reduces the overall hazard of this element.

Arsenic is a natural constituent of most soils. It is found in a number of foods, particularly shellfish. The average adult ingests somewhat less than 1 mg of arsenic per day through natural sources. Drinking water is a source of arsenic in some parts of the world. This was tragically illustrated in Bangladesh, where a United Nations program to develop water wells as a source of pathogen-free drinking water later resulted in perhaps millions of cases of arsenic poisoning from arsenic-containing well water. A directive by the U.S. Environmental Protection Agency in 2000 to lower the long-standing (since 1942) arsenic drinking water standard in the U.S. was overturned, pending further review by the newly elected administration in early 2001, causing a great deal of controversy (The new standard has since been reinstated).

10.5.3 Metabolism, Transport, and Toxic Effects of Arsenic

Biochemically, arsenic acts to coagulate proteins, forms complexes with coenzymes, and inhibits the production of adenosine triphosphate (ATP) (see Section 4.3). Like cadmium and mercury, arsenic is a sulfur-seeking element. Arsenic has some chemical similarities to phosphorus, and it substitutes for phosphorus in some biochemical processes, with adverse metabolic effects. Figure 10.7 summarizes one such effect. The top reaction in the figure illustrates the enzyme-catalyzed synthesis of 1,3-diphosphoglycerate from glyceraldehyde 3-phosphate. The product undergoes additional reactions to produce ATP, an essential energy-yielding substance in body metabolism. When arsenite AsO_3^{3-} is present, it bonds to glyceraldehyde 3-phosphate to yield a product that undergoes nonenzymatic spontaneous hydrolysis, thereby preventing ATP formation.

Symptoms of acute arsenic poisoning are many and may be severe — fatal at high doses. Fatal cases of arsenic poisoning have exhibited symptoms of fever, aversion to food, abnormal liver enlargement (hepatomegaly), cardiac arrhythmia, development of dark patches on skin and other tissue (melanosis), peripheral neuropathy, including sensory loss in the peripheral nervous system, gastrointestinal disorders, cardiovascular effects, and adverse effects on red blood cell formation, which can result in anemia. Mucous membranes may be irritated, form blisters, or slough off.

Chronic effects of arsenic poisoning include neurotoxic effects to the central and peripheral nervous systems. Symptoms include sensory changes, muscle sensitivity, prickling and tingling sensations (paresthesia), and muscle weakness. Liver injury is a common symptom of chronic arsenic poisoning. Studies of victims of chronic arsenic poisoning from contaminated drinking

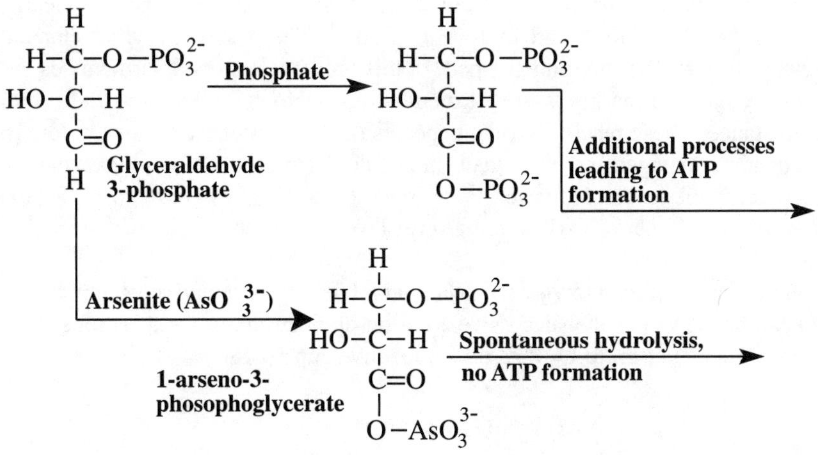

Figure 10.7 Interference of arsenic(III) with ATP production by phosphorylation.

water in Taiwan and Chile have exhibited blueness of the skin in extremities, a condition called acrocyanosis, the result of periphereal vascular disease. In extreme cases, this may progress to gangrene in the lower extremities, a condition called blackfoot disease.

There is now sufficient epidemiological evidence to classify arsenic as a human carcinogen and a cause of skin cancer. In people chronically exposed to toxic doses of arsenic, such cancers may be preceded by discolored skin (hyperpigmentation) and development of horny skin surfaces (hyperkeratosis). These areas may progress to locally invasive basal cell carcinomas or to squamous cell carcinomas capable of metastasis. Unlike skin cancers that develop on skin exposed to ultraviolet solar radiation, arsenic-induced skin cancer frequently develops in areas not commonly exposed to sunlight, such as the palms of hands or soles of feet.

Analysis of hair, fingernails, and toenails can serve as evidence of arsenic ingestion. Such analyses are complicated by the possible presence of arsenic contamination, particularly in a work environment in which the air and surroundings may be contaminated with arsenic. Levels of arsenic may be correlated with the growth of nails and hair so that careful analysis of segments of these materials can indicate time frames of exposure.

Antidotes to arsenic poisoning take advantage of the element's sulfur-seeking tendencies and contain sulfhydryl groups. One such antidote is 2,3-mercaptopropanol (BAL), discussed in the preceding section as an antidote for lead poisoning.

10.6 NONMETALS

10.6.1 Oxygen and Ozone

Molecular oxygen, O_2, is essential for life processes in both humans and other aerobic organisms and is potentially damaging to tissue. Exposure to excessive levels of O_2 can cause toxic responses. This was tragically illustrated by the use of oxygen to assist the respiration of premature infants, a procedure that caused many to become blind. Even at normal levels of oxygen, some toxicants can cause this essential element to cause toxic lesions. To understand why this is so, consider that aerobic organisms, including humans, derive their energy by mediating the oxidation of nutrient molecules such as glucose:

$$C_6H_{12}O_6 + 6O_2 \rightarrow 6CO_2 + 6H_2O + \text{energy} \tag{10.6.1}$$

In this aerobic respiration process, molecular oxygen is the oxidizing agent or electron receptor and is reduced to the –2 oxidation state in the H_2O product. The process by which elemental oxygen accepts electrons is complex and multistepped. In this process, reactive intermediates are produced that can seriously damage the lipids in cell membranes, DNA in cell nuclei, and proteins. Under normal circumstances, these reactive oxidant species undergo further reactions before they can do much harm, or are scavenged by antioxidant molecules or by the action of enzymes designed to keep them at acceptable levels. However, under conditions of excessive exposure to oxidants and by the action of some kinds of toxicants, harmful levels of reactive intermediate oxidant species can build up to harmful levels.

The metabolic conversion of oxygen(0) in elemental oxygen to bound oxygen in the –2 oxidation state in H_2O can be viewed as the successive addition of electrons (e⁻) and H^+ ions to O_2. The first step is addition of an electron to O_2 to produce reactive superoxide ion, $O_2^{-\cdot}$:

$$O_2 + e^- \rightarrow O_2^{-\cdot} \tag{10.6.2}$$

In formulas such as that of superoxide, the dot represents an unpaired electron. Species that have unpaired electrons are very reactive **free radicals**. Addition of H^+ ion to superoxide produces reactive hydroperoxyl radical, $HO_2\cdot$:

$$O_2^{-\cdot} + H^+ \rightarrow HO_2\cdot \tag{10.6.3}$$

Another electron and H^+ ion may be added to the hydroperoxyl radical, a process equivalent to adding an H atom, to produce hydrogen peroxide:

$$HO_2\cdot + e^- + H^+ \rightarrow H_2O_2 \tag{10.6.4}$$

Hydrogen peroxide may be produced from the superoxide radical anion by the action of superoxide dismutase enzyme. The catalase enzyme may act on hydrogen peroxide to produce O_2 and H_2O. Hydrogen peroxide may also be eliminated by the action of glutathione peroxidase, producing the oxidized form of glutathione (see below). In the presence of appropriate metal ion catalysts, hydrogen peroxide may undergo the Haber–Weiss reaction,

$$O_2^{-\cdot} + H_2O_2 + Fe^{2+} \rightarrow Fe^{3+} + O_2 + OH^- + HO\cdot \tag{10.6.5}$$

and the Fenton reaction,

$$H_2O_2 + Fe^{2+} \rightarrow Fe^{3+} + OH^- + HO\cdot \tag{10.6.6}$$

to produce hydroxyl radical, $HO\cdot$.

Superoxide, hydroperoxyl, and especially reactive hydroxyl radicals along with hydrogen peroxide attack tissue and DNA either directly or through their reaction products. The damage done is sometimes referred to as oxidative lesions. It is now recognized that some toxicants have the ability to promote the formation of reactive oxidizing species to the extent that defensive mechanisms against oxidants are overwhelmed, a condition called **oxidative stress**. Under conditions of oxidative stress, lipids, nucleic acids, and proteins may be damaged by reactive oxidants. Another very damaging effect of oxidant reactive intermediates is **lipid peroxidation**, in which polyunsaturated fatty acids on lipids are attacked and oxidized, as shown in Figure 10.8. This can be especially damaging to lipid-rich cell membranes.

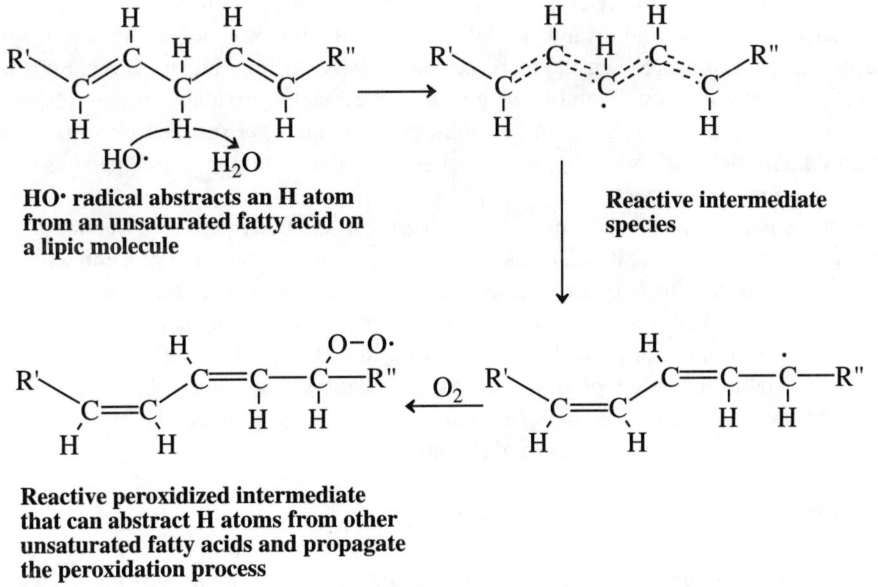

Figure 10.8 Peroxidation of lipid molecules by reactive radical species such as the hydroxyl radical, HO·.

Superoxide radical anion, hydroxyl radical, and hydrogen peroxide are known as prooxidants, whereas substances that neutralize their effects are called antioxidants. Oxidative stress occurs when the prooxidant–antioxidant balance becomes too favorable to the prooxidants. The effects of prooxidants can be neutralized by their direct reaction with small-molecule antioxidants, including glutathione, ascorbate, and tocopherols. In addition, oxidizing radicals are scavenged from a living system by several enzymes, including peroxidase, superoxide dismutase, and catalase. Oxidative lesions on DNA may be repaired by DNA repair enzymes.

Probably the most important antioxidant molecule is glutathione, a tripeptide formed from glutamic acid, cysteine, and glycine amino acids:

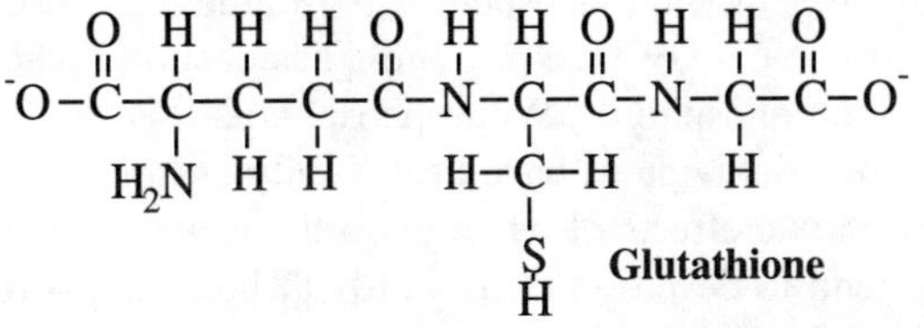

This substance reacts with oxidant radicals to produce H_2O and the oxidized form of glutathione, consisting of two of the molecules of this substance joined by an SS bridge.

A potent oxidizing form of elemental oxygen is **ozone**, O_3. This species is arguably the most toxic environmental pollutant to which the general population is exposed because of its presence in polluted atmospheres, especially under conditions where photochemical smog is present. It can be a pollutant of the workplace in locations where electrical discharges or ultraviolet radiation pass through air (from sources such as laser printers). The reactions for the production of ozone, beginning with the splitting of O_2 molecules to produce O atoms, are given in Section 2.8.

A deep lung irritant, ozone causes pulmonary edema, which can be fatal. It is also strongly irritating to the upper respiratory system and eyes and is largely responsible for the unpleasantness of photochemical smog. A level of 1 ppm of ozone in air has a distinct odor, and inhalation of such air causes severe irritation and headache. The primary toxicological concern with ozone involves the lungs. Exposure to ozone increases the activity of free-radical-scavenging enzymes in the lung, indicative of ozone's ability to generate the reactive oxidant species responsible for oxidative stress. Arterial lesions leading to pulmonary edema have resulted from ozone exposure by inhalation. Animal studies of ozone inhalation have shown injury of epithelial (surface) cells throughout the respiratory tract.

Like nitrogen dioxide and ionizing radiation, ozone in the body produces free radicals that can be involved in destructive oxidation processes, such as lipid peroxidation or reaction with sulfhydryl (–SH) groups. Exposure to ozone can cause chromosomal damage. Ozone also appears to have adverse immunological effects. Radical-scavenging compounds, antioxidants, and compounds containing sulfhydryl groups can protect organisms from the effects of ozone.

Ozone is notable for being **phytotoxic** (toxic to plants). Loss of crop productivity from the phytotoxic action of ozone is a major concern in areas afflicted with photochemical smog, of which ozone is the single most characteristic manifestation.

10.6.2 Phosphorus

The most common elemental form of phosphorus, white phosphorus, is highly toxic. White phosphorus (melting point (mp), 44°C; boiling point (bp), 280°C) is a colorless waxy solid, sometimes with a yellow tint. It ignites spontaneously in air to yield a dense fog of finely divided, highly deliquescent P_4O_{10}:

$$P_4 + 5O_2 \rightarrow P_4O_{10} \tag{10.6.7}$$

White phosphorus can be absorbed into the body, particularly through inhalation, as well as through the oral and dermal routes. It has a number of systemic effects, including anemia, gastrointestinal system dysfunction, and bone brittleness. Acute exposure to relatively high levels results in gastrointestinal disturbances and weakness due to biochemical effects on the liver. Chronic poisoning occurs largely through the inhalation of low concentrations of white phosphorus and through direct contact with this toxicant. Severe eye damage can result from chronic exposure to elemental white phosphorus. A number of cases of white phosphorus poisoning have resulted from exposure in the fireworks industry. At least one case of fatal poisoning has occurred when a child accidentally ate a firecracker containing white phosphorus. White phosphorus used to be a common ingredient of rat poisons, and some suicidal individuals have been fatally poisoned from ingesting rat poison.

The most characteristic toxic effects of white phosphorus are musculoskeletal effects. Victims of phosphorus poisoning tend to develop necrosis of both bone and soft tissue in the oral cavity. As a result, the jawbone may deteriorate and become brittle, a condition called **phossy jaw**. Instances of this malady have been reported among workers handling white phosphorus, and it is believed that direct exposure of the mouth and oral cavity have occurred as the result of poor hygiene practices. Those afflicted with phossy jaw tend to develop abscessed teeth, and the sockets remaining from the extraction of teeth heal poorly. Infections of the jaw around teeth accompanied by severe pain are common symptoms of phossy jaw.

10.6.3 The Halogens

The elemental **halogens** — fluorine, chlorine, bromine, and iodine — are all toxic. Both fluorine and chlorine are highly corrosive gases that are very damaging to exposed tissue. These elements are chemically and toxicologically similar to many of their compounds, such as the interhalogen compounds, discussed in Chapter 11. The toxicities of halogen compounds are discussed in the next two sections.

10.6.3.1 Fluorine

Fluorine, F_2 (mp, –218°C; bp, –187°C), is a pale yellow gas produced from calcium fluoride ore by first liberating hydrogen fluoride with sulfuric acid, then electrolyzing the HF in a 4:1 mixture with potassium fluoride, KF, as shown in the reaction

$$2HF(molten\ KF) \xrightarrow{\text{Direct current}} H_2(cathode) + F_2(anode) \tag{10.6.8}$$

Of all the elements, fluorine is the most reactive and the most electronegative (a measure of tendency to acquire electrons). In its chemically combined form, it always has an oxidation number of –1. Fluorine has numerous industrial uses, such as the manufacture of UF_6, a gas used to enrich uranium in its fissionable isotope, uranium-235. Fluorine is used to manufacture uranium hexafluoride, SF_6, a dielectric material contained in some electrical and electronic apparatus. A number of organic compounds contain fluorine, particularly the chlorofluorocarbons used as refrigerants and organofluorine polymers, such as DuPont's Teflon.

Given elemental fluorine's extreme chemical reactivity, it is not surprising that F_2 is quite toxic. It is classified as "a most toxic irritant." It strongly attacks skin and the mucous membranes of the nose and eyes.

10.6.3.2 Chlorine

Elemental chlorine, Cl_2 (mp, –101°C; bp, –34.5°C), is a greenish yellow gas that is produced industrially in large quantities for numerous uses, such as the production of organochlorine solvents (see Chapter 11) and water disinfection. Liquified Cl_2 is shipped in large quantities in railway tank cars, and human exposure to chlorine from transportation accidents is not uncommon.

Chlorine was the original poison gas used in World War I. It is a strong oxidant and reacts with water to produce an acidic oxidizing solution by the following reactions:

$$Cl_2 + H_2O \rightleftarrows HCl + HOCl \tag{10.6.9}$$

$$Cl_2 + H_2O \rightleftarrows 2HCl + \{O\} \tag{10.6.10}$$

where HOCl is oxidant hypochlorous acid and {O} stands for nascent oxygen (in a chemical sense regarded as freshly generated, highly reactive oxygen atoms). When chlorine reacts in the moist tissue lining the respiratory tract, the effect is quite damaging to the tissue. Levels of 10 to 20 ppm of chlorine gas in air cause immediate irritation to the respiratory tract, and brief exposure to 1000 ppm of Cl_2 can be fatal. Because of its intensely irritating properties, chlorine is not an insidious poison, and exposed individuals will rapidly seek to get away from the source if they are not immediately overcome by the gas.

10.6.3.3 Bromine

Bromine, Br_2 (mp, –7.3°C; bp, 58.7°C), is a dark red liquid prepared commercially from elemental chlorine and bromide ion in bromide brines by the reaction

$$Cl_2 + 2Br^- \rightarrow 2Cl^- + Br_2 \tag{10.6.11}$$

and the elemental bromine product is swept from the reaction mixture with steam. The major use of elemental bromine is for the production of organobromine compounds such as 1,1-dibromoethane, formerly widely used as a grain and soil fumigant for insect control and as a component of leaded gasoline for scavenging lead from engine cylinders.

Bromine is toxic when inhaled or ingested. Like chlorine and fluorine, it is an irritant to the respiratory tract and eyes because it attacks their mucous membranes. Pulmonary edema may result from severe bromine poisoning. The severely irritating nature of bromine causes a withdrawal response in its presence, thereby limiting exposure.

10.6.3.4 Iodine

Elemental iodine, I_2 (solid, sublimes at 184°C), consists of violet-black rhombic crystals with a lustrous metallic appearance. More irritating to the lungs than bromine or chlorine, its general effects are similar to the effects of these elements. Exposure to iodine is limited by its low vapor pressure, compared to liquid bromine or gaseous chlorine or fluorine.

10.6.4 Radionuclides

10.6.4.1 Radon

In Section 9.3 the toxicological effects of ionizing radiation are mentioned, and radon is cited as a source of such radiation. Radon can pose very distinct health risks.[8] Radon's toxicity is not the result of its chemical properties, because it is a noble gas and does not enter into any normal chemical reactions. However, it is a radioactive element (radionuclide) that emits positively charged alpha particles, the largest and — when emitted inside the body — the most damaging form of radioactivity. Furthermore, the products of the radioactive decay of radon are also alpha emitters. Alpha particles emitted from a radionuclide in the lung cause damage to cells lining the lung bronchi and other tissues, resulting in processes that can cause cancer.

Radon is a decay product of radium, which in turn is produced by the radioactive decay of uranium. During its brief lifetime, radon may diffuse upward through soil and into dwellings through cracks in basement floors. Radioactive decay products of radon become attached to particles in indoor air, are inhaled, and lodge in the lungs until they undergo radioactive decay, damaging lung tissue. Synergistic effects between radon and smoking appear to be responsible for most of the cases of cancer associated with radon exposure.

10.6.4.2 Radium

A second radionuclide to which humans are likely to be exposed is **radium**, Ra. Occupational exposure to radium is known to have caused cancers in humans, most tragically in the cases of a number of young women who were exposed to radium because of their employment in painting luminescent radium-containing paint on the dials of watches, clocks, and instruments.[9] These workers would touch their tongues with the very fine brushes used for the radioactive paint in order to "point" the brushes. Many eventually developed bone cancer and died from this malady.

The most likely route for human exposures to low doses of radium is through drinking water. Areas in the U.S. where significant radium contamination of water has been observed include the uranium-producing regions of the western U.S., Iowa, Illinois, Wisconsin, Missouri, Minnesota, Florida, North Carolina, Virginia, and the New England states.

The maximum contaminant level (MCL) for total radium (^{226}Ra plus ^{228}Ra) in drinking water is specified by the U.S. Environmental Protection Agency as 5 pCi/l, where a picocurie is 0.037 disintegrations per second. Perhaps as many as several hundred municipal water supplies in the U.S. exceed this level and require additional treatment to remove radium. Fortunately, conventional water-softening processes, which are designed to take out excessive levels of calcium, are relatively efficient in removing radium from water.

10.6.4.3 Fission Products

The anthropogenic radionuclides of most concern are those produced as fission products from nuclear weapons and nuclear reactors. The most devastating release from the latter source to date resulted from the April 26, 1986, explosion, partial meltdown of the reactor core, and breach of confinement structures by a power reactor at Chernobyl in the Ukraine. This disaster released 5×10^7 Ci of radionuclides from the site, which contaminated large areas of Soviet Ukraine and Byelorussia, as well as areas of Scandinavia, Italy, France, Poland, Turkey, and Greece. Radioactive fission products that are the same or similar to elements involved in life processes can be particularly hazardous. One of these is radioactive iodine, which tends to accumulate in the thyroid gland, which may develop cancer or otherwise be damaged as a result. Radioactive cesium exists as the Cs^+ ion and is similar to sodium and potassium in its physiological behavior. Radioactive strontium forms the Sr^{2+} ion and substitutes for Ca^{2+}, especially in bone.

REFERENCES

1. Blake, B.L., Lawler, C.P., and Mailman, R.B., Biochemical toxicology of the central nervous system, in *Introduction to Biochemical Toxicology*, 3rd ed., Hodgson, E. and Smart, R.C., Eds., Wiley Interscience, New York, 2001, pp. 453–486.
2. Sharma, P., Johri, S., and Shukla, S., Beryllium-induced toxicity and its prevention by treatment with chelating agents, *J. Appl. Toxicol.*, 20, 313–318, 2000.
3. Domingo, J.L., Vanadium: a review of the reproductive and developmental toxicity, *Reprod. Toxicol.*, 10, 175–182, 1996.
4. Goldwaser, I. et al., Insulin-like effects of vanadium: basic and clinical implications, *J. Inorg. Biochem.*, 80, 21–25, 2000.
5. Schaffer, A.W. et al., Increased blood cobalt and chromium after total hip replacement, *J. Toxicol. Clin. Toxicol.*, 37, 839–844, 1999.
6. Harada, E. et al., Transgenic tobacco plants expressing a rice *cysteine synthase* gene are tolerant to toxic levels of cadmium, *J. Plant Physiol.*, 158, 655–661, 2001.
7. Goyer, R.A. and Clarkson, T.W., Toxic effects of metals, in *Casarett and Doull's Toxicology: The Basic Science of Poisons*, 6th ed., Klaassen, C.D., Ed., McGraw-Hill, New York, 2001, chap. 23, pp. 811–867.
8. Committee on Health Risks of Exposure to Radon, *Health Effects of Exposure to Radon*, National Academy Press, Washington, D.C., 1999.
9. Mullner, R., *Deadly Glow: The Radium Dial Worker Tragedy*, American Public Health Association, Washington, D.C., 1999.

SUPPLEMENTARY REFERENCE

Manahan, S.E., *Environmental Chemistry*, 7th ed., CRC Press/Lewis Publishers, Boca Raton, FL, 2000.

QUESTIONS AND PROBLEMS

1. Why is it difficult to define what is meant by a toxic element? What are the major categories of toxic elements? Give an example of each.
2. Into which four main categories are elements divided in the periodic table? Why does one of these categories consist of elements of no toxicological chemical significance? What might be a toxicity characteristic of these "nontoxic" elements?
3. What has the Greenland ice pack revealed about the environmental chemistry and distribution of a toxicologically significant element?

4. List and explain the forms in which metals may occur in the body.
5. What is a metal chelate? How are metal complexes related to chelates? In what sense may water be regarded as a ligand and metal ions dissolved in water regarded as complex ions?
6. What is the distinguishing feature of organometallic compounds as related to metal complexes? How is methylation related to organometallic compounds?
7. Which two kinds of functional molecules in biomolecules are most available for bonding to metal ions by complexation? What other functional group forms especially strong bonds with some important toxic heavy metals? In what common biological compound produced as a defense against heavy metal poisoning is this functional group most abundant?
8. In what form are metals most likely to be taken in by the pulmonary route? What is one very special case of a toxic heavy metal taken in by this route?
9. What are the major toxic effects of beryllium? What may be said about the latent period for beryllium poisoning?
10. Although metal ions are generally not very soluble in hydrocarbons, vanadium occurs at high levels in some crude oil products. What is there about vanadium in crude oil that enables this to occur?
11. What are the most common oxidation states of chromium? Of these, why is chromium in the lower oxidation state generally insignificant in water?
12. In what respect does cadmium's chemical similarity to zinc possibly contribute to the toxicity of cadmium? Which organ in the body is most susceptible to cadmium poisoning?
13. What is a cumulative poison? In what sense is cadmium a cumulative poison? What might be a metabolic explanation for why a poison is cumulative?
14. Match the following:
    (a) $PbSO_4$ 1. Organometallic compound
    (b) $Pb(C_2H_5)_4$ 2. In sealed nickel–cadmium batteries
    (c) PbO_2 3. Basic salt
    (d) $Pb(OH)_2 \cdot 2PbCO_3$ 4. Strong oxidant
    (e) $Pb(OH)_2$ 5. Ionic lead(II) compound
15. Match the following:
    (a) Hg metal   1. In Ruben–Mallory dry cell
    (b) HgO   2. Very soluble in water
    (c) $Hg(C_2H_3O_2)_2$ 3. Explosives' detonator
    (d) $HgCl_2$ 4. Used in gauges
    (e) $Hg(ONC)_2$ 5. Soluble in a number of organic solvents
16. What is the predominant function of the blood–brain barrier? How is it affected by mercury?
17. What is the greatest single use for lead? How might this use lead to lead exposure?
18. What is the effect of calcium on the absorption of dietary lead? How might this effect be explained?
19. What is the major biochemical effect of lead, and how is this effect manifested?
20. What are the toxic effects of lead and cadmium on the kidney?
21. What is used as a therapeutic agent for lead poisoning? Why is this antidote always administered with calcium?
22. Explain what is shown by the illustration below:

23. What toxicological chemical effect is illustrated by the figure below?

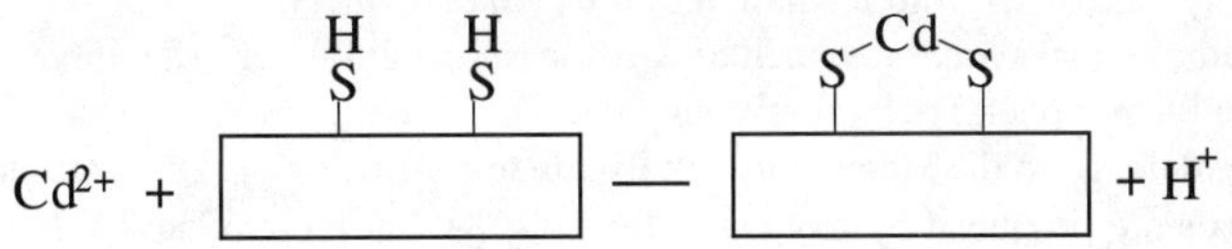

$$Cd^{2+} + \begin{array}{c} H \quad H \\ | \quad | \\ S \quad S \\ \hline \quad \quad \quad \end{array} \longrightarrow \begin{array}{c} \quad Cd \\ S \diagdown \diagup S \\ \hline \quad \quad \quad \end{array} + H^+$$

24. What are some of the uses of elemental arsenic and of arsenic compounds? How might these uses lead to human exposure?
25. Which of the oxidation states of arsenic is most likely to be toxic?
26. Explain what is shown by the following figure:

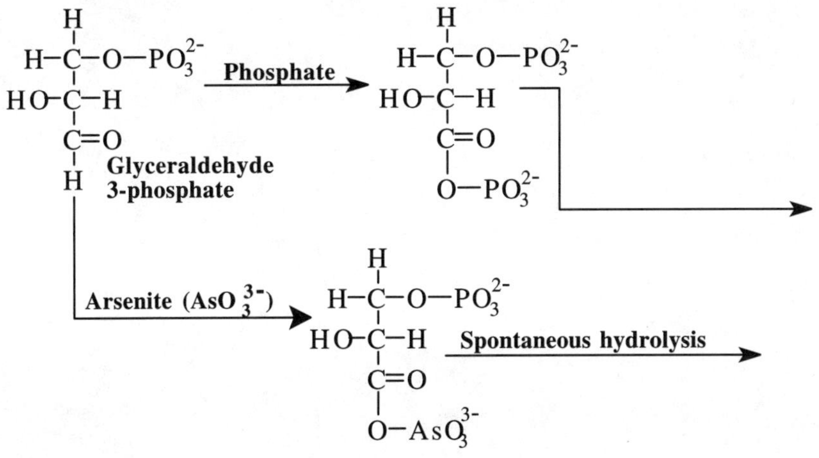

[Figure showing Glyceraldehyde 3-phosphate reacting with Phosphate and with Arsenite (AsO$_3^{3-}$), followed by spontaneous hydrolysis]

27. List the respects in which arsenic is similar to cadmium and mercury, as well as phosphorus. Why is its chemical similarity to phosphorus especially damaging?
28. In what respects do antidotes to arsenic poisoning take advantage of arsenic's sulfur-seeking tendencies? What is the name and chemical formula of one such antidote?
29. Explain what the following figure shows about toxicological chemistry:

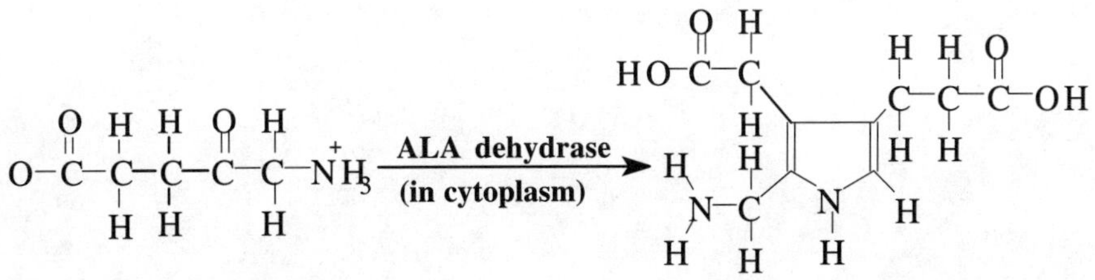

[Figure showing reaction catalyzed by ALA dehydrase (in cytoplasm)]

30. Phosphorus and arsenic are chemically similar. Compare the toxic effects of elemental and combined phosphorus and arsenic.
31. Although noble gases are chemically unreactive and cannot be toxic because of any chemical interactions, one such gas is particularly toxic by nonchemical mechanisms. Which noble gas is that, and why is it toxic?
32. Which metallic element, though chemically not similar to radon, operates through a similar mode of toxic action? What is the most likely route of exposure to this element?

33. Designate which of the following is **not** true of the toxicological hazards or effects of lead:
    (a) inhibition of the synthesis of hemoglobin
    (b) particularly hazardous from inhalation of the elemental metal
    (c) psychopathological symptoms, including restlessness, dullness, irritability, and memory loss
    (d) effects on the peripheral nervous system
    (e) reversible damage to the kidney through its adverse effect on proximal tubules
34. Which radicals are produced by oxygen in the body? What are radicals? Why are they toxic?

CHAPTER 11

Toxic Inorganic Compounds

11.1 INTRODUCTION

In Chapter 10 elements were discussed that as a rule tend to be toxic in their various forms. Chapter 11 covers toxic inorganic compounds of elements that are not themselves generally regarded as toxic. These elements include for the most part the lighter nonmetals located in the upper right of the periodic table (Figure 1.3) and exclude the heavy metals. Most of the elements involved in the inorganic compounds discussed in this chapter are those that are essential for life processes. Any division between "toxic" and "nontoxic" elements is by nature artificial in that most of the heavy metals have compounds of relatively low toxicity, and there are deadly compounds that contain elements essential for life.

11.1.1 Chapter Organization

In general, this chapter is organized in the order of increasing atomic number of the elements that are covered. Inorganic compounds of carbon, atomic number 6, are discussed first, followed by toxic inorganic compounds of nitrogen, atomic number 7. The next element, oxygen, occurs in so many different inorganic compounds that it is not discussed in a separate category. The halogens — fluorine, chlorine, bromine, and iodine — are discussed as a group because of their chemical similarities. The other major elements whose toxic inorganic compounds are discussed are silicon, phosphorus, and sulfur.

11.2 TOXIC INORGANIC CARBON COMPOUNDS

11.2.1 Cyanide

Cyanide, in the form of either gaseous **hydrogen cyanide** (HCN) or **cyanide ion** (CN^-) (present in cyanide salts such as KCN), is a notably toxic substance. Cyanide is a rapidly acting poison, and the fatal oral dose to humans is believed to be only 60 to 90 mg. Hydrogen cyanide and cyanide salts have numerous uses; examples are as ingredients of pest poisons, fumigants, metal (silver) polishes, and photographic chemical solutions. Therefore, exposure to cyanide is certainly possible. Hydrogen cyanide is used as a fumigant to kill pests such as rodents in warehouses, grain storage bins, greenhouses, and holds of ships, where its high toxicity and ability to penetrate obscure spaces are advantageous. Cyanide salt solutions are used to extract some metals such as gold from ores, in metal refining, in metal plating, and for salvaging silver from exposed photographic and x-ray film. Cyanide is used in various chemical syntheses. Polyacrylic polymers may evolve HCN during combustion, adding to the toxic gases that are usually responsible for deaths in fires. Sodium nitroprusside, $Na_2Fe(NO)(CN)_5$, used intravenously in humans to control hypertension, can hydrolyze in the body to release cyanide and cause cyanide poisoning.

Some plants contain cyanogenic glycosides, saccharidal substances that contain the –CN group and that may hydrolyze to release cyanide. Such substances, called **cyanogens**, include amygdalin, linamarin, and linseed cyanogens consisting of mixtures of linustatin and neolinustatin.[1] The release of cyanide by the enzymatic or acidic hydrolysis of amygdalin in the digestive tract is shown below:

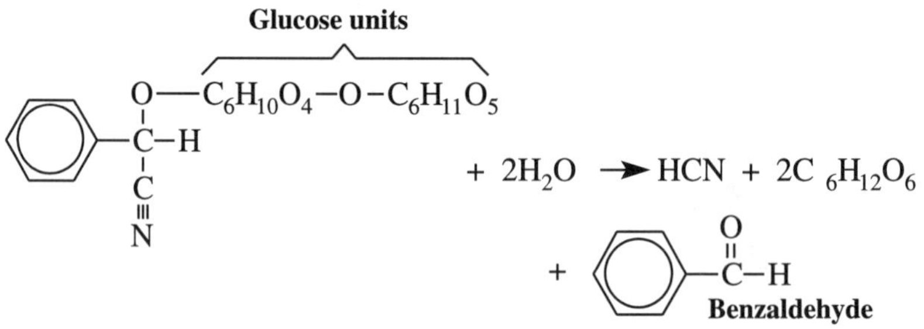

$$\text{Amygdalin} + 2H_2O \rightarrow HCN + 2\,C_6H_{12}O_6 + \text{Benzaldehyde} \quad (11.2.1)$$

The Romans used cyanide from natural seed sources, such as apple seeds, for executions and suicides. The seeds of apples, apricots, cherries, peaches, plums, and some other fruits contain sources of cyanide. Other natural sources of cyanide include arrowgrass, sorghum, flax, velvet grass, and white clover.

A potential source of cyanide poisoning is cassava, a starch from the root of *Manihot esculenta*, used as food in much of Africa. The root contains cyanogenic linamarin, which is normally removed in processing the root for food. Widespread cases of a spinal cord disorder called konzo and characterized by spastic paralysis have been attributed to ingestion of linamarin from inadequately processed cassava root.

11.2.1.1 Biochemical Action of Cyanide

Cyanide deprives the body of oxygen by acting as a **chemical asphyxiant** (in contrast to simple asphyxiants that simply displace oxygen in respired air). In acting as an asphyxiant, cyanide inhibits an enzyme (see enzyme inhibition, Section 7.6) involved in a key step in the oxidative phosphorylation pathway, by which the body utilizes oxygen in cell mitochondria. The inhibited enzyme is ferricytochrome oxidase (Fe(III)-oxid), an iron-containing metalloprotein that acts as an acceptor of electrons and is converted to ferrouscytochrome oxidase (Fe(II)-oxid) during the oxidation of glucose. The ferrouscytochrome oxidase that is formed transfers the electrons to molecular oxygen and produces energetic adenosine triphosphate (ATP) from adenosine diphosphate (ADP) (see Section 4.3), regenerating Fe(III)-oxid that can repeat the cycle. The overall process is represented as follows:

$$\text{Fe(III)-oxid} + \text{Reducing agent} \rightarrow \text{Fe(II)-oxid} + \text{Oxidized reducing agent} \quad (11.2.2)$$

$$\text{Fe(II)-oxid} + 2H^+ + \tfrac{1}{2}O_2 \xrightarrow{\text{ADP} \rightarrow \text{ATP}} \text{Fe(III)-oxid} + H_2O \quad (11.2.3)$$

Cyanide bonds to the iron(III) of the ferricytochrome enzyme, preventing its reduction to iron(II) in the first of the two reactions above. The result is that ferrouscytochrome oxidase, which is required to react with O_2, is not formed and utilization of oxygen in cells is prevented, leading to rapid cessation of metabolic processes. The decreased utilization of oxygen in tissue results in a buildup of oxyhemoglobin in venous blood, which gives the skin and mucous membranes a characteristic red color (flush).

The metabolic pathway for the detoxification of cyanide involves conversion to the less toxic thiocyanate by a reaction requiring thiosulfate or colloidal sulfur as a substrate:

$$CN^- + S_2O_3^{2-} \xrightarrow{\text{Rhodanase}} SCN^- + SO_3^{2-} \quad (11.2.4)$$

This reaction is catalyzed by *rhodanase* enzyme, also called *mitochondrial sulfur transferase*. Although not found in the blood, this enzyme does occur abundantly in liver and kidney tissue. Because of this reaction, thiosulfate can be administered as an antidote for cyanide poisoning.

Nitrite, NO_2^-, administered intravenously as sodium nitrite solution or inhaled as amyl nitrite, $C_5H_{11}NO_2$, an ester which hydrolyzes to NO_2^- in the blood, functions as an antidote to cyanide poisoning. This occurs because nitrite oxidizes iron(II) in blood hemoglobin (HbFe(II)) to methemoglobin (HbFe(III)), a brown substance that is ineffective in carrying oxygen to tissues. (This reaction is the mechanism of nitrite toxicity; excessive formation of methemoglobin causes oxygen deprivation that can be fatal.) Methemoblogin in the blood, however, has a high affinity for cyanide and removes it from ferricytochrome oxidase enzyme that has been inhibited by binding of cyanide (Fe(III)-oxid–CN),

$$HbFe(III) + Fe(III)\text{-oxid–CN} \rightarrow HbFe(III)\text{–CN} + Fe(III)\text{-oxid} \quad (11.2.5)$$

freeing the ferricytochrome oxidase enzyme so that it can participate in its normal metabolic functions. Additional treatment with thiosulfate results in elimination of the cyanide:

$$HbFe(III)\text{–CN} + S_2O_3^{2-} \rightarrow SCN^- + HbFe(III) + SO_3^{2-} \quad (11.2.6)$$

11.2.2 Carbon Monoxide

Carbon monoxide, CO, is a toxic industrial gas produced by the incomplete combustion of carbonaceous fuels. It is used as a reductant for metal ores, for chemical synthesis, and as a fuel. As an environmental toxicant, it is responsible for a significant number of accidental poisonings annually. Observable acute effects of carbon monoxide exposure in humans cover a wide range of symptoms and severity. These include impairment of judgment and visual perception at CO levels of 10 ppm in air; dizziness, headache, and weariness (100 ppm); loss of consciousness (250 ppm); and rapid death (1000 ppm). Chronic effects of long-term low-level exposure to carbon monoxide include disorders of the respiratory system and the heart. As evidence of the latter, cardiac dysfunctions, including arrhythmia and myocardia ischemia (blood deficiency in the heart muscles), have been reported in victims of carbon monoxide poisoning.[2] Autopsies of such victims have shown scattered hemorrhages throughout the heart.

11.2.3 Biochemical Action of Carbon Monoxide

Carbon monoxide enters the bloodstream through the lungs and reacts with oxyhemoglobin (O_2Hb) to produce carboxyhemoglobin (COHb):

$$O_2Hb + CO \rightarrow COHb + O_2 \quad (11.2.7)$$

Carboxyhemoglobin is several times more stable than oxyhemoglobin and ties up the hemoglobin so that it cannot carry oxygen to body tissues.

11.2.4 Cyanogen, Cyanamide, and Cyanates

Cyanogen, NCCN, is a colorless, violently flammable gas with a pungent odor. It may cause permanent injury or even death in exposed individuals. Fumes produced by the reaction of cyanogen with water or acids are highly toxic.

Cyanamide, H$_2$NCN, and calcium cyanamide, CaNCN, are used as fertilizers and raw materials. Calcium cyanamide is employed for the desulfurization and nitridation of steel. Inhalation or oral ingestion of cyanamide causes dizziness, lowers blood pressure, and increases rates of pulse and respiration. Calcium cyanamide acts as a primary irritant to the skin and to nose and throat tissues. The major metabolic product of cyanimide is N-acetylcyanamide, which is found in the urine of subjects exposed to cyanamide.

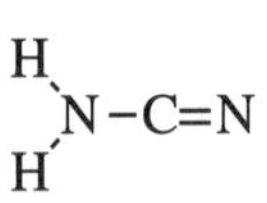

$$\underset{\text{Cyanamide}}{\overset{H}{\underset{H}{N}}-C=N} \qquad \underset{\text{N-acetylcyanamide}}{H-\overset{H}{\underset{H}{C}}-\overset{O}{\overset{\|}{C}}-\overset{H}{\underset{}{N}}-C\equiv N}$$

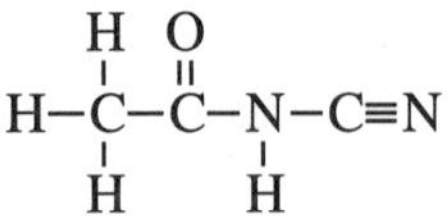

Cyanic acid, HOCN (boiling point (bp), 23.3°C; melting point (mp), –86°C), is a dangerously explosive liquid with an acrid odor. The acid forms cyanate salts, such as NaOCN and KOCN. During decomposition from heat or contact with strong acid, cyanic acid evolves very toxic fumes.

11.3 TOXIC INORGANIC NITROGEN COMPOUNDS

11.3.1 Ammonia

Ammonia, NH$_3$, is widely used as a gas for chemical synthesis, fertilizer, and other applications. It is also used as a solution of concentrated NH$_3$ in water as a chemical reagent and as a fertilizer. Tanks of liquified anhydrous ammonia are common targets for the operators of "meth labs" in rural areas, who steal this dangerous chemical to make illicit amphetamines. Undoubtedly, some of the thieves suffer injury in the process, though such injuries are rarely reported.

The evaporation of liquid ammonia in contact with flesh can cause frostbite. Ammonia is a potent skin corrosive and can damage eye tissue. When inhaled, ammonia causes constriction of the bronchioles. Because of its high water solubility, ammonia is absorbed by the moist tissues of the upper respiratory tract. Irritant damage to the lungs from ammonia can cause edema and changes in lung permeability.

11.3.2 Hydrazine

Hydrazine,

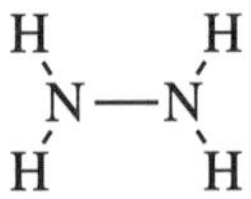

$$\overset{H}{\underset{H}{N}}-\overset{H}{\underset{H}{N}} \quad \text{Hydrazine}$$

is a common inorganic nitrogen compound. Hydrazine is hepatotoxic, causing accumulation of triglycerides in the liver, a condition commonly called fatty liver. These effects may be related to hydrazine's ability to increase the activity of enzymes required to produce diglycerides, depletion of ATP, or inhibition of protein synthesis. Hydrazine acting in the liver induces hydrolysis of glycogen (animal starch) to release glucose, causing excessive blood glucose levels, a condition called hyperglycemia. This can result in depletion of glycogen, leading to the opposite effect, hypoglycemia. Hydrazine inhibits some enzymes, including phosphoenol pyruvatecarboxykinase and some transaminases that are involved in intermediary metabolism. Swelling of cell mitochondria has been observed after exposure to hydrazine, and prolonged exposure can result in formation of large megamitochondria.

The most serious toxicologic effect of hydrazine is its ability to indirectly cause methylation of DNA, leading to cancer. Inhalation of hydrazine has been linked to lung cancer.

11.3.3 Nitrogen Oxides

The two most common oxides of nitrogen are **nitric oxide** (NO) and **nitrogen dioxide** (NO_2), designated collectively as NO_x. Nitric oxide is produced in combustion processes from organically bound nitrogen endogenous to fossil fuels (particularly coal, heavy fuel oil, and shale oil) and from atmospheric nitrogen under the conditions that exist in an internal combustion engine, as shown by the two following reactions:

$$2N(fossil\ fuel) + O_2 \rightarrow 2NO \quad (11.3.1)$$

$$N_2 + O_2 \xrightarrow{\text{Internal combustion engine}} 2NO \quad (11.3.2)$$

Under the conditions of photochemical smog formation, nitric oxide is converted to nitrogen dioxide by the following overall reaction:

$$2NO + O_2 \xrightarrow{\text{Organics, photochemical processes}} 2NO_2 \quad (11.3.3)$$

This conversion consists of complex chain reactions involving light energy and unstable reactive intermediate species. The conditions required are stagnant air, low humidity, intense sunlight, and the presence of reactive hydrocarbons, particularly those from automobile exhausts. Of the NO_x constituents, NO_2 is generally regarded as the more toxic, although all nitrogen oxides and potential sources thereof (such as nitric acid in the presence of oxidizable organic matter) should be accorded the same respect as nitrogen dioxide.

11.3.4 Effects of NO_2 Poisoning

The toxic effects of NO_2 have been summarized.[3] Inhalation of NO_2 causes severe irritation of the innermost parts of the lungs, resulting in pulmonary edema and fatal bronchiolitis fibrosa obliterans. Inhalation, for even very brief periods, of air containing 200 to 700 ppm of NO_2 can be fatal. The biochemical action of NO_2 includes disruption of some enzyme systems, such as lactic dehydrogenase. Nitrogen dioxide probably acts as an oxidizing agent similar to, though weaker than, ozone, which is discussed in Section 10.6.1. Included is the formation of free radicals, particularly the hydroxyl radical HO·. Like ozone, it is likely that NO_2 causes **lipid peroxidation**. This is a process in which the C=C double bonds in unsaturated lipids are attacked by free radicals and undergo chain reactions in the presence of O_2, resulting in their oxidative destruction.

11.3.5 Nitrous Oxide

Nitrous oxide, once commonly known as laughing gas, is used as an oxidant gas and in dental surgery as a general anesthetic. It is a central nervous system depressant and can act as an asphyxiant.

11.4 HYDROGEN HALIDES

Hydrogen halides are compounds with the general formula HX, where X is F, Cl, Br, or I. They are all gases, and all are relatively toxic. Because of their abundance and industrial uses, HF and HCl have the greatest toxicological significance of these gases.

11.4.1 Hydrogen Fluoride

Hydrogen fluoride, HF (mp, –83.1°C; bp, 19.5°C), may be in the form of either a clear, colorless liquid or gas. It forms corrosive fumes when exposed to the atmosphere. The major commercial application of hydrogen fluoride is as an alkylating catalyst in petroleum refining. Pot room workers in the primary aluminum industry are exposed to levels up to 5 mg/m^3 in the workplace atmosphere and exhibit elevated levels of F$^-$ ion in their blood plasma.[4] Hydrogen fluoride in aqueous solution is called **hydrofluoric acid**, which contains 30 to 60% HF by mass. Hydrofluoric acid must be kept in plastic containers because it vigorously attacks glass and other materials containing silica (SiO$_2$), producing gaseous silicon tetrafluoride, SiF$_4$. Hydrofluoric acid is used to etch glass and clean stone.

Both hydrogen fluoride and hydrofluoric acid, referred to collectively as HF, are extreme irritants to any tissue they contact. Exposed areas heal poorly, gangrene may develop, and ulcers can occur in affected areas of the upper respiratory tract.

The toxic nature of fluoride ion, F$^-$, is not confined to its presence in HF. It is toxic in soluble fluoride salts, such as NaF. At relatively low levels, such as about 1 ppm, used in some drinking water supplies, fluoride prevents tooth decay. At excessive levels, fluoride causes **fluorosis**, a condition characterized by bone abnormalities and mottled, soft teeth. Livestock are especially susceptible to poisoning from fluoride fallout on grazing land as a result of industrial pollution. In severe cases, the animals become lame and even die.

11.4.2 Hydrogen Chloride

Hydrogen chloride, HCl (mp, –114°C; bp, –84.8°C), may be encountered as a gas, pressurized liquid, or aqueous solution called **hydrochloric acid**, commonly denoted simply as HCl. This compound is colorless in the pure state and in aqueous solution. As a saturated solution containing 36% HCl, hydrochloric acid is a major industrial chemical, with U.S. production of about 2.3 million tons per year. It is used for chemical and food manufacture, acid treatment of oil wells to increase crude oil flow, and metal processing.

Hydrogen chloride is not nearly as toxic as HF, although inhalation can cause spasms of the larynx as well as pulmonary edema and even death at high levels. Because of its high affinity for water, HCl vapor tends to dehydrate tissue of the eyes and respiratory tract. Hydrochloric acid is a natural physiological fluid found as a dilute solution in the stomachs of humans and other animals.

11.4.3 Hydrogen Bromide and Hydrogen Iodide

Hydrogen bromide, HBr (mp, –87°C; bp, –66.5°C), and **hydrogen iodide**, HI (mp, –50.8°C; bp, –35.4°C), are both pale yellow or colorless gases, although contamination by their respective elements tends to impart some color to these compounds. Both are very dense gases, 3.5 g/l for HBr and 5.7 g/l for HI at 0°C and atmospheric pressure. These compounds are used much less than HCl. Both are irritants to the skin and eyes and to the oral and respiratory mucous membranes.

11.5 INTERHALOGEN COMPOUNDS AND HALOGEN OXIDES

Halogens form compounds among themselves and with oxygen. Some of these compounds are important in industry and toxicologically. Some of the more important such compounds are discussed below.

Table 11.1 Major Interhalogen Compounds

Compound Name and Formula	Physical Properties
Chlorine monofluoride, ClF	Colorless gas; mp, −154°C; bp, 101°C
Chlorine trifluoride, ClF_3	Colorless gas; mp, −83°C; bp, 12°C
Bromine monofluoride, BrF	Pale brown gas; bp, 20°C
Bromine trifluoride, BrF_3	Colorless liquid; mp, 8.8°C; bp, 127°C
Bromine pentafluoride, BrF_5	Colorless liquid; mp, −61.3°C; bp, 40°C
Bromine monochloride, BrCl	Red or yellow highly unstable liquid and gas
Iodine trifluoride, IF_3	Yellow solid decomposing at 28°C
Iodine pentafluoride, IF_5	Colorless liquid; mp, 9.4°C; bp, 100°C
Iodine heptafluoride, IF_7	Colorless sublimable solid; mp, 5.5°C
Iodine monobromide, IBr	Gray sublimable solid; mp, 42°C
Iodine monochloride, ICl	Red-brown solid alpha form; mp, 27°C; bp, 9°C
Iodine pentabromide, IBr_5	Crystalline solid
Iodine tribromide, IBr_3	Dark brown liquid
Iodine trichloride, ICl_3	Orange-yellow solid subliming at 64°C
Iodine pentachloride, ICl_5	—

Table 11.2 Major Oxides of the Halogens

Compound Name and Formula	Physical Properties
Fluorine monoxide, OF_2	Colorless gas; mp, −224°C; bp, −145°C
Chlorine monoxide, Cl_2O	Orange gas; mp, −20°C; bp, 2.2°C
Chlorine dioxide, ClO_2	Orange gas; mp, −59°C; bp, 9.9°C
Chlorine heptaoxide, Cl_2O_7	Colorless oil; mp, −91.5°C; bp, 82°C
Bromine monoxide, Br_2O	Brown solid; decomp. −18°C
Bromine dioxide, BrO_2	Yellow solid; decomp. 0°C
Iodine dioxide, IO_2	Yellow solid
Iodine pentoxide, I_2O_5	Colorless oil; decomp. 325°C

11.5.1 Interhalogen Compounds

Fluorine is a sufficiently strong oxidant to oxidize chlorine, bromine, and iodine, whereas chlorine can oxidize bromine and iodine. The compounds thus formed are called **interhalogen compounds**. The major interhalogen compounds are listed in Table 11.1.

The liquid interhalogen compounds are usually described as "fuming" liquids. For the most part, interhalogen compounds exhibit extreme reactivity. They react with water or steam to produce hydrohalic acid solutions (HF, HCl) and nascent oxygen {O}. They tend to be potent oxidizing agents for organic matter and oxidizable inorganic compounds. These chemical properties are reflected in the toxicities of the interhalogen compounds. Too reactive to enter biological systems in their original chemical state, they tend to be powerful corrosive irritants that acidify, oxidize, and dehydrate tissue. The skin, eyes, and mucous membranes of the mouth, throat, and pulmonary systems are susceptible to attack by interhalogen compounds. In some respects, the toxicities of the interhalogen compounds resemble the toxic properties of the elemental forms of the elements from which they are composed. The by-products of chemical reactions of the interhalogen compounds — such as HF from fluorine compounds — pose additional toxicological hazards.

11.5.2 Halogen Oxides

The oxides of the halogens tend to be unstable and reactive. Although these compounds are called oxides, it is permissible to call the ones containing fluorine fluorides because fluorine is more electronegative than oxygen. The major halogen oxides are listed in Table 11.2. Commercially, the most important of the halogen oxides is chlorine dioxide, which offers some advantages over chlorine as a water disinfectant. It is also employed for odor control and bleaching wood pulp. Because of its extreme instability, chlorine dioxide is manufactured on the site where it is used.

Investigations on human blood and on rodents suggest that ClO_2 and its metabolic product ClO_2^- cause formation of methemoglobin, decrease the activities of glucose-6-phosphate dehydrogenase and glutathione peroxidase enzymes, reduce levels of reduced glutathione (a protective agent against oxidative stress), increase levels of hydrogen peroxide, and cause breakdown of red blood cells releasing hemoglobin (hemolysis).[5] These effects would suggest an overall hematotoxicity of chlorine dioxide.

For the most part, the halogen oxides are highly reactive toxic substances. Their toxicity and hazard characteristics are similar to those of the interhalogen compounds, described previously in this section.

11.5.3 Hypochlorous Acid and Hypochlorites

The halogens form several oxyacids and their corresponding salts. Of these, the most important is hypochlorous acid (HOCl), formed by the following reaction:

$$Cl_2 + H_2O \rightleftharpoons HCl + HOCl \qquad (11.5.1)$$

Hypochlorous acid and hypochlorites are used for bleaching and disinfection. They produce active (nascent) oxygen, {O}, as shown by the reaction below, and the resulting oxidizing action is largely responsible for the toxicity of hypochlorous acid and hypochlorites as irritants to eye, skin, and mucous membrane tissue.

$$HClO \rightarrow H^+ + Cl^- + \{O\} \qquad (11.5.2)$$

11.5.4 Perchlorates

Perchlorates are the most oxidized of the salts of the chlorooxyacids. Although perchlorates are not particularly toxic, ammonium perchlorate (NH_4ClO_4) should be mentioned because it is a powerful oxidizer and reactive chemical produced in large quantities as a fuel oxidizer in solid rocket fuels. Each of the U.S. space shuttle booster rockets contains about 350,000 kg of ammonium perchlorate in its propellant mixture. By 1988, U.S. consumption of ammonium perchlorate for rocket fuel uses was of the order of 24 million kg/year. In May 1988, a series of massive explosions in Henderson, Nevada, demolished one of only two plants producing ammonium perchlorate for the U.S. space shuttle, MX missile, and other applications, so that supplies were severely curtailed. The plant has since been rebuilt.

The toxicological hazard of perchlorate salts may depend on the cation in the compound. In general, the salts should be considered as skin irritants and treated as such. Perchlorate ion, ClO_4^-, may compete physiologically with iodide ion, I^-. This can occur in the uptake of iodide by the thyroid, leading to the biosynthesis of thyroid hormones. As a consequence, perchlorate can cause symptoms of iodine deficiency.

11.6 NITROGEN COMPOUNDS OF THE HALOGENS

11.6.1 Nitrogen Halides

The general formula of the nitrogen halides is N_nX_x, where X is F, Cl, Br, or I. A list of nitrogen halides is presented in Table 11.3. The nitrogen halides are considered to be very toxic, largely as irritants to eyes, skin, and mucous membranes. Direct exposure to nitrogen halide compounds tends

Table 11.3 Nitrogen Halides

Compound Name and Formula	Physical Properties
Nitrogen trifluoride, NF_3	Colorless gas; mp, –209°C; bp, –129°C
Nitrogen trichloride, NCl_3	Volatile yellow oil; melting below –40°C; boiling below 71°C; exploding around 90°C
Nitrogen tribromide, NBr_3	Solid crystals
Nitrogen triiodide, NI_3	Black crystalline explosive substance
Tetrafluorohydrazine, N_2F_4	—

to be limited because of their reactivity, which may destroy the compound before exposure. Nitrogen triiodide is so reactive that even a "puff" of air can detonate it.[6]

11.6.2 Azides

Halogen azides are compounds with the general formula XN_3, where X is one of the halogens. These compounds are extremely reactive and can be spontaneously explosive. Their reactions with water can produce toxic fumes of the elemental halogen, acid (e.g., HCl), and NO_X. The compound vapors are irritants.

11.6.3 Monochloramine and Dichloramine

The substitution of Cl for H on ammonia can be viewed as a means of forming nitrogen trichloride (Table 11.3), monochloramine, and dichloramine. The formation of the last two compounds from ammonium ion in water is shown by the following reactions:

$$NH_4^+ + HOCl \rightarrow H^+ + H_2O + NH_2Cl \qquad (11.6.1)$$
$$\text{Monochloramine}$$

$$NH_2Cl + HOCl \rightarrow H_2O + NHCl_2 \qquad (11.6.2)$$
$$\text{Dichloramine}$$

The chloramines are disinfectants in water and are formed deliberately in the purification of drinking water to provide **combined available chlorine**. Although combined available chlorine is a weaker disinfectant than water, containing Cl_2, HOCl, and OCl^-, it is retained longer in the water distribution system, affording longer-lasting disinfection.

Since they work as disinfectants, the chloramines have to have some toxic effects. They have been shown to inhibit acetylcholinesterase activity.[7]

11.7 INORGANIC COMPOUNDS OF SILICON

Because of its use in semiconductors, silicon has emerged as a key element in modern technology. Concurrent with this phenomenon has been an awareness of the toxicity of silicon compounds, many of which, fortunately, have relatively low toxicities. This section covers the toxicological aspects of inorganic silicon compounds.

11.7.1 Silica

The silicon compound that has probably caused the most illness in humans is **silica**, SiO_2. Silica is a hard mineral substance known as quartz in the pure form and occurring in a variety of minerals, such as sand, sandstone, and diatomaceous earth. Because of silica's occurrence in a large number of common materials that are widely used in construction, sand blasting, refractories manufacture, and many other industrial applications, human exposure to silica dust is widespread. Such exposure causes a condition called **silicosis**, a type of pulmonary fibrosis, one of the most common disabling conditions that result from industrial exposure to hazardous substances. Silicosis causes fibrosis and nodules in the lung, lowering lung capacity and making the subject more liable to pulmonary diseases, such as pneumonia. A lung condition called silicotuberculosis may develop. Severe cases of silicosis can cause death from insufficient oxygen or from heart failure.

Silica exposure has been associated with increased incidences of **scleroderma**, a condition manifested by hardened, rigid connective tissue. In this respect, it is believed that silica acts by an adjuvant mechanism in which it enhances the autoimmune response caused by other agents, such as silicones or paraffin.[8]

11.7.2 Asbestos

Asbestos describes a group of silicate minerals, such as those of the serpentine group, approximate formula $Mg_3P(Si_2O_5)(OH)_4$, which occur as mineral fibers. Asbestos has many properties, such as insulating abilities and heat resistance, that have given it numerous uses. It has been used in structural materials, brake linings, insulation, and pipe manufacture. Unfortunately, inhalation of asbestos damages the lungs and results in a characteristic type of lung cancer in some exposed subjects. The toxic effects of asbestos are initiated when asbestos fibers in the lung act as local irritants and become phagocytosed by macrophages (large white blood cells). The bodies of phagocytosed asbestos are taken up by cellular lysosomes, which secrete hydrolytic enzymes, digesting the matter surrounding the asbestos particles and releasing them to start the process over. This process causes lymphoid tissue to aggregate in the vicinity of the insult, forming fibrotic lesions from the synthesis of excess collagen.[9]

The three major pathological conditions caused by the inhalation of asbestos are asbestosis (a pneumonia condition), mesothelioma (tumor of the mesothelial tissue lining the chest cavity adjacent to the lungs), and bronchogenic carcinoma (cancer originating with the air passages in the lungs). Because of these health effects, uses of asbestos have been severely curtailed and widespread programs have been undertaken to remove asbestos from buildings.

Lung cancer from asbestos exposure has a strong synergistic relationship with exposure to cigarette smoke.[10] Long-term exposure to asbestos, alone, increases the incidence of lung cancer about 5-fold, cigarette smoking roughly 10-fold, but the two together more than 50-fold.

11.7.3 Silanes

Compounds of silicon with hydrogen are called **silanes**. The simplest of these is silane, SiH_4. Disilane is H_3SiSiH_3. Numerous organic silanes exist in which alkyl moieties are substituted for H.

In addition to SiH_4, the inorganic silanes produced for commercial use are dichloro- and trichlorosilane, SiH_2Cl_2 and $SiHCl_3$, respectively. These compounds are used as intermediates in the synthesis of organosilicon compounds and in the production of high-purity silicon for semiconductors. Several kinds of inorganic compounds derived from silanes have potential uses in the manufacture of photovoltaic devices for the direct conversion of solar energy to electricity. In general, not much is known about the toxicities of silanes. Silane itself burns readily in air. Chlorosilanes are irritants to eye, nasal, and lung tissue. The toxicities of silane, dichlorosilane, and tetraethoxysilane, $Si(OC_2H_5)$, have been reviewed for their relevance in the semiconductor industry.[11] The major effects of silane and tetraethoxysilane appeared to be nephrotoxicity (kidney damage).

11.7.4 Silicon Halides and Halohydrides

Four **silicon tetrahalides**, with the general formula SiX_4, are known to exist. Of these, only silicon tetrachloride, $SiCl_4$, is produced in significant quantities. It is used to manufacture fumed silica (finely divided SiO_2). In addition, numerous **silicon halohydrides**, with the general formula $H_{4-x}SiX_x$, have been synthesized. The commercially important compound of this type is trichlorosilane, $HSiCl_3$, which is used to manufacture organotrichlorosilanes and elemental silicon for semiconductors.

Both silicon tetrachloride and trichlorosilane are fuming liquids with suffocating odors. They both react with water to give off HCl vapor.

11.8 INORGANIC PHOSPHORUS COMPOUNDS

11.8.1 Phosphine

Phosphine, PH_3 (mp, $-132°C$; bp, $-88°C$), is a colorless gas that undergoes autoignition at 100°C. It is used for the synthesis of organophosphorus compounds. Its inadvertent production in chemical syntheses involving other phosphorus compounds is a potential hazard in industrial processes and in the laboratory. Phosphine gas is a pulmonary tract irritant and central nervous system depressant that is very toxic when inhaled and can be fatal. Symptoms of acute exposure include headache, dizziness, burning pain below the sternum, nausea, vomiting, difficult, painful breathing, pulmonary irritation and edema, cough with fluorescent green sputum, tremors, and fatigue. Convulsions have appeared in some victims after they have apparently recovered from phosphine poisoning. Workers chronically exposed to phosphine have exhibited inflammation of the nasal cavity and throat, nausea, dizziness, weakness, and adverse gastrointestinal, cardiorespiratory, and central nervous system effects. Chronic effects have also included hepatotoxic symptoms, jaundice, nervous system abnormalities, and increased bone density.

Arsine gas, AsH_3, is mentioned here because of the position of arsenic directly below phosphorus in the periodic table, and hence the similarity between arsine and phosphine. Arsine may be generated by chemically reductive processes in the refining of various metals. It is a highly toxic substance that can cause fatal instances of poisoning. Its major effect is on the blood, and it may cause breakdown of red blood cells with liberation of hemoglobin (hemolysis).[12] Symptomatic of this effect is the presence of hemoglobin in urine (hemoglobinuria). Acute symptoms of arsine poisoning include headache, shortness of breath, nausea, and vomiting. Jaundice and anemia may also accompany arsine poisoning.

11.8.2 Phosphorus Pentoxide

The oxide most commonly formed by the combustion of elemental white phosphorus and many phosphorus compounds is P_4O_{10}. As an item of commerce, this compound is usually misnamed **phosphorus pentoxide**. When produced from the combustion of elemental phosphorus (see Reaction 10.6.7), it is a fluffy white powder that removes water from air to form syrupy orthophosphoric acid:

$$P_4O_{10} + 6H_2O \rightarrow 4H_3PO_4 \qquad (11.8.1)$$

Because of its dehydrating action and formation of acid, phosphorus pentoxide is a corrosive irritant to skin, eyes, and mucous membranes.

11.8.3 Phosphorus Halides

Phosphorus forms halides with the general formulas PX_3 and PX_5. Typical of such compounds are phosphorus trifluoride (PF_3), a colorless gas (mp, $-152°C$; bp, $-102°C$), and phosphorus pentabromide (PBr_5), a yellow solid that decomposes at approximately $100°C$. Of these compounds, the most important commercially is phosphorus pentachloride, used as a catalyst in organic synthesis, as a chlorinating agent, and as a raw material to make phosphorus oxychloride ($POCl_3$). Phosphorus halides react violently with water to produce the corresponding hydrogen halides and oxophosphorus acids, as shown by the following reaction of phosphorus pentachloride:

$$PCl_5 + 4H_2O \rightarrow H_3PO_4 + 5HCl \qquad (11.8.2)$$

Largely because of their acid-forming tendencies, the phosphorus halides are strong irritants to eyes, skin, and mucous membranes, and should be regarded as very toxic.

11.8.4 Phosphorus Oxyhalides

Phosphorus oxyhalides, with the general formula POX_3, are known for fluoride, chloride, and bromide. Of these, the one with commercial uses is phosphorus oxychloride ($POCl_3$). Its uses are similar to those of phosphorus trichloride, acting in chemical synthesis as a chlorinating agent and for the production of organic chemical intermediates. It is a faintly yellow fuming liquid (mp, $1°C$; bp, $105°C$). It reacts with water to form hydrochloric acid and phosphonic acid (H_3PO_3). The liquid evolves toxic vapors, and it is a strong irritant to the eyes, skin, and mucous membranes. Phosphorus oxychloride is metabolized to phosphorodichloridic acid,

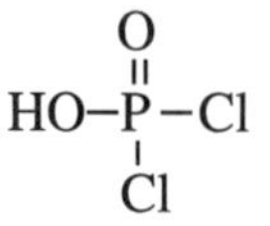

$$\begin{array}{c} O \\ \parallel \\ HO-P-Cl \\ | \\ Cl \end{array} \quad \text{Phosphorodichloridic acid}$$

a phosphorylating agent that phosphorylates acetylcholinesterase at the active site to form enzymically inactive (O-phosphoserine)acetylcholinesterase.[13]

11.9 INORGANIC COMPOUNDS OF SULFUR

One of the elements essential for life, sulfur is a constituent of several of the more important toxic inorganic compounds. The common elemental form of yellow crystalline or powdered sulfur, S_8, has a low toxicity, although chronic inhalation of it can irritate mucous membranes.

11.9.1 Hydrogen Sulfide

Hydrogen sulfide (H_2S) is a colorless gas (mp, $-86°C$; bp, $-61°C$) with a foul, rotten-egg odor. It is produced in large quantities as a by-product of coal coking and petroleum refining, and massive quantities are removed in the cleansing of sour natural gas. Hydrogen sulfide is released in large quantities from volcanoes and hydrothermal vents. Indeed, if Yellowstone National Park in the U.S. were an industrial enterprise, parts of it would be shut down because of release of hydrogen sulfide from geothermal sources. Hydrogen sulfide is a major source of elemental sulfur by a process that involves oxidation of part of the H_2S to SO_2, followed by the Claus reaction:

$$2H_2S(g) + SO_2(g) \rightarrow 2H_2O(l) + 3S(s) \tag{11.9.1}$$

Hydrogen sulfide is a very toxic substance, which in some cases can cause a fatal response more rapidly even than hydrogen cyanide, the toxic effects of which it greatly resembles. Like cyanide, hydrogen sulfide inhibits the cytochrome oxidase system essential for respiration. Hydrogen sulfide affects the central nervous system, causing symptoms that include headache, dizziness, and excitement. Rapid death occurs at exposures to air containing more than about 1000 ppm of H_2S, and somewhat lower exposures for about 30 min can be lethal. Death results from asphyxiation as a consequence of respiratory system paralysis. Sulfide can also cause localized toxic effects at the point of contact, one of which is pulmonary edema. Another localized effect is eye conjunctivitis, a condition called "gas eye," perhaps named after conditions suffered by gas works employees formerly exposed to hydrogen sulfide produced in the gasification of high-sulfur coal in the production of synthetic gas, once widely used for cooking and lighting.

Accidental poisonings by hydrogen sulfide are not uncommon. In the most notorious such case, 22 people (by some accounts many more) were killed in 1950 in Poza Rica, Mexico, when a flare used to "dispose" of hydrogen sulfide from natural gas by burning it to sulfur dioxide became extinguished, releasing large quantities of H_2S and asphyxiating victims as they slept. In 1975, at Denver City, Texas, nine people were killed from hydrogen sulfide blown out of a secondary petroleum recovery well. There are numerous effects of chronic H_2S poisoning, including general debility.

Hydrogen sulfide is acted on in the body by methylation with thiol S-methyl transferase. The initial product is methanethiol, $HSCH_3$, which is also quite toxic. A second methylation produces dimethylsulfide, H_2CSCH_3. (Of some interest is the fact that dimethylsulfide is the major volatile sulfur compound released to the atmosphere from oceans, where it is produced by the action of marine microorganisms.)

Bacteria acting anaerobically in the colon produce large quantities of hydrogen sulfide and methanethiol. There is evidence to suggest that the mucous membranes of the colon have enzymes that convert hydrogen sulfide and methanethiol to nontoxic thiosulfate, $S_2O_3^{2-}$.[14]

The nitrite-induced formation of blood methemoglobin has been used successfully to treat hydrogen sulfide poisoning. Like cyanide, hydrogen sulfide bonds to iron(III) in methemoglobin so that it is not available to inhibit cytochrome oxidase.[15]

11.9.2 Sulfur Dioxide and Sulfites

Sulfur dioxide (SO_2) is an intermediate in the production of sulfuric acid. It is a common air pollutant produced by the combustion of pyrite (FeS_2) in coal and organically bound sulfur in coal and fuel oil, as shown by the two following reactions:

$$4FeS_2 + 11O_2 \rightarrow 2Fe_2O_3 + 8SO_2 \tag{11.9.2}$$

$$S(organic,\ in\ fuel) + O_2 \rightarrow SO_2 \tag{11.9.3}$$

These sources add millions of tons of sulfur dioxide to the global atmosphere annually and are largely responsible for acid rain.

Sulfur dioxide is an irritant to the eyes, skin, mucous membranes, and respiratory system. As a water-soluble gas, it is largely removed in the upper respiratory tract. Its major effect is as a respiratory tract irritant, where it irritates the upper airways and causes bronchioconstriction, resulting in increased airflow resistance.[16] Subjects who are hyperresponsive to sulfur dioxide are especially at risk from it. Asthmatics may suffer bronchioconstriction after only a few breaths of sulfur dioxide-contaminated air. The degree of response of asthma sufferers to sulfur dioxide is highly variable.[17]

Dissolved in water, sulfur dioxide produces **sulfurous acid** (H_2SO_3), **hydrogen sulfite ion** (HSO_3^-), and **sulfite ion** (SO_3^{2-}). Sodium sulfite (Na_2SO_3) has been used as a chemical food preservative, although some individuals are hypersensitive to it.

11.9.3 Sulfuric Acid

Sulfuric acid is number one in synthetic chemical production. It is used to produce phosphate fertilizer, high octane gasoline, and a wide variety of inorganic and organic chemicals. Large quantities are consumed to pickle steel (cleaning and removal of surface oxides); disposal of spent pickling liquor can be a problem.

Sulfuric acid is of particular concern as an atmospheric pollutant. In times past, air polluted with unquestionably toxic levels of sulfuric acid aerosols (see Section 2.8.3), such as in the severe air pollution that occurred in London and around various smelters in the 1950s and early 1960s, produced toxic effects and even fatalities. At present, sulfuric acid is a major contributor to acid precipitation (see Section 2.8.1), and it may well be the most intense common irritant occurring in air polluted with acid substances. Most of the pollutant sulfur that becomes atmospheric H_2SO_4 is emitted to the atmosphere as SO_2 from the burning of sulfur-containing fuels (particularly coal). Sulfur dioxide emissions are almost always accompanied by emissions of particulate matter, which often contains metals, such as vanadium, iron, and manganese. These metals can catalyze the oxidation of SO_2 to H_2SO_4, either on particle surfaces or leached into aqueous solution in aerosol droplets:

$$SO_2 + \tfrac{1}{2}O_2 + H_2O \rightarrow H_2SO_4(aq) \qquad (11.9.4)$$

The result can be formation of an aerosol mist of droplets containing intensely irritating sulfuric acid.

Sulfuric acid is a severely corrosive poison and dehydrating agent in the concentrated liquid form. It readily penetrates skin to reach subcutaneous tissue and causes tissue necrosis, with effects resembling those of severe thermal burns. Sulfuric acid fumes and mists can act as irritants to eye and respiratory tract tissue. Industrial exposure has caused tooth erosion in workers.

At lower levels, inhalation of sulfuric acid from sources such as atmospheric precipitation is damaging to the pulmonary tract. Compared to sulfur dioxide, sulfuric acid is the much more potent lung tissue irritant. Animal studies and limited data from exposed humans indicate that inhalation of H_2SO_4 aerosol increases airway resistance and inhibits bronchial clearance of inhaled particles. Asthmatic subjects are sensitive to sulfuric acid inhalation, and the effect may be synergistic with sulfur dioxide. Therefore, particularly for sensitive individuals, exposure to air containing sulfuric acid, sulfur dioxide, and particles — all of which tend to occur together when one is present in a polluted atmosphere — may be particularly damaging to the lungs.

11.9.4 Carbon Disulfide

Carbon disulfide, CS_2, is a toxicologically important compound because of its widespread use in making rayon and cellophane from cellulose and its well-established toxic effects. Skin contact with carbon disulfide has caused skin disorders, including blisters in rayon plant workers. Very high levels of atmospheric carbon disulfide vapor in the workplace of the order of 10 ppt can cause life-threatening effects on the central nervous system. Epidemiologic studies of viscose rayon workers exposed to carbon disulfide have shown increased mortalities, including cardiovascular mortality. Vascular atherosclerotic changes have been observed in workers exposed to carbon disulfide for long periods of time.

Table 11.4 Inorganic Sulfur Compounds

Compound Name and Formula	Properties
Sulfur	
Monofluoride, S_2F_2	Colorless gas; mp, −104°C; bp, −99°C; toxicity similar to HF
Tetrafluoride, SF_4	Gas; mp, −124°C; bp, −40°C; powerful irritant
Hexafluoride, SF_6	Colorless gas; mp, −51°C; surprisingly nontoxic when pure, but often contaminated with toxic lower fluorides
Monochloride, S_2Cl_2	Oily; fuming orange liquid; mp, −80°C; bp, 138°C; strong irritant to eyes, skin, and lungs
Tetrachloride, SCl_4	Brownish yellow liquid or gas; mp, −30°C; decomp. below 0°C; irritant
Trioxide, SO_3	Solid anhydride of sulfuric acid (see toxic effects above); reacts with moisture or steam to produce sulfuric acid
Sulfuryl chloride, SO_2Cl_2	Colorless liquid; mp, −54°C; bp, 69°C; used for organic synthesis, corrosive toxic irritant
Thionyl chloride, $SOCl_2$	Colorless-to-orange fuming liquid; mp, −105°C; bp, 79°C; toxic corrosive irritant
Carbon oxysulfide, COS	Volatile liquid by-product of natural gas or petroleum refining; toxic narcotic

The most notable toxicological effects of carbon disulfide are on the nervous system, including damage to the peripheral nervous system.[18] Individuals exposed to carbon disulfide have lost consciousness. Decreased nerve conduction velocities have been observed in workers exposed to 10 to 20 ppm of carbon disulfide in the workplace over periods of 10 to 20 years. Brain abnormalities suggesting toxic encephalopathy have been observed in exposed workers. Some studies have suggested the possibility of mental performance and personality disorders in workers exposed to carbon disulfide, including heightened levels of anxiety, introversion, and depression.

11.9.5 Miscellaneous Inorganic Sulfur Compounds

A large number of inorganic sulfur compounds, including halides and salts, are widely used in industry. The more important of these are listed in Table 11.4.

REFERENCES

1. Lei, V., Amoa-Awua, W.K.A., and Brimer, L., Degradation of cyanogenic glycosides by *Lactobacillus plantarum* strains from spontaneous cassava fermentation and other microorganisms, *Int. J. Food Microbiol.*, 53, 169–184, 1999.
2. Gandini, C. et al., Carbon monoxide cardiotoxicity, *J. Toxicol. Clin. Toxicol.*, 39, 35–44, 2001.
3. Elsayed, N.M., Toxicity of nitrogen dioxide: an introduction, *Toxicology*, 89, 161–174, 1994.
4. Lund, K. et al., Exposure to hydrogen fluoride: an experimental study in humans of concentrations of fluoride in plasma, symptoms, and lung function, *Occup. Environ. Med.*, 54, 32–37, 1997.
5. Ueno, H., Sayato, Y., and Nakamuro, K., Hematological effects of chlorine dioxide on in vitro exposure in mouse, rat, and human blood and on subchronic exposure in mice, *J. Health Sci.*, 46, 110–116, 2000.
6. Sax, N.I., *Dangerous Properties of Industrial Materials*, 5th ed., Van Nostrand Reinhold, New York, 1979.
7. Wang, Z. and Minami, M., Effects of chloramine on neuronal cholinergic factors: further studies of toxicity mechanism suggested by an unusual case record, *Biogenic Amines*, 12, 213–223, 1996.
8. Burns, L.A., Meade, B.J., and Munson, A.E., Toxic responses of the immune system, in *Casarett and Doull's Toxicology: The Basic Science of Poisons*, 5th ed., Klaassen, C.D., Ed., McGraw-Hill, New York, 1996, chap. 12, pp. 355–402.
9. Timbrell, J.A., *Principles of Biochemical Toxicology*, 3rd ed., Taylor and Francis, London, 2000, p. 190.
10. Modifying factors of toxic effects, in *Basic Toxicology*, Lu, F.C., Ed., Hemisphere Publishing Corp., New York, 1991, chap. 5, pp. 61–73.

11. Nakashima, H. et al., Toxicity of silicon compounds in semiconductor industries, *J. Occup. Health*, 40, 270–275, 1998.
12. Goyer, R.A., Toxic effects of metals, in *Casarett and Doull's Toxicology: The Basic Science of Poisons*, 5th ed., Klaassen, C.D., Ed., McGraw-Hill, New York, 1996, chap. 23, pp. 691–736.
13. Quistad, G.B. et al., Toxicity of phosphorus oxychloride to mammals and insects that can be attributed to selective phosphorylation of acetylcholinesterase by phosphorodichloridic acid, *Chem. Res. Toxicol.*, 13, 652–657, 2000.
14. Furne, J. et al., Oxidation of hydrogen sulfide and methanethiol to thiosulfate by rat tissues: a specialized function of the colonic mucosa, *Biochem. Pharmacol.*, 62, 255–259, 2001.
15. Smith, R.P., Toxic responses of the blood, in *Casarett and Doull's Toxicology: The Basic Science of Poisons*, 5th ed., Klaassen, C.D., Ed., McGraw-Hill, New York, 1996, chap. 11, pp. 335–354.
16. Costa, D.L., Air pollution, in *Casarett and Doull's Toxicology: The Basic Science of Poisons*, 5th ed., Klaassen, C.D., Ed., McGraw-Hill, New York, 1996, chap. 28, pp. 857–882.
17. Trenga, C.A., Koenig, J.Q., and Williams, P.V., Sulphur dioxide sensitivity and plasma antioxidants in adult subjects with asthma, *Occup. Environ. Med.*, 56, 544–547, 1999.
18. Newhook, R., Meek, M.E., and Walker, M., Carbon disulfide: hazard characterization and exposure-response analysis, *Environ. Carcinog. Ecotoxicol. Rev.*, C19, 125–160, 2001.

QUESTIONS AND PROBLEMS

1. What are the two main toxic forms of cyanide? Which of these is most dangerous by inhalation? Which by ingestion?
2. What is a common natural source of cyanide? How is this form converted to toxic cyanide ion in the body? How did the Romans use this substance?
3. In what sense does cyanide deprive the body of oxygen? How does this differ from the way in which methane gas or nitrogen gas deprives the body of oxygen, or the way in which carbon monoxide does?
4. What is the biochemical action of carbon monoxide? What is the receptor with which carbon monoxide reacts? In what sense is this reaction reversible?
5. In general, how does NO_X enter the atmosphere? How does the more toxic form of NO_X form in the atmosphere?
6. Which organ is most affected by exposure to NO_2? What are the toxic effects, including bronchiolitis fibrosa obliterans? In general, what is the biochemical action of NO_2, and how does it involve free radical and lipid peroxidation?
7. What is the major toxic effect of nitrous oxide, N_2O? What might lead you to believe that it is much less toxic than NO_2?
8. What are the major toxicological effects of ozone? What kinds of groups does it attack in the body?
9. What is the main kind of reaction of chlorine in water? In what sense is this reaction tied to chlorine toxicity?
10. What may be said about the toxicity of fluorine compared to that of chlorine? Why is toxic exposure to bromine and iodine usually less of a problem than that to fluorine or chlorine?
11. Of the hydrogen halides, which is the most dangerous? How does this substance occur? What does it do to the body?
12. What are the nature and symptoms of fluorosis? How can this condition result from air pollution?
13. What kind of compound is ClF_3? What kind of compound is ClO_2? What are their chemical and toxicological similarities?
14. What kinds of compounds are NF_3 and NCl_3? What may be said about their chemical properties? What is their major toxicological effect?
15. What would lead you to believe that monochloramine and dichloramine are not regarded as very toxic, at least in an unconcentrated form? How are these compounds used? How are they related to combined available chlorine?
16. What is the chemical nature of silica? What is its major toxicological effect? What are the symptoms of this toxic effect?

TOXIC INORGANIC COMPOUNDS

17. In addition to silica, there is another silicon-containing mineral that is toxic. What is it? What are its toxic effects? How are its toxic effects synergistic with cigarette smoke?
18. What are silanes? How are they used? Is much known about their toxicities? What are the toxic effects of chlorosilanes?
19. What is the most commonly produced silicon tetrahalide? How is it used? Why might its toxicological properties be similar to those of HCl?
20. Why is PH_3 a particular hazard in the laboratory and in industrial chemical synthesis? Is it very toxic? What are its major toxic effects?
21. What role may be played by particulate matter and by metals, such as vanadium, iron, and manganese, in the production of toxic sulfuric acid?
22. What is the chemical nature of P_4O_{10}? What does it form when exposed to atmospheric moisture? What are its major toxicological effects? Suggest a sequence of reactions by which H_3PO_4 might be formed from PH_3.
23. What is PCl_5? How is it used? How does it react with water, and how is this reaction related to its toxic properties? What are some compounds that are related to PCl_5?
24. What is the most commonly used phosphorus oxyhalide, general formula POX_3? What are its industrial uses? How does it react with water, and how is this reaction related to the fact that it is a strong irritant to the eyes, skin, and mucous membranes?
25. In addition to the "miscellaneous" inorganic sulfur compounds listed in Table 11.4, four sulfur compounds were discussed separately for their toxicities. Of these, which is the most toxic? What is its mode of toxicity?
26. Why does exposure to fatal doses of H_2S still occur? What are some specific incidents in which such exposure has occurred?
27. What is SO_2 like chemically? What does it form in water? Why does it contribute to acid precipitation? In what sense is it less effective than H_2SO_4 as a constituent of acid precipitation?
28. Explain how the following reactions may lead to the occurrence of a major toxic air pollutant:

$$4FeS_2 + 11O_2 \rightarrow 2Fe_2O_3 + 8SO_2$$

$$S(organic,\ in\ fuel) + O_2 \rightarrow SO_2$$

$$SO_2 + \tfrac{1}{2}O_2 + H_2O \rightarrow H_2SO_4(aq)$$

29. What are the major toxic effects of sulfuric acid? How is exposure to sulfuric acid likely to occur?

CHAPTER 12

Organometallics and Organometalloids

12.1 THE NATURE OF ORGANOMETALLIC AND ORGANOMETALLOID COMPOUNDS

An **organometallic compound** is one in which the metal atom is bonded to at least one carbon atom in an organic group. An **organometalloid compound** is a compound in which a metalloid element is bonded to at least one carbon atom in an organic group. The metalloid elements are shown in the periodic table of elements in Figure 1.3 and consist of boron, silicon, germanium, arsenic, antimony, tellurium, and astatine (a very rare radioactive element). In subsequent discussions, *organometallic* will be used as a term to designate both organometallic and organometalloid compounds, and *metal* will refer to both metals and metalloids, unless otherwise indicated. Given the predominance of the metals among the elements, and the ability of most to form organometallic compounds, it is not surprising that there are so many organometallic compounds, and new ones are being synthesized regularly. Fortunately, only a small fraction of these compounds are produced in nature or for commercial use, which greatly simplifies the study of their toxicities.

A further clarification of the nature of organometallic compounds is based on the **electronegativities** of the elements involved, i.e., the abilities of covalently bonded atoms to attract electrons to themselves. Electronegativity values of the elements range from 0.86 for cesium to 4.10 for fluorine. The value for carbon is 2.50, and all organometallic compounds involve bonds between carbon and an element with an electronegativity value of less than 2.50. The value of the electronegativity of phosphorus is 2.06, but it is so nonmetallic in its behavior that its organic compounds are not classified as organometallic compounds.

Organometallic compounds are very important in environmental and toxicological chemistry. The formation of organometallic species in the environment, such as occurs with the methylation of mercury by anaerobic bacteria in sediments, is an important mode of mobilizing metals. Toxicologically, organometallic species often behave in an entirely different way from inorganic forms of metals and may be more toxic than the inorganic ions or compounds.

12.2 CLASSIFICATION OF ORGANOMETALLIC COMPOUNDS

The simplest way to classify organometallic compounds for the purpose of discussing their toxicology is the following:[1]

1. Those in which the organic group is an alkyl group, such as ethyl in tetraethyllead, $Pb(C_2H_5)_4$:

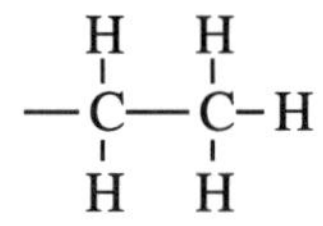

$$-\underset{H}{\overset{H}{\underset{|}{C}}}-\underset{H}{\overset{H}{\underset{|}{C}}}-H$$

2. Those in which the organic group is carbon monoxide:

$$:C\equiv O:$$

(In the preceding Lewis formula of CO, each dash represents a pair of bonding electrons, and each pair of dots represents an unshared pair of electrons.) Compounds with carbon monoxide bonded to metals, some of which are quite volatile and toxic, are called **carbonyls**.

3. Those in which the organic group is a π electron donor, such as ethylene or benzene:

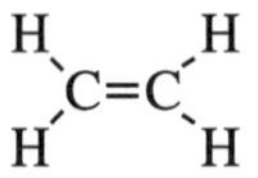

Ethylene    **Benzene**

Combinations exist of the three general types of compounds outlined above, the most prominent of which are arene carbonyl species, in which a metal atom is bonded to both an aromatic entity, such as benzene, and several carbon monoxide molecules. A more detailed discussion of the types of compounds and bonding follows.

12.2.1 Ionically Bonded Organic Groups

Negatively charged hydrocarbon groups are called **carbanions**. These can be bonded to group 1A and 2A metal cations, such as Na^+ and Mg^{2+}, by predominantly ionic bonds. In some carbanions the negative charge is localized on a single carbon atom. For species in which conjugated double bonds and aromaticity are possible, the charge may be delocalized over several atoms, thereby increasing the carbanions' stability (see Figure 12.1).

Ionic organic compounds involving carbanions react readily with oxygen. For example, ethylsodium, $C_2H_5^-Na^+$, self-ignites in air. Ionic organometallic compounds are extremely reactive in water, as shown by the following reaction:

$$C_2H_5^-Na^+ \xrightarrow{H_2O} \text{Organic products} + NaOH \quad (12.2.1)$$

One of the products of such a reaction is a strong base, such as NaOH, which is very corrosive to exposed tissue.

12.2.2 Organic Groups Bonded with Classical Covalent Bonds

A major group of organometallic compounds has carbon–metal covalent single bonds in which both the C and metal (or metalloid) atoms contribute one electron each to be shared in the bond (in contrast to ionic bonds, in which electrons are transferred between atoms). The bonds produced by this sharing arrangement are sigma-covalent bonds, in which the electron density is concentrated between the two nuclei. Since in all cases the more electronegative atom in this bond is carbon

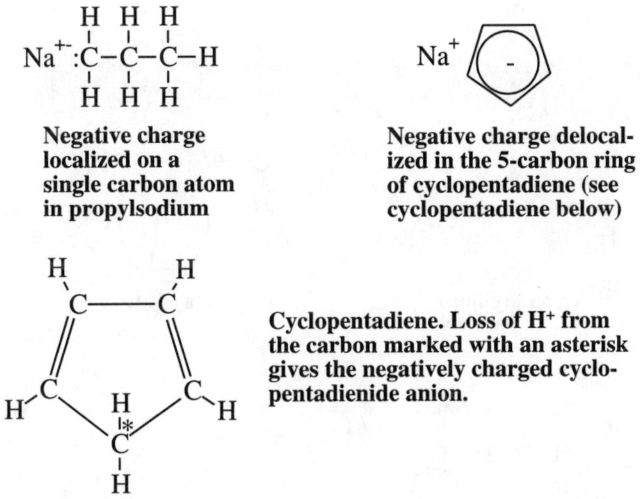

Na$^+$:C(H)(H)—C(H)(H)—C(H)(H)—H

Negative charge localized on a single carbon atom in propylsodium

Na$^+$ [cyclopentadienide ring with −]

Negative charge delocalized in the 5-carbon ring of cyclopentadiene (see cyclopentadiene below)

[cyclopentadiene structure with asterisk on CH$_2$ carbon]

Cyclopentadiene. Loss of H$^+$ from the carbon marked with an asterisk gives the negatively charged cyclopentadienide anion.

Figure 12.1 Carbanions showing localized and delocalized negative charges.

(see Section 12.1), the electrons in the bond tend to be more attracted to the more electronegative atom, and the covalent bond has a **polar** character, as denoted by the following:

$$\overset{\delta+}{M}\text{—}\overset{\delta-}{C}$$

When the electronegativity difference is extreme, such as when the metal atom is Na, K, or Ca, an ionic bond is formed. In cases of less extreme differences in electronegativity, the bond may be only partially ionic; i.e., it is intermediate between a covalent and ionic bond. Organometallic compounds with classical covalent bonds are formed with representative elements and with zinc, cadmium, and mercury, which have filled d orbitals. In some cases, these bonds are also formed with transition metals. Organometallic compounds with this kind of bonding comprise some of the most important and toxicologically significant organometallic compounds. Examples of such compounds are shown in Figure 12.2.

The two most common reactions of sigma-covalently bonded organometallic compounds are oxidation and hydrolysis (see Chapter 1). These compounds have very high heats of combustion because of the stabilities of their oxidation products, which consist of metal oxide, water, and carbon dioxide, as shown by the following reaction for the oxidation of diethyl zinc:

$$\text{Zn}(C_2H_5)_2 + 7O_2 \rightarrow \text{ZnO}(s) + 5H_2O(g) + 4CO_2(g) \tag{12.2.2}$$

Industrial accidents in which the combustion of organometallic compounds generates respirable, toxic metal oxide fumes can certainly pose a hazard.

The organometallic compounds most likely to undergo hydrolysis are those with ionic bonds, those with relatively polar covalent bonds, and those with vacant atomic orbitals (see Chapter 1) on the metal atom, which can accept more electrons. These provide sites of attack for the water molecules. For example, liquid trimethylaluminum reacts almost explosively with water or water and air:

$$\text{Al}(CH_3)_3 \xrightarrow[\{O_2\}]{H_2O} \text{Al}(OH)_3 + \text{Organic products} \tag{12.2.3}$$

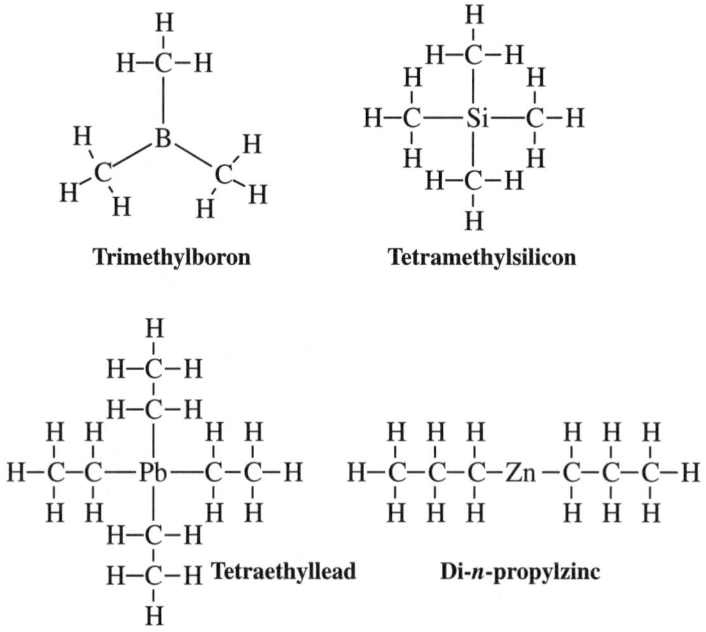

Figure 12.2 Some organometallic compounds with sigma-covalent metal–carbon bonds.

In addition to the dangers posed by the vigor of the reaction, it is possible that noxious organic products are evolved. Accidental exposure to air in the presence of moisture can result in the generation of sufficient heat to cause complete combustion of trimethylaluminum to the oxides of aluminum and carbon and to water.

12.2.3 Organometallic Compounds with Dative Covalent Bonds

Dative covalent bonds, or coordinate covalent bonds, are those in which electrons are shared (as in all covalent bonds), but in which both electrons involved in each bond are contributed from the same atom. Such bonds occur in organometallic compounds of transition metals having vacant d orbitals. It is beyond the scope of this book to discuss such bonding in detail; the reader needing additional information should refer to works on organometallic compounds.[1,2] The most common organometallic compounds that have dative covalent bonds are **carbonyl compounds**, which are formed from a transition metal and carbon monoxide, where the metal is usually in the –1, 0, or +1 oxidation state. In these compounds the carbon atom on the carbon monoxide acts as an electron-pair donor:

$$M + :CO: \rightarrow M:CO:$$
$$\uparrow$$
$$\textbf{Dative bond} \qquad (12.2.4)$$

Most carbonyl compounds have several carbon monoxide molecules bonded to a metal.

Many transition metal carbonyl compounds are known. The one that is the most significant toxicologically, because of its widespread occurrence and extremely poisonous nature, is the nickel carbonyl compound, $Ni(CO)_4$. Perhaps the next most abundant is $Fe(CO)_5$. Other examples are

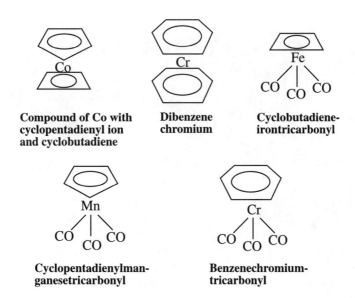

Figure 12.3 Compounds of metals with π-electron donor hydrocarbons and with carbon monoxide.

$V(CO)_6$ and $Cr(CO)_6$. In some cases, bonding favors compounds with two metal atoms per molecule, such as $(CO)_5Mn–Mn(CO)_5$ or $(CO)_4Co–Co(CO)_4$.

12.2.4 Organometallic Compounds Involving π-Electron Donors

Unsaturated hydrocarbons, such as ethylene, butadiene, cyclopentadiene, and benzene, contain π-electrons that occupy orbitals that are not in a direct line between the two atoms bonded together, but are above and below a plane through that line. These electrons can participate in bonds to metal atoms in organometallic compounds. Furthermore, the metal atoms in a number of organometallic compounds are bonded to both a π-electron donor organic species — most commonly the cyclopentadienyl anion with a –1 charge — and one or more CO molecules. A typical compound of this class is cyclopentadienylcobalt-dicarbonyl, $C_5H_5Co(CO)_2$. Examples of these compounds and of compounds consisting of metals bonded only to organic π-electron donors are shown in Figure 12.3.

12.3 MIXED ORGANOMETALLIC COMPOUNDS

So far in this chapter the discussion has centered on compounds in which all of the metal bonds are with carbon. A large number of compounds exist that have at least one bond between the metal and a C atom on an organic group, as well as other covalent or ionic bonds between the metal and atoms other than carbon. Because they have at least one metal–carbon bond, as well as properties, uses, and toxicological effects typical of organometallic compounds, it is useful to consider such compounds along with organometallic compounds. Examples are monomethylmercury chloride, CH_3HgCl, in which the organometallic CH_3Hg^+ ion is ionically bonded to the chloride anion. Another example is phenyldichloroarsine, $C_6H_5AsCl_2$, in which a phenyl group is covalently bonded to arsenic through an As–C bond and two Cl atoms are also covalently bonded to arsenic.

A number of compounds exist that consist of organic groups bonded to a metal atom through atoms other than carbon. Although they do not meet the strict definition thereof, such compounds can be classified as organometallics for the discussion of their toxicology and aspects of their chemistry. An example of such a compound is isopropyl titanate, $Ti(OC_3H_7)_4$,

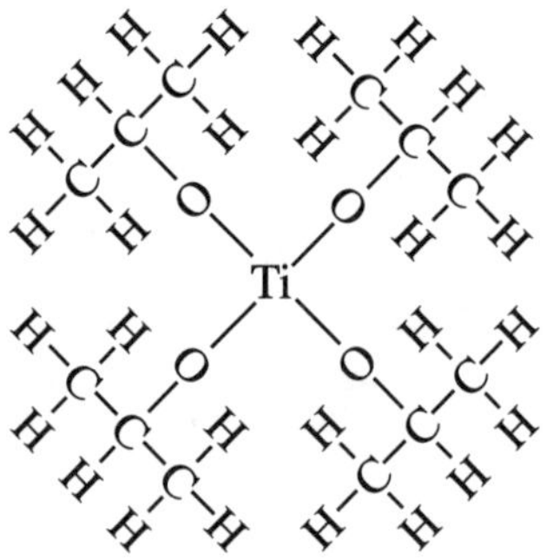

Isopropyl titanate

also called titanium isopropylate. This compound is a colorless liquid melting at 14.8°C and boiling at 104°C, low values that reflect the organic nature of the molecule, which is obvious even in the two-dimensional structural representation of the formula above. The behavior of isopropyl titanate is more that of an organometallic compound than that of an inorganic compound, and by virtue of its titanium content, it is not properly classified as an organic compound. The term *organometal* is sometimes applied to such a compound. For toxicological considerations, it may be regarded as an organometallic compound.

Several compounds are discussed in this chapter that have some organometallic character, but which also have formulas, structures, and properties of inorganic or organic compounds. These compounds could be called mixed organometallics. However, so long as the differences are understood, compounds such as isopropyl titanate (see above) that do not meet all the criteria of organometallic compounds can be regarded as such for the discussion of their toxicities.

12.4 ORGANOMETALLIC COMPOUND TOXICITY

Some organometallic compounds have been known and used for decades, so that their toxicological properties are rather well known. Prominent among these are organoarsenicals used as drugs, organomercury fungicides, and tetramethyl- and tetraethyllead, used as antiknock additives for gasoline. Since about 1950, there has been very substantial growth in chemical research devoted to organometallic compounds, and large numbers and varieties of these compounds have been synthesized. Although the applications of organoarsenicals and organomercury compounds as human drugs and pesticides have been virtually eliminated because of their toxicities, environmental effects, and the development of safer substitutes, a wide variety of new organometallic compounds has come into use for various purposes, such as catalysis and chemical synthesis. The toxicological properties of these compounds are very important, and they should be treated with great caution until proven safe. Many are very reactive chemically, so they are hazardous to directly exposed tissue, even if not toxic systemically.

12.5 COMPOUNDS OF GROUP 1A METALS

12.5.1 Lithium Compounds

Table 12.1 shows some organometallic lithium compounds. It is seen from their formulas that these compounds are ionic. As discussed in Section 12.2, 1A metals have low electronegativities and form ionic compounds with hydrocarbon anions. Of these elements, lithium tends to form metal–carbon bonds with the most covalent character; therefore, lithium compounds are more stable (though generally quite reactive) than other organometallic compounds of group 1A metals, most

Table 12.1 Some Organometallic Compounds of Lithium

Name	Formula	Properties and Uses
Methyllithium	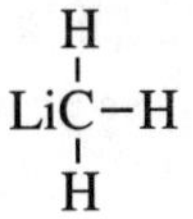 LiCH₃ (LiC with H, H, H)	Initiator for solution polymerization of elastomers
Ethyllithium	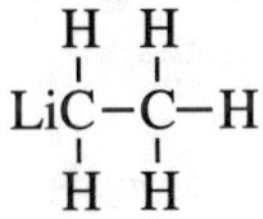 LiC₂H₅ (LiC-C with H's)	Transparent crystals melting at 95°C, pyrophoric,[a] decomposes in water
Tert-butyllithium	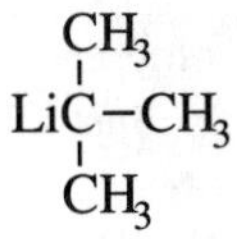 LiC(CH₃)₃	Colorless crystalline solid subliming at 70-80°C, used as synthesis reagent
Phenyllithium	 Li–C₆H₅	Colorless pyrophoric solid used in Grignard-type reactions to attach a phenyl group

[a] Pyrophoric means spontaneously flammable in air.

likely to exist as liquids or low-melting-point solids, and generally more soluble in organic solvents.[3] These compounds are moisture sensitive, both in the pure state and in solution, and can undergo spontaneous ignition when exposed to air.

The most widely used organolithium compound is *n*-butyllithium (see formulas of related compounds in Table 12.1), used as an initiator for the production of elastomers by solution polymerization, predominantly of styrene-butadiene.

Lithium forms a very unstable carbonyl, for which the toxicity is suspected of being high. The formula of this compound is LiCOCOLi, written in this manner to show that the two CO molecules form bridges between two Li atoms.

Unless otherwise known, the toxicities of lithium organometallic compounds should be regarded as those of lithium compounds and of organometallic compounds in general. The latter were discussed in Section 12.4. Lithium oxide and hydroxide are caustic bases, and they may be formed by the combustion of lithium organometallic compounds or by their reaction with water.

Lithium ion, Li⁺, is a central nervous system toxicant that causes dizziness, prostration, anorexia, apathy, and nausea. It can also cause kidney damage and, in large doses, coma and death.

12.5.2 Compounds of Group 1A Metals Other Than Lithium

As discussed in Section 12.2, group 1A metals form ionic metal–carbon bonds. Organometallic compounds of group 1A metals other than lithium have metal–carbon bonds with less of a covalent character than the corresponding bonds in lithium compounds and tend to be especially reactive. Compounds of rubidium and cesium are rarely encountered outside the laboratory, so their toxicological significance is relatively minor. Therefore, aside from lithium compounds, the toxicology of sodium and potassium compounds is of most concern.

Both sodium and potassium salts are natural constituents of body tissues and fluids as Na⁺ and K⁺ ions, respectively, and are not themselves toxic at normal physiological levels. The oxides and hydroxides of both these metals are very caustic, corrosive substances that damage exposed tissue. Oxides are formed by the combustion of sodium and potassium organometallics, and hydroxides are produced by the reaction of the oxides with water or by direct reaction of the organometallics with water, as shown below for cyclopentadienylsodium:

$$C_5H_5^- Na^+ + H_2O \rightarrow C_5H_6 + NaOH \qquad (12.5.1)$$

Both sodium and potassium form carbonyl compounds, NaCO and KCO, respectively. Both compounds are highly reactive solids prone to explode when exposed to water or air. Decomposition of the carbonyls gives off caustic oxides and hydroxides of Na and K, as well as toxic carbon monoxide.

Sodium and potassium form alkoxide compounds with the general formula M^+OR, in which R is a hydrocarbon group. Typically, sodium reacts with methanol:

$$2CH_3OH + 2Na \rightarrow 2Na^+OCH_3 + H_2 \qquad (12.5.2)$$

to yield sodium methoxide and hydrogen gas. The alkoxide compounds are highly basic and caustic, reacting with water to form the corresponding hydroxides, as illustrated by the following reaction:

$$K^+OCH_3 + H_2O \rightarrow KOH + CH_3OH \qquad (12.5.3)$$

12.6 COMPOUNDS OF GROUP 2A METALS

The organometallic compound chemistry of the 2A metals is similar to that of the 1A metals, and ionically bonded compounds predominate. As is the case with lithium in group 1A, the first 2A element, beryllium, behaves atypically, with a greater covalent character in its metal–carbon bonds.

Beryllium organometallic compounds should be accorded the respect due all beryllium compounds because of beryllium's extreme toxicity (see Section 10.4). Dimethylberyllium, $Be(CH_3)_2$, is a white solid having needle-like crystals. When heated to decomposition, it gives off highly toxic beryllium oxide fumes. Diethylberyllium, $Be(C_2H_5)_2$, with a melting point of 12°C and a boiling point of 110°C, is a colorless liquid at room temperature and is especially dangerous because of its volatility.

12.6.1 Magnesium

The organometallic chemistry of magnesium has been of the utmost importance for many decades because of **Grignard reagents**, the first of which was made by Victor Grignard around 1900 by the reaction

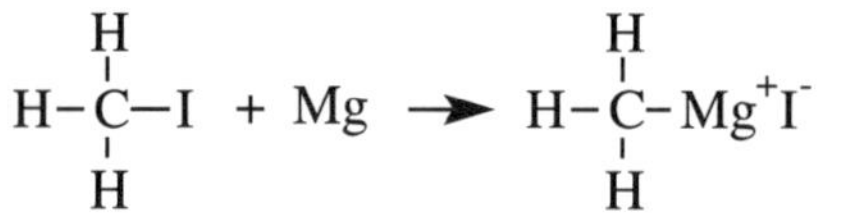

$$\underset{\textbf{Iodomethane}}{CH_3\text{-}I} + Mg \rightarrow \underset{\textbf{Methylmagnesium iodide}}{CH_3\text{-}Mg^+I^-} \qquad (12.6.1)$$

Grignard reagents are particularly useful in organic chemical synthesis for the attachment of their organic component ($-CH_3$ in the preceding example) to another organic molecule. The development of Grignard reagents was such an advance in organic chemical synthesis that in 1912 Victor Grignard received the Nobel Prize for his work.

Grignard reagents can cause damage to skin or pulmonary tissue in the unlikely event that they are inhaled. These reagents react rapidly with both water and oxygen, releasing a great deal of heat in the process. Ethyl ether solutions of methylmagnesium bromide (CH_3MgBr) are particularly hazardous because of the spontaneous ignition of the reagent and the solvent ether in which it is contained when the mixture contacts water, such as water on a moist laboratory bench top.

The simplest dialkyl magnesium compounds are dimethylmagnesium, $Mg(CH_3)_2$, and diethylmagnesium, $Mg(C_2H_5)_2$. Both are pyrophoric compounds that are violently reactive to water and steam and that self-ignite in air, the latter even in carbon dioxide (like the elemental form, magnesium in an organometallic compound removes O from CO_2 to form MgO and release elemental carbon). Diethylmagnesium has a melting point of 0°C and is a liquid at room temperature. Diphenylmagnesium, $Mg(C_6H_5)_2$, is a feathery solid, somewhat less hazardous than the dimethyl and diethyl compounds. It is violently reactive with water and is spontaneously flammable in humid air, but not dry air.

Unlike the caustic oxides and hydroxides of group 1A metals, magnesium hydroxide, $Mg(OH)_2$, formed by the reaction of air and water with magnesium organometallic compounds, is a relatively benign substance that is used as a food additive and ingredient of milk of magnesia.

12.6.2 Calcium, Strontium, and Barium

It is much more difficult to make organometallic compounds of Ca, Sr, and Ba than it is to make those of the first two group 2A metals. Whereas organometallic compounds of beryllium and magnesium have metal–carbon bonds with a significant degree of covalent character, the Ca, Sr, and Ba organometallic compounds are much more ionic. These compounds are extremely reactive to water, water vapor, and atmospheric oxygen. There are relatively few organometallic compounds of calcium, strontium, and barium; their industrial uses are few, so their toxicology is of limited concern. Grignard reagents in which the metal is calcium rather than magnesium (general formula RCa^+X^-) have been prepared, but are not as useful for synthesis as the corresponding magnesium compounds.

12.7 COMPOUNDS OF GROUP 2B METALS

It is convenient to consider the organometallic compound chemistry of the group 2B metals immediately following that of the 2A metals because both have two $2s$ electrons and no partially filled d orbitals. The group 2B metals — zinc, cadmium, and mercury — form an abundance of organometallic compounds, many of which have significant uses. Furthermore, cadmium and mercury (both discussed in Chapter 10) are notably toxic elements, so the toxicological aspects of their organometallic compounds are of particular concern. Therefore, the organometallic compound chemistry of each of the 2B metals will be discussed separately.

12.7.1 Zinc

Organozinc compounds are widely used as reagents.[4] A typical synthesis of a zinc organometallic compound is given by the reaction below, in which the Grignard-type compound CH_3ZnI is an intermediate:

$$H_3C-I + 2Zn \rightarrow H_3C-Zn-CH_3 + 2ZnI_2 \quad (12.7.1)$$

Dimethylzinc has a rather low melting temperature of –40°C, and it boils at 46°C. At room temperature, it is a mobile, volatile liquid that undergoes self-ignition in air and reacts violently with water. The same properties are exhibited by diethylzinc, $(C_2H_5)_2Zn$, which melts at –28°C and boils at 118°C. Both dimethylzinc and diethylzinc are used in organometallic chemical vapor

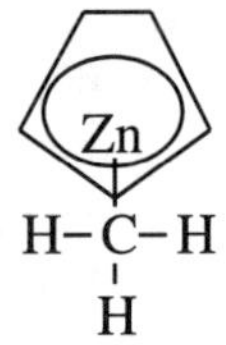

Figure 12.4 Methylcyclopentadienylzinc. The monomer shown exists in the vapor phase. In the solid phase, a polymeric form exists.

deposition of zinc and zinc oxide in fabrication of semiconductors and light-emitting diodes. Diphenylzinc, $(C_6H_5)Zn$, is considerably less reactive than its methyl and ethyl analogs; it is a white crystalline solid melting at 107°C. Zinc organometallics are similar in many respects to their analogous magnesium compounds (see Section 12.6), but do not react with carbon dioxide, as do some of the more reactive magnesium compounds. An example of an organozinc compound involving a π-bonded group is that of methylcyclopentadienylzinc, shown in Figure 12.4.

Zinc forms a variety of Grignard-type compounds, such as ethylzinc chloride, ethylzinc bromide, butylzinc chloride, and butylzinc iodide.

Zinc organometallic compounds should be accorded the same caution in respect to toxicology as that given to organometallic compounds in general. The combustion of highly flammable organozinc compounds such as dimethyl and diethyl compounds produces very finely divided particles of zinc oxide fumes, as illustrated by the reaction

$$2(CH_3)_2Zn + 8O_2 \rightarrow 2ZnO + 4CO_2 + 6H_2O \qquad (12.7.2)$$

Although zinc oxide is used as a healing agent and food additive, inhalation of zinc oxide fume particles causes zinc **metal fume fever**, characterized by elevated temperature and chills. The toxic effect of zinc fume has been attributed to its flocculation in lung airways, which prevents maximum penetration of air to the alveoli and perhaps activates endogenous pyrogen in blood leukocytes. An interesting aspect of this discomfiting but less-than-deadly affliction is the immunity that exposed individuals develop to it, but which is lost after only a day or two of nonexposure. Thus workers exposed to zinc fume usually suffer most from the metal fume fever at the beginning of the work week, and less with consecutive days of exposure as their systems adapt to the metal fume.

Diphenylzinc illustrates the toxicity hazard that may obtain from the organic part of an organometallic compound upon decomposition. Under some conditions, this compound can react to release toxic phenol (see Chapter 14):

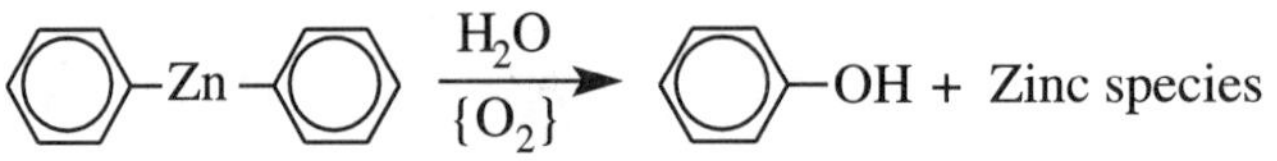

$$\text{C}_6\text{H}_5-Zn-\text{C}_6\text{H}_5 \xrightarrow[\{O_2\}]{H_2O} \text{C}_6\text{H}_5-OH + \text{Zinc species} \qquad (12.7.3)$$

A number of zinc compounds with organic constituents (e.g., zinc salts of organic acids) have therapeutic uses. These include antidandruff zinc pyridinethione, antifungal zinc undecylenate used to treat athlete's foot, zinc stearate and palmitate (zinc soap), and antibacterial zinc bacitracin. Zinc naphthenate is used as a low-toxicity wood preservative, and zinc phenolsulfonate has insecticidal properties and was once used as an intestinal antiseptic. The inhalation of zinc soaps by infants has been known to cause acute fatal pneumonitis characterized by lung lesions similar to, but more serious than, those caused by talc. Zinc pyridine thione (zinc 2-pyridinethiol-1-oxide) has been shown to cause retinal detachment and blindness in dogs; this is an apparently species-specific effect because laboratory tests at the same and even much higher dosages in monkeys and rodents do not show the same effect.

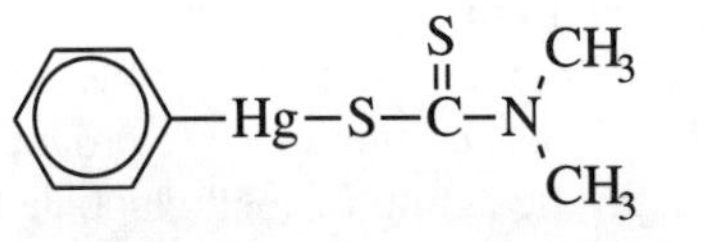

Phenylmercurydimethyldithiocarbamate (slimicide for wood pulp and mold retardant for paper)

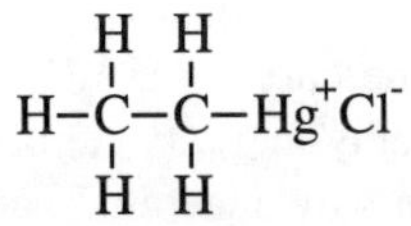

Ethylmercury chloride (seed fungicide)

Figure 12.5 Two organomercury compounds that have been used for fungicidal purposes.

12.7.2 Cadmium

In the absence of water, cadmium halides, CdX_2, react with organolithium compounds, as shown by the following example:

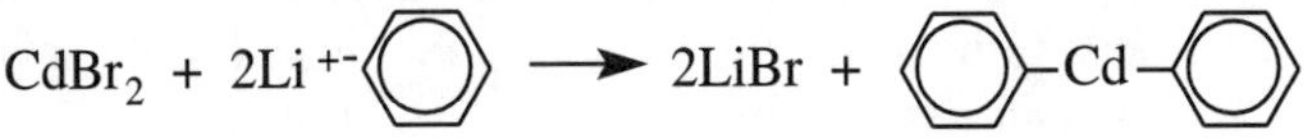

$$CdBr_2 + 2Li^+\text{-}C_6H_5 \rightarrow 2LiBr + C_6H_5\text{-}Cd\text{-}C_6H_5 \quad (12.7.4)$$

Dimethylcadmium, $(CH_3)_2Cd$, is an oily liquid at room temperature and has a very unpleasant odor. The compound melts at $-4.5°C$ and boils at $106°C$. It decomposes in contact with water. Diethylcadmium is likewise an oil; it melts at $-21°C$, boils at $64°C$, and reacts explosively with oxygen in air. Dipropylcadmium, $(C_3H_7)_2Cd$, is an oil that melts at $-83°C$, boils at $84°C$, and reacts with water. The dialkyl cadmium compounds are distillable, but decompose above about $150°C$, evolving toxic cadmium fume.

The toxicology of cadmium organometallic compounds is of particular concern because of the high toxicity of cadmium. The organometallic compounds of cadmium form vapors that can be inhaled and that can cross membranes because of their lipid solubility. The reaction of cadmium organometallic compounds with water can release highly toxic fumes of cadmium and CdO. Inhalation of these fumes can cause chronic cadmium poisoning and death. The toxicological aspects of cadmium are discussed in Section 10.4.

Evidence has been detected of the biomethylation of cadmium. Studies with differential pulse anodic stripping voltammetry have shown detectable amounts of monomethylcadmium ion, H_3CCd^+, in surface water of the South Atlantic.[5] Examination of water from some Arctic meltwater ponds showed that up to half of the cadmium present in the water was in the monomethylcadmium ion form.

12.7.3 Mercury

In 1853, E. Frankland made the first synthetic organomercury compound by the photochemical reaction below:

$$2Hg + 2CH_3I + h\nu(\text{sunlight}) \rightarrow (CH_3)_2Hg + HgI_2 \quad (12.7.5)$$

Numerous synthetic routes are available for the preparation of a variety of mercury organometallic compounds.

In the late 1800s and early 1900s, numerous organomercury pharmaceutical compounds were synthesized and used. These have since been replaced by more effective and safe nonmercury substitutes. Organomercury compounds have been widely used as pesticidal fungicides (see Figure 12.5), but these applications have been phased out because of the adverse effects of mercury in the environment. Mercury levels in organs of wildlife, such as white-tailed eagles in Germany and Austria, have decreased significantly with the phaseout of organomercury seed-treating chemicals.[6]

The most notorious mercury compounds in the environment are monomethyl mercury (CH_3Hg^+) salts and dimethylmercury ($(CH_3)_2Hg$). The latter compound is both soluble and volatile, and the salts of the monomethylmercury cation are soluble. These compounds are produced from inorganic mercury in sediments by anaerobic bacteria through the action of methylcobalamin, a vitamin B_{12} analog and intermediate in the synthesis of methane:

$$HgCl_2(s) \xrightarrow{\text{Methylcobalamin}} CH_3Hg^+(aq) + 2Cl^- \quad (12.7.6)$$

The preceding reaction is favored in somewhat acidic water in which anaerobic decay, which often produces CH_4, is occurring. If the water is neutral or slightly alkaline, dimethylmercury formation is favored; this volatile compound may escape to the atmosphere. Discovered around 1970, the biosynthesis of the methylmercury species in sediments was an unpleasant surprise, in that it provides a means for otherwise insoluble inorganic mercury compounds to get into natural waters. Furthermore, these species are lipid soluble, so that they undergo bioaccumulation and biomagnification in aquatic organisms. Fish tissue often contains more than 1000 times the concentration of mercury as does the surrounding water.

The toxicity of mercury is discussed in Section 10.4. Some special considerations apply to organomercury compounds, the foremost of which is their lipid solubility and resulting high degree of absorption and facile distribution through biological systems. The lipid solubilities and high vapor pressures of the methylmercuries favor their absorption by the pulmonary route. These compounds also can be absorbed through the skin, and their uptake approaches 100% (compared to less than 10% for inorganic mercury compounds) in the gastrointestinal tract.

With respect to distribution in the body, the methylmercury species behave more like mercury metal, $Hg(0)$, than inorganic mercury(II), Hg^{2+}. Like elemental mercury, methylmercury compounds traverse the blood–brain barrier and affect the central nervous system. However, the psychopathological effects of methylmercury compounds (laughing, crying, impaired intellectual abilities) are different from those of elemental mercury (irritability, shyness).

Both dimethylmercury and salts of monomethylmercury, such as H_3CHgCl, are extraordinarily dangerous. Most of what is known about their toxicities has been learned from exposure to monomethylmercury chloride on treated seed grains consumed by people and by exposure of people in Japan to seafood contaminated with methylmercury compounds. (Early investigators of volatile dimethylmercury, a liquid that readily penetrates skin, died from its toxic effects within months of making the compound.) Fetuses of pregnant women who consumed seafood contaminated with methylmercury have suffered grievous damage. The major effects of exposure of adults to methylmercury compounds are neurotoxic effects on the brain. Victims exhibit a variety of devastating symptoms, the earliest of which are numbnesss and tingling of the mouth, lips, fingers, and toes. Swallowing and word pronunciation become difficult, and the victim staggers while attempting to walk. Symptoms of weakness and extreme fatigue are accompanied by loss of hearing, vision, and ability to concentrate. Ultimately, spasticity, coma, and death occur.

The extreme toxicity of dimethylmercury was demonstrated tragically by the 1997 death of Professor Karen Wetterhahn of Dartmouth College. Dr. Wetterhahn was exposed to dimethylmercury from an accidental spill of about two drops of this liquid onto the latex rubber gloves she was wearing for protection. The lipid-soluble compound permeates latex and skin, and Dr. Wetterhahn died less than a year later from neurotoxic effects to the brain.

12.8 ORGANOTIN AND ORGANOGERMANIUM COMPOUNDS

Global production of organotin compounds has reached levels around 40,000 metric tons per year, consuming about 7 to 8% of the tin used each year. Of all the metals, tin has the greatest

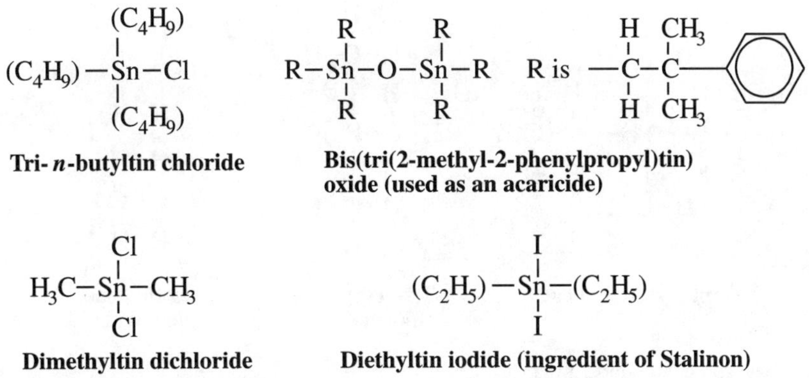

Figure 12.6 Examples of organotin compounds.

number of organometallic compounds in commercial use.[7] Major industrial uses include applications of tin compounds in fungicides, acaricides, disinfectants, antifouling paints, stabilizers to lessen the effects of heat and light in polyvinyl chloride (PVC) plastics, catalysts, and precursors for the formation of films of SnO_2 on glass. Tributyl tin (TBT) chloride and related TBT compounds have bactericidal, fungicidal, and insecticidal properties and are of particular environmental significance because of their once widespread use as industrial biocides. In addition to tributyl tin chloride, other tributyl tin compounds used as biocides include hydroxide, naphthenate, bis(tributyltin) oxide, and tris(tributylstannyl) phosphate. A major use of TBT has been in boat and ship hull coatings to prevent the growth of fouling organisms. Other applications have included preservation of wood, leather, paper, and textiles. Because of their antifungal activity, TBT compounds have been used as slimicides in cooling tower water.

In addition to synthetic organotin compounds, methylated tin species can be produced biologically in the environment. Figure 12.6 gives some examples of the many known organotin compounds.

12.8.1 Toxicology of Organotin Compounds

Many organotin compounds have the general formula R_nSnX_{4-n}, where R is a hydrocarbon group and X is an inorganic entity, such as a chlorine atom, or an organic group bonded to tin through a noncarbon atom (for example, acetate bonded to Sn through an O atom). As a general rule, in a series of these compounds, toxicity is at a maximum value for n = 3. Furthermore, the toxicity is generally more dependent on the nature of the R groups than on X.

Organotin compounds are readily absorbed through the skin, and skin rashes may result. Organotin compounds, especially those of the R_3SnX type, bind to proteins, probably through the sulfur on cysteine and histidine residues. Interference with mitochondrial function by several mechanisms appears to be the mode of biochemical action leading to toxic responses.

Much of what is known of organotin toxicity to humans was learned in the 1950s from exposure of humans in France to Stalinon, used to treat skin disorders, osteomyelitis, and anthrax. The active ingredient of this formulation was diethyltin iodide, although the toxic agent may have been impurity triethyltin iodide. Neural tissue was most susceptible to damage. Victims exhibited swelling of brain tissue, edema of white matter, and cerebral hemorrhages. Tragically, approximately 100 people died from taking Stalinon in France.

Although human exposure to organotin compounds is not believed to cause many cases of poisoning, the ecotoxicological effects of organotins may be quite significant. This is because of exposure to sediment-dwelling organisms to organotins leached from ship and boat hulls treated with biocidal organotins. Increasingly stringent regulation of this application of organotin compounds should continue to reduce the ecotoxicological problems resulting from these compounds.

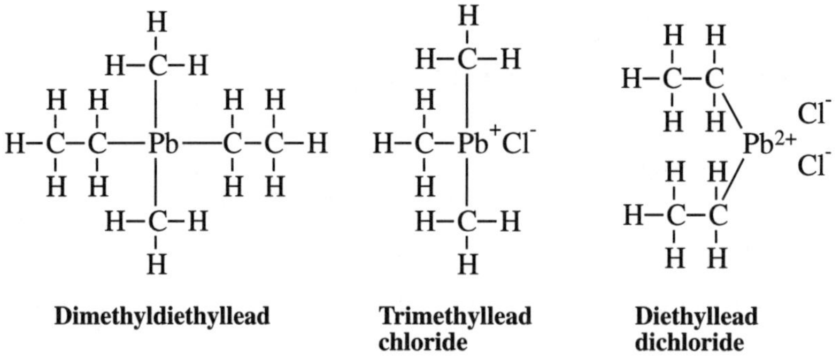

Figure 12.7 Alkyllead compounds and salts.

12.8.2 Organogermanium Compounds

Organogermanium compounds, including tetramethyl- and tetraethylgermanium, are used in the semiconductor industry to prepare deposits of germanium. Spirogermanium,

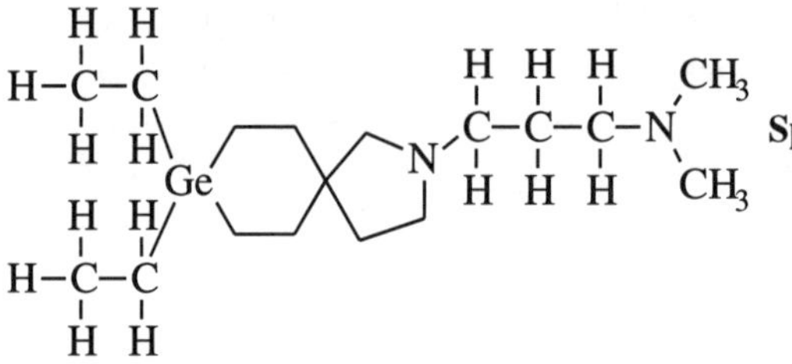

has been tested for antitumor activity. Not much is known about the toxicities of organogermanium compounds, although spirogermanium was of some interest for chemotherapy because it is reputed to be only moderately toxic.

12.9 ORGANOLEAD COMPOUNDS

The toxicities and environmental effects of organolead compounds are particularly noteworthy because of the former widespread use and distribution of tetraethyllead as a gasoline additive (see structure in Figure 12.2).[8] Although more than 1000 organolead compounds have been synthesized, those of commercial and toxicological importance are largely limited to the alkyl (methyl and ethyl) compounds and their salts, examples of which are shown in Figure 12.7.

In addition to manufactured organolead compounds, the possibility exists of biological methylation of lead, such as occurs with mercury (see Section 12.7). However, there is a great deal of uncertainty regarding biological methylation of lead in the environment.

12.9.1 Toxicology of Organolead Compounds

Because of the large amounts of tetraethyllead used as a gasoline additive, the toxicology of this compound has been investigated much more extensively than that of other organolead compounds and is discussed briefly here. Tetraethyllead is a colorless, oily liquid with a strong affinity for lipids and is considered highly toxic by inhalation, ingestion, and absorption through the skin. Most commonly, exposure is through inhalation, and around 70% of inhaled tetraethyllead is absorbed. Numerous cases of poisoning have been reported in individuals sniffing leaded gasoline for "recreational" purposes.

In common with other organollead compounds, tetraethyllead has a strong affinity for lipid and nerve tissue and is readily transported to the brain. Symptoms of tetraethyllead poisoning reflect effects on the central nervous system. Among these symptoms are fatigue, weakness, restlessness, ataxia, psychosis, and convulsions. Victims may also experience nausea, vomiting, and diarhhea. In cases of fatal tetraethyllead poisoning, victims may experience convulsions and coma; death has occurred as soon as one or two days after exposure. Almost one third of victims acutely exposed to tetraethyllead die, although fatalities from chronic exposure have been comparatively rare, considering the widespread use of tetraethyllead. Recovery from poisoning by this compound tends to be slow.

The toxicological action of tetraethyllead is different from that of inorganic lead. As one manifestation of this difference, chelation therapy is ineffective for the treatment of tetraethyllead poisoning. The toxic action of tetraethyllead appears to involve its metabolic conversion to the triethyl form.

12.10 ORGANOARSENIC COMPOUNDS

There are two major sources of organoarsenic compounds: those produced for commercial applications and those produced from the biomethylation of inorganic arsenic by microorganisms. Many different organoarsenic compounds have been identified.

12.10.1 Organoarsenic Compounds from Biological Processes

The reactions that follow illustrate the production of organoarsenic compounds by bacteria. In a reducing environment, arsenic(V) is reduced to arsenic(III):

$$H_3AsO_4 + 2H^+ + 2e^- \rightarrow H_3AsO_3 + H_2O \tag{12.10.1}$$

Through the action of methylcobalamin in bacteria, arsenic(III) is methylated to methyl, and then to dimethylarsinic acid:

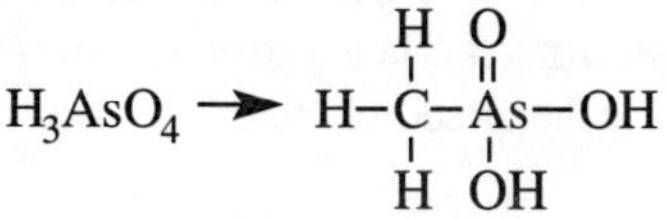

$$H_3AsO_4 \rightarrow \underset{\substack{|\\HOH}}{\overset{\substack{HO\\|\|}}{H-C-\overset{}{As}-OH}} \tag{12.10.2}$$

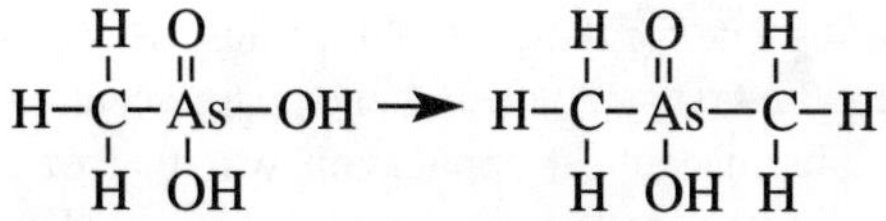

$$\underset{\substack{|\\HOH}}{\overset{\substack{HO\\|\|}}{H-C-\overset{}{As}-OH}} \rightarrow \underset{\substack{|||\\HOHH}}{\overset{\substack{HOH\\|\||}}{H-C-\overset{}{As}-C-H}} \tag{12.10.3}$$

Dimethylarsinic acid can be reduced to volatile dimethylarsine:

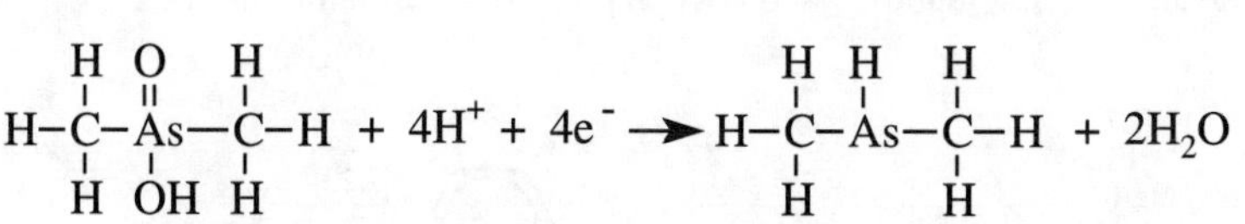

$$\underset{\substack{|||\\HOHH}}{\overset{\substack{HOH\\|\||}}{H-C-\overset{}{As}-C-H}} + 4H^+ + 4e^- \rightarrow \underset{\substack{|||\\HH}}{\overset{\substack{HHH\\|||}}{H-C-\overset{}{As}-C-H}} + 2H_2O \tag{12.10.4}$$

Methylarsinic acid and dimethylarsinic acid are the two organoarsenic compounds that are most likely to be encountered in the environment.

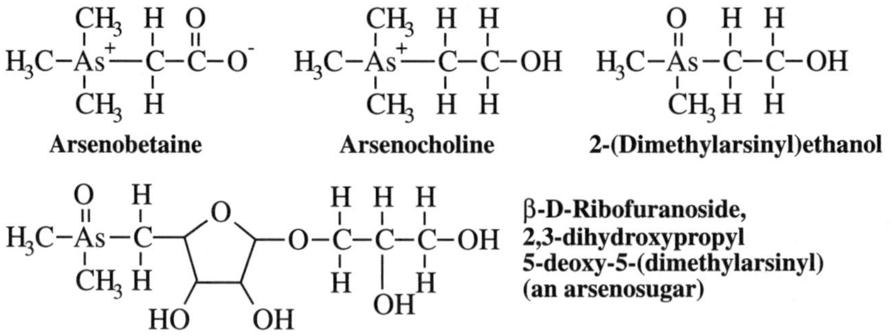

Arsenobetaine, **Arsenocholine**, **2-(Dimethylarsinyl)ethanol**

β-D-Ribofuranoside, 2,3-dihydroxypropyl 5-deoxy-5-(dimethylarsinyl) (an arsenosugar)

Figure 12.8 Examples of organoarsenic compounds found in seaweed and in sheep feeding on the seaweed.

Biomethylated arsenic was responsible for numerous cases of arsenic poisoning in Europe during the 1800s. Under humid conditions, arsenic in plaster and wallpaper pigments was converted to biomethylated forms, as manifested by the strong garlic odor of the products, and people sleeping and working in the rooms became ill from inhaling the volatile organoarsenic compounds.

Foods from marine sources may have high levels of arsenic. An interesting study of arsenic in sheep that live off seaweed detected 15 different organoarsenic compounds in the seaweed and in the blood, urine, liver, kidney, muscle, and wool of the sheep.[9] The rare breed of North Ronaldsay sheep studied live on the beach of Orkney Island off Northern Scotland, eating up to 3 kg per day of seaweed washed ashore. The seaweed consumed is predominantly brown algae containing 20 to 100 mg of arsenic per milligram dry mass, giving each adult sheep an intake of approximately 50 kg/day of arsenic. The arsenic in the seaweed is present predominantly as four kinds of dimethylarsinoylribosides known as arsenosugars. In addition to the arsenosugars, the organoarsenic species detected in either the seaweed or samples from the sheep include dimethylarsinic acid, monomethyl arsonic acid, trimethylarsine oxide, tetramethylarsonium ion, arsenobetaine, arsenocholine, and dimethylarsinyl ethanol. Examples of these compounds are shown in Figure 12.8. Around 95% of the arsenic excreted from the sheep was in the form of dimethylarsinic acid, shown in reaction 12.10.3. The blood, urine, and tissue arsenic concentrations in these sheep were approximately 100 times those of grass-fed sheep that did not eat the arsenic-laden seaweed. However, the arsenic levels in the meat from the sheep did not exceed U.K. guidelines of a maximum of 1 mg/kg fresh weight. The fact that sheep have been kept on Orkney Island beach and feeding on arsenic-laden seaweed for several centuries suggests that arsenic tolerance has developed in this particular breed of sheep.

12.10.2 Synthetic Organoarsenic Compounds

Although now essentially obsolete for the treatment of human diseases because of their toxicities, organoarsenic compounds were the first synthetic organic pharmaceutical agents and were widely used in the early 1900s. The first pharmaceutical application was that of atoxyl (the sodium salt of 4-aminophenylarsinic acid), which was used to treat sleeping sickness. The synthesis of Salvarsan by Dr. Paul Ehrlich in 1907 was a development that may be considered the beginning of modern **chemotherapy** (chemical treatment of disease). Salvarsan was widely used for the treatment of syphilis. Toxic effects of Salvarsan included jaundice and encephalitis (brain inflammation).

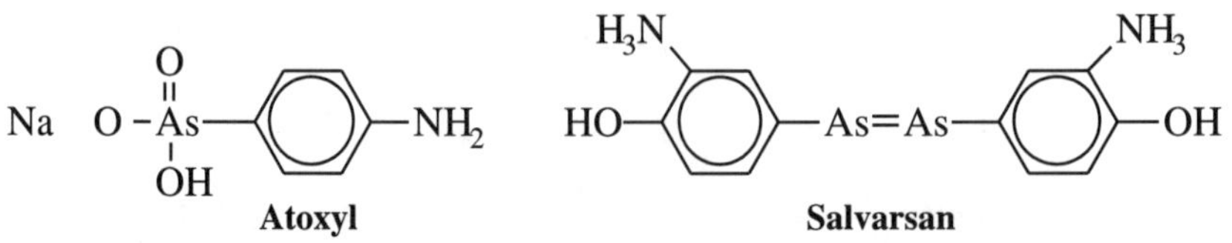

Atoxyl, **Salvarsan**

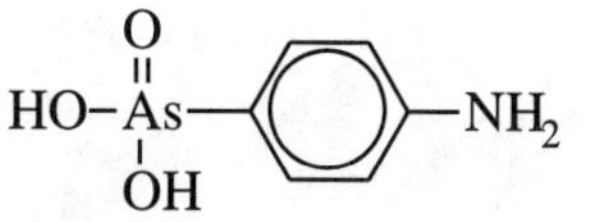

Arsanilic acid

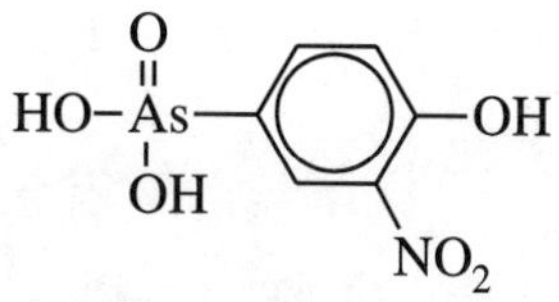

3-Nitro-4-hydroxyphenyl-arsinic acid (Roxarsone)

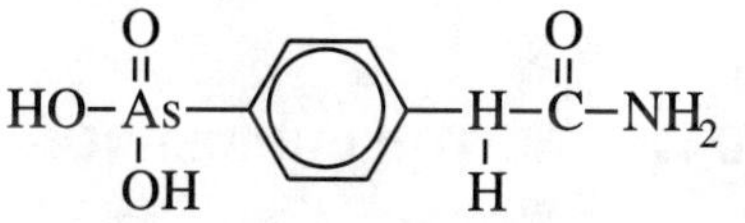

N-carbamoylarsinic acid (Carbarsone)

Figure 12.9 Major organoarsenic animal feed additives. Arsanilic acid and Roxarsone are used to control swine dysentery and increase the rate of gain relative to the amount of feed in swine and chickens. Carbarsone and nitarsone (4-nitrophenylarsanilic acid) act as antihistomonads in chickens.

Some organoarsenic compounds that are cytotoxic (toxic to tissue) have been found to have antitumor activity. One of these, which is active against breast cancer and leukemic cells, is 2-methylthio-4-(2'-phenylarsenic acid)-aminopyrimidine:

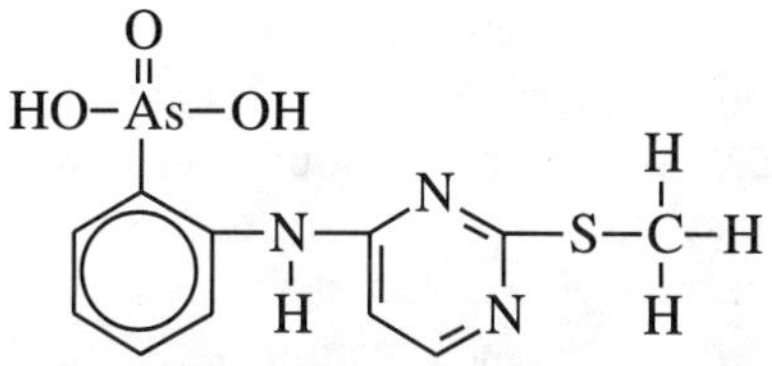

2-Methylthio-4-(2'-phenylarsenic acid)-aminopyrimidine

Organoarsenic compounds are used as animal feed additives. The major organoarsenic feed additives and their uses are summarized in Figure 12.9.

12.10.3 Toxicities of Organoarsenic Compounds

The toxicities of organoarsenic compounds vary over a wide range. In general, the toxicities are less for those compounds that are not metabolized in the body and that are excreted in an unchanged form. Examples of such compounds are the animal feed additives shown in Figure 12.9. Metabolic breakdown of organoarsenic compounds to inorganic forms is correlated with high toxicity. This is especially true when the product is inorganic arsenic(III), which, for the most part, is more toxic than arsenic(V). The toxicity of arsenic(III) is related to its strong affinity for sulfhydryl (–SH) groups. Detrimental effects are especially likely to occur when sulfhydryl groups are adjacent to each other on the active sites of enzymes, enabling chelation of the arsenic and inhibition of the enzyme.

To a certain extent, toxic effects of dimethylarsinic acid (cacodylic acid) have been observed because of its applications as an herbicide and the former uses of its sodium salt for the treatment of human skin disease and leukemia. It is most toxic via ingestion because the acidic medium in the stomach converts the compound to inorganic arsenic(III). A portion of inorganic arsenic in the body is converted to dimethylarsinic acid, which is excreted in urine, sweat, and exhaled air, accompanied by a strong garlic odor. Roxarsone has a relatively high acute toxicity to rats and dogs. Among the effects observed in these animals are internal hemorrhage, kidney congestion, and gastroenteritis. Rats fed fatal doses of about 400 ppm in the diet exhibited progressive weakness prior to death.

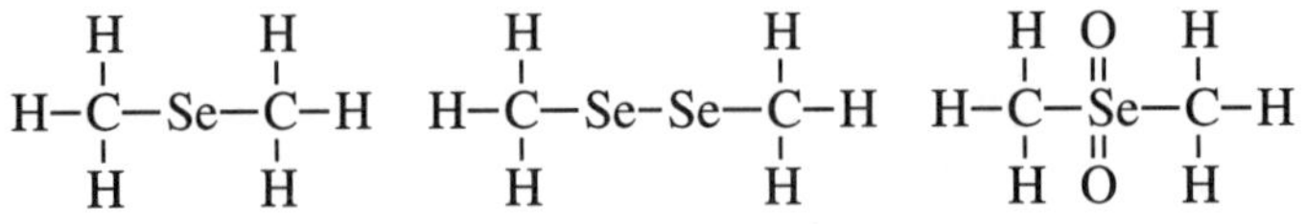

$$\text{H-}\underset{\underset{H}{|}}{\overset{\overset{H}{|}}{C}}\text{-Se-}\underset{\underset{H}{|}}{\overset{\overset{H}{|}}{C}}\text{-H} \qquad \text{H-}\underset{\underset{H}{|}}{\overset{\overset{H}{|}}{C}}\text{-Se-Se-}\underset{\underset{H}{|}}{\overset{\overset{H}{|}}{C}}\text{-H} \qquad \text{H-}\underset{\underset{H}{|}}{\overset{\overset{H}{|}}{C}}\text{-}\underset{\underset{O}{||}}{\overset{\overset{O}{||}}{Se}}\text{-}\underset{\underset{H}{|}}{\overset{\overset{H}{|}}{C}}\text{-H}$$

        Dimethylselenide   **Dimethyldiselenide**   **Dimethylselenone**

Figure 12.10 Example organoselenium compounds.

12.11 ORGANOSELENIUM AND ORGANOTELLURIUM COMPOUNDS

Organo compounds of the two group 6A elements, selenium and tellurium, are of considerable environmental and toxicological importance. Organoselenium and organotellurium compounds are produced both synthetically and by microorganisms. The selenium compounds are the more significant because of the greater abundance of this element.

12.11.1 Organoselenium Compounds

The structures of three common organoselenium compounds produced by organisms are given in Figure 12.10. Some organisms convert inorganic selenium to dimethylselenide. Several genera of fungi are especially adept at this biomethylation process, and their activities are readily detected from the very strong ultragarlic odor of the product. The bioconversion of inorganic selenium(II) and selenium(VI) to dimethylselenide and dimethyldiselenide, respectively, occurs in animals such as rats, and the volatile compounds are evolved with exhaled air. Another organoselenium compound produced by bacteria is dimethylselenone. Some synthetic organoselenium compounds have selenium as part of a ring, such as is the case with the cyclic ether 1,4-diselenane.

Inorganic selenium compounds are rather toxic, and probably attach to protein sulfhydryl groups, much like inorganic arsenic. In general, organoselenium compounds are regarded as less toxic than inorganic selenium compounds.

12.11.2 Organotellurium Compounds

Inorganic tellurium is used in some specialized alloys, to color glass, and as a pigment in some porcelain products. The breath of workers exposed to inorganic tellurium has a garlic odor, perhaps indicative of bioconversion to organotellurium species. Dimethyltelluride can be produced by fungi from inorganic tellurium compounds. Tellurium is a rather rare element in the geosphere and in water, so that biomethylation of this element is unlikely to be a major environmental problem. In general, the toxicities of tellurium compounds are less than those of their selenium analogs.

REFERENCES

1. Elschenbroich, C. and Salzer, A., *Organometallics: A Concise Introduction*, 2nd ed., Vch Verlagsgesellschaft Mbh, Berlin, 1992.
2. Crabtree, R.H., *The Organometallic Chemistry of the Transition Metals*, 3rd ed., John Wiley & Sons, New York, 2000.
3. Bach, R. et al., Lithium and lithium compounds, in *Kirk–Othmer Concise Encyclopedia of Chemical Technology*, Wiley Interscience, New York, 1985, pp. 706–707.
4. Knochel, P. and Jones, P., Eds., *Organozinc Reagents: A Practical Approach*, Oxford University Press, Oxford, 1999.
5. Pongratz, R. and Heumann, K., Determination of monomethyl cadmium in the environment by differential pulse anodic stripping voltammetry, *Anal. Chem.*, 68, 1262–1266, 1996.

6. Kenntner, N., Tataruch, F., and Krone, O., Heavy metals in soft tissue of white-tailed eagles found dead or moribund in Germany and Austria from 1993 to 2000, *Environ. Toxicol. Chem.*, 20, 1831–1837, 2001.
7. Hoch, M., Organotin compounds in the environment: an overview, *Appl. Geochem.*, 16, 719–743, 2001.
8. Hernberg, S., Lead poisoning in a historical perspective, *Am. J. Industrial Med.*, 38, 244–254, 2000.
9. Feldmann, J. et al., An appetite for arsenic: the seaweed-eating sheep from Orkney, *Spec. Publ. R. Soc. Chem.*, 267, 380–386, 2001.

SUPPLEMENTARY REFERENCES

Bochmann, M., *Organometallics*, Oxford University Press, New York, 1994.
Craig, P.J., Ed., *Organometallic Compounds in the Environment*, John Wiley & Sons, New York, 1986.
Crompton, T.R., *Occurrence and Analysis of Organometallic Compounds in the Environment*, John Wiley & Sons, New York, 1998.
Kirchner, K. and Weissensteiner, W., *Organometallic Chemistry and Catalysis*, Springer, New York, 2001.
Spessard, G.O. and Miessler, G.L., *Organometallic Chemistry*, Prentice Hall, Upper Saddle River, 1997.
Thayer, J.S., *Environmental Chemistry of the Heavy Elements: Hydrido and Organo Compounds*, VCH, New York, 1995.

QUESTIONS AND PROBLEMS

1. How is carbon involved in defining what an organometallic compound is? How is electronegativity involved in this definition? How does an organometalloid differ from an organometallic?
2. What are the three major kinds of organic groups, or ligands, bonded to a metal in an organometallic compound? How might the bonding of an alkyl ligand to an element with a very low electronegativity, such as potassium, differ from the bonding to an element with a higher electronegativity, such as arsenic?
3. What is a carbanion? How are carbanions involved in organometallic compounds? How can neutral cyclopentadiene form a carbanion?
4. Match the following pertaining to bonding in organometallic compounds:

    (a) Sigma-covalent   1. Mixed organometallic
    (b) Dative covalent   2. Formed by benzene, cyclopentadiene
    (c) Bonds with π-electrons 3. Shared electrons all contributed by one atom
    (d) CH_3HgCl 4. Electron density is concentrated between the two nuclei

5. What would be the expected reactions of $C_2H_5^-Na^+$ with water? How might this species react with oxygen in air? What toxic effects might result from these kinds of reactions?
6. Discuss the historical aspects of organometallic compound toxicity, including organoarsenicals used as pharmaceutical agents, gasoline antiknock additives, and compounds used in applications such as catalysis and chemical synthesis.
7. Which organometallic compounds of group 1A are more stable than other organometallic compounds of this group, most likely to exist as liquids or low-melting-point solids, and generally more soluble in organic solvents?
8. In general, how should the toxicities of lithium organometallic compounds be regarded? Do they have any unique toxicity characteristics?
9. What are alkoxide compounds? In what sense are they organometallic compounds? In what respects are they not organometallic compounds? What does the reaction

$$K^+{}^-OCH_3 + H_2O \rightarrow KOH + CH_3OH$$

    show about alkoxides?
10. What are Grignard reagents? In what sense are they mixed organometals?

11. Diethylmagnesium, $Mg(C_2H_5)_2$, is described as a pyrophoric compound that is violently reactive to water and steam and that self-ignites in air, burning even in a carbon dioxide atmosphere. Describe the significance of this description in terms of reactivity, susceptibility to hydrolysis or oxidation, and potential toxic effects.
12. Describe what is shown by the following reaction:

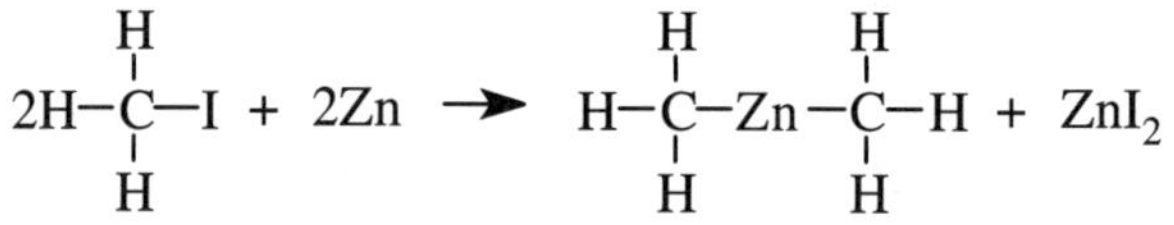

$$2\,H{-}\underset{H}{\overset{H}{\underset{|}{\overset{|}{C}}}}{-}I + 2Zn \rightarrow H{-}\underset{H}{\overset{H}{\underset{|}{\overset{|}{C}}}}{-}Zn{-}\underset{H}{\overset{H}{\underset{|}{\overset{|}{C}}}}{-}H + ZnI_2$$

13. Describe a specific toxic reaction that may result from the following combustion reaction:

$$2(CH_3)_2Zn + 8O_2 \rightarrow 2ZnO + 4CO_2 + 6H_2O$$

14. Why is the toxicology of cadmium organometallic compounds of particular concern?
15. Describe one chemical and one biochemical means of synthesis of $(CH_3)_2Hg$. In what sense was the discovery of biosynthesis of methylmercury species an unpleasant surprise in environmental chemistry?
16. List some special considerations that apply to organomercury compounds. How do their properties and pathways in the body compare to $Hg(0)$ and Hg^{2+}?
17. Describe the biocidal properties and uses of tributyl tin chloride and related tributyl tin compounds.
18. What are some of the biocidal uses of tributyl tin compounds?
19. In what sense are the toxicities and environmental effects of organolead compounds particularly noteworthy?
20. What is some of the evidence that the toxicological action of tetraethyllead is different from that of inorganic lead? What are some of the symptoms of tetraethyllead poisoning?
21. What are the two major sources of organoarsenic compounds? Give some examples of organoarsenic compounds produced by these two routes.
22. What may be said about the range of toxicities of organoarsenic compounds? How do these toxicities vary with organoarsenic compounds that are readily metabolized in the body, compared to those that are excreted in an unchanged form?
23. Why are organoselenium compounds of more concern than organotellurium compounds despite the close chemical similarity of selenium and tellurium?

CHAPTER 13

Toxic Organic Compounds and Hydrocarbons

13.1 INTRODUCTION

The fundamentals of organic chemistry are reviewed in Chapter 1. The present chapter is the first of seven that discuss the toxicological chemistry of organic compounds that are largely of synthetic origin. Since the vast majority of the several million known chemical compounds are organic — most of them toxic to a greater or lesser degree — the toxicological chemistry of organic compounds covers an enormous area. Specifically, this chapter discusses hydrocarbons, which are organic compounds composed only of carbon and hydrogen and are in a sense the simplest of the organic compounds. Hydrocarbons occur naturally in petroleum, natural gas, and tar sands, and they can be produced by pyrolysis of coal and oil shale or by chemical synthesis from H_2 and CO.

13.2 CLASSIFICATION OF HYDROCARBONS

For purposes of discussion of hydrocarbon toxicities in this chapter, hydrocarbons will be grouped into the five categories: (1) **alkanes**, (2) **unsaturated nonaromatic** hydrocarbons, (3) **aromatic** hydrocarbons (understood to have only one or two linked aromatic rings in their structures), (4) **polycyclic** aromatic hydrocarbons with multiple rings, and (5) **mixed** hydrocarbons containing combinations of two or more of the preceding types. These classifications are summarized in Figure 13.1.

13.2.1 Alkanes

Alkanes, also called **paraffins** or **aliphatic hydrocarbons**, are hydrocarbons in which the C atoms are joined by single covalent bonds (sigma bonds) consisting of two shared electrons (see Section 1.3). As shown by the examples in Figure 13.1 and Section 1.7, alkanes may exist as straight chains or branched chains. They may also exist as cyclic structures, for example, as in cyclohexane (C_6H_{12}). Each cyclohexane molecule consists of six carbon atoms (each with two H atoms attached) in a ring. The general molecular formula for straight- and branched-chain alkanes is C_nH_{2n+2}, and that of a cyclic alkane is C_nH_{2n}. The names of alkanes having from one to ten carbon atoms per molecule are respectively (1) methane, (2) ethane, (3) propane, (4) butane, (5) pentane, (6) hexane, (7) heptane, (8) octane, (9) nonane, and (10) decane. These names may be prefixed by n- to denote a straight-chain alkane. The same base names are used to designate substituent groups on molecules; for example, a straight-chain four-carbon alkane group (derived from butane) attached by an end carbon to a molecule is designated as an n-butyl group.

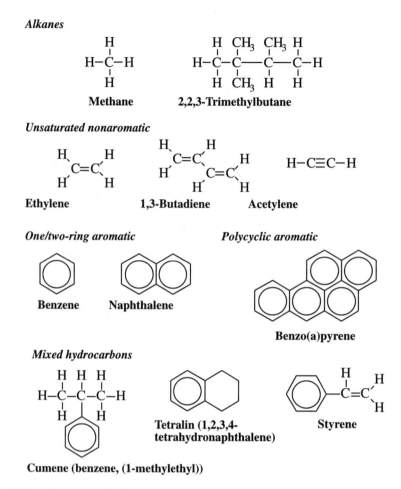

Figure 13.1 Hydrocarbons classified for discussion of their toxicological chemistry.

Alkanes undergo a number of chemical reactions, two classes of which should be mentioned here. The first of these is **oxidation** with molecular oxygen in air, as shown for the following combustion reaction of propane:

$$C_3H_8 + 5O_2 \rightarrow 3CO_2 + 4H_2O + \text{heat} \qquad (13.2.1)$$

Such reactions can pose flammability and explosion hazards. Another hazard occurs during combustion in an oxygen-deficient atmosphere or in an automobile engine, in which significant quantities of toxic carbon monoxide (CO) are produced.

The second major type of alkane reaction that should be considered here consists of **substitution reactions**, in which one or more H atoms on an alkane are replaced by atoms of another element. Most commonly, the H is replaced by a halogen, usually chlorine, to yield **organohalide** compounds; when chlorine is the substituent, the product is called an **organochlorine** compound. An example of this kind of reaction is that of methane with chlorine to give carbon tetrachloride, reaction 13.2.2. Organohalide compounds are of great toxicological significance and are discussed in Chapter 16.

$$CH_4 + 4Cl_2 \rightarrow CCl_4 + 4HCl \qquad (13.2.2)$$

$$\underset{Cl}{\overset{H}{\diagdown}}C=C\underset{Cl}{\overset{H}{\diagup}} \qquad \underset{Cl}{\overset{H}{\diagdown}}C=C\underset{H}{\overset{Cl}{\diagup}}$$

Cis -1,2-dichloroethylene, mp -80.5°C, bp 59°C

Trans -1,2-dichloroethylene, mp -50°C, bp 48°C

Figure 13.2 The two geometrical isomers of 1,2-dichloroethane.

13.2.2 Unsaturated Nonaromatic Hydrocarbons

Unsaturated hydrocarbons are those that have multiple bonds, each involving more than two shared electrons, between carbon atoms. Such compounds are usually **alkenes** or **olefins** that have double bonds consisting of four shared electrons, as shown for ethylene and 1,3-butadiene in Figure 13.1. Triple bonds consisting of six shared electrons are also possible, as illustrated by acetylene in the same figure.

Alkenes may undergo **addition reactions**, in which pairs of atoms are added across unsaturated bonds, as shown in the following reaction of ethylene with hydrogen to give ethane:

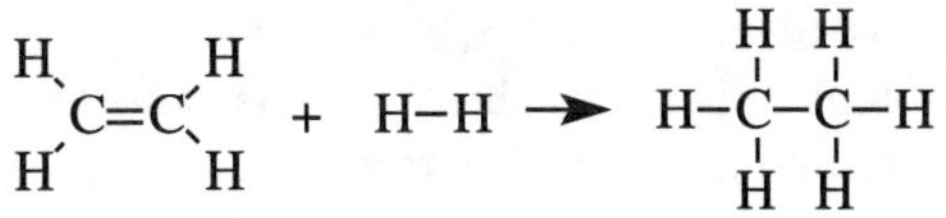

$$\underset{H}{\overset{H}{\diagdown}}C=C\underset{H}{\overset{H}{\diagup}} \;+\; H-H \;\rightarrow\; H-\underset{H}{\overset{H}{\underset{|}{\overset{|}{C}}}}-\underset{H}{\overset{H}{\underset{|}{\overset{|}{C}}}}-H \tag{13.2.3}$$

This kind of reaction, which is not possible with alkanes, adds to the chemical and metabolic, as well as toxicological, versatility of compounds containing unsaturated bonds.

Another example of an addition reaction is that of a molecule of HCl gas to one of acetylene to yield vinyl chloride:

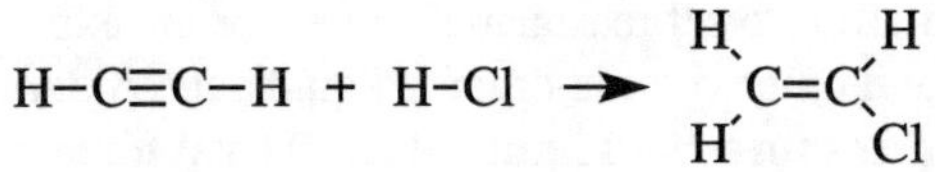

$$H-C\equiv C-H \;+\; H-Cl \;\rightarrow\; \underset{H}{\overset{H}{\diagdown}}C=C\underset{Cl}{\overset{H}{\diagup}} \tag{13.2.4}$$

The vinyl chloride product is the monomer used to manufacture polyvinylchloride plastic and is a carcinogen known to cause a rare form of liver cancer among exposed workers.

As discussed in Section 1.7, compounds with double bonds can exist as geometrical isomers exemplified by the two isomers of 1,2-dichloroethylene in Figure 13.2. Although both of these compounds have the molecular formula $C_2H_2Cl_2$, the orientations of their H and Cl atoms relative to each other are different, and their properties, such as melting and boiling points, are not the same. Their toxicities are both relatively low, but significantly different. The *cis*- isomer is an irritant and narcotic known to damage the liver and kidneys of experimental animals. The *trans*- isomer causes weakness, tremor, and cramps due to its effects on the central nervous system, as well as nausea and vomiting, resulting from adverse effects on the gastrointestinal tract.

13.2.3 Aromatic Hydrocarbons

Aromatic compounds were discussed briefly in Section 1.7. The characteristics of **aromaticity** of organic compounds are numerous and are discussed at length in works on organic chemistry. These characteristics include a low hydrogen:carbon atomic ratio, C–C bonds that are quite strong and of intermediate length between such bonds in alkanes and those in alkenes, a tendency to

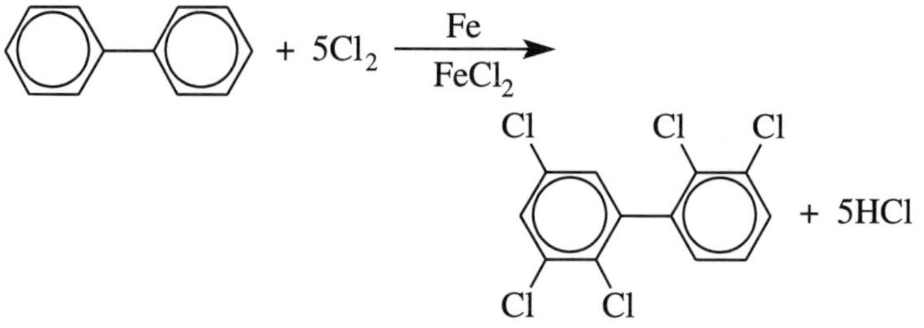

Figure 13.3 An example of a substitution reaction of an aromatic hydrocarbon compound (biphenyl) to produce an organochlorine product (2,3,5,2',3'-pentachlorobiphenyl, a PCB compound). The product is 1 of 210 possible congeners of PCBs, widespread and persistent pollutants found in the fat tissue of most humans and of considerable environmental and toxicological concern.

undergo substitution reactions (see Reaction 13.2.2) rather than the addition reactions characteristic of alkenes, and delocalization of π-electrons over several carbon atoms, resulting in resonance stabilization of the molecule. For more detailed explanations of these concepts, refer to standard textbooks on organic chemistry. For purposes of discussion here, most of the aromatic compounds discussed are those that contain single benzene rings or fused benzene rings, such as those in naphthalene or benzo(a)pyrene, shown in Figure 13.1.

An example reaction of aromatic compounds with considerable environmental and toxicological significance is the chlorination of biphenyl. Biphenyl gets its name from the fact that it consists of two **phenyl** groups (where a phenyl group is a benzene molecule less a hydrogen atom) joined by a single covalent bond. In the presence of an iron(II) chloride catalyst, this compound reacts with chlorine to form a number of different molecules of polychlorinated biphenyls (PCBs), as shown in Figure 13.3. These environmentally persistent compounds are discussed in Chapter 16.

13.3 TOXICOLOGY OF ALKANES

Worker exposure to alkanes, especially the lower-molecular-mass compounds, is most likely to come from inhalation. In an effort to set reasonable values for the exposure by inhalation of vapors of solvents, hydrocarbons, and other volatile organic liquids, the American Conference of Governmental Industrial Hygienists sets **threshold limit values** (TLVs) for airborne toxicants.[1,2] The **time-weighted average exposure** (E) is calculated by the formula

$$E = \frac{C_a T_a + C_b T_b + \cdots + C_n T_n}{8} \quad (13.3.1)$$

where C is the concentration of the substance in the air for a particular time T (hours), such as a level of 3.1 ppm by volume for 1.25 h. The 8 in the denominator is for an 8-h day. In addition to exposures calculated by this equation, there are short-term exposure limits (STELs) and ceiling (C) recommendations applicable to higher exposure levels for brief periods of time, such as 10 min once each day.

"Safe" levels of air contaminants are difficult to set based on systemic toxicologic effects. Therefore, TLVs often reflect nonsystemic effects of odor, narcosis, eye irritation, and skin irritation. Because of this, comparison of TLVs is often not useful in comparing systemic toxicological effects of chemicals in the workplace.

13.3.1 Methane and Ethane

Methane and ethane are **simple asphyxiants**, which means that air containing high levels of these gases does not contain sufficient oxygen to support respiration. Table 13.1 shows the levels of asphyxiants in air at which various effects are observed in humans. Simple asphyxiant gases are

TOXIC ORGANIC COMPOUNDS AND HYDROCARBONS

Table 13.1 Effects of Simple Asphyxiants in Air

Percent Asphyxiant[a]	Percent Oxygen, O_2[a]	Effect on Humans
0–33	21–14	No major adverse symptoms
33–50	14–10.5	Discernible effects beginning with air hunger and progressing to impaired mental alertness and muscular coordination
50–75	10.5–5.3	Fatigue, depression of all sensations, faulty judgment, emotional instability; in later phases, nausea, vomiting, prostration, unconsciousness, convulsions, coma, death
75–100	5.3–0	Death within a few minutes

[a] Percent by volume on a "dry" (water vapor-free) basis.

not known to have major systemic toxicological effects, although subtle effects that are hard to detect should be considered as possibilities.

13.3.2 Propane and Butane

Propane has the formula C_3H_8 and butane C_4H_8. There are two isomers of butane, *n*-butane and isobutane (2-methylpropane). Propane and the butane isomers are gases at room temperature and atmospheric pressure; like methane and ethane, all three are asphyxiants. A high concentration of propane affects the central nervous system. There are essentially no known systemic toxicological effects of the two butane isomers; behavior similar to that of propane might be expected.

13.3.3 Pentane through Octane

The alkanes with five to eight carbon atoms consist of *n*-alkanes, and there is an increasing number of branched-chain isomers with higher numbers of C atoms per molecule. For example, there are nine isomers of heptane C_7H_{16}. These compounds are all volatile liquids under ambient conditions; the boiling points for the straight-chain isomers range from 36.1°C for *n*-pentane to 125.8°C for *n*-octane. In addition to their uses in fuels, such as in gasoline, these compounds are employed as solvents in formulations for a number of commercial products, including varnishes, glues, and inks. They are also used for the extraction of fats.

Once regarded as toxicologically almost harmless, the C_5–C_8 aliphatic hydrocarbons are now recognized as having some significant toxic effects. Exposure to the C_5–C_8 hydrocarbons is primarily via the pulmonary route, and high levels in air have killed experimental animals. Humans inhaling high levels of these hydrocarbons have become dizzy and have lost coordination as a result of central nervous system depression.

Of the C_5–C_8 alkanes, the one most commonly used for nonfuel purposes is *n*-hexane. It acts as a solvent for the extraction of oils from seeds, such as cottonseed and sunflower seed. This alkane serves as a solvent medium for several important polymerization processes and in mixtures with more polar solvents, such as furfural,

Furfural

for the separation of fatty acids. **Polyneuropathy** (multiple disorders of the nervous system) has been reported in several cases of human exposure to *n*-hexane, such as Japanese workers involved in home production of sandals using glue with *n*-hexane solvent. The workers suffered from muscle weakness and impaired sensory function of the hands and feet. Biopsy examination of nerves in leg muscles of the exposed workers showed loss of myelin (a fatty substance constituting a sheath around certain nerve fibers) and degeneration of axons (part of a nerve cell through which nerve impulses are transferred out of the cell). The symptoms of polyneuropathy were reversible, with recovery taking several years after exposure was ended.

Exposure of the skin to C_5–C_8 liquids causes dermatitis. This is the most common toxicological occupational problem associated with the use of hydrocarbon liquids in the workplace, and is a consequence of the dissolution of the fat portions of the skin. In addition to becoming inflamed, the skin becomes dry and scaly.

13.3.4 Alkanes above Octane

Alkanes higher than C_8 are contained in kerosene, jet fuel, diesel fuel, mineral oil, and fuel oil distilled from crude oil as middle distillate fuels with a boiling range of approximately 175 to 370°C. Kerosene, also called fuel oil no. 1, is a mixture of primarily C_8–C_{16} hydrocarbons, predominantly alkanes. Diesel fuel is called fuel oil no. 2. The heavier fuel oils, no. 3 to 6, are characterized by increasing viscosity, darker color, and higher boiling temperatures with increasing fuel oil number. Mineral oil is a carefully selected fraction of petroleum hydrocarbons with density ranges of 0.83 to 0.86 g/ml for light mineral oil and 0.875 to 0.905 g/ml for heavy mineral oil.

The higher alkanes are not regarded as very toxic, although there are some reservations about their toxicities. Inhalation is the most common route of occupational exposure and can result in dizziness, headache, and stupor. In cases of extreme exposure, coma and death have occurred. Inhalation of mists or aspiration of vomitus containing higher alkane liquids has caused a condition known as aspiration pneumonia. They are not regarded as carcinogenic, although experimental mice have shown weak tumorigenic responses with long latency periods upon prolonged skin exposure to middle distillate fuels. The observed effects have been attrributed to chronic skin irritation, and these substances do not produce tumors in the absence of skin irritation.[3] Middle distillate fuels can be effective carriers of known carcinogens, especially polycyclic aromatic hydrocarbons.

13.3.5 Solid and Semisolid Alkanes

Semisolid petroleum jelly is a highly refined product commonly known as vaseline, a mixture of predominantly C_{16}–C_{19} alkanes. Carefully controlled refining processes are used to remove nitrogen and sulfur compounds, resins, and unsaturated hydrocarbons. Paraffin wax is a similar product, behaving as a solid. Neither petroleum jelly nor paraffin is digested or absorbed by the body.

13.3.6 Cyclohexane

Cyclohexane, the six-carbon ring hydrocarbon with the molecular formula C_6H_{12}, is the most significant of the cyclic alkanes. Under ambient conditions it is a clear, volatile, highly flammable liquid. It is manufactured by the hydrogenation of benzene and is used primarily as a raw material for the synthesis of cyclohexanol and cyclohexanone through a liquid-phase oxidation with air in the presence of a dissolved cobalt catalyst.

Cyclohexanol   **Cyclohexanone**

Like *n*-hexane, cyclohexane has a toxicity rating of 3, moderately toxic (see Table 6.1 for toxicity ratings). Cyclohexane acts as a weak anesthetic similar to, but more potent than, *n*-hexane. Systemic effects have not been shown in humans.

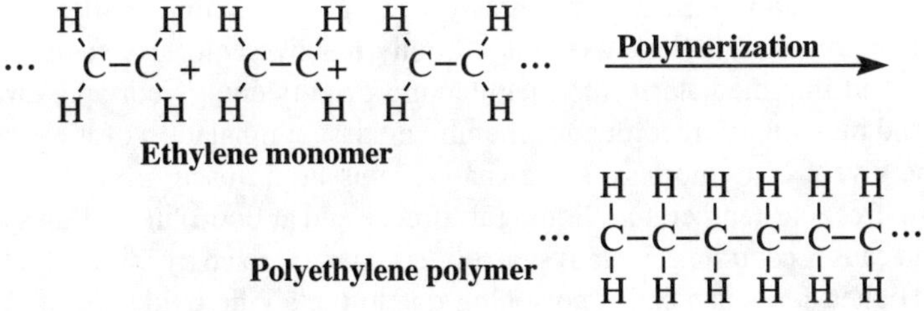

$$\cdots \overset{H}{\underset{H}{C}}-\overset{H}{\underset{H}{C}} + \overset{H}{\underset{H}{C}}-\overset{H}{\underset{H}{C}} + \overset{H}{\underset{H}{C}}-\overset{H}{\underset{H}{C}} \cdots \xrightarrow{\text{Polymerization}}$$

Ethylene monomer

$$\cdots \overset{H\ H\ H\ H\ H\ H}{\underset{H\ H\ H\ H\ H\ H}{C-C-C-C-C-C}} \cdots$$

Polyethylene polymer

Figure 13.4 Polymerization of ethylene to produce polyethylene.

13.4 TOXICOLOGY OF UNSATURATED NONAROMATIC HYDROCARBONS

Ethylene (structure in Figure 13.1) is the most widely used organic chemical. Almost all of it is consumed as a chemical feedstock for the manufacture of other organic chemicals. Polymerization of ethylene to produce polyethylene is illustrated in Figure 13.4. In addition to polyethylene, other polymeric plastics, elastomers, fibers, and resins are manufactured with ethylene as one of the ingredients. Ethylene is also the raw material for the manufacture of ethylene glycol antifreeze, solvents, plasticizers, surfactants, and coatings.

The boiling point (bp) of ethylene is –105°C, and under ambient conditions it is a colorless gas. It has a somewhat sweet odor, is highly flammable, and forms explosive mixtures with air. Because of its double bond (unsaturation), ethylene is much more active than the alkanes. It undergoes addition reactions, as shown in the following examples, to form a number of important products:

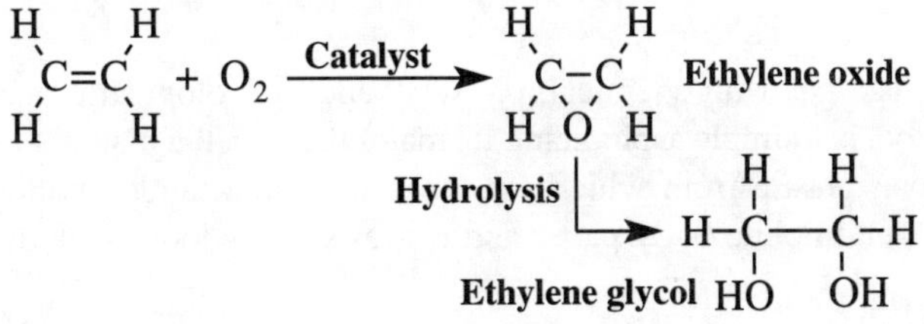

$$\overset{H}{\underset{H}{C}}=\overset{H}{\underset{H}{C}} + O_2 \xrightarrow{\text{Catalyst}} \overset{H}{\underset{H}{C}}-\overset{H}{\underset{H}{C}} \text{ Ethylene oxide} \tag{13.4.1}$$

$$\xrightarrow{\text{Hydrolysis}} H-\overset{H}{\underset{HO}{C}}-\overset{H}{\underset{OH}{C}}-H \text{ Ethylene glycol}$$

$$\overset{H}{\underset{H}{C}}=\overset{H}{\underset{H}{C}} + Br_2 \rightarrow Br-\overset{H}{\underset{H}{C}}-\overset{H}{\underset{H}{C}}-Br \tag{13.4.2}$$

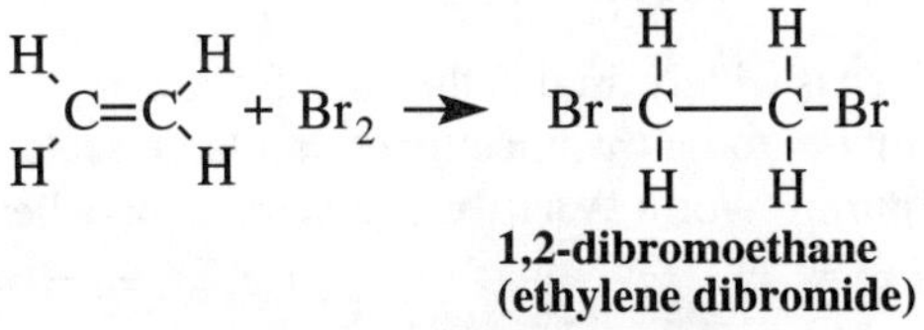

1,2-dibromoethane (ethylene dibromide)

$$\overset{H}{\underset{H}{C}}=\overset{H}{\underset{H}{C}} + Cl_2 \rightarrow Cl-\overset{H}{\underset{H}{C}}-\overset{H}{\underset{H}{C}}-Cl \tag{13.4.3}$$

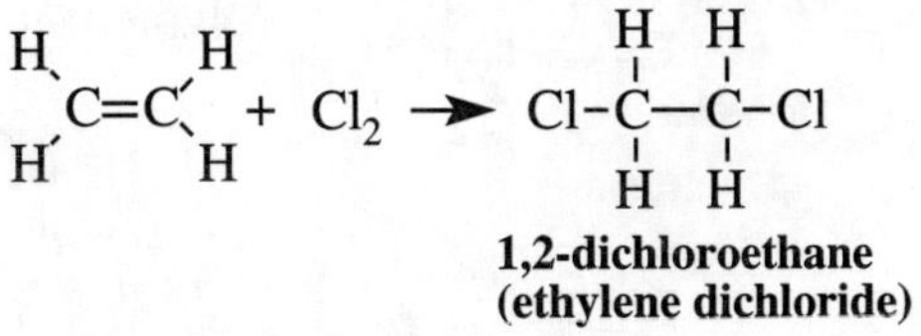

1,2-dichloroethane (ethylene dichloride)

$$\overset{H}{\underset{H}{C}}=\overset{H}{\underset{H}{C}} + HCl \rightarrow H-\overset{H}{\underset{H}{C}}-\overset{H}{\underset{H}{C}}-Cl \tag{13.4.4}$$

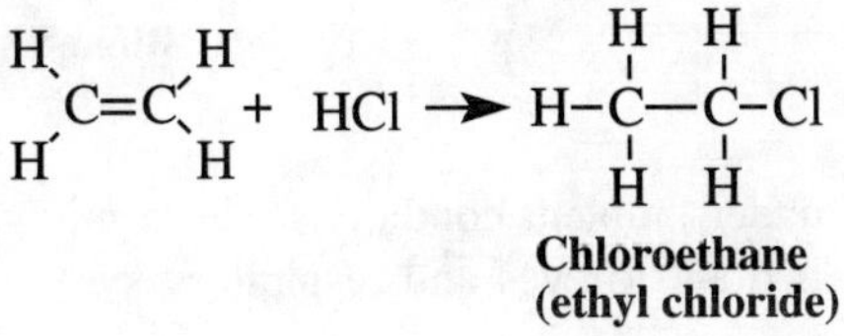

Chloroethane (ethyl chloride)

The products of the addition reactions shown above are all commercially, toxicologically, and environmentally important. Ethylene oxide is a highly reactive colorless gas used as a sterilizing agent, fumigant, and intermediate in the manufacture of ethylene glycol and surfactants. It is an irritant to eyes and pulmonary tract mucous membrane tissue; inhalation of it can cause pulmonary edema. Ethylene glycol is a colorless, somewhat viscous liquid used in mixtures with water as a high-boiling, low-freezing-temperature liquid (antifreeze and antiboil) in cooling systems. Ingestion of this compound causes central nervous system effects characterized by initial stimulation, followed by depression. Higher doses can cause poisoning due to metabolic oxidation of ethylene glycol to glycolic acid, glyoxylic acid, and oxalic acid. Glycolic acid causes acidosis, and oxalate forms insoluble calcium oxalate, which clogs the kidneys, as discussed in Section 14.2.

Ethylene dibromide has been used as an insecticidal fumigant and additive to scavenge lead from leaded gasoline combustion. During the early 1980s, there was considerable concern about residues of this compound in food products, and it was suspected of being a carcinogen, mutagen, and teratogen. Ethylene dichloride (bp, 83.5°C) is a colorless, volatile liquid with a pleasant odor that is used as a soil and foodstuff fumigant. It has a number of toxicological effects, including adverse effects on the eye, liver, and kidneys, and a narcotic effect on the central nervous system. Ethyl chloride seems to have similar, but much less severe, toxic effects.

A highly flammable compound, ethylene forms dangerously explosive mixtures with air. It is phytotoxic (toxic to plants). Ethylene, itself, is not very toxic to animals, but it is a simple asphyxiant (see Section 13.3 and Table 13.1). At high concentrations, it acts as an anesthetic to induce unconsciousness. The only significant pathway of human exposure to ethylene is through inhalation. This exposure is limited by the low blood–gas solubility ratio of ethylene, which applies at levels below saturation of blood with the gas. This ratio for ethylene is only 0.14, compared, for example, with the very high value of 15 for chloroform.[4]

13.4.1 Propylene

Propylene (C_3H_6) is a gas with chemical, physical, and toxicological properties very similar to those of ethylene. It, too, is a simple asphyxiant. Its major use is in the manufacture of polypropylene polymer, a hard, strong plastic from which are made injection-molded bottles, as well as pipes, valves, battery cases, automobile body parts, and rot-resistant indoor–outdoor carpet.

13.4.2 1,3-Butadiene

The dialkene 1,3-butadiene is widely used in the manufacture of polymers, particularly synthetic rubber. The first synthetic rubber to be manufactured on a large scale and used as a substitute for unavailable natural rubber during World War II was a styrene–butadiene polymer:

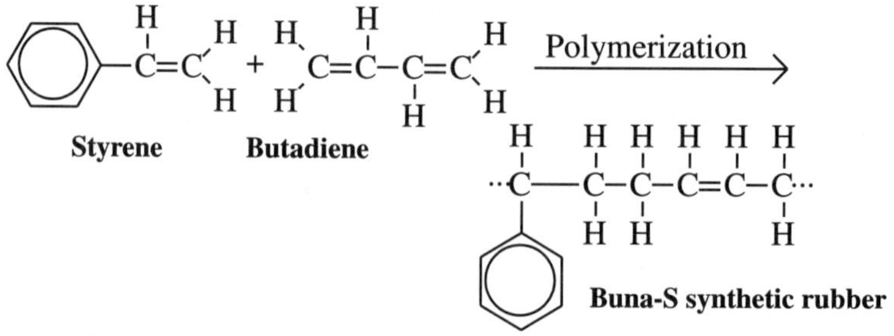

$$\text{Styrene} + \text{Butadiene} \xrightarrow{\text{Polymerization}} \text{Buna-S synthetic rubber} \quad (13.4.5)$$

Butadiene is a colorless gas under ambient conditions with a mild, somewhat aromatic odor. At lower levels, the vapor is an irritant to eyes and respiratory system mucous membranes, and at

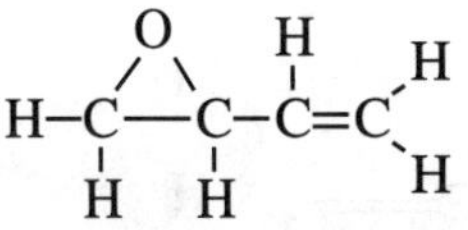

1,2-Epoxybutene-3

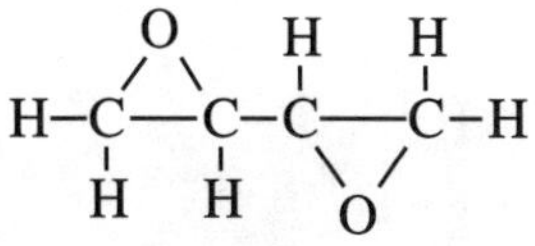

Diepoxybutane

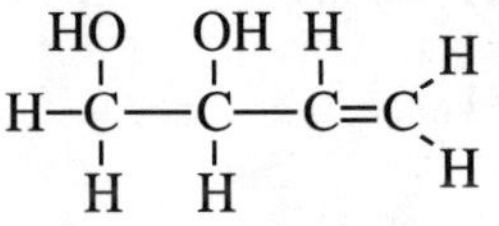

3-Butene-1,2-diol

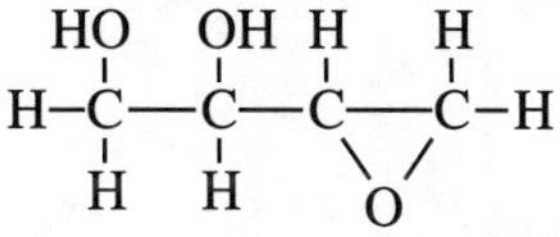

3,4-Epoxy-1,2-butane diol

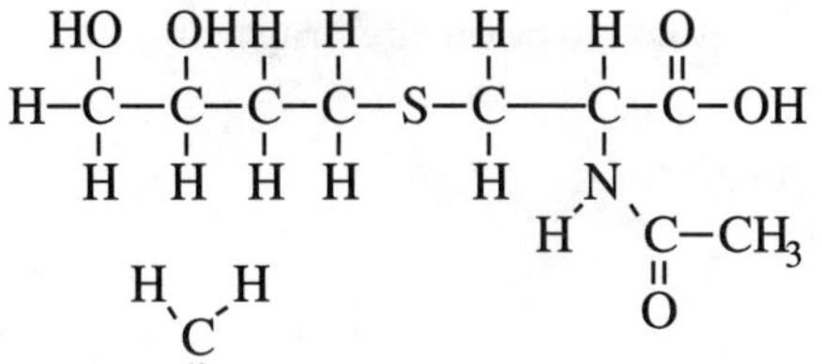

L-Cysteine, N-acetyl-S-(3,4-dihydroxybutyl) mercapturic acid conjugate

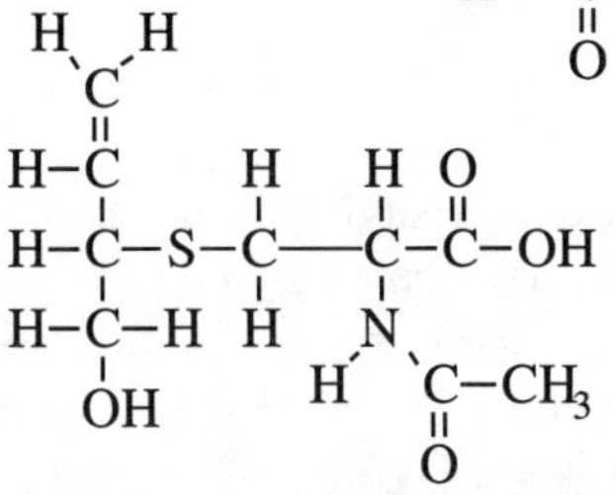

L-Cysteine, N-acetyl-S-[1-(hydroxymethyl)-2-propenyl] mercapturic acid conjugate

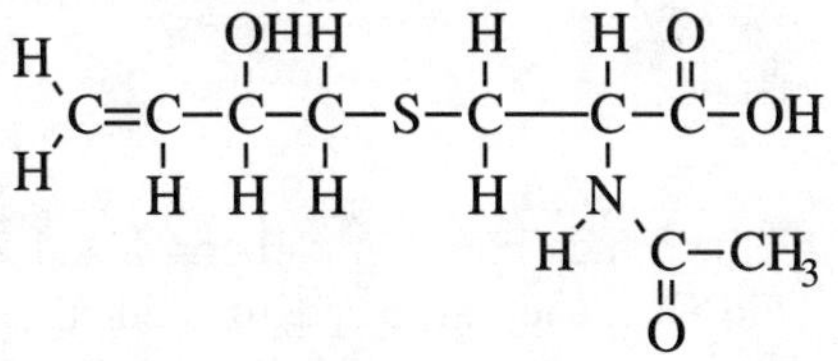

L-Cysteine, N-acetyl-S-(2-hydroxy-3-butenyl) mercapturic acid conjugate

Figure 13.5 Common metabolites of 1,3-butadiene.

higher levels, it can cause unconsciousness and even death. Symptoms of human exposure include, initially, blurred vision, nausea, and paresthesia, accompanied by dryness of the mouth, nose, and throat. In cases of severe exposure, fatigue, headache, vertigo, and decreased pulse rate and blood pressure may be followed by unconsciousness. Fatal exposures have occurred only as the result of catastrophic releases of 1,3-butadiene gas. The compound boils at –4.5°C and is readily stored and handled as a liquid. Release of the liquid can cause frostbite-like burns on exposed flesh.

The aspect of 1,3-butadiene of greatest toxicological concern is its potential carcinogenicity. Butadiene is a known carcinogen to rats and mice and is more likely to cause cancer in the latter. Although it is a suspected carcinogen to humans, epidemiological studies of exposed workers in the synthetic rubber and plastics industries suggest that normal worker exposures are insufficient to cause cancer. Butadiene is acted on by P-450 isoenzymes to produce genotoxic metabolites, most prominently epoxybutene and diepoxybutene.[5] In addition, microsomal metabolic processes in rats produce the two possible stereoisomers of diepoxybutane, 3-butene-1,2-diol, and the two stereoisomers of 3,4-epoxy-1,2-butanediol (Figure 13.5). The production of mercapturic acid derivatives of the oxidation products of 1,3-butadiene (see Figure 13.5) results in detoxication of this compound and serves as a biomarker of exposure to it. Other useful biomarkers consist of the hemoglobin adducts 1- and 2-hydroxy-3-butenylvaline.[6]

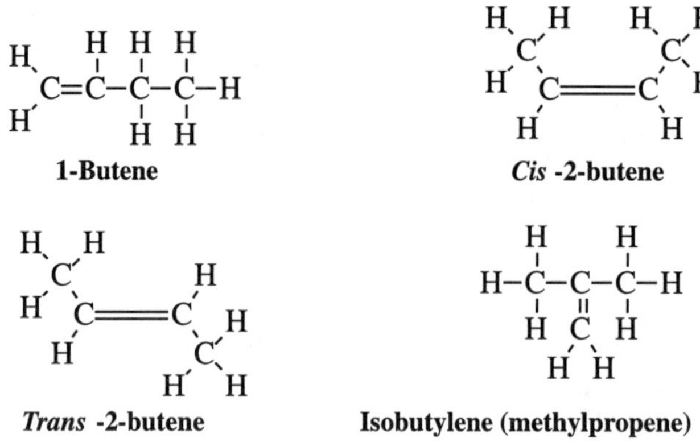

282 TOXICOLOGICAL CHEMISTRY AND BIOCHEMISTRY

Figure 13.6 The four butylene compounds, formula C_4H_8.

13.4.3 Butylenes

There are four monoalkenes with the formula C_4H_8 (butylenes), as shown in Figure 13.6. All gases under ambient conditions, these compounds have boiling points ranging from –6.9°C for isobutylene to 3.8°C for *cis*-2-butene. The butylenes readily undergo isomerization (change to other isomers). They participate in addition reactions and form polymers. Their major hazard is extreme flammability. Though not regarded as particularly toxic, they are asphyxiants and have a narcotic effect when inhaled.

13.4.4 Alpha-Olefins

Alpha-olefins are linear alkenes with double bonds between carbons 1 and 2 in the general range of carbon chain length C_6 through about C_{18}. They are used for numerous purposes. The C_6–C_8 compounds are used as comonomers to manufacture modified polyethylene polymer, and the C_{12}–C_{18} alpha-olefins are used as raw materials in the manufacture of detergents. The compounds are also used to manufacture lubricants and plasticizers. Worldwide consumption of the alpha-olefins was around 1 million metric tons. With such large quantities involved, due consideration needs to be given to the toxicological and occupational health aspects of these compounds.

13.4.5 Cyclopentadiene and Dicyclopentadiene

The cyclic dialkene cyclopentadiene has the structural formula shown below:

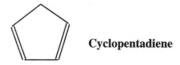
Cyclopentadiene

Two molecules of cyclopentadiene readily and spontaneously join together to produce dicyclopentadiene, widely used to produce polymeric elastomers, polyhalogenated flame retardants, and polychlorinated pesticides. Dicyclopentadiene mp, 32.9°C; bp, 166.6°C) exists as colorless crystals. It is an irritant and has narcotic effects. It is considered to have a high oral toxicity and to be moderately toxic through dermal absorption.

TOXIC ORGANIC COMPOUNDS AND HYDROCARBONS

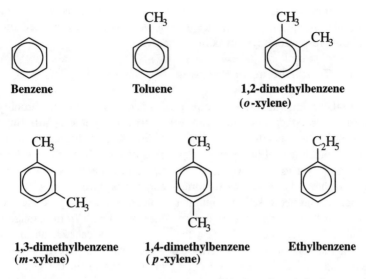

Figure 13.7 Benzene and its most common methyl-substituted hydrocarbon derivatives.

13.4.6 Acetylene

Acetylene (Figure 13.1) is widely used as a chemical raw material and fuel for oxyacetylene torches. It was once the principal raw material for the manufacture of vinyl chloride (see reaction 13.2.4), but other synthetic routes are now used. Acetylene is a colorless gas with an odor resembling garlic. Though not notably toxic, it acts as an asphyxiant and narcotic and has been used for anesthesia. Exposure can cause headache, dizziness, and gastric disturbances. Some adverse effects from exposure to acetylene may be due to the presence of impurities in the commercial product.

13.5 BENZENE AND ITS DERIVATIVES

Figure 13.7 shows the structural formulas of benzene and its major hydrocarbon derivatives. These compounds are very significant in chemical synthesis, as solvents, and in unleaded gasoline formulations.

13.5.1 Benzene

Benzene (C_6H_6) is chemically the single most significant hydrocarbon. It is used as a starting material for the manufacture of numerous products, including phenolic and polyester resins, polystyrene plastics and elastomers (through intermediate styrene, Figure 13.1), alkylbenzene surfactants, chlorobenzene compounds, insecticides, and dyes. Benzene (bp, 80.1°C) is a volatile, colorless, highly flammable liquid with a characteristic odor.

13.5.1.1 Acute Toxic Effects of Benzene

Benzene has been in commercial use for over a century, and toxic effects of it have been suspected since about 1900. Benzene has both acute and chronic toxicological effects.[7] It is usually absorbed as a vapor through the respiratory tract, although absorption of liquid through the skin and intake through the gastrointestinal tract are also possible. Benzene is a skin irritant, and progressively higher local exposures can cause skin redness (erythema), burning sensations, fluid accumulation (edema), and blistering. Inhalation of air containing about 64 g/m³ of benzene can be fatal within a few minutes; about one tenth that level of benzene causes acute poisoning within an hour, including a narcotic effect on the central nervous system manifested progressively by excitation, depression, respiratory system failure, and death.

13.5.1.2 Chronic Toxic Effects of Benzene

Of greater overall concern than the acute effects of benzene exposure are chronic effects, which are still subject to intense study. As with many other toxicants, subjects suffering from chronic benzene exposure suffer nonspecific symptoms, including fatigue, headache, and appetite loss. More specifically, blood abnormalities appear in people suffering chronic benzene poisoning. The most common of these is a lowered white cell count. More detailed examination may show an abnormal increase in blood lymphocytes (colorless corpuscles introduced to the blood from the lymph glands), anemia, and decrease in the number of blood platelets required for clotting (thrombocytopenia). Some of the observed blood abnormalities may result from damage by benzene to bone marrow. Epidemiological studies suggest that benzene may cause acute melogenous (from bone marrow) leukemia. Because of concerns that long-term exposure to benzene may cause preleukemia, leukemia, or cancer, the allowable levels of benzene in the workplace have been greatly reduced, and substitutes such as toluene and xylene are used wherever possible.

13.5.1.3 Metabolism of Benzene

For a hydrocarbon, the water solubility of benzene is a moderately high 1.80 g/l at 25°C. The vapor is readily absorbed by blood, from which it is strongly taken up by fatty tissues. For nonmetabolized benzene, the process is reversible and benzene is excreted through the lungs. Benzene metabolism occurs largely in the liver. Initially, benzene is oxidized by the action of cytochrome P-450 enzymes to benzene oxepin and benzene oxide, which are interchangeable through the action of cytochrome P-450 enzymes:

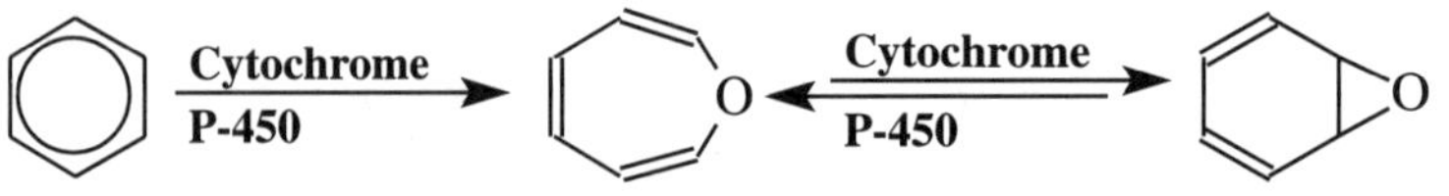

$$\text{Benzene} \xrightarrow{\text{Cytochrome P-450}} \text{Benzene oxepin} \xleftrightarrow{\text{Cytochrome P-450}} \text{Benzene oxide} \quad (13.5.1)$$

Benzene oxide may be hydrated through the action of epoxide hydrolase enzyme,

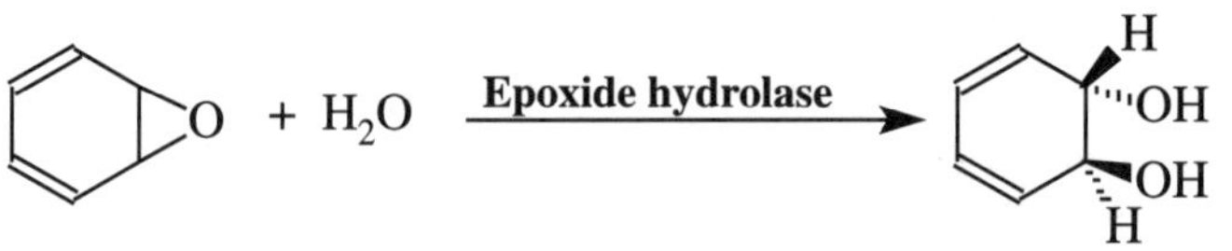

$$\text{Benzene oxide} + H_2O \xrightarrow{\text{Epoxide hydrolase}} \text{Benzene } trans\text{-1,2-dihydrodiol} \quad (13.5.2)$$

to produce benzene *trans*-1,2-dihydrodiol. This product is acted on by dihydrodiol dehydrogenase enzyme,

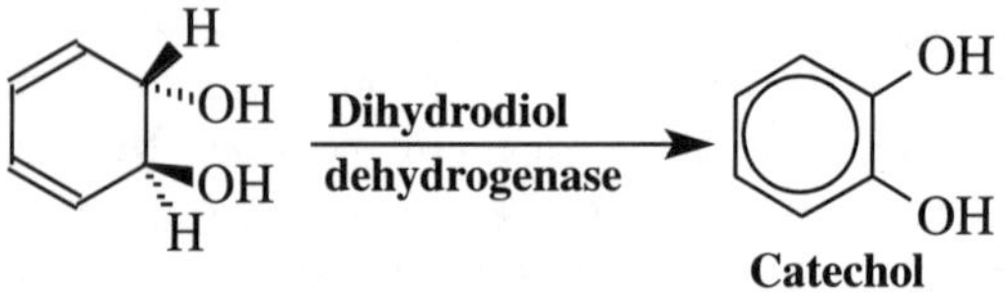

$$\text{Benzene } trans\text{-1,2-dihydrodiol} \xrightarrow{\text{Dihydrodiol dehydrogenase}} \text{Catechol} \quad (13.5.3)$$

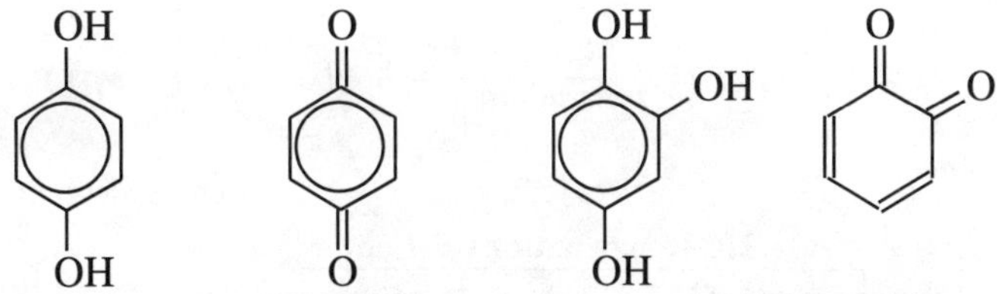

p-Benzoquinone *o*-Benzoquinone

Hydroquinone 1,2,4-trihydroxybenzene

Figure 13.8 Products of phenol and catechol produced by the metabolic oxidation of benzene.

to produce catechol. Benzene oxepin or oxide may also react to produce muconaldehyde and muconic acid:

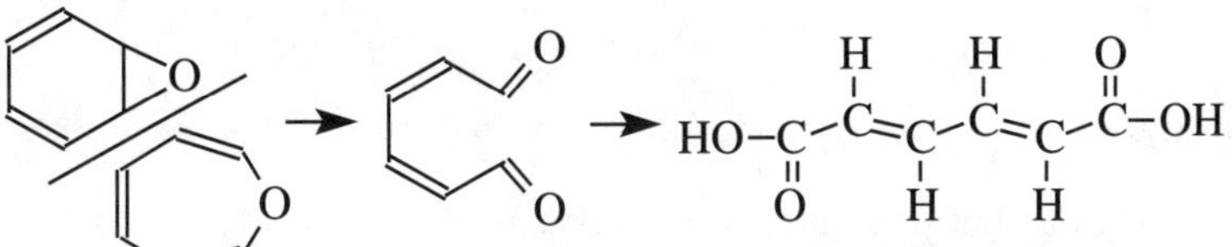

Muconaldehyde ***Trans*, *trans*-muconic acid** (13.5.4)

Benzene oxepin or oxide may form a glutathione conjugate or undergo nonenzymatic rearrangement to produce phenol. Phenol and catechol produce several oxyaryl species, shown in Figure 13.8.

Phase 1 oxidation products of benzene, including phenol, hydroquinone, catechol, 1,2,4-trihydroxybenzene, and *trans*,*trans*-muconic acid in urine, are evidence of exposure to benzene. Another substance observed in urine of individuals exposed to benzene is S-phenylmercapturic acid,

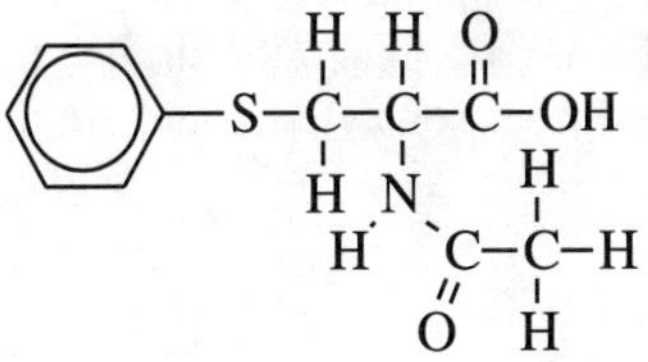

S-phenylmercapturic acid (L-cysteine, N-acetyl-S-phenyl-)

which is formed as a result of the phase 2 conjugation of benzene oxide by glutathione and subsequent reactions. Hemoglobin and albumin adducts of benzene oxide are commonly detected in the blood of workers exposed to benzene.

The oxidized metabolites of benzene, including reactive benzene oxide intermediate, are known to bind with DNA, RNA, and proteins. This can result in cell destruction, alteration of cell growth, and inhibition of enzymes involved in the processes of forming blood cells. This phenomenon is probably responsible for the bone marrow damage, aplastic anemia (lowered production of blood cells due to damage to bone marrow), and, in severe cases, leukemia associated with benzene exposure.

13.5.2 Toluene, Xylenes, and Ethylbenzene

Toluene is a colorless liquid boiling at 101.4°C. Gasoline is 5 to 7% toluene and is the most common source of human exposure to toluene. Toluene is one of the most common solvents inhaled by solvent abusers. It is classified as moderately toxic through inhalation or ingestion and has a low toxicity by dermal exposure. Concentrations in ambient air up to 200 ppm usually do not result

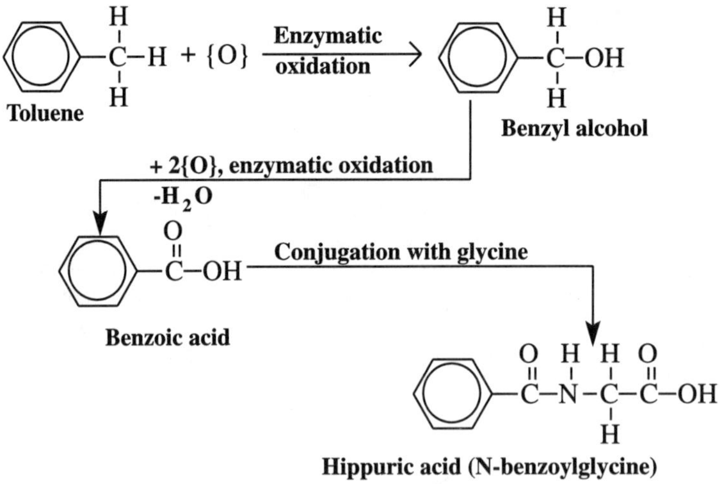

Figure 13.9 Metabolic oxidation of toluene with conjugation to hippuric acid, which is excreted with urine.

in significant symptoms, but exposure to 500 ppm may cause headache, nausea, lassitude, and impaired coordination without detectable physiological effects. At massive exposure levels, toluene acts as a narcotic, which can lead to coma.

Toluene tends to enter brain tissue, which it affects, and accumulates in adipose tissue. Unlike benzene, toluene possesses an aliphatic side chain that can be oxidized enzymatically, leading to products that are readily excreted from the body. The metabolism of toluene is thought to proceed via oxidation of the methyl group and formation of the conjugate compound hippuric acid, as shown in Figure 13.9.

Xylenes and ethylbenzene (Figure 13.7) are common gasoline constituents, industrial solvents, and reagents, so human exposure to these materials is common. The absorption (primarily through inhalation), metabolism, and effects of these solvents are generally similar to those of toluene. Effects are largely on the central nervous system. Effects of xylenes and ethylbenzene on organs other than the central nervous system appear to be limited.

13.5.3 Styrene

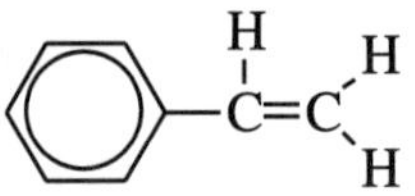

Styrene,

is widely used to make various kinds of rubber (see styrene–butadiene polymer in reaction 13.4.5), polystyrene plastics, resins, and insulators. As a consequence, human exposure to this substance in the workplace has been quite high. As with the other volatile aromatic hydrocarbons discussed in this section, styrene is readily absorbed by inhalation, is lipid soluble, and is readily metabolized in the liver. The presence of the C=C group in styrene provides an active site for biochemical attack, and styrene is readily oxidized metabolically to styrene oxide:

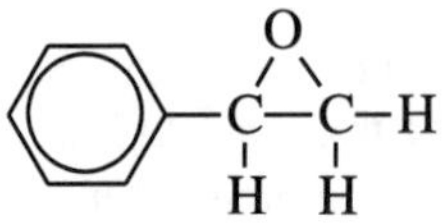

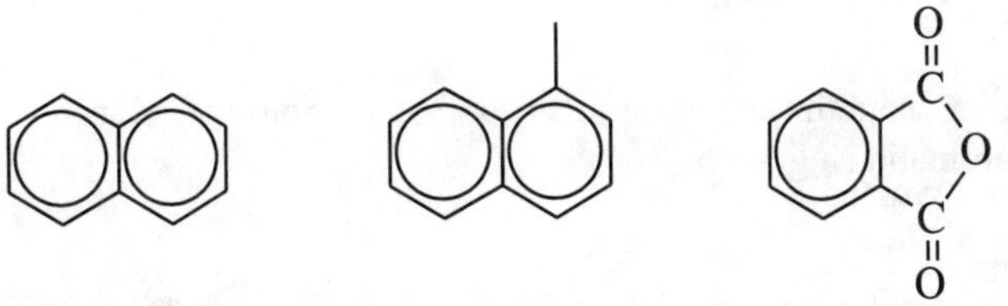

Naphthalene   **1-(2-propyl)naphthalene**   **Phthalic anhydride**

Figure 13.10 Naphthalene and two of its derivatives.

The major toxicological concern with styrene has to do with its potential role as a procarcinogen in producing carcinogenic styrene oxide, itself an industrial chemical to which workers may be exposed. Styrene oxide that is inhaled directly is distributed in the body by systemic circulation. However, styrene oxide that is produced by the metabolic oxidation of styrene in the liver is rapidly hydrolyzed in the liver by the action of epoxide hydrolase, leading to the formation of mandelic acid and phenylgloxylic acid, probably making the carcinogenicity hazard of styrene much lower than that of styrene oxide:[8]

Mandelic acid   **Glyoxylic acid**

The albumin adduct of styrene oxide, S-(2-hydroxyl-1-phenylethyl)cysteine,

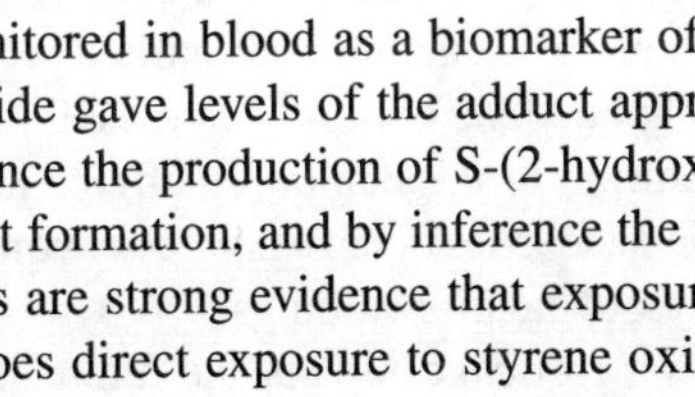

S-(2-hydroxy-1-phenylethyl)cysteine

has been monitored in blood as a biomarker of exposure to styrene and styrene oxide.[9] Exposures to styrene oxide gave levels of the adduct approximately 2000 times that of comparable exposure to styrene. Since the production of S-(2-hydroxyl-1-phenylethyl)cysteine is a measure of tendency toward adduct formation, and by inference the formation of nucleic acid adducts leading to cancer, these findings are strong evidence that exposure to styrene poses a much lower risk of carcinogenicity than does direct exposure to styrene oxide.

13.6 NAPHTHALENE

Naphthalene, also known as tar camphor, and its alkyl derivatives, such as 1-(2-propyl)naphthalene (Figure 13.10), are important industrial chemicals. Used to make mothballs, naphthalene is a volatile white crystalline solid with a characteristic odor. Coal tar and petroleum are the major sources of naphthalene. Numerous industrial chemical derivatives are manufactured from it. The most important of these is phthalic anhydride (Figure 13.10), used to make phthalic acid plasticizers, which are discussed in Chapter 14.

13.6.1 Metabolism of Naphthalene

The metabolism of naphthalene is similar to that of benzene, starting with an enzymatic epoxidation of the aromatic ring:

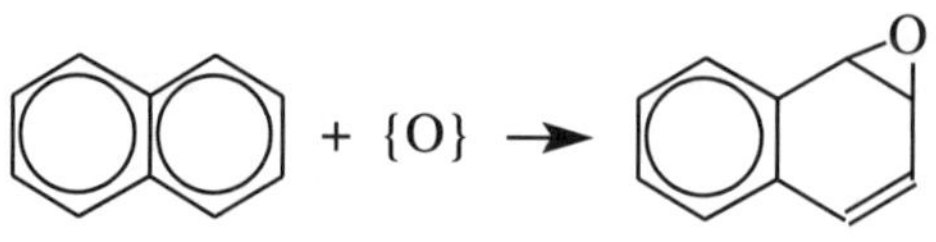

$$\text{naphthalene} + \{O\} \rightarrow \text{naphthalene-1,2-oxide} \quad (13.6.1)$$

followed by a nonenzymatic rearrangement to 1-naphthol:

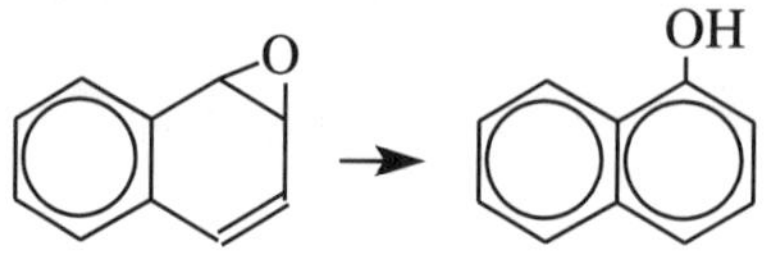

$$\text{naphthalene-1,2-oxide} \rightarrow \text{1-naphthol} \quad (13.6.2)$$

or addition of water to produce naphthalene-1,2-dihydrodiol through the action of epoxide hydrase enzyme:

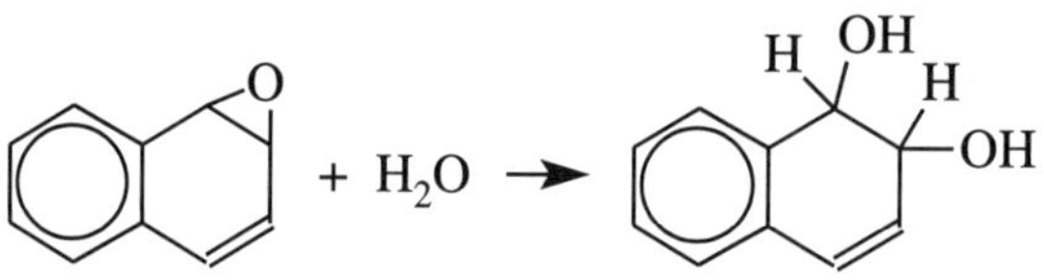

$$\text{naphthalene-1,2-oxide} + H_2O \rightarrow \text{naphthalene-1,2-dihydrodiol} \quad (13.6.3)$$

Elimination of the metabolized naphthalene from the body may occur as a mercapturic acid, preceded by the glutathione S-transferase-catalyzed formation of a glutathione conjugate.

13.6.2 Toxic Effects of Naphthalene

Exposure to naphthalene can cause a severe hemolytic crisis in some individuals with a genetically linked metabolic defect associated with insufficient activity of the glucose-6-phosphate dehydrogenase enzyme in red blood cells.[10] Effects include anemia and marked reductions in red cell count, hemoglobin, and hematocrit. Contact of naphthalene with skin can result in skin irritation or severe dermatitis in sensitized individuals. In addition to the hemolytic effects just noted, both inhalation and ingestion of naphthalene can cause headaches, confusion, and vomiting. Kidney failure is usually the ultimate cause of death in cases of fatal poisonings.

Naphthalene may adversely affect the eye, causing cortical cataracts and retinal degeneration.[11] These affects are attributed to the naphthalene dihydrodiol metabolite (see the product of reaction 13.6.3).

13.7 POLYCYCLIC AROMATIC HYDROCARBONS

Benzo(a)pyrene (Figure 13.1) is the most studied of the polycyclic aromatic hydrocarbons (PAHs). These compounds are formed by the incomplete combustion of other hydrocarbons so that hydrogen is consumed in the preferential formation of H_2O. The condensed aromatic ring system of the PAH compounds produced is the thermodynamically favored form of the hydrogen-deficient, carbon-rich residue. To cite an extreme example, the H:C ratio in methane (CH_4) is 4:1, whereas in benzo(a)pyrene ($C_{20}H_{12}$) it is only 3:5.

There are many conditions of partial combustion and pyrolysis that favor production of PAH compounds, and they are encountered abundantly in the atmosphere, soil, and elsewhere in the environment. Sources of PAH compounds include engine exhausts, wood stove smoke, cigarette smoke, and charbroiled food. Coal tars and petroleum residues have high levels of PAHs.

13.7.1 PAH Metabolism

The metabolism of PAH compounds is mentioned here with benzo(a)pyrene as an example. Several steps lead to the formation of the carcinogenic metabolite product of benzo(a)pyrene. After an initial oxidation to form the 7,8-epoxide, the 7,8-diol is produced through the action of epoxide hydrase enzyme, as shown by the following reaction:

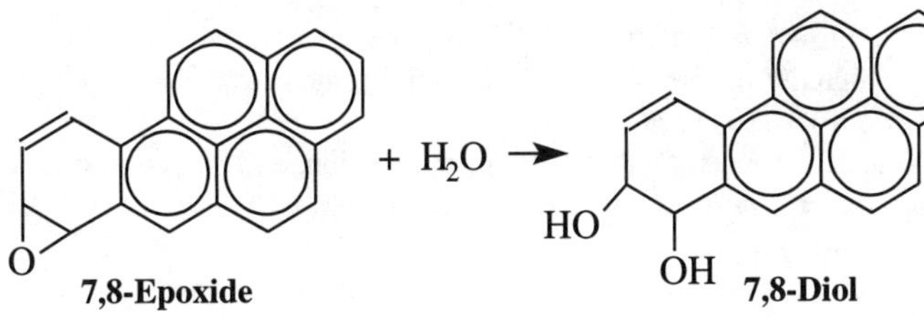

$$\text{7,8-Epoxide} + H_2O \rightarrow \text{7,8-Diol} \tag{13.7.1}$$

The microsomal mixed-function oxidase enzyme system further oxidizes the diol to the carcinogenic 7,8-diol-9,10-epoxide:

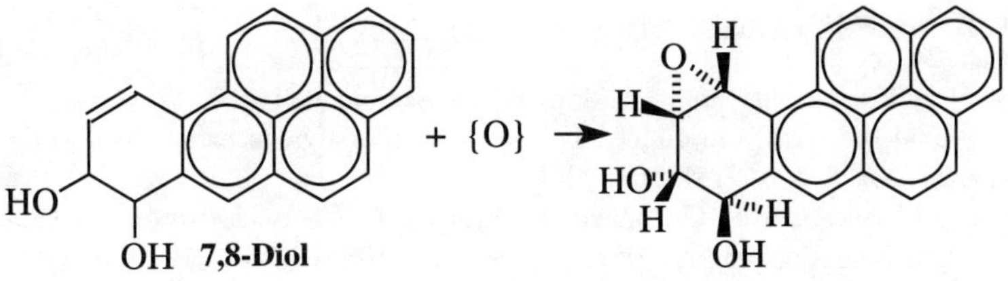

$$\text{7,8-Diol} + \{O\} \rightarrow \text{7,8-Diol-9,10-epoxide, carcinogenic (+)anti-isomer}$$

Several isomers of the 7,8-diol-9,10-epoxide are formed, depending on the orientations of the epoxide and OH groups relative to the plane of the molecule. The (+)antiisomer is the one that is regarded as carcinogenic based on its demonstrated mutagenicity, ability to bind with DNA, and extreme pulmonary carcinogenicity to newborn mice.[12]

Because of inhalation of smoke, especially tobacco smoke, the lungs are the most likely sites of cancer from exposure to PAH compounds. However, these compounds are also found in foods cooked under direct exposure to pyrolysis conditions and are suspected of causing cancer in the alimentary canal. Extraordinarily high rates of esophageal cancer have been observed in Linxian, China, and may be attributable to PAHs from unvented cookstoves.[13] In this study, the glucuronide conjugate of 1-hydroxypyrene was monitored as a biomarker of exposure to PAH compounds (Figure 13.11).

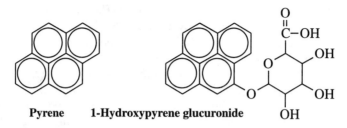

Figure 13.11 Pyrene, a common PAH compound, and the 1-hydroxypyrene glucuronide conjugate that may serve as a biomarker of exposure to pyrene.

REFERENCES

1. Threshold limit values (TLVs) for chemical substances and physical agents and biological exposure indices (BEIs) 2002, American Conference of Governmental Industrial Hygienists, Cincinnati, OH, 2002.
2. Kennedy, G.L., Setting a threshold limit value (TLV): the process, *Chem. Health Safety*, 8, 13–15, 2001.
3. Nessel, C.S., A comprehensive evaluation of the carcinogenic potential of middle distillate fuels, *Drug Chem. Toxicol.*, 22, 165–180, 1999.
4. Rozman, K.K. and Klaassen, C.D., Adsorption, distribution, and excretion of toxicants, in *Casarett and Doull's Toxicology: The Basic Science of Poisons*, 6th ed., Klaassen, C.D., Ed., McGraw-Hill, New York, 2001, chap. 8, pp. 107–132.
5. Bond, J.A. and Medinsky, M.A., Insights into the toxicokinetics and toxicodynamics of 1,3-butadiene, *Chem. Biol. Interact.*, 135/136, 599–614, 2001.
6. Boogaard, P.J., van Sittert, N.J., and Megens, H.J.J.J., Urinary metabolites and hemoglobin adducts as biomarkers of exposure to 1,3-butadiene: a basis for 1,3-butadiene cancer risk assessment, *Chem. Biol. Interact.*, 135/136, 695–701, 2001.
7. Bruckner, J.V. and Warren, D.A., Toxic effects of solvents and vapors, in *Casarett and Doull's Toxicology: The Basic Science of Poisons*, 6th ed., Klaassen, C.D., Ed., McGraw-Hill, New York, 2001, chap. 24, pp. 869–916.
8. Tornero-Velez, R. and Rappaport, S.M., Physiological modeling of the relative contributions of styrene-7,8-oxide derived from direct inhalation and from styrene metabolism to the systemic dose in humans, *Toxicol. Sci.*, 64, 151–161, 2001.
9. Rappaport, S.M. and Yeowell-O'Connell, K., Protein adducts as dosimeters of human exposure to styrene, styrene-7,8-oxide, and benzene, *Toxicol. Lett.*, 108, 117–126, 1999.
10. Gosselin, R.E., Smith, R.P., and Hodge, H.C., Naphthalene, in *Clinical Toxicology of Commercial Products*, 5th ed., Williams & Wilkins, Baltimore, 1984, pp. III-307–III-311.
11. Fox, D.A. and Boyes, W.K., Toxic responses of the ocular and visual system, in *Casarett and Doull's Toxicology: The Basic Science of Poisons*, 6th ed., Klaassen, C.D., Ed., McGraw-Hill, New York, 2001, chap. 17, pp. 565–595.
12. Rubin, H., Synergistic mechanisms in carcinogenesis by polycyclic aromatic hydrocarbons and by tobacco smoke: a biohistorical perspective with updates, *Carcinogenesis*, 22, 1903–1930, 2001.
13. Roth, M.J. et al., High urine 1-hydroxpyrene glucuronide concentrations in Linxian, China, an area of high risk for squamous esophageal cancer, *Biomarkers*, 6, 381–386, 2001.

QUESTIONS AND PROBLEMS

1. Using compounds other than those shown in Figure 13.1, give examples of each of the following kinds of hydrocarbons: (1) alkanes, (2) unsaturated nonaromatic hydrocarbons, (3) aromatic hydrocarbons, (4) polycyclic aromatic hydrocarbons with multiple rings, and (5) mixed hydrocarbons.
2. What kind of carbon–carbon bond characterizes alkanes? What kind of carbon–carbon bond characterizes other types of hydrocarbons?

TOXIC ORGANIC COMPOUNDS AND HYDROCARBONS

3. Give examples of hydrocarbons having the following general formulas: C_nH_{2n+2}, C_nH_{2n}, C_nH_n, and C_nH_x, where x is a number less than n.
4. What are the two most important reactions of alkanes? What kind of additional reaction is possible with alkenes? What may the latter have to do with the toxicological chemistry of alkenes?
5. What kind of reaction is shown below? What is the organic reactant? What is the product? What is the special toxicological significance of the product?

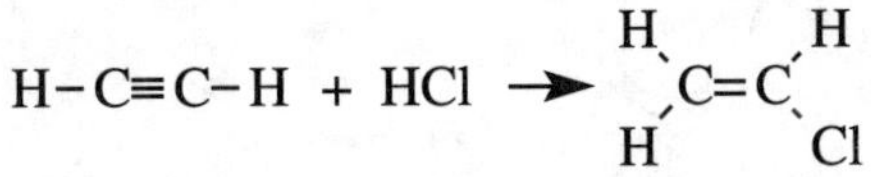

$$H-C\equiv C-H + HCl \rightarrow \underset{H}{\overset{H}{C}}=\underset{Cl}{\overset{H}{C}}$$

6. What structural phenomenon may be shown by the following formulas? What is its toxicological significance?

$$\underset{Cl}{\overset{H}{C}}=\underset{Cl}{\overset{H}{C}} \qquad \underset{Cl}{\overset{H}{C}}=\underset{H}{\overset{Cl}{C}}$$

7. Describe the special characteristics of aromaticity.
8. Explain the significance of the following formula:

$$E = \frac{C_a T_a + C_b T_b + \cdots + C_n T_n}{8}$$

9. What is the main toxicological characteristic of low-molecular-mass alkanes? What condition may be caused by exposure to somewhat higher-molecular-mass alkanes, such as *n*-hexane? How is this condition caused?
10. Consider the following reactions:

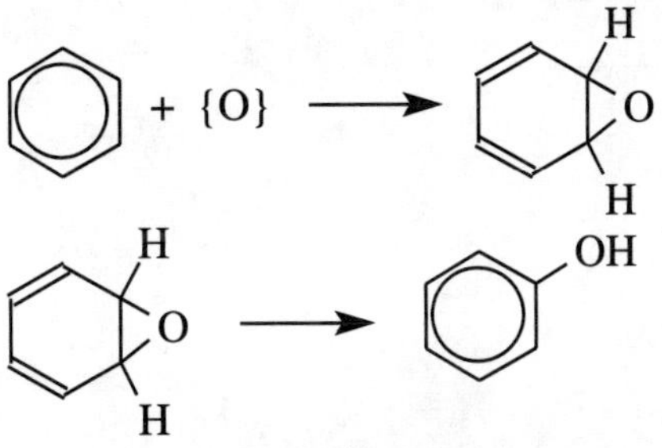

Discuss these reactions in terms of their significance for benzene toxicity and toxicological chemistry, phase I reactions, phase II reactions, and other aspects pertinent to benzene's effects on the body.
11. What is the formula of acetylene? What are its main toxicological effects?
12. What are the major acute toxicological effects of benzene? How does benzene exposure usually occur? How does benzene affect the central nervous system? At what levels of exposure are the acute toxicological effects manifested?
13. What are the chronic toxicological effects of benzene? What kinds of blood abnormalities are caused by benzene exposure? How does benzene toxicity affect white cell count? How does it affect bone marrow?
14. What may be said about the vapor pressure and water solubilities of benzene as they influence its toxicity?

15. In what important respects are the toxicological chemistry and toxicity of toluene quite different from those of benzene? How is hippuric acid formed from toluene?
16. What are the major toxicological chemical and toxicological aspects of naphthalene?
17. Discuss what the following shows about the toxicological chemistry and toxicity of some important polycyclic aromatic hydrocarbons:

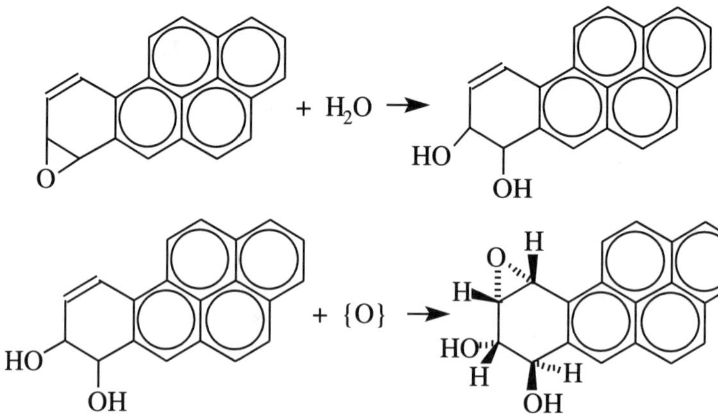

CHAPTER 14

Organooxygen Compounds

14.1 INTRODUCTION

A very large number of organic compounds and natural products, many of which are toxic, contain oxygen in their structures. This chapter concentrates on organic compounds that have oxygen covalently bonded to carbon. Organic compounds in which oxygen is bonded to nitrogen, sulfur, phosphorus, and the halogens are discussed in Chapters 15 to 18.

14.1.1 Oxygen-Containing Functional Groups

As shown in Table 1.4 and Figure 14.1, there are several kinds of oxygen-containing functional groups in organic compounds. In general, the organooxygen compounds can be classified according to the degree of oxygenation, location of oxygen on the hydrocarbon moiety, presence of unsaturated bonds in the hydrocarbon structure, and presence or absence of aromatic rings. Some of the features of organooxygen compounds listed above can be seen from an examination of some of the oxidation products of propane in Figure 14.1. Some organooxygen compounds discussed in this chapter are made from the bonding together of two of the many molecules shown in Figure 14.1.

14.2 ALCOHOLS

This section discusses the toxicological chemistry of the **alcohols**, oxygenated compounds in which the hydroxyl functional group is attached to an aliphatic or olefinic hydrocarbon skeleton. The phenols, which have –OH bonded to an aromatic ring, are covered in Section 14.3. The three lightest alcohols — methanol, ethanol, and ethylene glycol (shown in Figure 14.2) — are discussed individually in some detail because of their widespread use and human exposure to them. The higher alcohols, defined broadly as those containing three or more carbon atoms per molecule, are discussed as a group.

14.2.1 Methanol

Methanol, also called methyl alcohol and once commonly know as wood alcohol, is a clear, volatile liquid mp, –98°C; bp, 65°C). Until the early 1900s, the major commercial source of methanol was the destructive distillation (pyrolysis) of wood, a process that yields a product contaminated with allyl alcohol, acetone, and acetic acid. Now methanol is synthesized by the following reaction of hydrogen gas and carbon monoxide, both readily obtained from natural gas or coal gasification:

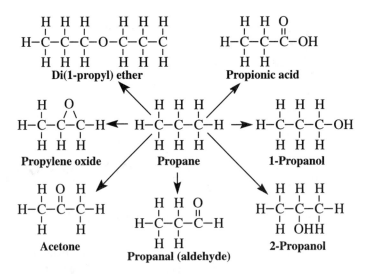

Figure 14.1 Oxygenated derivatives of propane.

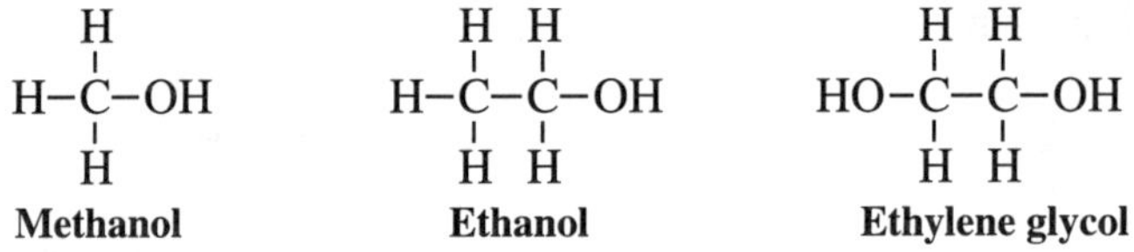

$$\underset{\text{Methanol}}{\text{H–C–OH}} \quad \underset{\text{Ethanol}}{\text{H–C–C–OH}} \quad \underset{\text{Ethylene glycol}}{\text{HO–C–C–OH}}$$

Figure 14.2 Three lighter alcohols with particular toxicological significance.

$$CO + 2H_2 \xrightarrow[\text{catalyst}]{\text{Metal}} CH_3OH \qquad (14.2.1)$$

The greatest use for methanol is in the manufacture of formaldehyde (see Section 14.5). Additional uses include the synthesis of other chemicals, including acetic acid, applications as an organic solvent, and addition to unleaded gasoline for fuel, antifreeze, and antiknock properties.

Methanol has been responsible for the deaths of many humans who ingested it accidentally or as a substitute for beverage ethanol. The fatal human dose is believed to lie between 50 and 250 g. In the body, methanol undergoes metabolic oxidation to formaldehyde and formic acid:[1]

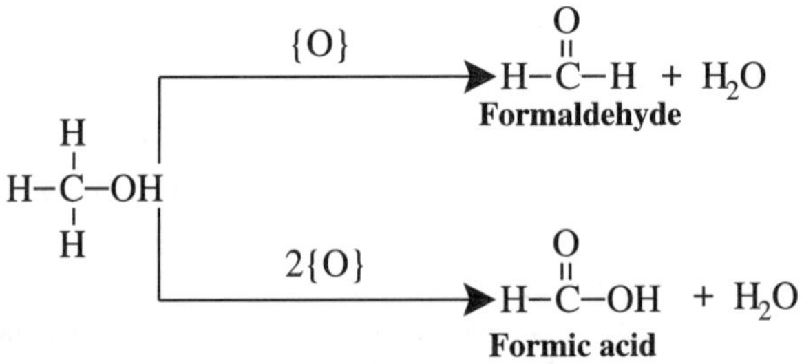

$$\text{H–C–OH} \begin{cases} \xrightarrow{\{O\}} \text{H–C–H (Formaldehyde)} + H_2O \\ \xrightarrow{2\{O\}} \text{H–C–OH (Formic acid)} + H_2O \end{cases} \qquad (14.2.2)$$

The formic acid product of this reaction causes acidosis, with major adverse effects on the central nervous system, retina, and optic nerve.[2] In cases of acute exposure, an initially mild inebriation is followed in about 10 to 20 h by unconsciousness and cardiac depression; death may occur. For sublethal doses, initial symptoms of visual dysfunction may clear temporarily, followed by blindness from deterioration of the optic nerve and retinal ganglion cells. Chronic exposures to lower levels of methanol may result from fume inhalation.

Methanol occurs in some foods. Distilled fruit spirits such as those from the fermentation of Bartlett pears contain some methanol. This has led to European standards for methanol limits in distilled fruit spirits. The levels of methanol can be reduced by appropriate adjustment of fermentation conditions and the distillation processes used.[3]

14.2.2 Ethanol

Ethanol, ethyl alcohol (mp, –114°C; bp, 78°C), is a clear, colorless liquid widely used as a beverage ingredient, synthetic chemical, solvent, germicide, antifreeze, and gasoline additive. It is produced by the fermentation of carbohydrates or by the hydration of ethylene, as shown by the following two reactions:

$$C_6H_{12}O_6 \xrightarrow{\text{Yeasts}} 2C_2H_5OH + 2CO_2 \qquad (14.2.3)$$

$$\underset{H}{\overset{H}{>}}C=C\underset{H}{\overset{H}{<}} + H_2O \xrightarrow{\text{Mixed-bed catalyst}} H-\underset{H}{\overset{H}{\underset{|}{C}}}-\underset{H}{\overset{H}{\underset{|}{C}}}-OH \qquad (14.2.4)$$

Ethanol misused in beverages is responsible for more deaths than any other chemical when account is taken of chronic alcoholism, vehicle fatalities caused by intoxicated drivers, and alcohol-related homicides. Chronic alcoholism is a distinct disease arising from generally long-term systemic effects of the ingestion of alcohol. Often the most damaging manifestation of chronic alcohol toxicity consists of adverse effects on the liver (alcohol-induced hepatotoxicity).[4] Some of these adverse effects are due to oxidative stress and lipid peroxidation. Other effects may result from the formation of protein adducts of acetaldehyde and 1-hydroxyethyl radical, leading to immunogenic processes that damage the liver.

Ethanol has a range of acute effects, normally expressed as a function of percent blood ethanol. In general, these effects are related to central nervous system depression. Mild effects such as decreased inhibitions and slowed reaction times begin to appear at about 0.05% blood ethanol. Most individuals are clinically intoxicated at a level of 0.15 to 0.3% blood ethanol; in the 0.3 to 0.5% range, stupor may be produced; and at 0.5% and above, coma and often death occur.

Metabolically, ethanol is oxidized first to acetaldehyde (Section 14.6), then to CO_2. The overall oxidation rate is faster than that for methanol.

In addition to absorption through the gastrointestinal tract, ethanol can be absorbed by the alveoli of the lungs. Symptoms of intoxication can be observed from inhalation of air containing more than 1000 ppm ethanol.

One of the more serious toxic effects of ethanol is its role as a teratogen when ingested during pregnancy, causing **fetal alcohol syndrome**. Fetal alcohol syndrome is manifested by a number of effects, with perhaps more to be discovered. One of the more obvious of these is the occurrence of defects in the head and face structure. Fetal alcohol syndrome is also manifested by central nervous system abnormalities, and it is one of the leading causes of nongenetic mental retardation. It also retards growth, both prenatally and postnatally. Ethanol and its first metabolite, acetaldehyde, rapidly cross the placenta and have adverse effects on its function. Both of these compounds are teratogens, and both are toxic to embryos.

14.2.3 Ethylene Glycol

Although used in cosmetics, chemical synthesis, and other applications, most ethylene glycol is consumed as the major ingredient of antifreeze and antiboil formulations for automobile radiators. Ethylene glycol (mp, –13°C; bp, 198°C) is synthesized by the oxidation of ethylene to ethylene oxide, followed by hydrolysis of the latter compound:

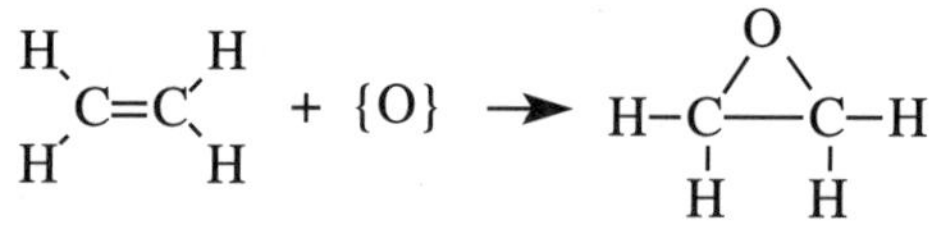

$$\begin{array}{c}H\\H\end{array}C=C\begin{array}{c}H\\H\end{array} + \{O\} \rightarrow H-\underset{\underset{H}{|}}{C}\overset{O}{-}\underset{\underset{H}{|}}{C}-H \qquad (14.2.5)$$

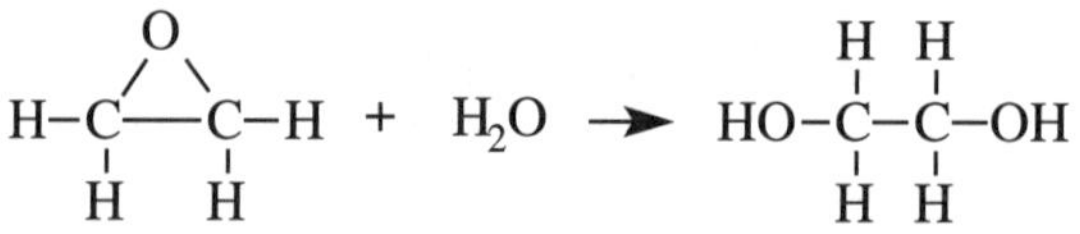

$$H-\underset{\underset{H}{|}}{C}\overset{O}{-}\underset{\underset{H}{|}}{C}-H + H_2O \rightarrow HO-\underset{\underset{H}{|}}{\overset{H}{|}}{C}-\underset{\underset{H}{|}}{\overset{H}{|}}{C}-OH \qquad (14.2.6)$$

Toxic exposures to ethylene glycol are rare because of its low vapor pressure, but inhalation of droplets can be very dangerous. Significant numbers of human fatalities attributable to ethylene glycol poisoning have been documented.[5] From the limited amount of information available, the LD_{50} for humans has been estimated to be about 110 g. Ingested ethylene glycol initially stimulates the central nervous system, and then depresses it. Victims may suffer acidemia from the presence of the intermediate metabolite glycolic acid. Kidney damage occurs later, and it can be fatal. The kidneys are harmed because of the deposition of solid calcium oxalate, resulting from the following overall process:

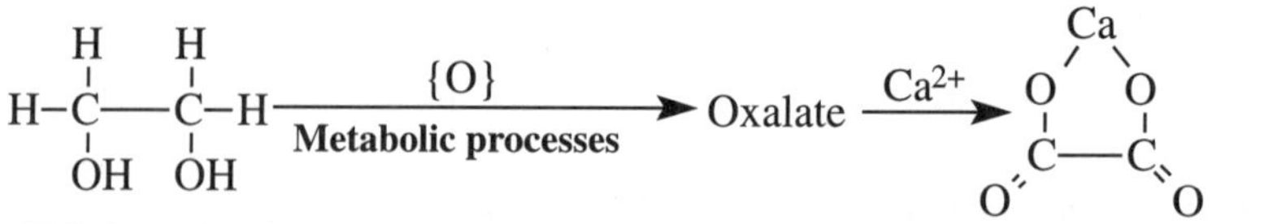

$$\underset{\text{Ethylene glycol}}{H-\underset{\underset{OH}{|}}{\overset{H}{|}}{C}-\underset{\underset{OH}{|}}{\overset{H}{|}}{C}-H} \xrightarrow[\text{Metabolic processes}]{\{O\}} \text{Oxalate} \xrightarrow{Ca^{2+}} \underset{\text{Calcium oxalate (solid)}}{\text{structure}} \qquad (14.2.7)$$

Important intermediates in this process are glycoaldehyde, glycolate, and glyoxalate:

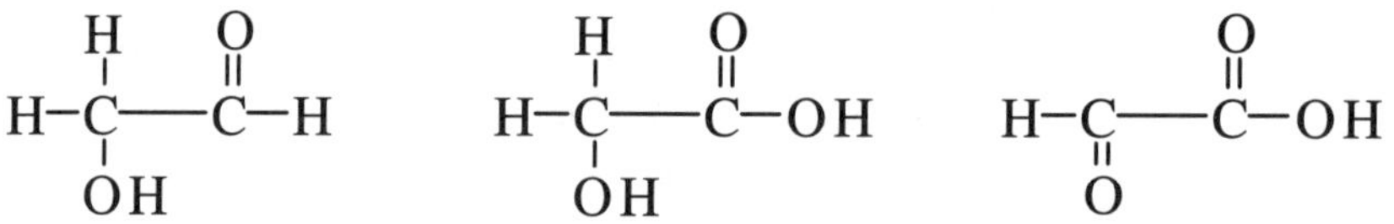

Glycoaldehyde   **Glycolic acid**   **Glyoxylic acid**

Kidney failure from the metabolic formation of calcium oxalate has been especially common in cat species, which have voracious appetites for ethylene glycol in antifreeze. Deposits of solid calcium oxalate have also been observed in the liver and brain tissues of victims of ethylene glycol poisoning.

14.2.4 The Higher Alcohols

Numerous alcohols containing three or more carbon atoms are used as solvents and chemical intermediates and for other purposes. Exposure to these compounds can occur, and their toxicities are important. Some of the more significant of these alcohols are listed in Table 14.1.

ORGANOOXYGEN COMPOUNDS

Table 14.1 Some Alcohols with Three or More Carbons

Alcohol Name and Formula	Properties
2-Propanol, $CH_3CHOHCH_3$	Isopropyl alcohol; used as rubbing alcohol and food additive; irritant; narcotic; relatively low toxicity
Allyl alcohol, $CH_2=CHCH_2OH$	Olefinic alcohol; pungent odor; strongly irritating to eyes, mouth, lungs
1-Butanol, $CH_3(CH_2)_2CH_2OH$	Butyl alcohol or *n*-butanol; irritant; limited toxicity because of low vapor pressure
1-Pentanol, $CH_3(CH_2)_3CH_2OH$	Amyl alcohol; liquid; bp, 138°C; irritant, causes headache and nausea; low vapor pressure and low water solubility reduce toxicity hazard
1-Decanol, $CH_3(CH_2)_8CH_2OH$	Viscous liquid; bp, 233°C; low acute toxicity
2-Ethylhexanol, $CH_3(CH_2)_3CH-(C_2H_5)CH_2OH$	2–Ethylhexyl alcohol; important industrial solvent; toxicity similar to those of butyl alcohols

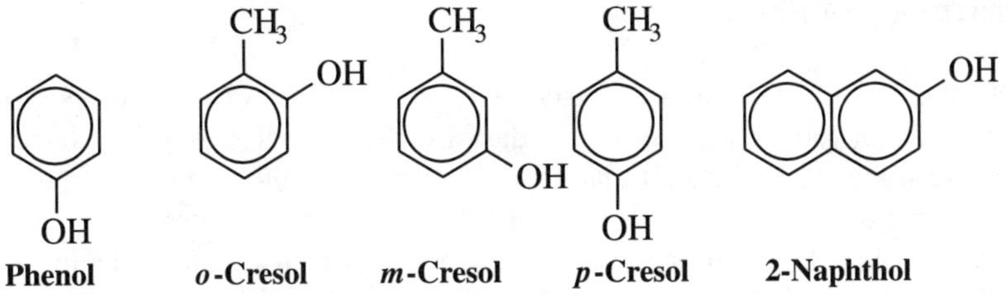

Figure 14.3 Major phenolic compounds.

An important alcohol in toxicology studies is ***n*-octanol**, $CH_3(CH_2)_6CH_2OH$. This compound is applied to the measurement of the **octanol–water partition coefficient**, which is used to estimate how readily organic toxicants are transferred from water to lipids, a tendency usually associated with ability to cross cell membranes and cause toxic effects. As just one example, the octanol–water partition coefficient can be used to estimate the tendency of organic compounds to be taken up from water to the lipid gill tissue of fish.

14.3 PHENOLS

Phenols are aryl alcohols in which the –OH group is bonded to an aromatic hydrocarbon moiety. The simplest of these compounds is phenol, in which the hydrocarbon portion is the phenyl group. Figure 14.3 shows some of the more important phenolic compounds. Phenols have properties that are quite different from those of the aliphatic and olefinic alcohols. Many important phenolic compounds have nitro groups (–NO_2) and halogen atoms (particularly Cl) bonded to the aromatic rings. These substituents may strongly affect chemical and toxicological behavior; such compounds are discussed in Chapters 15 and 16.

14.3.1 Properties and Uses of Phenols

The physical properties of the phenols listed in Figure 14.3 are summarized briefly in Table 14.2. These phenolic compounds are weak acids that ionize to phenolate ions in the presence of base:

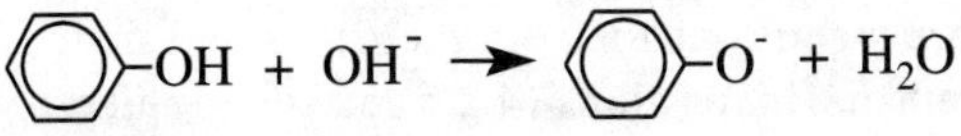

$$\text{C}_6\text{H}_5\text{-OH} + \text{OH}^- \rightarrow \text{C}_6\text{H}_5\text{-O}^- + \text{H}_2\text{O} \tag{14.3.1}$$

Table 14.2 Properties of Major Phenolic Compounds

Compound	Properties
Phenol	Carbolic acid; white solid; characteristic odor; mp, 41°C; bp, 182°C
m-Cresol	Often occurs mixed with *ortho-* and *para-* analogs as cresol or cresylic acid; light yellow liquid; mp, 11°C; bp, 203°C
o-Cresol	Solid; mp, 31°C; bp, 191°C
p-Cresol	Crystalline solid with phenolic odor; mp, 36°C; bp, 202°C
1-Naphthol	Alpha-naphthol; colorless solid; mp, 96°C; bp, 282°C
2-Naphthol	Beta-naphthol; mp, 122°C; bp, 288°C

Phenols are extracted commercially from coal tar into aqueous base as the phenolate ions. The major commercial use of phenol is in the manufacture of phenolic resin polymers, usually with formaldehyde. Phenols and cresols are used as antiseptics and disinfectants in areas such as barns where the phenol odor can be tolerated. Phenol was the original antiseptic used on wounds and in surgery, starting with the work of Lord Lister in 1885.

14.3.2 Toxicology of Phenols

Generally, the phenols have similar toxicological effects. Phenol is a protoplasmic poison, so it damages all kinds of cells. Early medical studies that demonstrated asepsis with phenol revealed its toxicity as well. Phenol is alleged to have caused "an astonishing number of poisonings" since it came into general use.[6]

Fatal doses of phenol may be absorbed through the skin. Its acute toxicological effects are predominantly on the central nervous system. Death can occur as early as a half hour after exposure. Key organs damaged by chronic exposure to phenol include the spleen, pancreas, and kidneys. Lung edema can also occur.

Some phenol is eliminated from the body as the unchanged molecular compound, although most is metabolized prior to excretion. As noted in Section 7.2.1, phase II reactions in the body result in the conjugation of phenol to produce sulfates and glucuronides. These water-soluble metabolic products are eliminated via the kidneys. Urinary phenyl glucuronide may be measured to monitor exposure to phenol.[7]

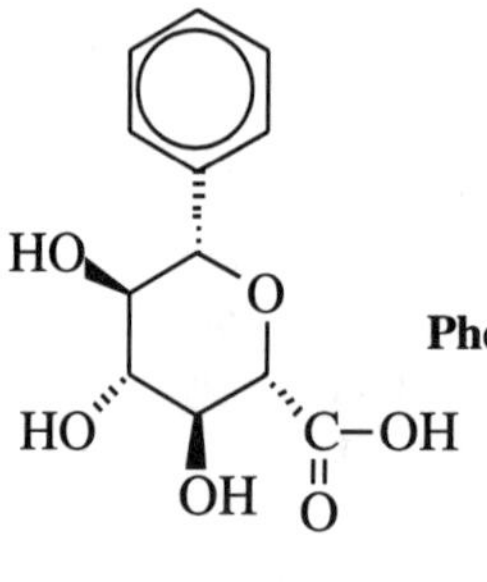

Phenyl glucuronide

Oral doses of naphthols can be fatal. Acute poisoning by these compounds can cause severe gastrointestinal disturbances, kidney malfunction, circulatory system failure, and convulsions. Naphthols can be absorbed through the skin, one effect of which can be eye damage involving the cornea and lens.

14.4 OXIDES

Hydrocarbon **oxides** are significant for both their uses and their toxic effects.[8] As shown for ethylene oxide (1,2-epoxyethane) in reactions 14.2.5 and 14.2.6 and propylene oxide (1,2-epoxypro-

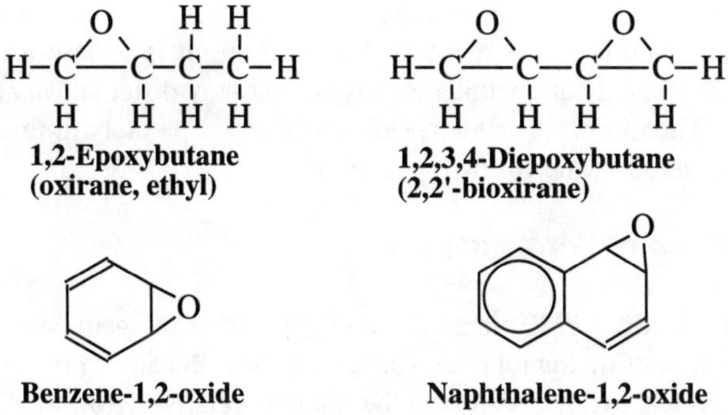

Figure 14.4 Some common epoxide compounds.

1,2-Epoxybutane (oxirane, ethyl)

1,2,3,4-Diepoxybutane (2,2'-bioxirane)

Benzene-1,2-oxide

Naphthalene-1,2-oxide

pane) in Figure 14.1, these compounds are characterized by an **epoxide** functional group consisting of an oxygen atom bridging two adjacent C atoms. As discussed in Section 4.2, the metabolic formation of such a group is called epoxidation and is a major type of the phase I reactions of xenobiotic compounds. In addition to ethylene and propylene oxides, four other common hydrocarbon oxides are shown in Figure 14.4.

Ethylene oxide (mp, –111°C; bp, 11°C) is a colorless, sweet-smelling, flammable, explosive gas. It is used as a chemical intermediate, sterilant, and fumigant. It has a moderate to high toxicity, is a mutagen, and is carcinogenic to experimental animals. When inhaled, ethylene oxide causes respiratory tract irritation, headache, drowsiness, and dyspnea. At higher levels, cyanosis, pulmonary edema, kidney damage, peripheral nerve damage, and death can result from inhalation of this compound. Animal studies have shown that inhalation of ethylene oxide causes a variety of tumors, raising concerns that it may be a human carcinogen.[8]

Propylene oxide (mp, –104°C; bp, 34°C) is a colorless, reactive, volatile liquid with uses similar to those of ethylene oxide. Its toxic effects are like those of ethylene oxide, though less severe. The properties of butylene oxide (liquid; bp, 63°C) are also similar to those of ethylene oxide. The oxidation product of 1,3-butadiene, 1,2,3,4-butadiene epoxide, is a direct-acting (primary) carcinogen.

As discussed in Section 13.5, benzene-1,2-oxide is an intermediate in the biochemical oxidation of benzene. It is probably responsible for the toxicity of benzene. It is hydrolyzed by the action of epoxide hydratase to the dihydrodiol shown below:

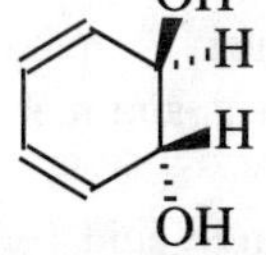

Benzene *trans*-1,2-dihydrodiol

Naphthalene-1,2-oxide is a metabolic intermediate in the oxidation of naphthalene mediated by cytochrome P-450.

14.5 FORMALDEHYDE

Aldehydes and ketones are compounds that contain the carbonyl (C=O) group. Of these compounds, **formaldehyde**,

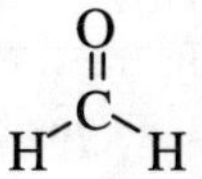

Formaldehyde

is uniquely important for several reasons. Among these are that its physical and chemical properties are atypical of aldehydes in some important respects. Furthermore, it is widely used in a number of applications and exhibits toxicological chemical behavior that may differ substantially from that of other common aldehydes. Therefore, formaldehyde is discussed separately in this section. Other aldehydes and ketones are covered in the following section.

14.5.1 Properties and Uses of Formaldehyde

Formaldehyde (mp, $-118°C$; bp, $-19°C$) is a colorless gas with a pungent, suffocating odor. It is manufactured by the oxidation of methanol over a silver catalyst. Because it undergoes a number of important reactions in chemical synthesis and can be made at relatively low cost, formaldehyde is one of the most widely used industrial chemicals. In the pure form it polymerizes with itself to give hydroxyl compounds, ketones, and other aldehydes. Because of this tendency, commercial formaldehyde is marketed as a 37 to 50% aqueous solution containing some methanol called **formalin**. Formaldehyde is a synthesis intermediate in the production of resins (particularly phenolic resins), as well as a large number of synthetic organic compounds, such as chelating agents. Formalin is employed in antiseptics, fumigants, tissue and biological specimen preservatives, and embalming fluid.

14.5.2 Toxicity of Formaldehyde and Formalin

The fact that formaldehyde is produced by natural processes in the environment and in the body would suggest that it might not be very toxic. However, such is not the case in that formaldehyde exhibits a number of toxic effects.

Exposure to inhaled formaldehyde via the respiratory tract is usually to molecular formaldehyde vapor, whereas exposure by other routes is usually to formalin. Exposure to formaldehyde vapor can occur in industrial settings. In recent years, a great deal of concern has arisen over the potential for exposure in buildings to formaldehyde vapor evolved from insulating foams that were not properly formulated and cured or when these foams burn. Hypersensitivity can result from prolonged, continuous exposure to formaldehyde. Furthermore, animal experiments have shown formaldehyde to be a lung carcinogen.

The human LD_{50} for the ingestion of formalin has been estimated at around 45 g. Deaths have been caused by as little as about 30 g, and individuals have survived ingestion of about 120 g, although in at least one such case removal of the stomach was required. Ingestion results in violent gastrointestinal disturbances, including vomiting and diarrhea. Formaldehyde attacks the mucous membrane linings of both the respiratory and alimentary tracts and reacts strongly with functional groups in molecules.

Metabolically, formaldehyde is rapidly oxidized to formic acid (see Section 14.7), which is responsible in large part for its toxicity. Formaldehyde reacts by addition and condensation reactions with a variety of biocompounds, including DNA and proteins, and in so doing forms adducts and DNA–protein cross-links.[9] Formaldehyde is incorporated into proteins and nucleic acids as the $-CH_3$ group. Reactive formaldehyde has a short systemic lifetime of only about 1 min; its formic acid metabolic product has a longer metabolic lifetime.

14.6 ALDEHYDES AND KETONES

In **aldehydes** the carbonyl group, C=O, is attached to a C and H atom at the end of a hydrocarbon chain, and in a **ketone** it is bonded to two C atoms in the middle of a hydrocarbon chain or ring. The hydrocarbon portion of aldehydes and ketones may consist of saturated or unsaturated straight

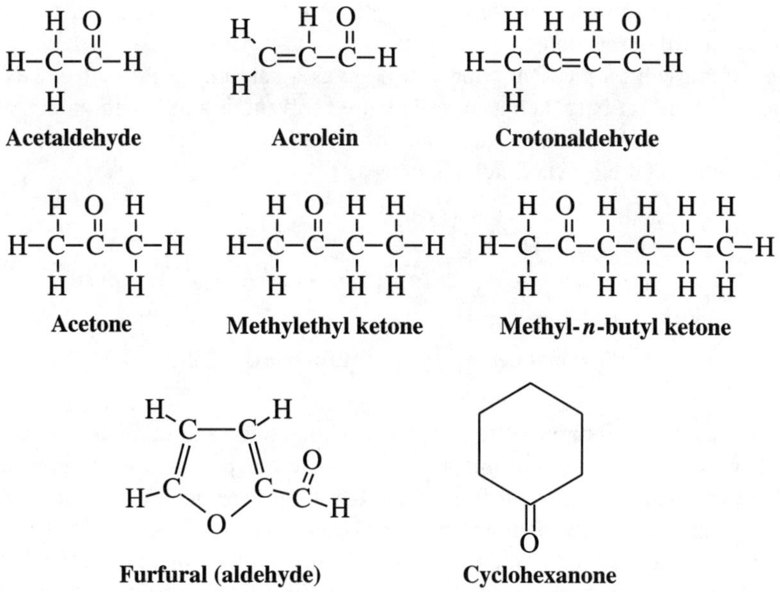

ORGANOOXYGEN COMPOUNDS

Figure 14.5 Aldehydes and ketones that are significant for their commercial uses and toxicological importance.

chains, branched chains, or rings. The structures of some important aldehydes and ketones are shown in Figure 14.5.

Both aldehydes and ketones are industrially important classes of chemicals. Aldehydes are reduced to make the corresponding alcohols and are used in the manufacture of resins, dyes, plasticizers, and alcohols. Some aldehydes are ingredients in perfumes and flavors. Several ketones are excellent solvents and are widely used for that purpose to dissolve gums, resins, laquers, nitrocellulose, and other substances.

14.6.1 Toxicities of Aldehydes and Ketones

In general, because of their water solubility and intensely irritating qualities, the lower aldehydes attack exposed moist tissue, particularly tissue in the eyes and mucous membranes of the upper respiratory tract. Because of their lower water solubility, the lower aldehydes can penetrate further into the respiratory tract and affect the lungs.

The toxicity of formaldehyde was discussed in the preceding section. The next higher aldehyde, acetaldehyde, is a colorless, volatile liquid (bp, 21°C). Toxicologically it acts as an irritant, and systemically as a narcotic to the central nervous system. Acrolein, a highly reactive alkenic aldehyde, is a colorless to light yellow liquid (bp, 52°C). It is a very reactive chemical that polymerizes readily. It is quite toxic by all routes of contact and ingestion. It has a choking odor and is extremely irritating to respiratory tract membranes. It is classified as an extreme lachrymator (substance that causes eyes to water). Because of this property, acrolein serves to warn of its own exposure. It can produce tissue necrosis, and direct contact with the eye can be especially hazardous. Crotonaldehyde is similarly dangerous and can cause burns to the eye cornea.

Metabolically, aldehydes are converted to the corresponding organic acids, as shown by the following general reaction:

$$R-\underset{\underset{H}{\|}}{\overset{O}{C}}-H + O \rightarrow R-\underset{\underset{H}{\|}}{\overset{O}{C}}-OH \qquad (14.6.1)$$

In mammals, the liver enzymes aldehyde dehydrogenase and aldehyde oxidase appear to be mainly responsible for this reaction.

Acetone is a liquid with a pleasant odor. It can act as a narcotic and dissolves fats from skin, causing dermatitis. Methyl-*n*-butyl ketone, a widely used solvent, is a mild neurotoxin. Methylethyl ketone is suspected of having caused neuropathic disorders in shoe factory workers.

Methylvinyl ketone and ethylvinyl ketone,

$$\underset{\text{Methylvinyl ketone}}{H_3C-\overset{O}{\overset{\|}{C}}-\overset{H}{\overset{|}{C}}=C\overset{H}{\underset{H}{\diagdown}}} \qquad \underset{\text{Ethylvinyl ketone}}{H_3C-\overset{H}{\underset{H}{\overset{|}{C}}}-\overset{O}{\overset{\|}{C}}-\overset{H}{\overset{|}{C}}=C\overset{H}{\underset{H}{\diagdown}}}$$

are both classified as α,β-unsaturated ketones. These compounds and α,β-unsaturated aldehydes, of which acrolein is an example, are mutagenic and therefore potentially carcinogenic. Human exposure to these compounds can result from a number of sources, including industrial chemicals (a purpose for which methyvinyl ketone is widely used), metabolites of industrial chemicals, pesticide metabolites, natural products, and pollutants. Ethylvinyl ketone is an especially common contaminant of foods, having been detected in meat, dairy products, fruit juices, kiwi fruit, and other foods. Both of these ketones have been found to form adducts with the guanine moiety in deoxyguanosine nucleoside and in 2'-deoxyguanosine 5'-monophosphate nucleotide (see Section 3.7). When inhaled, methylvinyl ketone is classified as a reactive, direct-acting gaseous irritant.[10]

14.7 CARBOXYLIC ACIDS

Carboxylic acids contain the –C(O)OH functional group bound to an aliphatic, olefinic, or aromatic hydrocarbon moiety. This section deals with those carboxylic acids that contain only C, H, and O. Carboxylic acids that contain other elements, such as trichloroacetic acid (a strong acid) or deadly poisonous monofluoroacetic acid, are discussed in later chapters. Some of the more significant carboxylic acids are shown in Figure 14.6.

Carboxylic acids are the oxidation products of aldehydes and are often synthesized by that route. Some of the higher carboxylic acids are constituents of oil, fat, and wax esters, from which they are prepared by hydrolysis. Carboxylic acids have many applications. Formic acid is used as a relatively inexpensive acid to neutralize base, in the treatment of textiles, and as a reducing agent. Acetic and propionic acids are added to foods for flavor and as preservatives. Among numerous other applications, these acids are also used to make cellulose plastics. Stearic acid acts as a dispersive agent and accelerator activator in rubber manufacture. Sodium stearate is a major ingredient of most soaps. Many preservative and antiseptic formulations contain benzoic acid. Large quantities of phthalic acid are used to make phthalate ester plasticizers (see Section 14.10). Acrylic acid and methacrylic acid (acrylic acid in which the alpha-hydrogen has been replaced with a –CH$_3$ group; see Figure 14.6) are used in large quantities to make acrylic polymers.

14.7.1 Toxicology of Carboxylic Acids

Concentrated solutions of formic acid are corrosive to tissue, much like strong mineral acids. In Europe, decalcifier formulations containing about 75% formic acid have been marketed for removing mineral scale. Children ingesting this material have suffered corrosive lesions to mouth and esophageal tissue. Although acetic acid is widely used in food preparation as a 4 to 6% solution in vinegar, pure acetic acid (glacial acetic acid) is extremely corrosive to tissue that it contacts.

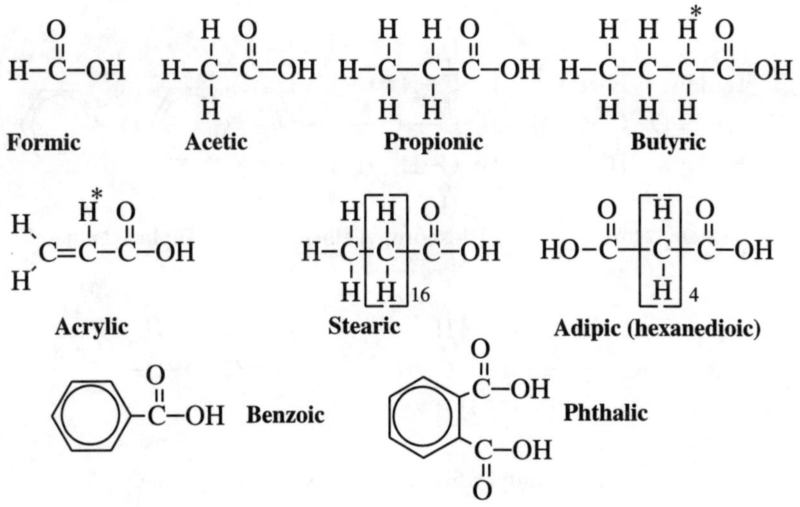

ORGANOOXYGEN COMPOUNDS

Figure 14.6 Some common carboxylic acids. The positions of the alpha-hydrogens have been marked with an asterisk for butyric and acrylic acids.

Acrylic and methacrylic acids are considered to be relatively toxic, both orally and by skin contact. In general, the presence of more than one carboxylic acid group per molecule, unsaturated bonds in the carbon skeleton, or the presence of a hydroxide group on the alpha-carbon position (see Figure 14.6) increases corrosivity and toxicity of carboxylic acids.

14.8 ETHERS

Three important classes of oxygenated organic compounds can be regarded as products of condensation of compounds containing the –OH group accompanied by a loss of H_2O, as shown by the following reaction:

$$R–OH + HO–R' \rightarrow R–O–R' + H_2O \qquad (14.8.1)$$

In this reaction, R–OH and HO–R' are either alcohols or carboxylic acids. When both are alcohols, R–O–R' is an ether; when one is an acid and the other an alcohol, the product is an ester; and when both are acids, an acid anhydride is produced. Ethers are discussed in this section, and the other two classes of products are discussed in the two sections that follow.

14.8.1 Examples and Uses of Ethers

An ether consists of two hydrocarbon moieties linked by an oxygen atom, as shown in Figure 14.7. Although diethyl ether is highly flammable, ethers are generally not very reactive. This property enables their uses in applications where an unreactive organic solvent is required. Some ethers form explosive peroxides when exposed to air, as shown by the example of diethyl ether peroxide in Figure 14.7.

Ethers are prominent members of a class of organic substances widely used as solvents, including hydrocarbons, chlorinated hydrocarbons, and alcohols, as well as ethers. Because of the widespread use of such solvents, human exposure is particularly likely.

Diethyl ether (mp, –116°C; bp, 34.6°C) is the most commercially important ether. It is used as a reaction medium, solvent, and extractant. The production of methyl *tert*-butyl ether increased markedly during the 1990s because of its application as an antiknock ingredient of unleaded

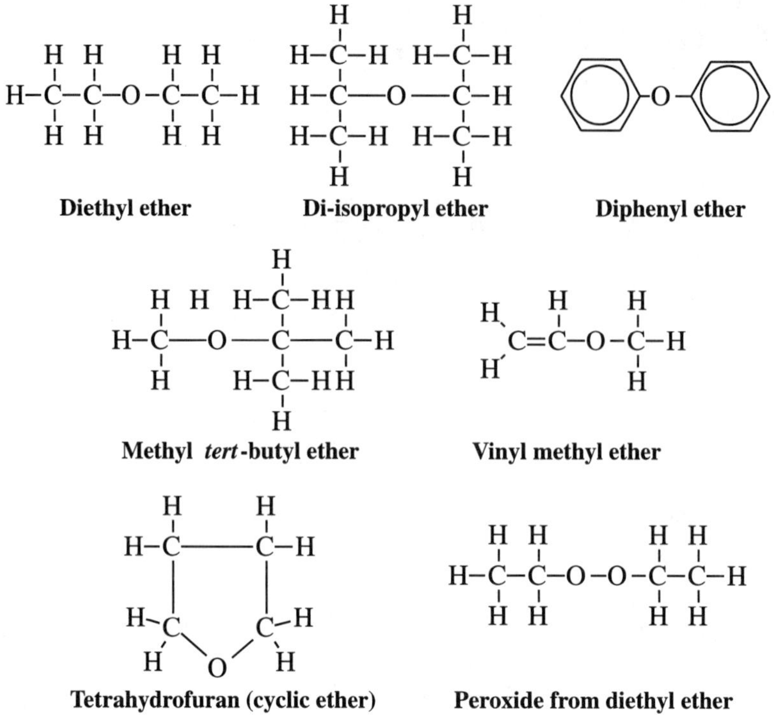

Figure 14.7 Structures of some common ethers.

gasoline, but its uses in this application are now being curtailed because it has become a troublesome water pollutant.

14.8.2 Toxicities of Ethers

Because of its volatility, the most likely route of exposure to diethyl ether is by inhalation. About 80% of this compound that gets into the body is eliminated unmetabolized as the vapor through the lungs. Diethyl ether is a central nervous system depressant, and for many years was the anesthetic of choice for surgery. At low doses, it causes drowsiness, intoxication, and stupor. Higher exposures result in unconsciousness and even death.

Compared to other classes of organic compounds, ethers have relatively low toxicities. This characteristic can be attributed to the low reactivity of the C–O–C functional group arising from the high strength of the carbon–oxygen bond. Like diethyl ether, several of the more volatile ethers affect the central nervous system. Hazards other than their toxicities tend to be relatively more important for ethers. These hazards are flammability and formation of explosive peroxides (especially with di-isopropyl ether).

14.9 ACID ANHYDRIDES

The most important carboxylic **acid anhydride** is acetic anhydride, the structure of which is

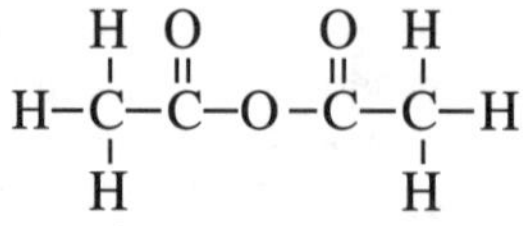

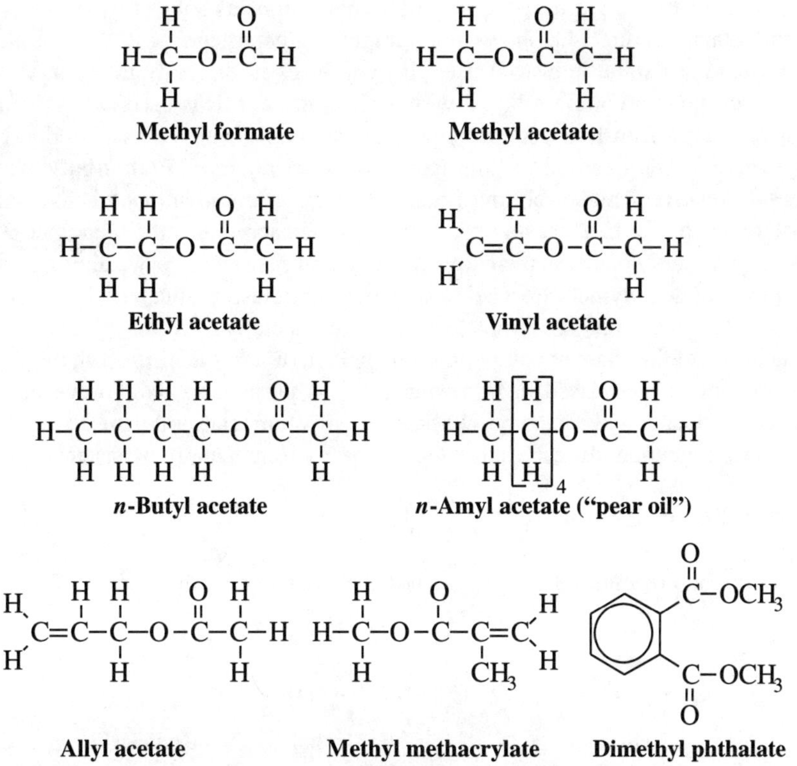

Figure 14.8 Some typical esters.

Annual world production of this chemical compound is on the order of a million metric tons. In chemical synthesis it functions as an acetylating agent (addition of $CH_3C(O)$ moiety). Its greatest single use is to make cellulose acetate, and it has additional applications in manufacturing textile sizing agents, the synthesis of salicylic acid (for aspirin manufacture), electrolytic polishing of aluminum, and the processing of semiconductor components.

14.9.1 Toxicological Considerations

In contrast to the relative safety of many ethers and esters, acetic anhydride is a systemic poison and especially corrosive to the skin, eyes, and upper respiratory tract. Levels in the air as low as 0.4 mg/m^3 adversely affect eyes, and contamination should be kept to less than one tenth that level in the workplace atmosphere. Blisters and burns that heal slowly result from skin exposure. Acetic anhydride has a very strong acetic acid odor that causes an intense burning sensation in the nose and throat that is accompanied by coughing. It is a powerful lachrymator. Fortunately, these unpleasant symptoms elicit a withdrawal response in exposed individuals.

14.10 ESTERS

Esters, such as those shown in Figure 14.8, are formed from an alcohol and acid, the reverse of reaction 14.10.1. Esters exhibit a wide range of biochemical diversity, and large numbers of them occur naturally. Fats, oils, and waxes are esters, as are many of the compounds responsible for odors and flavors of fruits, flowers, and other natural products. It follows that many esters are not toxic. Synthetic versions of many of the esters that occur naturally are produced for purposes such as flavoring ingredients. A number of esters that are not natural products have been synthesized for various purposes. Esters are used in industrial applications as solvents, plasticizers, lacquers, soaps, and surfactants. Figure 14.8 shows some representative esters.

Methyl formate has some industrial uses. It hydrolyzes in the body to methanol and formic acid.[11] Methyl acetate is a colorless liquid with a pleasant odor. It is used as a solvent and as an additive to give foods a fruit-like taste. Ethyl acetate is a liquid with a pleasant odor. Liquid vinyl acetate polymerizes when exposed to light to yield a solid polymer. Both n-butyl acetate and n-amyl acetate are relatively higher-boiling liquids than the esters mentioned above. Amyl acetate has a characteristic odor of bananas and pears. Methyl methacrylate is the monomer used to make some kinds of polymers noted for their transparency and resistance to weathering. Among their other applications, these polymers are used as substitutes for glass, particularly in automobile lights. Dimethyl phthalate is the simplest example of the environmentally important phthalate esters. Other significant members of this class of compounds are diethyl, di-n-butyl, di-n-octyl, bis(2-ethylhexyl), and butyl benzyl phthalates. Used in large quantities as plasticizers to improve the qualities of plastics, these compounds have become widespread environmental pollutants. The higher-molecular-mass phthalate compounds, especially, tend to be environmentally persistent.

14.10.1 Toxicities of Esters

The most common reaction of esters in exposed tissues is hydrolysis:

$$\underset{\textbf{Ester}}{R-O-\overset{\overset{O}{\|}}{C}-R'} + H_2O \longrightarrow \underset{\textbf{Alcohol}}{R-OH} + \underset{\textbf{Carboxylic acid}}{HO-\overset{\overset{O}{\|}}{C}-R'} \qquad (14.10.1)$$

To a large extent, therefore, the toxicities of esters tend to be those of their hydrolysis products. Two physical characteristics of many esters that affect their toxicities are relatively high volatility, which promotes exposure by the pulmonary route, and good solvent action, which affects penetration and tends to dissolve body lipids. Many volatile esters exhibit asphyxiant and narcotic action. As expected for compounds that occur naturally in foods, some esters are nontoxic (in reasonable doses). However, some of the synthetic esters, such as allyl acetate, have relatively high toxicities. As an example of a specific toxic effect, vinyl acetate acts as a skin defatting agent.

Although environmentally persistent, most of the common phthalates have low toxicity ratings of 2 or 3, based on acute toxic effects. There is particular concern with regard to di-2-ethylhexyl phthalate used as a plasticizer in polyvinyl chloride plastic medical devices.[12] Dialysis patients and hemophiliacs who receive frequent blood transfusions are especially likely to receive potentially harmful levels of di-2-ethylhexyl phthalates from contact of fluids with such devices.

REFERENCES

1. Lanigan, R.S., Final report on the safety assessment of methyl alcohol, *Int. J. Toxicol.*, 20 (Suppl. 1), 57–85, 2001.
2. Eells, J. et al., Development and characterization of a rodent model of methanol-induced retinal and optic nerve toxicity, *Neurotoxicology*, 21, 321–330, 2000.
3. Glatthar, J., Seen, T., and Pieper, H.J., Investigations on reducing the methanol content in distilled spirits made of bartlett pears, *Deutsche Lebensmittel-Rundschau*, 97, 209–216, 2001.
4. Lumeng, L. and Crabb, D.W., Alcoholic liver disease, *Curr. Opin. Gastroenterol.*, 17, 211–220, 2001.
5. Brent, J., Current management of ethylene glycol poisoning, *Drugs*, 61, 979–988, 2001.
6. Gosselin, R.E., Smith, R.P., and Hodge, H.C., Phenol, in *Clinical Toxicology of Commercial Products*, 5th ed., Williams & Wilkins, Baltimore, 1984, pp. III-344–III-348.

7. Staimer, N., Gee, S.J., and Hammock, B.D., Development of a class-selective enzyme immunoassay for urinary phenolic glucuronides, *Anal. Chim. Acta,* 441, 27–36, 2001.
8. Liteplo, R.G., Meek, M.E., and Bruce, W., Ethylene oxide: hazard characterization and exposure–response analysis, *Environ. Carcinog. Ecotoxicol. Rev.,* C19, 219–265, 2001.
9. Thrasher, J.D. and Kilburn, K.H., Embryo toxicity and teratogenicity of formaldehyde, *Arch. Environ. Health,* 56, 300–311, 2001.
10. Morgan, D.L. et al., Upper respiratory tract toxicity of inhaled methylvinyl ketone in F344 rats and B6C3F1 mice, *Toxicol. Sci.,* 58, 182–194, 2000.
11. Nihlen, A. and Droz, P.-O., Toxicokinetic modeling of methyl formate exposure and implications for biological monitoring, *Int. Arch. Occup. Environ. Health,* 73, 479–487, 2000.
12. Tickner, J.A. et al., Health risks posted by use of di-2-ethylhexyl phthalate (DEHP) in PVC medical devices: a critical review, *Am. J. Ind. Med.,* 39, 100–111, 2001.

QUESTIONS AND PROBLEMS

1. What are several of the bases for classifying organooxygen compounds?
2. In what respects are the chemical and toxicological chemical characteristics of methanol unique? What are some of the particular toxicological hazards of methanol?
3. What is the metabolic pathway of methanol degradation? How does this result in acidosis?
4. What are the major acute toxicological effects of ethanol? How is ethanol exposure usually measured or expressed? What is a particular chronic toxicological effect of long-term ethanol ingestion?
5. What are the metabolic products of ethanol oxidation in the body? How does the rate of ethanol metabolism compare to that of methanol metabolism?
6. What is the name of the long-chain alcohol $CH_3(CH_2)_6CH_2OH$? What is its water solubility? How is this alcohol used to describe bioaccumulation effects? What is the name of the parameter obtained using this alcohol to describe such effects?
7. In general, what are the toxicological characteristics of esters. Why is it reasonable to believe that many esters are not particularly toxic? What does the reaction below imply about the toxicities of esters?

$$R-O-\underset{\underset{\text{Ester}}{}}{\overset{\overset{O}{\|}}{C}}-R' + H_2O \rightarrow \underset{\text{Alcohol}}{R-OH} + \underset{\text{Carboxylic acid}}{HO-\overset{\overset{O}{\|}}{C}-R'}$$

8. What toxicological effect may result from the reaction below? Which organ is most susceptible to damage as a result?

$$\underset{\text{Ethylene glycol}}{\overset{HH}{\underset{OHOH}{H-C-C-H}}} \xrightarrow[\text{Metabolic processes}]{\{O\}} \text{Oxalate} \xrightarrow{Ca^{2+}} \underset{\text{Calcium oxalate (solid)}}{\text{Ca oxalate structure}}$$

9. Match the following pertaining to organooxygen compounds:
    (a) $CH_3CHOHCH_3$ 1. Olefinic alcohol
    (b) $CH_2=CHCH_2OH$ 2. *n*-Butanol
    (c) $CH_3(CH_2)_2CH_2OH$ 3. Used in bioaccumulation studies
    (d) $CH_3(CH_2)_6CH_2OH$ 4. Rubbing alcohol, food additive

10. What is shown by the following reaction? To what extent does this reaction occur?

$$\text{C}_6\text{H}_5\text{-OH} + \text{OH}^- \rightarrow \text{C}_6\text{H}_5\text{-O}^- + \text{H}_2\text{O}$$

11. Discuss the toxicology of phenol. Is it known to have many toxic effects? Why were so many people exposed around 100 years ago? What is meant by phenol being a protoplasmic poison?
12. What are epoxides? In what sense might they be regarded as ethers? Is there any way that epoxides may be formed from other kinds of compounds in the body? How might this occur?
13. What are the toxicological characteristics of formaldehyde? In what sense is the toxicological chemistry of formaldehyde unique? What is formalin? How is it related to formaldehyde? What metabolic phenomenon suggests that formaldehyde is not very toxic? Is this true?
14. What distinguishes an aldehyde from a ketone? From the material given in this chapter, can one conclude that there are any substantial differences in toxicities between aldehydes and ketones?
15. In large part because of the water solubility and intensely irritating qualities of the lower aldehydes, which kinds of tissue are these compounds most prone to attack?
16. Explain what is shown by the following general reaction in terms of the metabolism of an important class of toxic compounds:

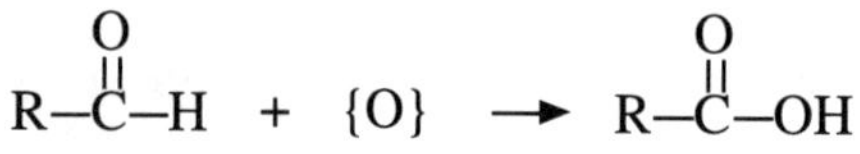

$$\underset{\text{R-C-H}}{\overset{\text{O}}{\|}} + \{O\} \rightarrow \underset{\text{R-C-OH}}{\overset{\text{O}}{\|}}$$

17. Why is it reasonable to believe that many carboxylic acids have only limited toxicities? Give some examples of carboxylic acids that are quite toxic.
18. Ethers are often used in applications where an unreactive organic solvent is required. In what sense are ethers unreactive? How is this reflected in their toxicological chemistry?
19. What is the most likely route of exposure to diethyl ether? How is much of the diethyl ether that enters the body by this route subsequently eliminated?
20. What are some of the important chemical and toxicological characteristics of the compound shown below:

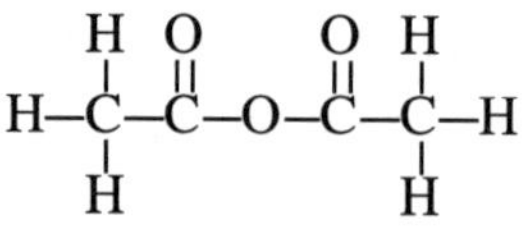

$$\text{H}_3\text{C-C(=O)-O-C(=O)-CH}_3$$

CHAPTER 15

Organonitrogen Compounds

15.1 INTRODUCTION

Nitrogen occurs in a wide variety of organic compounds of both synthetic and natural origin. This chapter discusses organic compounds that contain carbon, hydrogen, and nitrogen. Many significant organonitrogen compounds contain oxygen as well, and these are covered in later parts of the chapter. Not the least of the concerns regarding organonitrogen compounds is that a significant number of these compounds (including some aromatic amines and nitrosamines) are carcinogenic.

15.2 NONAROMATIC AMINES

15.2.1 Lower Aliphatic Amines

Amines may be regarded as derivatives of ammonia, NH_3, in which one to three of the H atoms have been replaced by hydrocarbon groups. When these groups are aliphatic groups of which none contains more than six C atoms, the compound may be classified as a **lower aliphatic amine**. Among the more commercially important of these amines are mono-, di-, and trimethylamine; mono-, di-, and triethylamine; dipropylamine; isopropylamine; butylamine; dibutylamine; diisobutylamine; cyclohexylamine; and dicyclohexylamine. Example structures are given in Figure 15.1.

The structures in Figure 15.1 indicate some important aspects of amines. Methylamine, methyl-2-propylamine, and triethylamine are primary, secondary, and tertiary amines, respectively. A primary amine has one hydrocarbon group substituted for H on NH_3, a secondary amine has two, and a tertiary amine has three. Dicyclohexylamine has two cycloalkane substituent groups attached and is a secondary amine. All of the aliphatic amines have strong odors. Of the compounds listed above as commercially important aliphatic amines, the methylamines and monoethylamine are gases under ambient conditions, whereas the others are colorless volatile liquids. The lower aliphatic amines are highly flammable. They are used primarily as intermediates in the manufacture of other chemicals, including polymers (rubber, plastics, textiles), agricultural chemicals, and medicinal chemicals.

The lower aliphatic amines are generally among the more toxic substances in routine, large-scale use. One of the reasons for their toxicity is that they are basic compounds and raise the pH of exposed tissue by hydrolysis with water in tissue, as shown by the following reaction:

$$R_3N + H_2O \rightarrow R_3NH^+ + OH^- \qquad (15.2.1)$$

Furthermore, these compounds are rapidly and easily taken into the body by all common exposure routes. The lower amines are corrosive to tissue and can cause tissue necrosis at the point of contact.

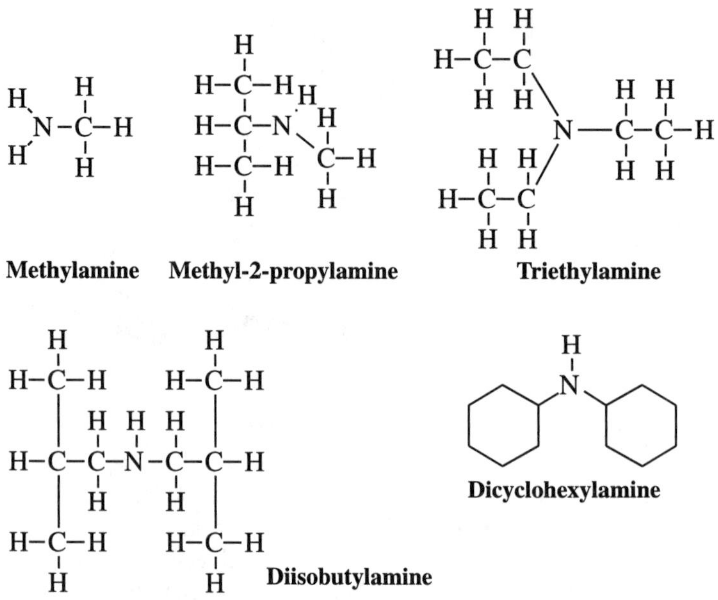

Figure 15.1 Examples of lower aliphatic amines.

Sensitive eye tissue is vulnerable to amines. These compounds can have systemic effects on many organs in the body. Necrosis of the liver and kidneys can occur, and exposed lungs can exhibit hemorrhage and edema. The immune system may become sensitized to amines.

Of the lower aliphatic amines, cyclohexylamine and dicyclohexylamine appear to have received the most attention for their toxicities. In addition to its caustic effects on eyes, mucous membranes, and skin, cyclohexylamine acts as a systemic poison. In humans the symptoms of systemic poisoning by this compound include nausea to the point of vomiting, anxiety, restlessness, and drowsiness. It adversely affects the female reproductive system. Dicyclohexylamine produces similar symptoms, but is considered to be more toxic. It is appreciably more likely to be absorbed in toxic levels through the skin, probably because of its less polar, more lipid-soluble nature.

15.2.2 Fatty Amines

Fatty amines are those containing alkyl groups having more than six carbon atoms. The commercial fatty amines are synthesized from fatty acids that occur in nature and are used as chemical intermediates. Other major uses of fatty amines and their derivatives include textile chemicals (particularly fabric softeners), emulsifiers for petroleum and asphalt, and flotation agents for ores.

Some attention has been given to the toxicity of octadecylamine, which contains a straight-chain, 18-carbon alkane group, because of its use as an anticorrosive agent in steam lines. There is some evidence to suggest that the compound is a primary skin sensitizer.

15.2.3 Alkyl Polyamines

Alkyl polyamines are those in which two or more amino groups are bonded to alkane moieties. The structures of the four most significant of these are shown in Figure 15.2. These compounds have a number of commercial uses, such as for solvents, emulsifiers, epoxy resin hardeners, stabilizers, and starting materials for dye synthesis. They also act as chelating agents; triethylenetetramine is especially effective for this purpose. Largely as a result of their strong alkalinity, the alkyl polyamines tend to be skin, eye, and respiratory tract irritants. The lower homologues are relatively stronger irritants.

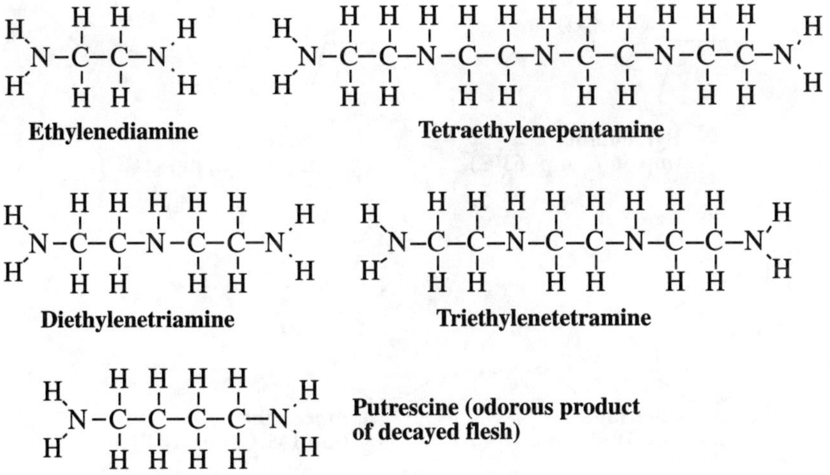

Ethylenediamine H₂N-CH₂-CH₂-NH₂

Tetraethylenepentamine H₂N-CH₂-CH₂-NH-CH₂-CH₂-NH-CH₂-CH₂-NH-CH₂-CH₂-NH₂

Diethylenetriamine H₂N-CH₂-CH₂-NH-CH₂-CH₂-NH₂

Triethylenetetramine H₂N-CH₂-CH₂-NH-CH₂-CH₂-NH-CH₂-CH₂-NH₂

Putrescine (odorous product of decayed flesh) H₂N-CH₂-CH₂-CH₂-CH₂-NH₂

Figure 15.2 Alkyl polyamines in which two or more amino groups are bonded to an alkane group.

Of the common alkyl polyamines, ethylenediamine is the most notable because of its widespread use and toxicity. Although it has a toxicity rating of only three, it can be very damaging to the eyes and is a strong skin sensitizer. The dihydrochloride and dihydroiodide salts have some uses as human and veterinary pharmaceuticals. The former is administered to acidify urine, and the latter as an iodine source. Putrescine is a notoriously odorous naturally occurring substance produced by bacteria in decaying flesh.

15.2.4 Cyclic Amines

Four simple amines in which N atoms are contained in a ring structure are shown in Figure 15.3. Of the compounds shown in Figure 15.3, the first three are liquids under ambient conditions and have the higher toxicity hazards expected of liquid toxicants. All four compounds are colorless in the pure form, but pyrrole darkens upon standing. All are considered to be toxic via the oral, dermal, and inhalation routes. There is little likelihood of inhaling piperazine, except as a dust, because of its low volatility.

15.3 CARBOCYCLIC AROMATIC AMINES

Carbocyclic aromatic amines are those in which at least one substituent group is an aromatic ring containing only C atoms as part of the ring structure, and with one of the C atoms in the ring bonded directly to the amino group. There are numerous compounds with many industrial uses in this class of amines. They are of particular toxicological concern because several have been shown to cause cancer in the human bladder, ureter, and pelvis, and are suspected of being lung, liver, and prostate carcinogens.

15.3.1 Aniline

Aniline,

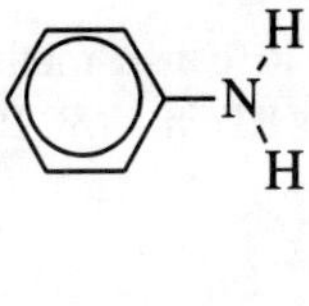

C₆H₅-NH₂ Aniline

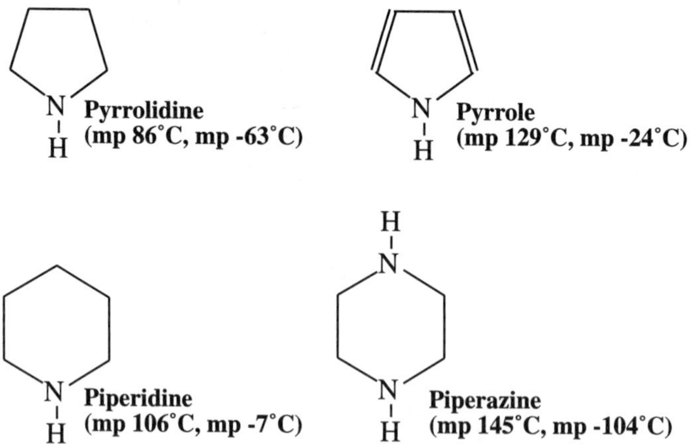

Pyrrolidine (mp 86°C, mp -63°C)

Pyrrole (mp 129°C, mp -24°C)

Piperidine (mp 106°C, mp -7°C)

Piperazine (mp 145°C, mp -104°C)

Figure 15.3 Some common cyclic amines.

has been an important industrial chemical for many decades. Currently, it is most widely used for the manufacture of polyurethanes and rubber, with lesser amounts consumed in the production of pesticides (herbicides, fungicides, insecticides, animal repellants), defoliants, dyes, antioxidants, antidegradants, and vulcanization accelerators. It is also an ingredient of some household products, such as polishes (stove and shoe), paints, varnishes, and marking inks. Aniline is a colorless liquid with an oily consistency and distinct odor; it freezes at –6.2°C and boils at 184.4°C.

Aniline is considered to be very toxic, with a toxicity rating of 4. It readily enters the body by inhalation, by ingestion, and through the skin. In its absorption and toxicological characteristics, aniline resembles nitrobenzene, which is discussed in Section 15.6. Aniline was the toxic agent responsible for affecting more than 20,000 people and killing 300 in Spain in 1981. Known as the Spanish toxic oil syndrome, this tragic epidemic was due to aniline-contaminated olive oil.[1]

The most common effect of aniline in humans is methemoglobinemia, caused by the oxidation of iron(II) in hemoglobin to iron(III), with the result that the hemoglobin can no longer transport oxygen in the body. This condition is characterized by cyanosis and a brown–black color of the blood. Unlike the condition caused by reversible binding of carbon monoxide to hemoglobin, oxygen therapy does not reverse the effects of methemoglobinemia. The effects can be reversed by the action of the methemoglobin reductase enzyme, as shown by the following reaction:

$$HbFe(III) \xrightarrow{\text{Methemoglobin reductase}} HbFe(II) \quad (15.3.1)$$

Rodents (mice, rats, rabbits) have a higher activity of this enzyme than do humans, so that extrapolation of rodent experiments with methemoglobinemia to humans is usually inappropriate. Methylene blue can also bring about the reduction of HbFe(III) to HbFe(II) and is used as an antidote for aniline poisoning.

Methemoglobinemia has resulted from exposure to aniline used as a vehicle in indelible laundry-marking inks, particularly those used to mark diapers. This condition was first recognized in 1886, and cases were reported for many decades thereafter. Infants who develop methemoglobinemia from this source suffer a 5 to 10% mortality rate. The skin of infants (particularly in the genital area; see Section 6.4) is more permeable to aniline than that of adults, and infant blood is more susceptible to methemoglobinemia.

Aniline must undergo biotransformation to cause methemoglobinemia because pure aniline does not oxidize iron(II) in hemoglobin to iron(III) *in vitro*. It is believed that the actual toxic agents

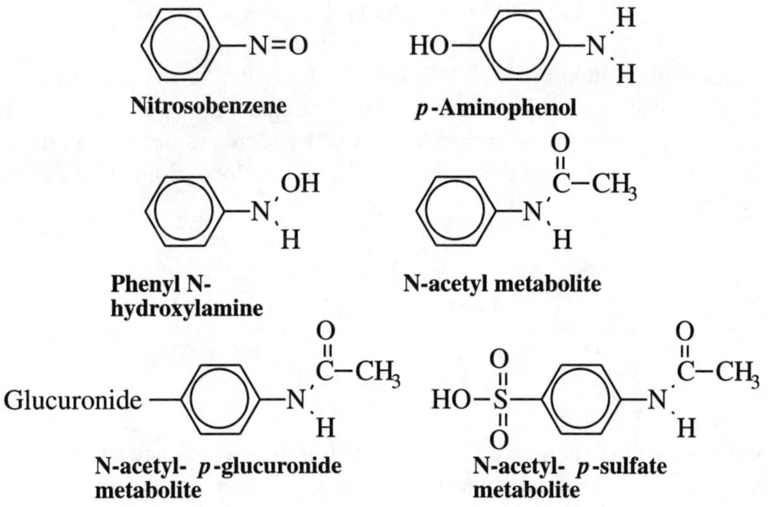

ORGANONITROGEN COMPOUNDS

Figure 15.4 Metabolites of aniline that are toxic or excreted.

formed from aniline are nitrosobenzene, aminophenol, and phenyl N-hydroxylamine, shown in Figure 15.4. The hepatic detoxification mechanisms for aniline are not very effective. The metabolites of aniline excreted from the body are N-acetyl, N-acetyl-*p*-glucuronide, and N-acetyl-*p*-sulfate products, also shown in Figure 15.4.

15.3.2 Benzidine

Benzidine, *p*-aminodiphenyl, is a solid compound that can be extracted from coal tar. It is highly toxic by oral ingestion, inhalation, and skin sorption and is one of the few proven human carcinogens. Its systemic effects include blood hemolysis, bone marrow depression, and kidney and liver damage.

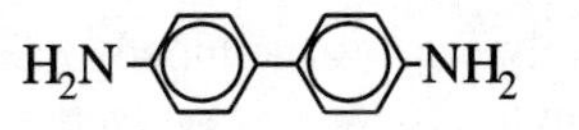

15.3.3 Naphthylamines

The two derivatives of naphthalene having single amino substituent groups are **1-naphthylamine** (alpha-naphthylamine) and **2-naphthylamine** (beta-naphthylamine). Both of these compounds are solids (lump, flake, dust) under normal conditions, although they may be encountered as liquids and vapors. Exposure can occur through inhalation, the gastrointestinal tract, or skin. Both compounds are highly toxic and are proven human bladder carcinogens.

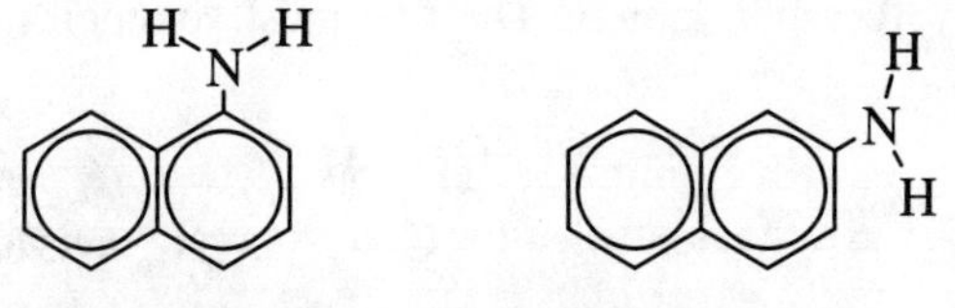

15.4 PYRIDINE AND ITS DERIVATIVES

Pyridine is a colorless liquid mp, –42°C; bp, 115°C) with a sharp, penetrating odor that can perhaps best be described as terrible. It is an aromatic compound in which an N atom is part of a six-membered ring. The most important derivatives of pyridine are the mono-, di-, and trimethyl derivatives; the 2-vinyl and 4-vinyl derivatives; 5-ethyl-2-methylpyridine (MEP); and piperidine, also called hexahydropyridine (below):

Pyridine and its substituted derivatives are recovered from coal tar. They tend to react like benzene and its analogous derivatives because of the aromatic ring. The major use of pyridine is as an initiator in the process by which rubber is vulcanized. Although considered moderately toxic, with a toxicity rating of three, pyridine has caused fatalities. Symptoms of acute pyridine poisoning from inhalation of the vapor have included eye irritation, nose and throat irritation, dizziness, abdominal discomfort, nausea, palpitations, and light-headedness.[2] Longer-term symptoms include diarrhea, anorexia, and fatigue. The major psychopathological effect of pyridine poisoning is mental depression.

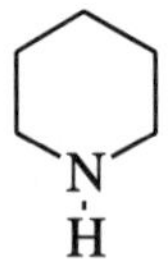

A notably toxic pyridine derivative is 1,2,3,6-tetrahydro-1-methyl-4-phenylpyridine (MPTP), which has the structural formula shown below:

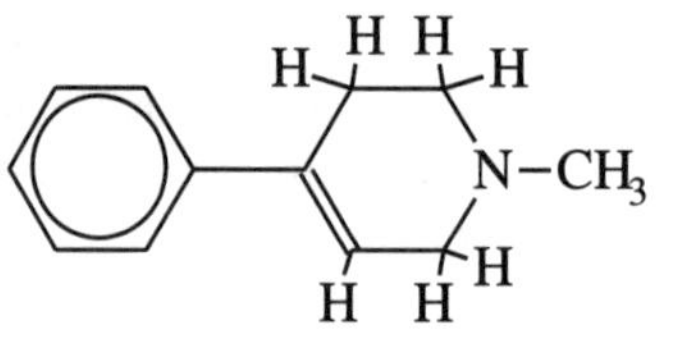

This compound is a protoxicant that readily crosses the blood–brain barrier, where it is acted on by the monoamine oxidase enzyme system to produce a positively charged neurotoxic species that cannot readily cross the blood–brain barrier to leave the brain. The result has been described as "selective neuronal death of the dopaminergic neurons in the zona compacta of the substantia nigra."[3] The symptoms of this disorder are very similar to Parkinson's disease, one of several common and devastating neurodegenerative diseases.

15.5 NITRILES

Nitriles are organic analogs of highly toxic hydrogen cyanide, HCN (see Section 11.2), where the H is replaced by a hydrocarbon moiety. The two most common nitriles are acetonitrile and acrylonitrile:

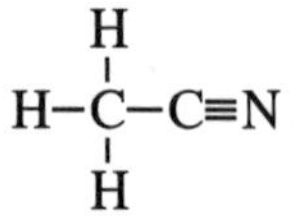

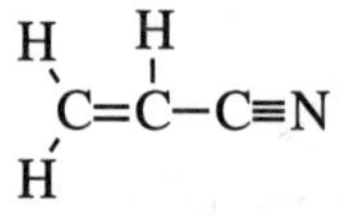

Acetonitrile (mp, –45°C; bp, 81°C) is a colorless liquid with a mild odor. Because of its good solvent properties for many organic and inorganic compounds and its relatively low boiling point, it has numerous industrial uses, particularly as a reaction medium that can be recovered. It is used as an organic solvent for lipophilic substances used in *in vitro* studies of metabolism of pharmaceutical agents.[4] Acetonitrile has a toxicity rating of 3 or 4; exposure can occur via the oral, pulmonary, and dermal routes. Although it is considered relatively safe, it is capable of causing human deaths, perhaps by metabolic release of cyanide.

Acrylonitrile is a colorless liquid with a peach-seed odor that is used in large quantities in the manufacture of acrylic fibers, dyes, and pharmaceutical chemicals. Containing both nitrile and C=C groups, acrylonitrile is a highly reactive compound with a strong tendency to polymerize. It has a toxicity rating of five, with a mode of toxic action resembling that of HCN. In addition to ingestion, it can be absorbed through the skin or by inhalation of the vapor. It causes blisters and arythema on exposed skin.

Because of its widespread industrial use and consequent worker exposure, the metabolism of acrylonitrile has been studied extensively.[5] There are two major pathways of acrylonitrile metabolism in humans. The first of these produces a glutathione conjugate and is considered to be detoxification. The second pathway produces cyanoethylene oxide,

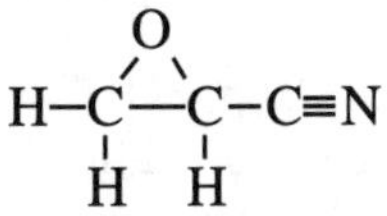

$$H-\overset{\overset{\displaystyle O}{\diagup\,\diagdown}}{\underset{H}{C}}-\underset{H}{C}-C\equiv N \quad \text{Cyanoethylene oxide}$$

followed by release of toxic cyanide, which inhibits enzymes responsible for respiration in tissue, thereby preventing tissue cells from utilizing oxygen. Acrylonitrile is a suspect carcinogen.

Acetone cyanohydrin (structure below) is an oxygen-containing nitrile that should be mentioned because of its extreme toxicity and widespread industrial applications. It is used to initiate polymerization reactions and in the synthesis of foaming agents, insecticides, and pharmaceutical compounds. A colorless liquid readily absorbed through the skin, it decomposes in the body to hydrogen cyanide, to which it should be considered toxicologically equivalent (toxicity rating, six) on a molecule-per-molecule basis.

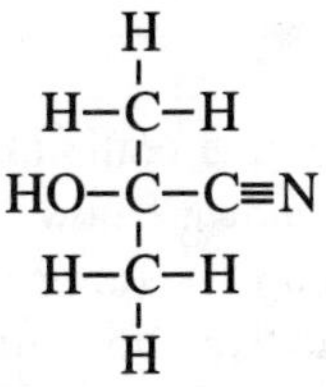

$$\begin{array}{c} H \\ | \\ H-C-H \\ | \\ HO-C-C\equiv N \\ | \\ H-C-H \\ | \\ H \end{array} \quad \text{Acetone cyanohydrin}$$

Nitriles are cyanogenic substances — substances that produce cyanide when metabolized. It is likely that nitriles are teratogens because of maternal production of cyanide in pregnant females. A study of the teratogenic effects on rats of saturated nitriles, including acetonitrile, propionitrile, and *n*-butyronitrile, and of unsaturated nitriles, including acrylonitrile, methacrylonitrile, allylnitrile, *cis*-2-pentenenitrile, and 2-chloroacrylonitrile, has shown a pattern of abnormal embryos similar to those observed from administration of inorganic cyanide.[6]

15.6 NITRO COMPOUNDS

The structures of three significant **nitro compounds**, which contain the $-NO_2$ functional group, are given in Figure 15.5.

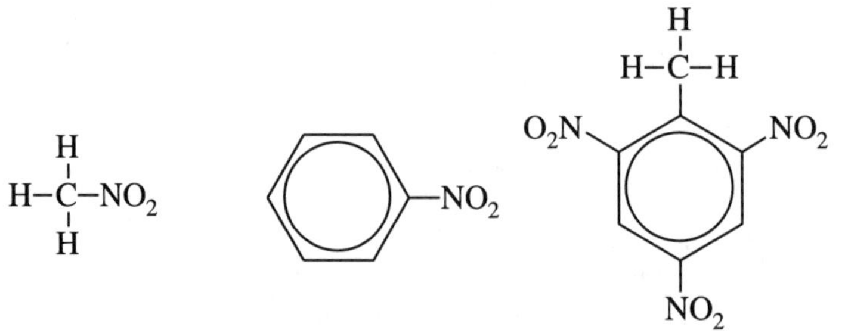

Nitromethane   **Nitrobenzene**   **Trinitrotoluene (TNT)**

Figure 15.5 Some of the more important nitro compounds.

The lightest of the nitro compounds is **nitromethane**, an oily liquid (mp, –29°C; bp, 101°C). It has a toxicity rating of three. Symptoms of poisoning include anorexia, diarrhea, nausea, and vomiting. The organs that are most susceptible to damage from it are the kidneys and liver. Severe peripheral neuropathy has been reported in two workers strongly exposed to nitromethane for several weeks.[7]

Nitrobenzene is a pale yellow oily liquid (mp, 5.7°C; bp, 211°C) with an odor of bitter almonds or shoe polish. It is produced mainly for the manufacture of aniline. It can enter the body through all routes and has a toxicity rating of five. Its toxic action is much like that of aniline, including the conversion of hemoglobin to methemoglobin, which deprives tissue of oxygen. Cyanosis is a major symptom of nitrobenzene poisoning.

Trinitrotoluene (TNT) is a solid material widely used as a military explosive. It has a toxicity rating of three or four. It can damage the cells of many kinds of tissue, including those of bone marrow, kidney, and liver. Extensive knowledge of the toxicity of TNT was obtained during World War II in the crash program to manufacture huge quantities of it. Toxic hepatitis developed in some workers under age 30 exposed to TNT systemically, whereas aplastic anemia was observed in some older victims of exposure. In the United States during World War II, 22 cases of fatal TNT poisoning were documented (many more people were blown up during manufacture and handling).

15.6.1 Nitro Alcohols and Nitro Phenols

Nitro alcohols are nonaromatic compounds containing both –OH and –NO$_2$ groups. A typical example of such a compound is **2-nitro-1-butanol**, shown below. These compounds are used in chemical synthesis to introduce nitro functional groups or (after reduction) amino groups onto molecules. They tend to have low volatilities and moderate toxicities. The aromatic nitrophenol, ***p*-nitrophenol**, is an industrially important compound with toxicological properties resembling those of phenol and nitrobenzene.

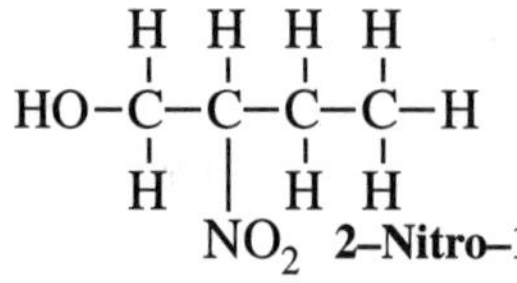

2–Nitro–1–butanol 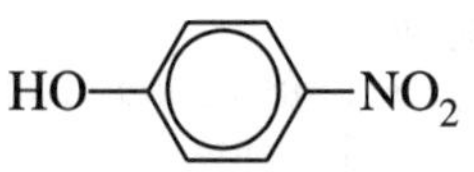   *p*-Nitrophenol

15.6.2 Dinoseb

Dinoseb is a nitrophenolic compound, once widely used as an herbicide and plant desiccant, that is noted for its toxic effects. The chemical name of this compound is 4,6-dinitro-2-*sec*-butylphenol, and its structure is

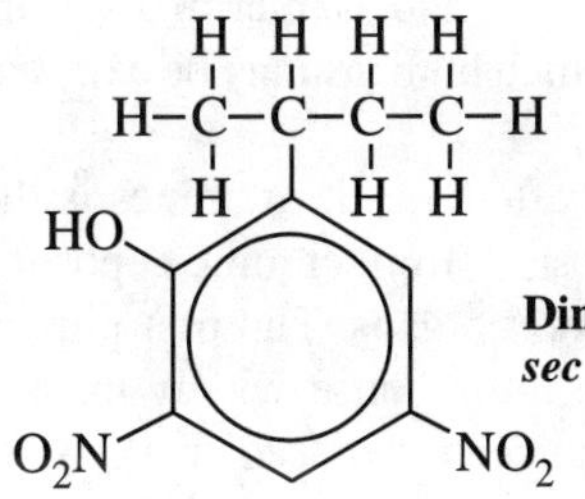

Dinoseb (4,6-dinitro-2-*sec* -butylphenol)

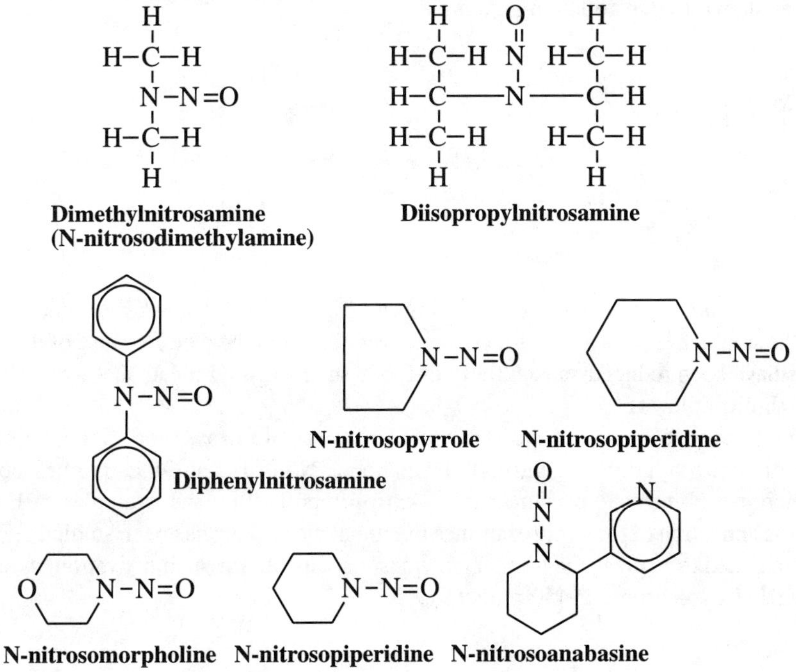

Figure 15.6 Examples of some important nitrosamines.

Dinoseb has a toxicity rating of five and is strongly suspected of causing birth defects in the children of women exposed to it early in pregnancy, as well as sterility in exposed men. In October 1986, the Environmental Protection Agency imposed an emergency ban on the use of the chemical, which was partially rescinded for the northwestern U.S. by court order early in 1987, although some uses were permitted, primarily in the northwestern U.S., through 1989. More than 10 years later, there were still controversies involving the cleanup of dinoseb-contaminated water in Washington State.[8]

15.7 NITROSAMINES

N-nitroso compounds, commonly called **nitrosamines**, are a class of compounds containing the N–N=O functional group. They are of particular toxicological significance because most that have been tested have been shown to be carcinogenic. The structural formulas of some nitrosamines are shown in Figure 15.6.

Some nitrosamines have been used as solvents and as intermediates in chemical synthesis. They have been found in a variety of materials to which humans may be exposed, including beer, whiskey, and cutting oils used in machining.

By far the most significant toxicological effect of nitrosamines is their carcinogenicity, which may result from exposure to a single large dose or from chronic exposure to relatively small doses. Different nitrosamines cause cancer in different organs. The first nitrosamine extensively investigated for carcinogenicity was dimethylnitrosamine, once widely used as an industrial solvent. It was known to cause liver damage and jaundice in exposed workers, and studies starting in the 1950s subsequently revealed its carcinogenic nature. Dimethylnitrosamine was found to alkylate DNA, which is the mechanism of its carcinogenicity (the alkylation of DNA as a cause of cancer is noted in the discussion of biochemistry of carcinogesis in Section 7.8).

The common means of synthesizing nitrosamines is the low-pH reaction of a secondary amine and nitrite, as shown by the following example:

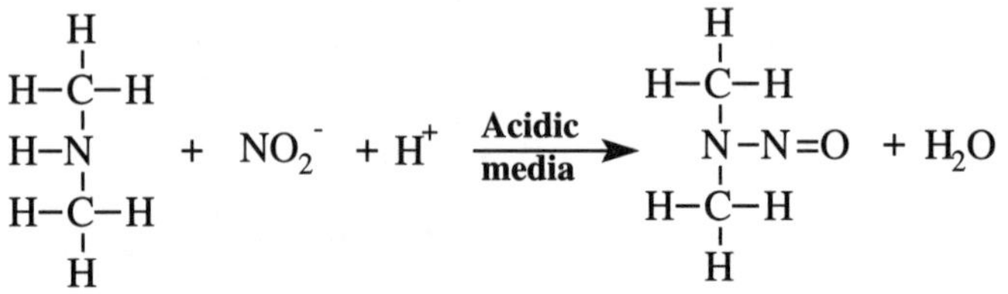

$$(CH_3)_2NH + NO_2^- + H^+ \xrightarrow{\text{Acidic media}} (CH_3)_2N-N=O + H_2O \quad (15.7.1)$$

The possibility of this kind of reaction occurring *in vivo* and producing nitrosamines in the acidic medium of the stomach is some cause for concern over nitrites in the diet. Because of this possibility, nitrite levels have been reduced substantially in foods such as cured meats that formerly contained relatively high nitrite levels.

Tobacco (chewing tobacco and snuff) contains a variety of nitrosamines, including N-nitrosatabine, 4-(methylnitrosamino)-1-(3-pyridyl)-1-butanone, N-nitrosanabasine, N-nitrosopyrrolidine, N-nitronornicotine, N-nitrosopiperidine, and N-nitrosomorpholine (see examples in Figure 15.6). The enzymatic activation of these nitrosamines to mutagenic species has been studied using bacteria genetically activated to express the human enzymes responsible for such activation, cytochrome P-450 and NADPH–cytochrome P-450 reductase.[9]

15.8 ISOCYANATES AND METHYL ISOCYANATE

Isocyanates are compounds with the general formula R–N=C=O. They have numerous uses in chemical synthesis, particularly in the manufacture of polymers with carefully tuned specialty properties. Methyl isocyanate is a raw material in the manufacture of carbaryl insecticide. Methyl isocyanate (like other isocyanates) can be synthesized by the reaction of a primary amine with phosgene in a moderately complex process, represented by reaction 15.8.1. Structures of three significant isocyanates are given in Figure 15.7.

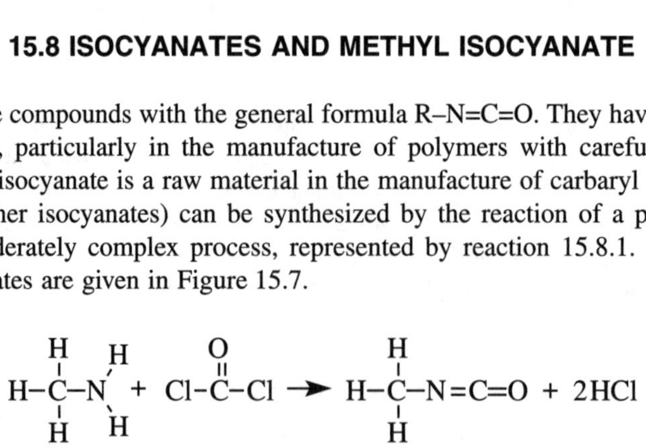

$$CH_3NH_2 + Cl-CO-Cl \rightarrow CH_3-N=C=O + 2HCl \quad (15.8.1)$$

Methylamine   Phosgene   Methyl isocyanate

Both chemically and toxicologically, the most significant property of isocyanates is the high chemical reactivity of the isocyanate functional group. Industrially, the most significant such reaction is with alcohols to yield urethane (carbamate) compounds, as shown by reaction 15.8.2. Multiple

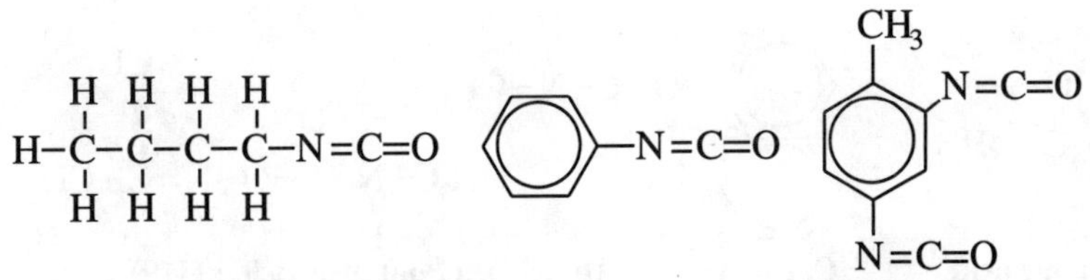

![n-Butyl isocyanate, Phenyl isocyanate, 2,4-Toluene diisocyanate structures]

Figure 15.7 Examples of isocyanate compounds.

isocyanate and –OH groups in the reactant molecules enable formation of polymers. The chemical versatility of isocyanates and the usefulness of the products — such as polymers and pesticides — from which they are made have resulted in their widespread industrial production and consumption.

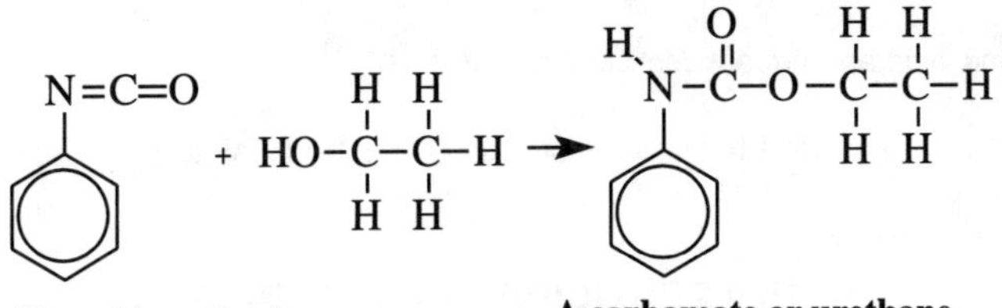

$$\text{Phenyl isocyanate} + \text{ethanol} \rightarrow \text{A carbamate or urethane compound} \tag{15.8.2}$$

Methyl isocyanate was the toxic agent involved in the most catastrophic industrial accident of all time, which took place in Bhopal, India, on December 2, 1984. This accident occurred when water got into a tank of methyl isocyanate, causing an exothermic reaction that built up pressure and ruptured a safety valve. This resulted in the release to the atmosphere of 30 to 40 tons of the compound over an approximately 3-h period. Subsequent exposure of people resulted in approximately 3,500 deaths and almost 100,000 injuries.

Most of the deaths at Bhopal resulted from devastating pulmonary edema, which caused respiratory failure, leading to cardiac arrest. The major debilitating effects of methyl isocyanate on the Bhopal victims were on the lungs, with survivors suffering long-term shortness of breath and weakness from lung damage. However, victims also suffered symptoms of nausea and bodily pain, and numerous toxic effects have been observed in the victims. Changes in the immune systems (effects on numbers of T cells, T-helper cells, and lymphocyte mitogenesis responses) of victims exposed to methyl isocyanate were also observed. The tendency of the compound to function as a systemic poison was somewhat surprising in view of its chemical reactivity with water — its half-life is only about 2 min in aqueous solution — and appears to be the result of its ability to bind with small-molecule proteins and peptides. The most prominent among these is glutathione, a tripeptide described as a conjugating agent in Section 7.4.2; binding to hemoglobin may also be possible. Isocyanate reacts reversibly with –SH groups on glutathione, probably to form S-(N-methylcarbamoyl)glutathione:

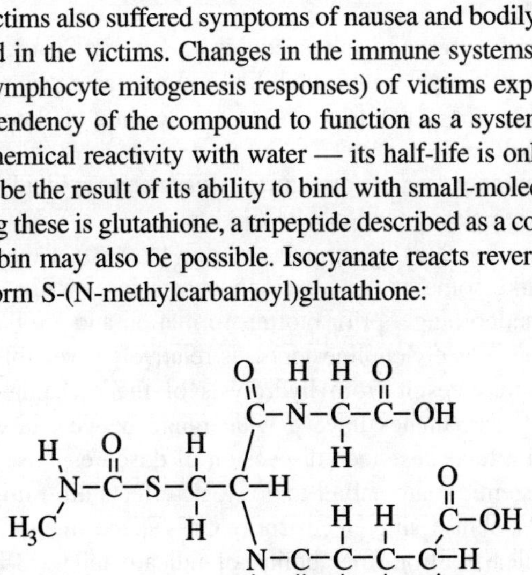

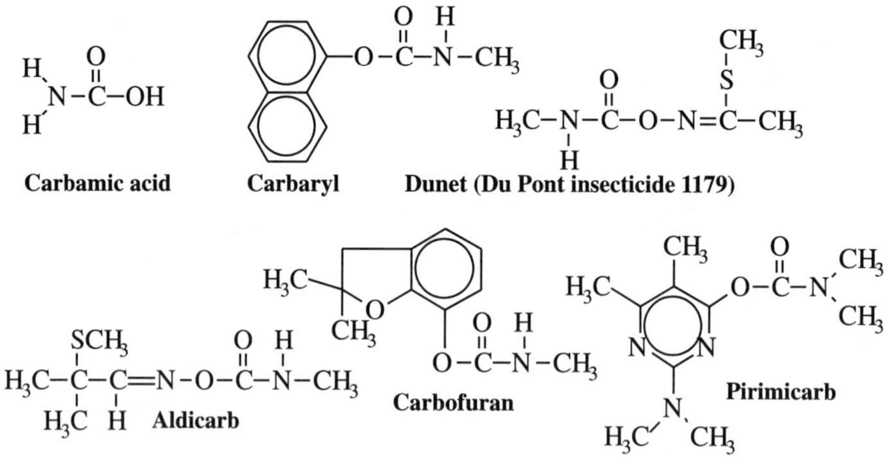

Figure 15.8 Carbamic acid and three insecticidal carbamates.

This complex can be transported to various organs in the body, where it releases isocyanate.

15.9 PESTICIDAL COMPOUNDS

A large number of organic compounds used as pesticides contain nitrogen. Space does not permit a detailed discussion of such compounds, but two general classes of them are cited here.

15.9.1 Carbamates

Pesticidal organic derivatives of carbamic acid, for which the formula is shown in Figure 15.8, are known collectively as **carbamates**. Some carbamate insecticides such as carbaryl, carbofuran, and pirimicarb have been in use for many years; others, including Dunet, are relatively recent. Carbamate pesticides have been widely used because some are more biodegradable than the formerly popular organochlorine insecticides and have lower dermal toxicities than most common organophosphate pesticides.

Carbaryl has been widely used as an insecticide on lawns or gardens. It has a low toxicity to mammals. **Carbofuran** has a high water solubility and acts as a plant systemic insecticide. It is taken up by the roots and leaves of plants so that insects feeding on the plant material are poisoned by the carbamate compound in it.

Pirimicarb has been widely used in agriculture as a systemic aphicide. Unlike many carbamates, it is rather persistent, with a strong tendency to bind to soil.

The toxic effects of carbamates to animals are due to the fact that these compounds inhibit acetylcholinesterase. Unlike some of the organophosphate insecticides (see Chapter 18), they do so without the need for undergoing a prior biotransformation and are therefore classified as direct inhibitors. Their inhibition of acetylcholinesterase is relatively reversible. Loss of acetylcholinesterase inhibition activity may result from hydrolysis of the carbamate ester, which can occur metabolically. In general, carbamates have a wide range between a dose that causes onset of poisoning symptoms and a fatal dose (see discussion of dose–response in Section 6.5). Although pirimicarb has a high systemic mammalian toxicity, its effects are mitigated by its low tendency to be absorbed through the skin. Using electrospray mass spectrometric analysis of urine samples, aldicarb sulfoxide and aldicarb sulfone metabolites of aldicarb and the 3-hydroxycarbofuran metabolite of carbofuran can be monitored as evidence of exposure to these insecticides:[10]

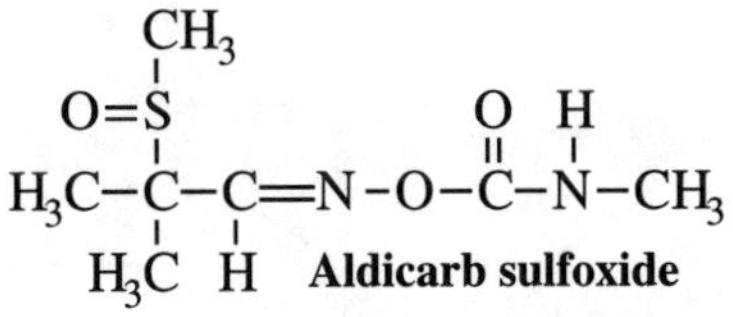

$$\underset{\text{Aldicarb sulfoxide}}{\overset{\overset{\displaystyle CH_3}{|}}{\underset{\underset{\displaystyle H_3C\ \ H}{|\ \ \ |}}{H_3C-\overset{\overset{\displaystyle O=S}{|}}{C}-C}}=N-O-\overset{\overset{\displaystyle O}{\|}}{C}-\overset{\overset{\displaystyle H}{|}}{N}-CH_3} \qquad \underset{\text{Aldicarb sulfone}}{\overset{\overset{\displaystyle CH_3}{|}}{\underset{\underset{\displaystyle H_3C\ \ H}{|\ \ \ |}}{H_3C-\overset{\overset{\displaystyle O=S=O}{|}}{C}-C}}=N-O-\overset{\overset{\displaystyle O}{\|}}{C}-\overset{\overset{\displaystyle H}{|}}{N}-CH_3}$$

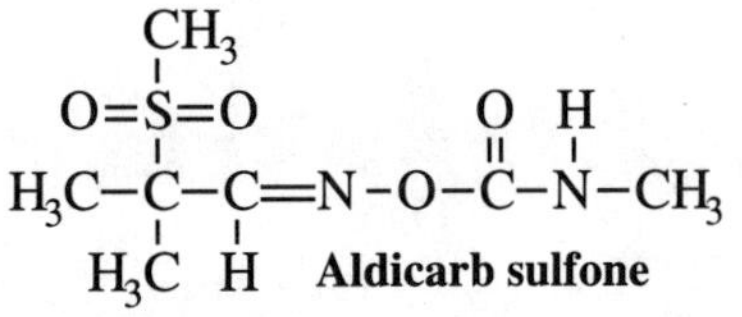

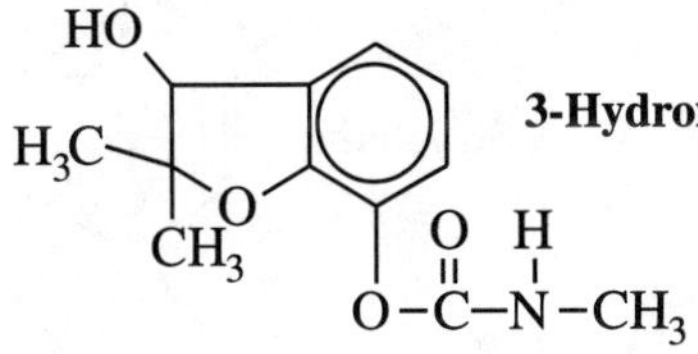

3-Hydroxycarbofuran

15.9.2 Bipyridilium Compounds

As shown by the structures in Figure 15.9, a bipyridilium compound contains two pyridine rings per molecule. The two important pesticidal compounds of this type are the herbicides **diquat** and **paraquat**; other members of this class of herbicides include chlormequat, morfamquat, and difenzoquat. Applied directly to plant tissue, these compounds rapidly destroy plant cells and give the plant a frostbitten appearance. However, they bind tenaciously to soil, especially the clay mineral fraction, which results in rapid loss of herbicidal activity so that sprayed fields can be planted within a day or two of herbicide application.

Paraquat, which was registered for use in 1965, is the most used of the bipyridilium herbicides. With a toxicity rating of five, it is reputed to have "been responsible for hundreds of human deaths."[11] Exposure to fatal or dangerous levels of paraquat can occur by all pathways, including inhalation of spray, skin contact, ingestion, and even suicidal hypodermic injections. Chronic health effects from long-term exposure are reputed to include pulmonary effects, skin cancer, and Parkinson's disease.[12] Despite these possibilities and its widespread application, paraquat is used safely without ill effects when proper procedures are followed.

Because of its widespread use as a herbicide, the possibility exists of substantial paraquat contamination of food. Drinking water contamination by paraquat has also been observed. The chronic effects of exposure to low levels of paraquat over extended periods of time are not well known. Acute exposure of animals to paraquat aerosols causes pulmonary fibrosis, and the lungs are affected even when exposure is through nonpulmonary routes. Paraquat affects enzyme activity. Acute exposure may cause variations in the levels of catecholamine, glucose, and insulin.

Although paraquat can be corrosive at the point of contact, it is a systemic poison that is devastating to a number of organs. The most prominent initial symptom of poisoning is vomiting, sometimes followed by diarrhea. Within a few days, dyspnea, cyanosis, and evidence of impairment of the kidneys, liver, and heart become obvious. In fatal cases, the lungs develop pulmonary fibrosis, often with pulmonary edema and hemorrhaging.

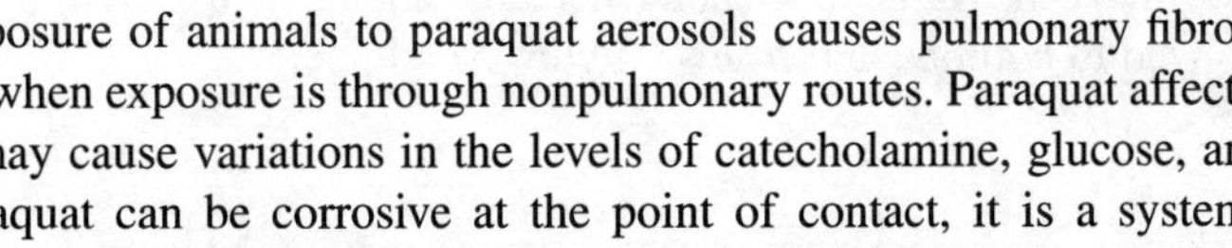

Diquat   **Paraquat**

Figure 15.9 The two major bipyridilium herbicides (cation forms).

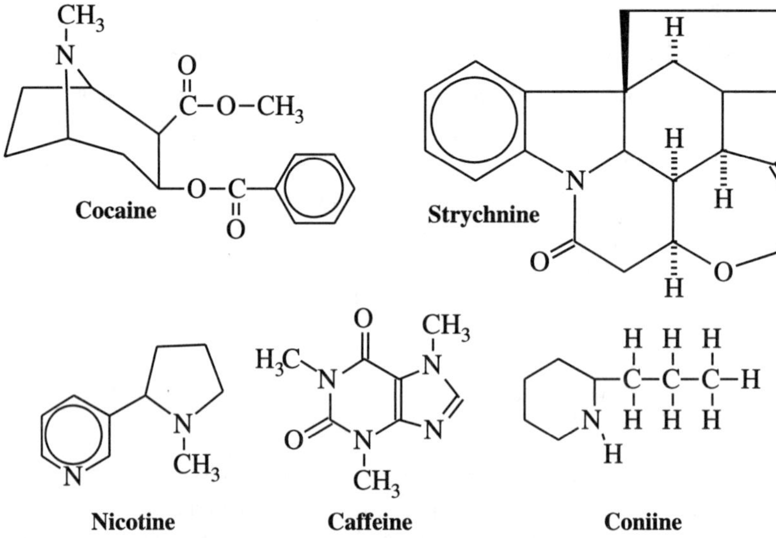

Figure 15.10 Structural formulas of typical alkaloids.

15.10 ALKALOIDS

Alkaloids are compounds of biosynthetic origin that contain nitrogen, usually in a heterocyclic ring. These compounds are produced by plants in which they are usually present as salts of organic acids. They tend to be basic and to have a variety of physiological effects. One of the more notorious alkaloids is cocaine, and alkaloidal strychnine is a deadly poison. The structural formulas of these compounds and three other alkaloids are given in Figure 15.10.

Among the alkaloids are some well-known (and dangerous) compounds. Nicotine is an agent in tobacco that has been described as "one of the most toxic of all poisons and (it) acts with great rapidity."[13] In 1988, the U.S. Surgeon General declared nicotine to be an addictive substance. Nicotine is metabolized to cotinine and *trans*-3'-hydroxycotinine,

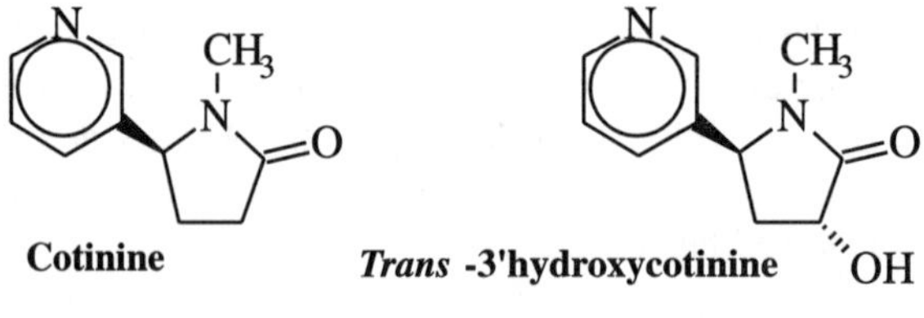

which may be detected in the urine of tobacco users. Coniine is the major toxic agent in poison hemlock (see Chapter 19). Alkaloidal strychnine is a powerful, fast-acting convulsant. Quinine and sterioisomeric quinidine are alkaloids that are effective antimalarial agents. Like some other alkaloids, caffeine contains oxygen. It is a stimulant that can be fatal to humans in a dose of about 10 g. Cocaine is currently the illicit drug of greatest concern. It is metabolized to benzoylecgonine, a compound in which the ester-linked –OCH$_3$ group in cocaine is replaced by the –OH group, which is detected in the urine of cocaine abusers.[14]

Benzoylecgonine

Alkaloids in feed crops, particularly grasses, can cause potentially fatal poisoning of livestock. Phalaris grasses used for pasture and forage in Australia have caused neurological and sudden death intoxication syndromes in livestock. Alkaloids similar to tryptamine and β-carboline have been implicated in these incidents.[15]

Tryptamine   **β-Carboline**

REFERENCES

1. Ladona, M.G. et al., Pharmacogenetic profile of xenobiotic enzyme metabolism in survivors of the Spanish toxic oil syndrome, *Environ. Health Perspect.*, 109, 369–375, 2001.
2. Pattanaik, U., Laa, S.B., and Bano, R., Accidental pyridine exposure: a workplace hazard, *Indian J. Occup. Health*, 42, 47–50, 1999.
3. Calne, D.B., Neurotoxins and degeneration in the central nervous system, *Neurotoxicology*, 12, 335–340, 1991.
4. Tang, C., Shou, M., and Rodrigues, D.A., Substrate-dependent effect of acetonitrile on human liver microsomal cytochrome P450 2C9 (CYP2C9) activity, *Drug Metab. Dispos.*, 28, 567–572, 2000.
5. Thier, R., Lewalter, J., and Bolt, H.M., Species differences in acrylonitrile metabolism and toxicity between experimental animals and humans based on observations in human accidental poisonings, *Arch. Toxicol.*, 74, 184–189, 2000.
6. Saillenfait, A.M. and Sabate, J.P., Comparative developmental toxicities of aliphatic nitriles: *in vivo* and *in vitro* observations, *Toxicol. Appl. Toxicol.*, 163, 149–163, 2000.
7. Page, E.H. et al., Peripheral neuropathy in workers exposed to nitromethane, *Am. J. Ind. Med.*, 40, 107–113, 2001.
8. Dinoseb cleanup: who pays the bill? Associated Press Newswires, February 24, 1999.
9. Fujita, K.-I. and Kamataki, T., Predicting the mutagenicity of tobacco-related N-nitrosamines in humans using 11 strains of *Salmonella typhimurium*, each coexpressing a form of human cytochrome P450 along with NADPH–cytochrome P450 reductase, *Environ. Mol. Mutag.*, 38, 339–346, 2001.
10. Fernandez, J.M., Vazquez, P.P., and Vidal, J.L.M., Analysis of N-methylcarbamate insecticides and some of their main metabolites in urine with liquid chromatography using diode array detection and electrospray mass spectrometry, *Anal. Chim. Acta*, 412, 131–139, 2000.
11. Gosselin, R.E., Smith, R.P., and Hodge, H.C., Paraquat, in *Clinical Toxicology of Commercial Products*, 5th ed., Williams & Wilkins, Baltimore, 1984, pp. III-328–III-336.
12. Wesseling, C. et al., Paraquat in developing countries, *Int. J. Occup. Environ. Health*, 7, 275–286, 2001.
13. Gosselin, R.E., Smith, R.P., and Hodge, H.C., Nicotine, in *Clinical Toxicology of Commercial Products*, 5th ed., Williams & Wilkins, Baltimore, 1984, pp. III-311–III-314.

14. Cone, E.J. et al., Cocaine metabolism and urinary excretion after different routes of administration, *Ther. Drug Monit.*, 20, 556–560, 1998.
15. Anderton, N. et al., New alkaloids from Phalaris spp.: a cause for concern? in *Book of Abstracts, 216th ACS National Meeting*, American Chemical Society, Washington, D.C., AGFD-168, 1998.

QUESTIONS AND PROBLEMS

1. Describe the sense in which amines may be regarded as derivatives of ammonia, NH_3. Distinguish among primary, secondary, and tertiary amines.
2. How are the compounds shown in the following figure characterized or described? What are their main toxicological characteristics?

3. What is the structural formula of aniline? What are its major uses? Why is human exposure to aniline likely to be relatively common? How is aniline taken into the body?
4. Which other nitrogen-containing nonamine organonitrogen compound does aniline most resemble in its toxicological characteristics? What is its most common manifestation of toxicity? How does this affect the subject?
5. What are fatty amines? From which raw materials that occur in nature are they commonly synthesized?
6. What are alkyl polyamines?
7. Of the following, the statement that is **not** true is:
   (a) The lower amines are corrosive to tissue and can cause tissue necrosis at the point of contact.
   (b) The most common toxic effect of the lower aliphatic amines is that they cause methemoglobinemia.
   (c) Sensitive eye tissue is vulnerable to amines.
   (d) Necrosis of the liver and kidneys can occur from exposure to amines, and exposed lungs can exhibit hemorrhage and edema.
   (e) The immune system may become sensitized to amines.
8. Explain what the reaction below shows about the toxicity of amines.

$$R_3N + H_2O \rightarrow R_3NH^+ + OH^-$$

9. Consider the compounds with the structural formulas shown below. Which of these are believed to be the actual toxic agents involved in aniline poisoning? Which are the forms eliminated from the body?

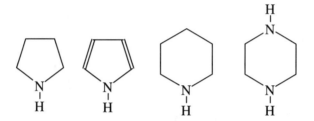

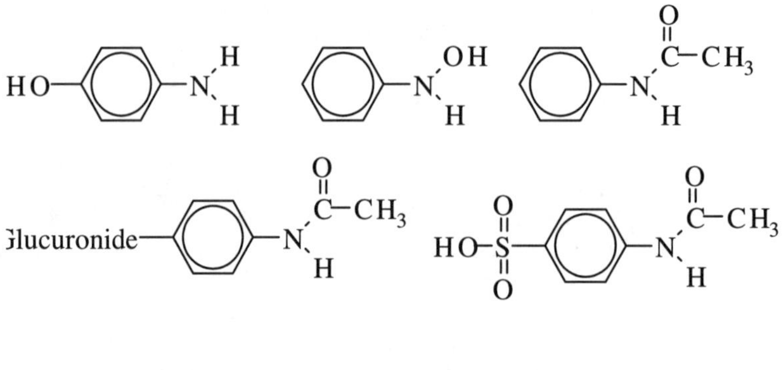

ORGANONITROGEN COMPOUNDS 325

10. What are the two naphthylamines? How does exposure to these compounds occur? What is their toxic effect of most concern?
11. What is the structural formula of pyridine. Is it highly toxic? In what respect is it like benzene?
12. Of which common inorganic compound are nitriles analogs? Which common natural product produces this highly toxic inorganic compound? How does this occur?
13. Acetonitrile is not highly toxic. What does this say about its toxicological chemistry and metabolism in the body?
14. What are two reasons that the compound below is of particular concern?

```
        H
        |
      H–C–H
        |
   HO–C–C≡N
        |
      H–C–H
        |
        H
```

15. Of which class of compounds is the N–N=O functional group characteristic? What is their most important toxicological characteristic?
16. Which class of compounds has the general formula R–N=C=O? Which of these is most notorious for an incident of poisoning? What happened?
17. What is the general formula of carbamates? Of which inorganic compound are they derivatives? How are carbamates used?
18. What does the reaction below illustrate?

```
    H                                      H
    |                                      |
  H–C–H                                  H–C–H
    |              Acidic                  |
  H–N    + NO₂⁻ + H⁺  ──────▶            N–N=O  + H₂O
    |              media                   |
  H–C–H                                  H–C–H
    |                                      |
    H                                      H
```

19. For what purposes are the compounds below used? What are their toxicity characteristics?

20. What kinds of compounds are the following? What are their sources? What may be said about their toxicities?

CHAPTER 16

Organohalide Compounds

16.1 INTRODUCTION

Organohalide compounds are halogen-substituted hydrocarbons with a wide range of physical and chemical properties produced in large quantities as solvents, heat transfer fluids, chemical intermediates, and for other applications. They may be saturated (alkyl halides), unsaturated (alkenyl halides), or aromatic (aryl halides). The major means of synthesizing organohalide compounds are shown by examples in Chapter 13 and include substitution halogenation, addition halogenation, and hydrohalogenation reactions, illustrated in reactions 13.2.2, 13.2.3, and 13.2.4, respectively. Most organohalide compounds are chlorides (chlorocarbons and chlorohydrocarbons), but they also include compounds of fluorine, bromine, and iodine, as well as mixed halides, such as the chlorofluorocarbons.

The chemical reactivities of organohalide compounds vary over a wide range. The alkyl halides are generally low in reactivity, but may undergo pyrolysis in flames to liberate noxious products, such as HCl gas. Alkenyl halides may be oxidized, which in some cases produces highly toxic phosgene, as shown by the following example:

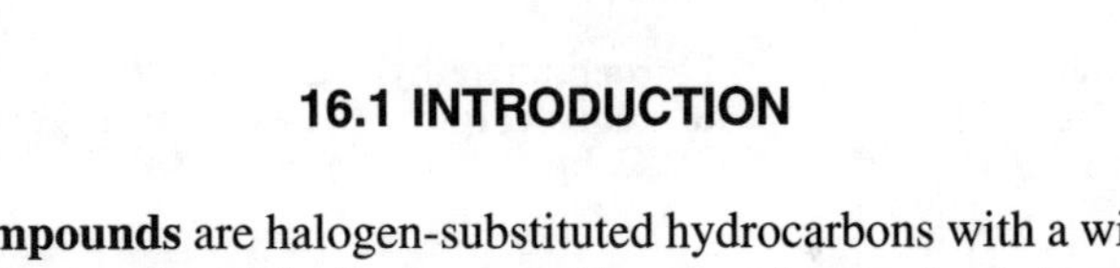

$$\underset{\text{Trichloroethylene}}{\overset{Cl}{\underset{H}{>}}C=C\overset{Cl}{\underset{Cl}{<}}} + O_2 \rightarrow HCl + \underset{\text{Phosgene}}{Cl-\overset{\overset{O}{\|}}{C}-Cl} + CO \qquad (16.1.1)$$

The toxicities of organohalide compounds vary widely. For example, dichlorodifluoromethane (Freon-12) is generally regarded as having a low toxicity, except for narcotic effects and the possibility of asphyxiation at high concentration. Vinyl chloride (see Section 16.3), however, is a known human carcinogen. The polychlorinated biphenyls (PCBs) are highly resistant to biodegradation and are extremely persistent in the environment.

16.1.1 Biogenic Organohalides

Organohalides were once regarded as being produced exclusively by human activities. However, more recent investigations have shown that organisms including algae and fungi release a variety of organohalides, and more than 2000 compounds from these biogenic sources have now been identified.[1] Most of these compounds are organochlorine and organobromine species. Bottom-ice microalgae and *Agarium cribrosum* kelp in the Arctic have been shown to be significant producers of environmental organobromine as bromoform, $HCBr_3$.[2]

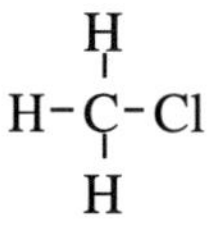

```
        H                    H                    Cl
        |                    |                    |
    H - C - Cl          Cl - C - Cl          Cl - C - Cl
        |                    |                    |
        H                    H                    Cl

   Chloromethane        Dichloromethane      Carbon tetrachloride
  (fp -98°C, bp -24°C)  (methylene chloride, (fp -23°C, bp 77°C)
                         fp -97°C, bp 40°C)

        F                   H  H                 Cl H
        |                   |  |                 |  |
    Cl - C - Cl         H - C- C - Cl       Cl - C- C - H
        |                   |  |                 |  |
        F                   H  H                 Cl H

  Dichlorodifluoro-     Chloroethane(ethyl-  1,1,1-Trichloroethane
  methane ("Freon 12,"  ene chloride, fp     (methyl chloroform,
  fp -158°C, bp -29°C)  -139°C, bp 12°C)     fp -33°C, bp 74°C)

         H  H
         |  |         1,2-Dibromoethane (ethylene dibromide,
    Br - C- C - Br    fp 9.3°C, bp 131°C)
         |  |
         H  H
```

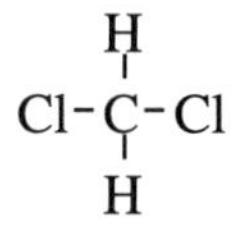

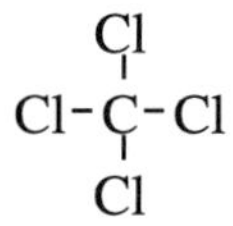

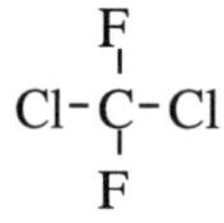

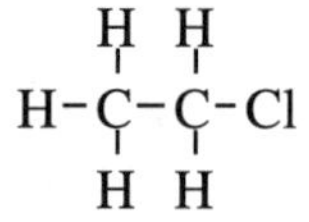

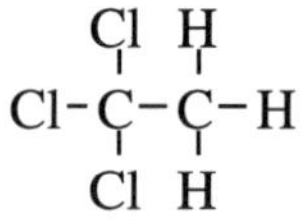

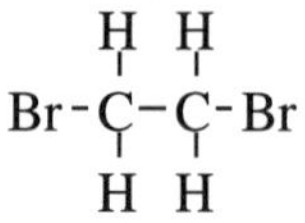

Figure 16.1 Some typical low-molecular-mass alkyl halides.

16.2 ALKYL HALIDES

Alkyl halides are compounds in which halogen atoms are substituted for hydrogen on an alkyl group. The structural formulas of some typical alkyl halides are given in Figure 16.1. Most of the commercially important alkyl halides are derivatives of alkanes of low molecular mass. A brief discussion of the uses of the compounds listed in Figure 16.1 will provide an idea of the versatility of the alkyl halides. Volatile chloromethane (methyl chloride) was once widely used as a refrigerant fluid and aerosol propellant; most of it now is consumed in the manufacture of silicones. Dichloromethane is a volatile liquid with excellent solvent properties for nonpolar organic solutes. It has been applied as a solvent for the decaffeination of coffee and in paint strippers, as a blowing agent in urethane polymer manufacture, and to depress vapor pressure in aerosol formulations. Once commonly sold as a solvent and stain remover, carbon tetrachloride is now severely curtailed. Chloroethane is an intermediate in the manufacture of tetraethyllead (now virtually discontinued in motor fuel) and is an ethylating agent in chemical synthesis. Methyl chloroform (1,1,1-trichloroethane) used to be one of the more common industrial chlorinated solvents. Insecticidal 1,2-dibromethane has been used in large quantities to fumigate soil, grain, and fruit and as a lead scavenger in leaded gasoline. It is an effective solvent for resins, gums, and waxes and serves as a chemical intermediate in the syntheses of some pharmaceutical compounds and dyes.

16.2.1 Toxicities of Alkyl Halides

The toxicities of alkyl halides vary a great deal with the compound. Although some of these compounds have been considered to be almost completely safe in the past, there is a marked tendency to regard each with more caution as additional health and animal toxicity study data become available. Perhaps the most universal toxic effect of alkyl halides is depression of the central nervous system. Chloroform, $CHCl_3$, was the first widely used general anesthetic, although many surgical patients were accidentally killed by it.

ORGANOHALIDE COMPOUNDS

16.2.2 Toxic Effects of Carbon Tetrachloride on the Liver

Of all the alkyl halides, carbon tetrachloride has the most notorious record of human toxicity, especially for its toxic effects on the liver. For many years it was widely used in consumer products as a degreasing solvent, in home fire extinguishers, and in other applications. However, numerous toxic effects, including some fatalities, were observed, and in 1970, the U.S. Food and Drug Administration (FDA) banned the sale of carbon tetrachloride and formulations containing it for home use.

Carbon tetrachloride is toxic through both inhalation and ingestion. Toxic symptoms from inhalation tend to be associated with nervous system, whereas those from ingestion often involve the gastrointestinal tract and liver. Both the liver and kidney may be substantially damaged by carbon tetrachloride.

The biochemical mechanism of carbon tetrachloride toxicity has been investigated in detail. The cytochrome P-450-dependent monooxygenase system acts on CCl_4 in the liver to produce the $Cl_3C\cdot$ free radical:

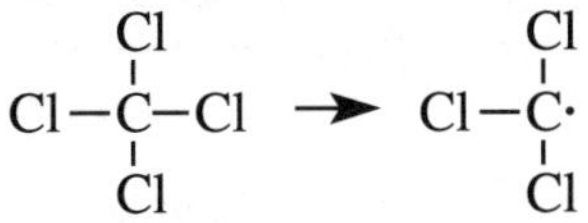

$$Cl_4C \rightarrow Cl_3C\cdot \qquad (16.2.1)$$

There are two major processes that this radical may initiate.[3] The radical can bind with liver cell components, the ultimate effect of which is inhibition of lipoprotein secretion. This causes fatty tissue to accumulate in the liver, leading to fatty liver or steatosis. Formation of DNA adducts with the $Cl_3C\cdot$ radical may initiate carcinogenesis. Another process that the $Cl_3C\cdot$ radical may undergo is combination with molecular oxygen to yield the highly reactive $Cl_3COO\cdot$ radical:

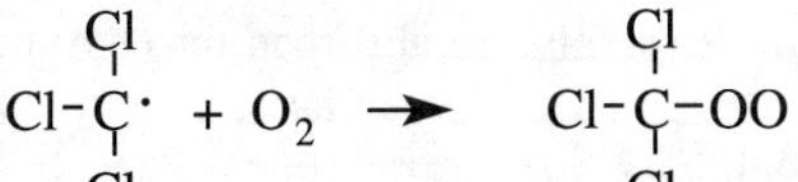

$$Cl_3C\cdot + O_2 \rightarrow Cl_3C-OO \qquad (16.2.2)$$

These radical species, along with others produced from their subsequent reactions, can react with biomolecules, such as proteins and DNA. The most damaging such reaction is **lipid peroxidation**, a process that involves the attack of chemically active species on unsaturated lipid molecules, followed by oxidation of the lipids through a free radical mechanism. It occurs in the liver and is a main mode of action of some hepatotoxicants, which can result in major cellular damage. The mechanism of lipid peroxidation may involve abstraction of the methylene hydrogens attached to doubly bonded carbon atoms in lipid molecules:

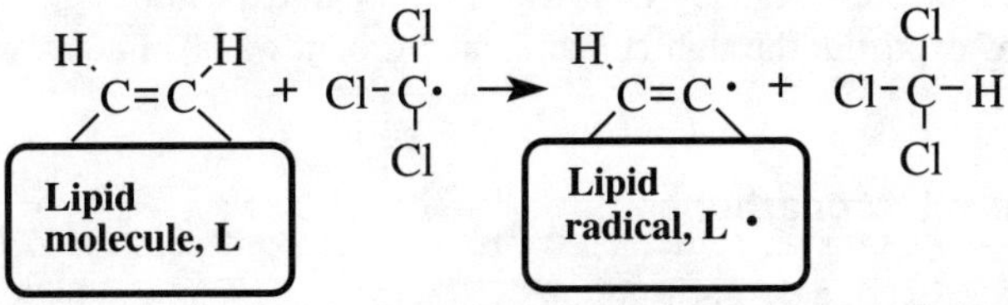

$$\underset{\text{Lipid molecule, L}}{H_2C=CH_2} + Cl_3C\cdot \rightarrow \underset{\text{Lipid radical, L}\cdot}{HC=CH\cdot} + Cl_3C-H \qquad (16.2.3)$$

Reaction of the lipid radical with molecular oxygen yields peroxy radical species:

$$\underset{\substack{\text{Lipid}\\\text{radical, L}\cdot}}{\overset{H}{\underset{}{>}}C=C\overset{\cdot}{<}} + O_2 \rightarrow \underset{\substack{\text{Lipid peroxy}\\\text{radical, LOO}\cdot}}{\overset{H}{\underset{}{>}}C=C\overset{OO\cdot}{<}} \qquad (16.2.4)$$

This species can initiate chain reaction sequences with other molecules as follows:

$$\underset{\substack{\text{Lipid peroxy}\\\text{radical, LOO}\cdot}}{\overset{H}{\underset{}{>}}C=C\overset{OO\cdot}{<}} + \underset{\substack{\text{Lipid}\\\text{molecule, L}}}{\overset{H}{\underset{}{>}}C=C\overset{H}{<}} \rightarrow \underset{\substack{\text{Lipid hydroper-}\\\text{oxide, LOOH}}}{\overset{H}{\underset{}{>}}C=C\overset{OOH}{<}} + \underset{\substack{\text{Lipid}\\\text{radical, L}\cdot}}{\overset{H}{\underset{}{>}}C=C\overset{\cdot}{<}} \qquad (16.2.5)$$

Once inititated, chain reactions such as these continue and cause massive alteration of the lipid molecules. The LOOH molecules are unstable and decompose to yield additional free radicals. The process terminates when free radical species combine with each other to form stable species.

16.2.3 Other Alkyl Halides

Dichloromethane has long been regarded as one of the least acutely toxic alkyl halides. This compound has been used in large quantities as a degreasing solvent, paint remover, aerosol propellant additive, and grain fumigant. Because of the high volatility of dichloromethane, its most common route of exposure is through air. It can also be absorbed through the skin or ingested with food or water. As a result of its properites and widespread uses, human exposure to dichloromethane has been relatively high. Human fatalities have occurred from very high exposures to methylene chloride in paint-stripping operations. It is not known to be a human carcinogen, although there is concern that it may possibly be carcinogenic.

Generally considered to be among the least toxic of the alkyl halides, 1,1,1-trichloroethane was once produced at levels of several hundred million kilograms per year. However, it is persistent in the atmosphere and is a strong stratospheric ozone-depleting chemical, so production has been severely curtailed and there is now little reason for concern over its toxicity.

A much more toxic alkyl halide is 1,2-dibromoethane. It is a severe irritant, damaging the lungs when inhaled in high concentrations, and a potential human carcinogen. It was widely used until the early 1980s to kill insects and worms on grain, vegetables, and fruits such as mangoes, papayas, and citrus. As a result, human exposure was relatively high. But these uses were banned by the U.S. Environmental Protection Agency in 1984, and human exposure is now negligible. There is still the possibility of exposure through contaminated groundwater in areas where dibromoethane has been used.

16.2.4 Hydrochlorofluorocarbons

Hydrochlorofluorocarbons (HCFCs) are now being produced in very large quantities as substitutes for ozone-depleting chlorofluorocarbons (CFCs). The two most common HCFCs are 1,1-dichloro-2,2,2-trifluoroethane (HCFC-123) and 1,1-dichloro-1-fluoroethane (HCFC-141b):

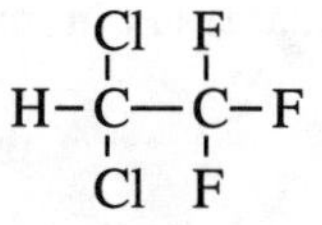

ORGANOHALIDE COMPOUNDS

$$\begin{array}{cc} \text{Cl} & \text{F} \\ | & | \\ \text{H-C-C-F} \\ | & | \\ \text{Cl} & \text{F} \end{array} \qquad \begin{array}{cc} \text{Cl} & \text{H} \\ | & | \\ \text{F-C-C-H} \\ | & | \\ \text{Cl} & \text{H} \end{array}$$

1,1-Dichloro-2,2,2-trifluoroethane (HCFC-123) 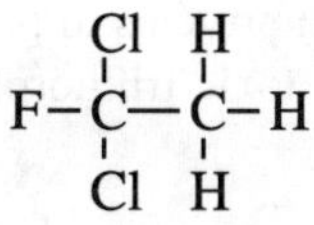   **1,1-Dichloro-1-fluoroethane (HCFC-141b)**

As a result of their increased production and the consequent exposure of organisms to HCFCs, these compounds have been subjected to intense scrutiny for their potential toxicological effects. Studies with rat liver tissue indicate that both HCFC-123 and HCFC-141b are metabolized by cytochrome P-450 enzymes to produce reactive metabolites, probably with intermediate production of free radical species. Studies of human volunteers who had inhaled levels of 250, 500, and 1000 ppm HCFC-141b showed that the major metabolite excreted in urine was 2,2-dichloro-2-fluoroethyl glucuronide, hydrolyzed by the action of β-glucuronidase enzyme to give 2,2-dichloro-2-fluoroethanol, which was measured by gas chromatography[4]:

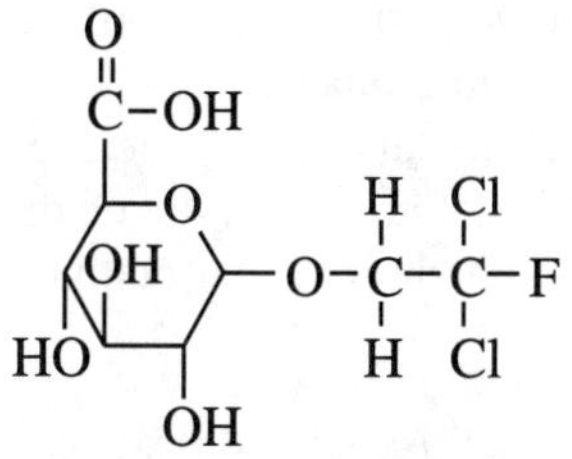

2,2-Dichloro-2-fluoroethyl glucuronide 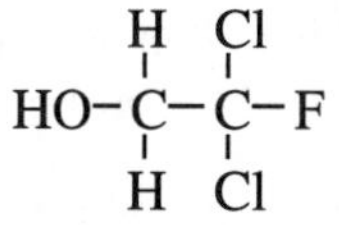   **2,2-Dichloro-2-fluoroethanol**

16.2.5 Halothane

Extensive toxicological studies have been performed on halothane,

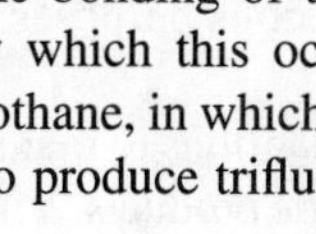

$$\begin{array}{cc} \text{F} & \text{Cl} \\ | & | \\ \text{F-C-C-Br} \\ | & | \\ \text{F} & \text{Cl} \end{array}$$

because it is a commonly used anesthetic. In rare cases, repeated exposure to halothane has caused liver cell necrosis, resulting in fatal halothane hepatitis in humans. This condition has been classified as an immune hepatitis resulting from the bonding of trifluoroacetyl metabolite of halothane to proteins.[5] The sequence of processes by which this occurs begins with the cytochrome P-450 catalyzed oxidative dehalogenation of halothane, in which the bromine atom, the best leaving group of the halogens on the molecule, is lost to produce trifluoroacetylchloride:

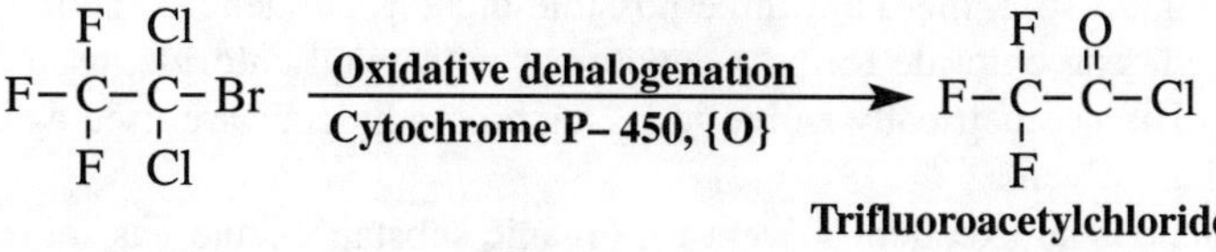

$$\begin{array}{cc} \text{F} & \text{Cl} \\ | & | \\ \text{F-C-C-Br} \\ | & | \\ \text{F} & \text{Cl} \end{array} \xrightarrow[\text{Cytochrome P-450, \{O\}}]{\text{Oxidative dehalogenation}} \begin{array}{cc} \text{F} & \text{O} \\ | & \| \\ \text{F-C-C-Cl} \\ | \\ \text{F} \end{array} \qquad (16.2.6)$$

Trifluoroacetylchloride

This product can bind with proteins to produce neo-antigens that result in immune hepatitis. It can also hydrolyze to produce toxic trifluoroacetic acid:

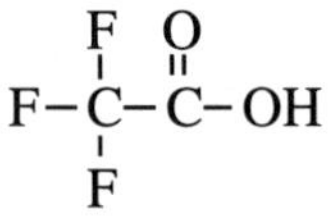

$$\underset{\underset{F}{|}}{\overset{\overset{F}{|}}{F-C}}-\overset{\overset{O}{\|}}{C}-OH \quad \text{Trifluoroacetic acid}$$

Halothane can also undergo reductive dehalogenation,

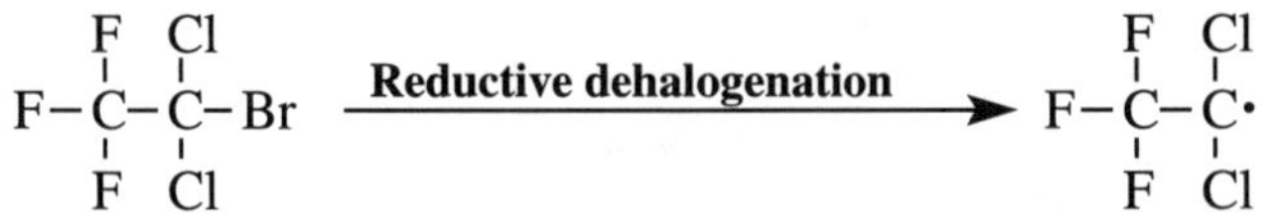

$$\underset{\underset{F}{|}\;\underset{Cl}{|}}{\overset{\overset{F}{|}\;\overset{Cl}{|}}{F-C-C}}-Br \quad \xrightarrow{\textbf{Reductive dehalogenation}} \quad \underset{\underset{F}{|}\;\underset{Cl}{|}}{\overset{\overset{F}{|}\;\overset{Cl}{|}}{F-C-C\cdot}} \qquad (16.2.7)$$

to generate a carbon-centered radical. As shown above for the $Cl_3C\cdot$ radical generated from carbon tetrachloride, this radical species may be involved with lipid peroxidation and protein binding, resulting in liver damage to rats and presumably to humans.

Evidence for the halothane metabolism outlined above has been found in products recovered from the breath and urine of humans subjected to halothane anesthetic during surgery.[6] As evidence of reductive metabolites, chlorotrifluoroethane and chlorodifluoroethylene,

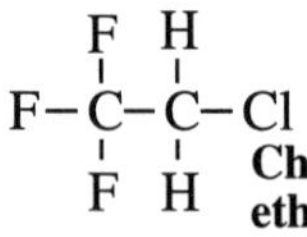

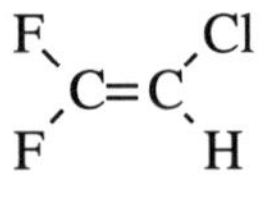

$$\underset{\underset{F}{|}\;\underset{H}{|}}{\overset{\overset{F}{|}\;\overset{H}{|}}{F-C-C}}-Cl \quad \text{Chlorotrifluoro-ethane} \qquad \overset{F}{\underset{F}{\diagdown}}C=C\overset{Cl}{\underset{H}{\diagup}} \quad \text{Chlorodifluoro-ethylene}$$

were found in breath samples and F⁻ in urine. Trifluoroacetic acid and Br⁻ in urine were evidence of oxidative metabolism.

16.3 ALKENYL HALIDES

The **alkenyl**, or **olefinic organohalides**, contain at least one halogen atom and at least one carbon–carbon double bond. The most significant of these are the lighter chlorinated compounds, such as those illustrated in Figure 16.2.

16.3.1 Uses of Alkenyl Halides

The alkenyl halides are used for numerous purposes. Some of the more important applications are discussed here.

Vinyl chloride is consumed in large quantities to manufacture polyvinyl chloride plastic, a major polymer in pipe, hose, wrapping, and other products. Vinyl chloride is a highly flammable volatile gas with a sweet, not unpleasant odor.

As shown in Figure 16.2, there are three possible dichloroethylene compounds, all clear, colorless liquids. Vinylidene chloride forms a copolymer with vinyl chloride, used in some kinds of coating materials. The geometrically isomeric 1,2-dichloroethylenes are used as organic synthesis intermediates and as solvents.

Trichloroethylene is an excellent solvent for organic substances and has some other properties that are favorable for a solvent. It is a clear, colorless, nonflammable, volatile liquid. It is an excellent

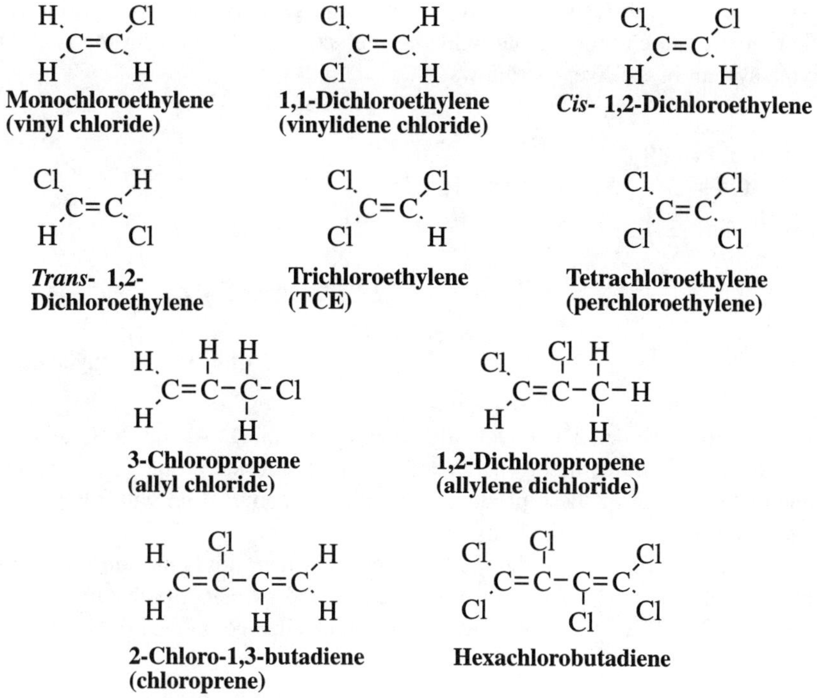

H Cl	Cl H	Cl Cl
C=C	C=C	C=C
H H	Cl H	H H
Monochloroethylene	1,1-Dichloroethylene	*Cis*- 1,2-Dichloroethylene
(vinyl chloride)	(vinylidene chloride)	

Cl H	Cl Cl	Cl Cl
C=C	C=C	C=C
H Cl	Cl H	Cl Cl
Trans- 1,2-Dichloroethylene	Trichloroethylene (TCE)	Tetrachloroethylene (perchloroethylene)

3-Chloropropene (allyl chloride)

1,2-Dichloropropene (allylene dichloride)

2-Chloro-1,3-butadiene (chloroprene)

Hexachlorobutadiene

Figure 16.2 The more common low-molecular-mass alkenyl chlorides

degreasing and dry-cleaning solvent and has been used as a household solvent and for food extraction (for example, in decaffeination of coffee).

Tetrachloroethylene is a colorless, nonflammable liquid with properties similar to those of trichloroethylene. Its major use is for dry cleaning, and it has some applications for degreasing metals.

The two chlorinated propene compounds shown are colorless liquids with pungent, irritating odors. Allyl chloride is an intermediate in the manufacture of allyl alcohol and other allyl compounds, including pharmaceuticals, insecticides, and thermosetting varnish and plastic resins. Dichloropropene compounds have been used as soil fumigants, as well as solvents for oil, fat, dry cleaning, and metal degreasing.

Large quantities of chloroprene, a colorless liquid with an ethereal odor, are used to make neoprene rubber. Hexachlorobutadiene is a colorless liquid with an odor somewhat like that of turpentine. It is used as a solvent for higher hydrocarbons and elastomers, as a hydraulic fluid, in transformers, and for heat transfer.

16.3.2 Toxic Effects of Alkenyl Halides

Because of their widespread use and disposal in the environment, the toxicities of the alkenyl halides are of considerable concern. They exhibit a wide range of acute and chronic toxic effects.

Many workers have been exposed to vinyl chloride because of its use in polyvinyl chloride plastic manufacture. The central nervous system, respiratory system, liver, and blood and lymph systems are all affected by exposure to vinyl chloride. Among the symptoms of poisoning are fatigue, weakness, and abdominal pain. Cyanosis may also occur. Vinyl chloride was abandoned as an anesthetic when it was found to induce cardiac arrhythmias.

The most notable effect of vinyl chloride is its carcinogenicity. It causes a rare angiosarcoma of the liver in chronically exposed individuals, observed particularly in those who cleaned autoclaves in the polyvinyl chloride fabrication industry. The carcinogenicity of vinyl chloride results from its metabolic oxidation to chloroethylene oxide by the action of the cytochrome P-450 monooxygenase enzyme system in the liver as follows:

$$\underset{H}{\overset{H}{>}}C=C\underset{H}{\overset{Cl}{<}} + \{O\} \rightarrow H-\underset{H}{\overset{O}{\underset{|}{C}}}-\underset{H}{\overset{|}{C}}-Cl$$

Rearrangement → $Cl-\underset{H}{\overset{H}{\underset{|}{C}}}-\overset{O}{\overset{\|}{C}}-H$

Chloroacetaldehyde (16.3.1)

The epoxide has a strong tendency to covalently bond to protein, DNA, and RNA, and it rearranges to chloroacetaldehyde, a known mutagen. Therefore, vinyl chloride produces two potentially carcinogenic metabolites. Both of these products can undergo conjugation with glutathione to yield products that are eliminated from the body.

It has been suggested that one of the mechanisms by which vinyl chloride causes liver cancer is by the addition of etheno (C_2H_4) adducts to adenine and cytosine, both nitrogenous bases in DNA (see Section 3.7). The addition of an etheno group (shaded below) to adenine produces 1,N6-ethenoadenine:

Adenine → Addition of C_2H_4 → **1,N6-Ethenoadenine** (16.3.2)

and its addition to cytosine produces 3,N4-ethenocytosine:[7]

Cytosine → Addition of C_2H_4 → **3,N4-Ethenocytosine** (16.3.3)

Based on animal studies and its structural similarity to vinyl chloride, 1,1-dichloroethylene is a suspect human carcinogen. Although both 1,2-dichloroethylene isomers have relatively low toxicities, their modes of action are different. The *cis* isomer is an irritant and narcotic, whereas the *trans* isomer affects both the central nervous system and the gastrointestinal tract, causing weakness, tremors, cramps, and nausea.

Trichloroethylene has caused liver carcinoma in experimental animals and is a suspect human carcinogen, although a recent review of the literature has concluded that "it would be wholly inappropriate to classify trichloroethylene as a human carcinogen."[8] Numerous body organs are affected by it. As with other organohalide solvents, skin dermatitis can result from dissolution of skin lipids by trichloroethylene. Exposure to it can affect the central nervous and respiratory systems, liver, kidneys, and heart. Symptoms of exposure include disturbed vision, headaches, nausea, cardiac arrhythmias, and burning or tingling sensations in the nerves (paresthesia). Trichloroacetate ion,

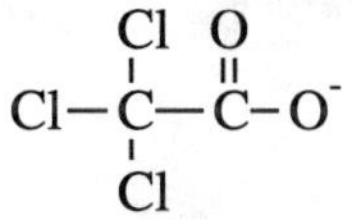

$$\text{Cl}-\underset{\underset{\text{Cl}}{|}}{\overset{\overset{\text{Cl}}{|}}{\text{C}}}-\overset{\overset{\text{O}}{\|}}{\text{C}}-\text{O}^- \quad \textbf{Trichloroacetate ion}$$

is a metabolite of trichloroethylene and may be toxicologically important.[9]

Tetrachloroethylene damages the liver, kidneys, and central nervous system. Because of its hepatotoxicity and experimental evidence of carcinogenicity in mice, it is a suspect human carcinogen.

The chlorinated propenes are obnoxious compounds. Unlike other compounds discussed so far in this section, their pungent odors and irritating effects lead to an avoidance response in exposed subjects. They are irritants to the eyes, skin, and respiratory tract. Contact with the skin can result in rashes, blisters, and burns. Chronic exposure to allyl chloride is manifested by aching muscles and bones; it damages the liver, lungs, and kidney and causes pulmonary edema.

Chloroprene is an eye and respiratory system irritant. It causes dermatitis to the skin and alopecia, a condition characterized by hair loss in the affected skin area. Affected individuals are often nervous and irritable.

Ingestion and inhalation of hexachlorobutadiene inhibits cells in the liver and kidney. Animal tests have shown both acute and chronic toxicities. The compound is a suspect human carcinogen.

16.3.3 Hexachlorocyclopentadiene

As shown by the structure below, hexachlorocyclopentadiene is a cyclic alkenyl halide with two double bonds:

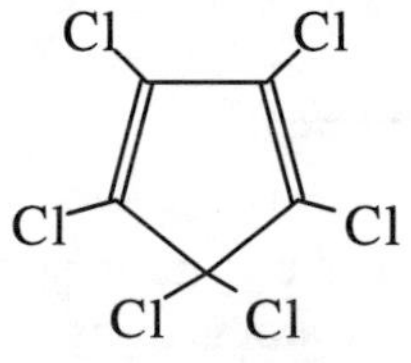

Hexachlorocyclopentadiene

It was once an important industrial chemical used directly as an agricultural fumigant and as an intermediate in the manufacture of insecticides. Hexachlorocyclopentadiene and still bottoms from its manufacture are found in hazardous waste chemical sites, and large quantities were disposed at the Love Canal site. The pure compound is a light yellow liquid (fp, 11°C; bp, 239°C) with a density of 1.7 g/cm^3 and a pungent, somewhat musty odor. With two double bonds, it is a very reactive compound and readily undergoes substitution and addition reactions. Its photolytic degradation yields water-soluble products.

Hexachlorocyclopentadiene is considered to be very toxic, with a toxicity rating of 4. Its fumes are strongly lacrimating, and it is a skin, eye, and mucuous membrane irritant. In experimental animals it has been found to damage most major organs, including the kidney, heart, brain, adrenal glands, and liver.

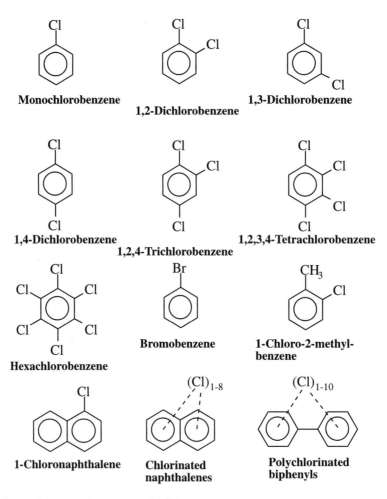

Figure 16.3 Some of the more important aryl halides.

16.4 ARYL HALIDES

Figure 16.3 gives the structural formulas of some of the more important aryl halides. These compounds are made by the substitution chlorination of aromatic hydrocarbons, as shown, for example, by the reaction below for the synthesis of a polychlorinated biphenyl:

$$\text{C}_6\text{H}_5\text{-C}_6\text{H}_5 + 5\text{Cl}_2 \xrightarrow{\text{Fe}/\text{FeCl}_2} \text{C}_{12}\text{H}_5\text{Cl}_5 + 5\text{HCl} \qquad (16.4.1)$$

16.4.1 Properties and Uses of Aryl Halides

Aryl halides have many uses, which have resulted in substantial human exposure and environmental contamination. Some of their major applications are summarized here.

Monochlorobenzene is a flammable clear liquid (fp, −45°C; bp, 132°C) used as a solvent, solvent carrier for methylene diisocyanate, pesticide, heat transfer fluid, and in the manufacture of aniline, nitrobenzene, and phenol. The 1,2- isomer of dichlorobenzene (*ortho*-dichlorobenzene) has been used as a solvent for degreasing hides and wool and as a raw material for dye manufacture. The 1,4- isomer (*para*-dichlorobenzene) is also used in dye manufacture and as a moth repellant and germicide. All three isomers have been used as fumigants and insecticides. The 1,2- and 1,3- (*meta*) isomers are liquids under ambient conditions, whereas the 1,4- isomer is a white sublimable solid. Used as a solvent, lubricant, dielectric fluid, chemical intermediate, and formerly as a termiticide, 1,2,4-trichlorobenzene is a liquid (fp, 17°C; bp, 213°C).

Hexachlorobenzene (perchlorobenzene) is a high-melting solid consisting of white needles and used as a seed fungicide, wood preservative, and intermediate for organic synthesis. Bromobenzene (fp, −31°C; bp, 156°C) serves as a solvent and motor oil additive, as well as an intermediate for organic synthesis. Most 1-chloro-2-methylbenzene is consumed in the manufacture of 1-chlorobenzotrifluoride.

There are two major classes of halogenated aryl compounds containing two benzene rings. One class is based on naphthalene and the other on biphenyl, as shown by the examples in Figure 16.3. For each class of compounds, the individual members range from liquids to solids, depending on the degree of chlorination. These compounds are manufactured by chlorination of the parent compounds and have been sold as mixtures with varying degrees of chlorine content. The desirable properties of the chlorinated naphthalenes, polychlorinated biphenyls, and polybrominated biphenyls, including their physical and chemical stabilities, have led to many uses, such as for heat transfer, and hydraulic fluids, dielectrics, and flame retardants. However, for environmental and toxicological reasons, these uses have been severely curtailed.

16.4.2 Toxic Effects of Aryl Halides

Exposure to monochlorobenzene usually occurs by inhalation or skin contact. It is an irritant and affects the respiratory system, liver, skin, and eyes. Ingestion of this compound has caused incoordination, pallor, cyanosis, and eventual collapse, effects similar to those of aniline poisoning (see Section 15.3). Workers exposed to chlorobenzene have complained of headaches, numbness, sleepiness, and digestive symptoms, including nausea and vomiting. In general, most of these workers were exposed to other substances as well, so it is uncertain that their symptoms were due to chlorobenzene alone.

Exposure to the dichlorobenzenes is also most likely to occur through inhalation or contact. These compounds are irritants and tend to damage the same organs as monochlorobenzene. The 1,4- isomer has been known to cause profuse rhinitis (running nose), nausea, weight loss associated with anorexia, jaundice, and liver cirrhosis. The di- and tetrachlorobenzenes are considered to be moderately toxic by inhalation and ingestion.

Hexachlorobenzene is a notorious compound in the annals of toxicology because of a massive poisoning incident involving 3000 people in Turkey during the period of 1955 to 1959. The victims ate seed wheat that had been treated with 10% hexachlorobenzene to deter fungal growth. As a consequence, they developed **porphyria cutanea tarda**, a condition in which the skin becomes blistered, fragile, photosensitive, and subject to excessive hair growth. In addition to the skin damage, the victims' eyes were damaged in severe cases, and many suffered weight loss associated with anorexia. Wasting of skeletal muscles was also observed. The possibility exists that many of these effects were due to the presence of manufacturing by-product impurity polychlorinated dibenzodioxins (see Section 16.6).

Bromobenzene can enter the body through the respiratory tract, gastrointestinal tract, or skin. Little information is available regarding its human toxicity. It has been shown to damage the livers of rats used in animal tests.

Wide variations have been reported in the toxicities of the chlorinated naphthalenes, raising the possibility that some of the effects observed were due to impurities introduced during manufacture. Humans exposed to the more highly halogenated fractions by inhaling the vapors have developed chloracne rash and have suffered from debilitating liver necrosis. In the 1940s and early 1950s, several hundred thousand cattle died from polychlorinated naphthalene-contaminated feed.

Polychlorinated biphenyls have received special attention as environmental pollutants and toxicants because of their widespread manufacture and use and extreme persistence. With highly sensitive electron capture detectors for gas chromatography, PCBs and their metabolites are routinely detected in the blood of people from the general population. A likely route of exposure to PCBs is through contaminated food, especially fish and wild game. Infants may be exposed through breast milk. Acute exposure of workers to PCBs has caused nose and lung irritation and is alleged to have caused limb weakness and numbness, liver damage, and altered immune systems. Although the carcinogenicity of PCBs to humans is subject to a great deal of uncertainty, both the International Agency for Research on Cancer (IARC) and the U.S. Environmental Protection Agency classify PCBs as probable carcinogens. The toxicology and health effects of PCBs have been summarized in the proceedings of a workshop on that topic.[10]

Although PCBs are poorly biodegradable and tend to accumulate in adipose tissue, they are metabolized to a certain extent. Prominent among their metabolic products found in blood are phenolic derivatives with at least one –OH group attached to the aromatic rings of PCBs.[11]

The polybrominated biphenyl (PBB) analogs of PCBs were the cause of massive livestock poisoning in Michigan in 1973 because of the addition of PBB flame retardant to livestock feed during its formulation.

16.5 ORGANOHALIDE INSECTICIDES

Organohalide compounds were the first of the widely used synthetic organic pesticides. In this section organohalide insecticides are discussed, and in Section 16.6 other pesticides of the organohalide chemical type are covered.

Figure 16.4 shows the structural formulas of some of the more common organohalide insecticides, now discontinued and of historical interest, and of concern in some old hazardous waste sites. Most of the insecticidal organohalide compounds contain chlorine as the only halogen. Ethylene dibromide and dichlorobromopropane are insecticidal, but are more properly classified as fumigants and nematocides.

As seen from the structural formulas in Figure 16.4, the organochlorine insecticides are of intermediate molecular mass and contain at least one aromatic or nonaromatic ring. They can be placed in four major chemical classes. The first of these consists of the chloroethylene derivatives, of which DDT and methoxychlor are the prime examples. The second major class is composed of chlorinated cyclodiene compounds, including aldrin, dieldrin, and heptachlor. The most highly chlorinated members of this class, such as chloredecone, are manufactured from hexachlorocyclopentadiene (see Section 16.3). The benzene hexachloride stereoisomers make up a third class of organochlorine insecticides, and the third group, known collectively as toxaphene, constitutes a fourth.

16.5.1 Toxicities of Organohalide Insecticides

Organohalide insecticides exhibit a wide range of toxic effects and varying degrees of toxicity. Many of these compounds are neuropoisons, and their most prominent acute effects are on the central nervous system, manifested by symptoms of central nervous system poisoning, including tremor, irregular jerking of the eyes, changes in personality, and loss of memory. Some of the toxic effects of specific organohalide insecticides and classes of these compounds are discussed below.

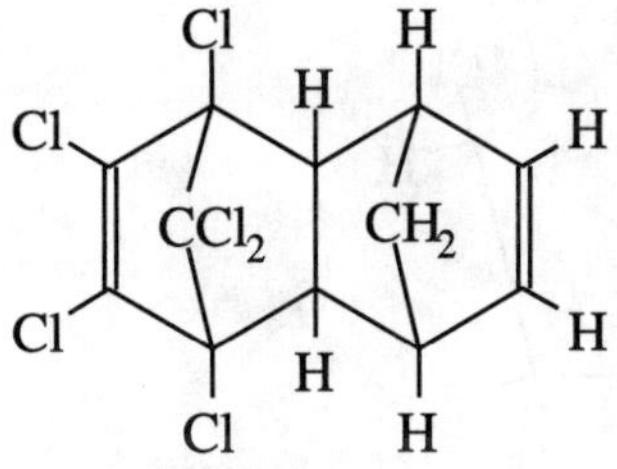

Aldrin

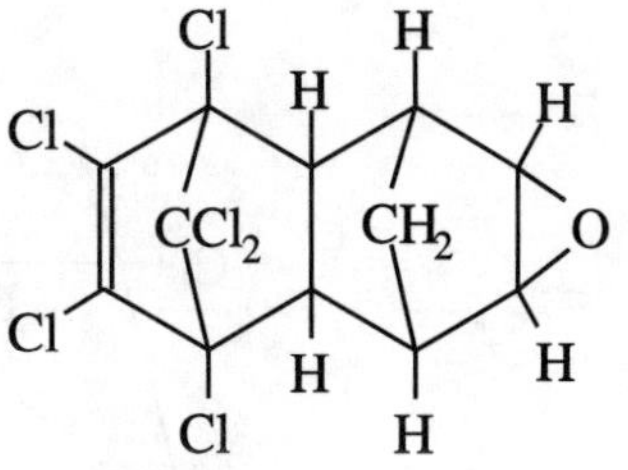

Dieldrin

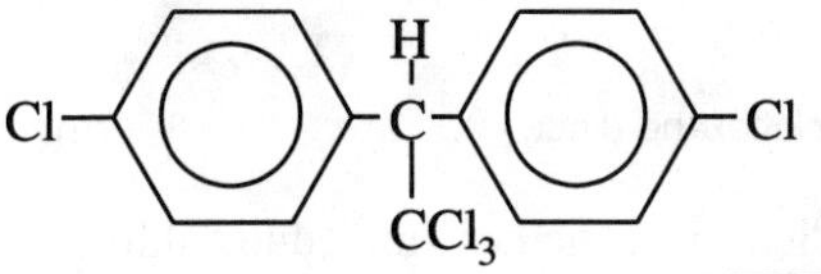

DDT

Methoxychlor

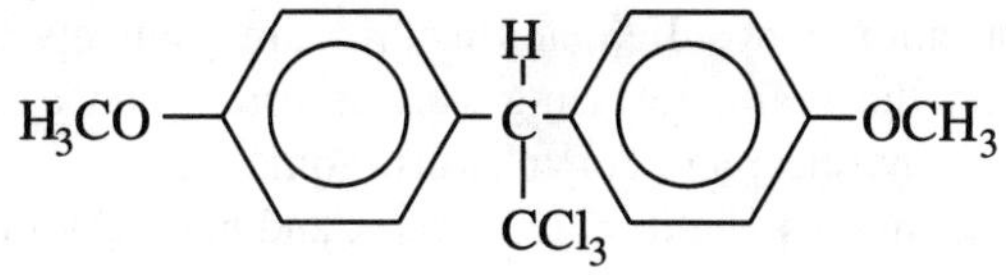

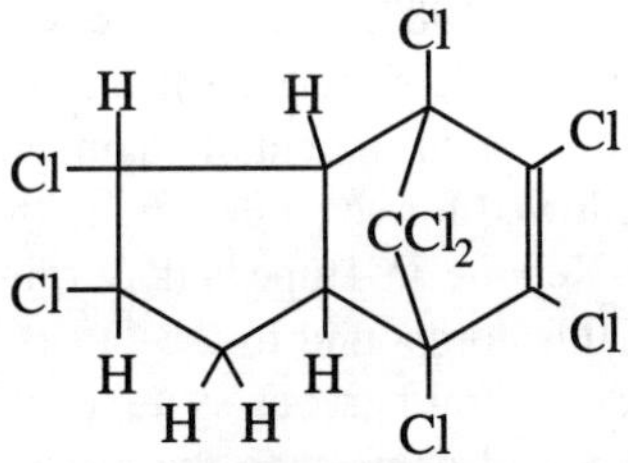

Chlordane

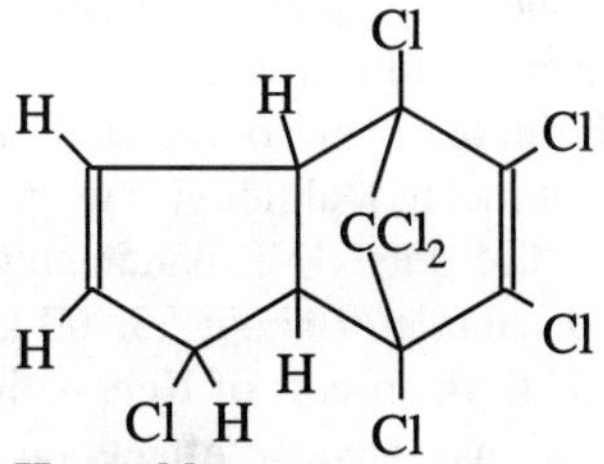

Heptachlor

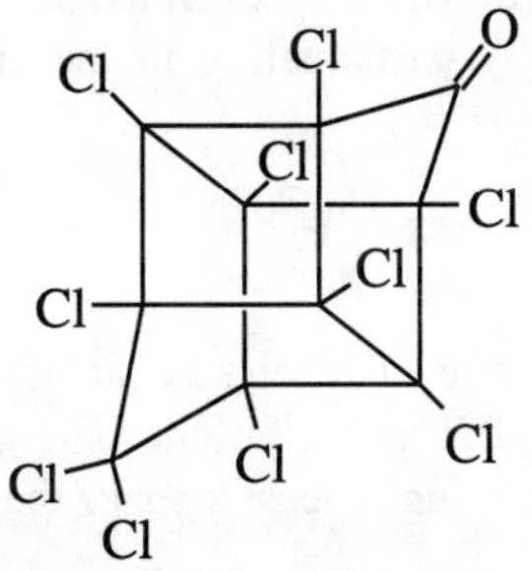

Chlordecone (Kepone); Mirex is structurally similar with 2 Cls in place of =O

Figure 16.4 Some typical organohalide insecticides.

Despite its role in the establishment of the modern environmental movement in Rachel Carson's classic book, *Silent Spring*, DDT's acute toxicity to humans is very low. It was applied directly to people on a large scale during World War II for the control of typhus and malaria. Symptoms of acute DDT poisoning are much the same as those described previously for organohalide insecticides in general and are, for the most part, neurotoxic in nature. In the environment, DDT undergoes bioaccumulation in the food chain, with animals at the top of the chain most affected. The most vulnerable of these are predator birds, which produce thin-shelled, readily broken eggs from ingestion of DDT through the food chain. The other major insecticidal chloroethane-based compound, methoxychlor, is a generally more biodegradable, less toxic compound than DDT, and has been used as a substitute for it.

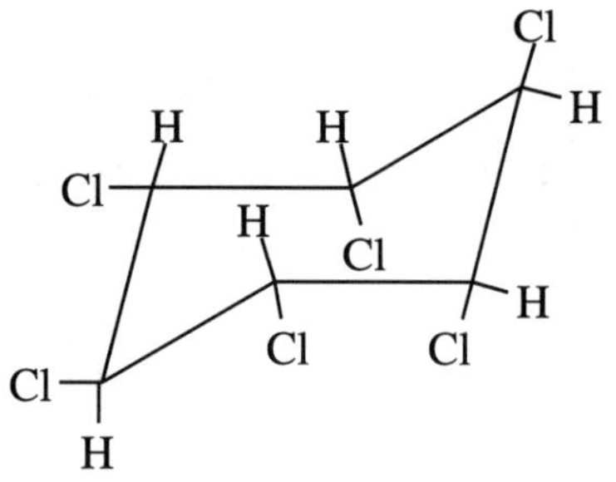

Figure 16.5 The gamma isomer of hexachlorocyclohexane (lindane).

The toxicities of the chlorinated cyclodiene insecticides, including aldrin, dieldrin, endrin, chlordane, heptachlor, endosulfan, and isodrin, are relatively high and similar to each other. They appear to act on the brain, releasing betaine esters and causing headaches, dizziness, nausea, vomiting, jerking muscles, and convulsions. Some members of this group are teratogenic or toxic to fetuses. In test animals, dieldrin, chlordane, and heptachlor cause liver cancer. The use of aldrin, dieldrin, and heptachlor has long been prohibited in the United States. The use of chlordane was continued for underground applications for termite control. In 1987, even this use was discontinued.

A significant number of human exposures to the insecticides derived from hexachlorocyclopentadiene (Mirex and Kepone) have occurred. Use of these environmentally damaging compounds was discontinued some time ago, although they were allowed for several years in the southeastern U.S. for eradication of fire ants. The manufacture of Kepone in Hopewell, Virginia, during the 1970s resulted in the discharge of about 53,000 kg of this compound to the James River through the city sewage system. Toxic effects of Kepone include central nervous system symptoms (irritability, tremor, hallucinations), adverse effects on sperm, and damage to the nerves and muscles. The compound causes liver cancer in rodents and is teratogenic in test animals. Studies of exposed workers have shown that Kepone absorbed by the liver is excreted through the bile and then reabsorbed from the gastrointestinal tract, thereby participating in the enterohepatic circulation system, as illustrated in Section 7.4.

16.5.2 Hexachlorocyclohexane

Hexachlorocyclohexane, once confusingly called benzene hexachloride (BHC), consists of several stereoisomers with different orientations of H and Cl atoms. The gamma isomer is shown in Figure 16.5. It is an effective insecticide, constituting at least 99% of the commercial insecticide **lindane**.

The toxic effects of lindane are very similar to those of DDT. Degeneration of kidney tubules, liver damage associated with fatty tissue, and hystoplastic anemia have been observed in individuals poisoned by lindane.

16.5.3 Toxaphene

Toxaphene is insecticidal chlorinated camphene and consists of a mixture of more than 170 compounds containing 10 C atoms and 6 to 10 Cl atoms per molecule and often represented by the empirical formula $C_{10}H_{10}Cl_8$. The structural formula of one of the molecules contained in toxaphene, 8-octachlorobornane, is given below. Toxaphene was once the most used insecticide in the U.S., with annual consumption of about 40 million kg.

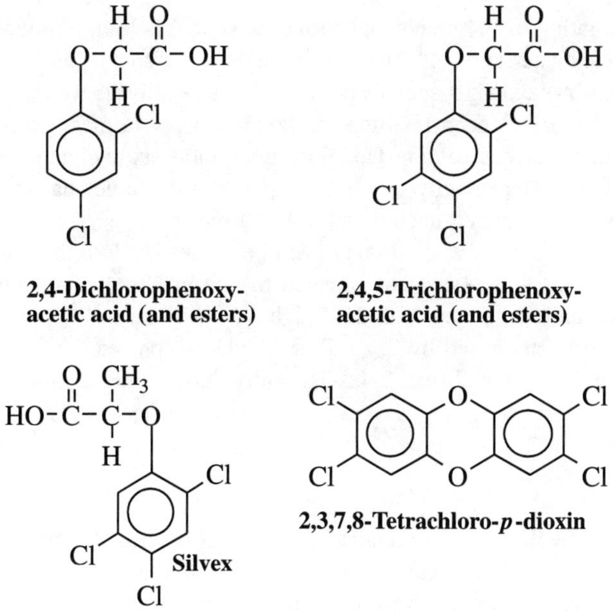

Figure 16.6 Herbicidal chlorophenoxy compounds and TCDD manufacturing by-product.

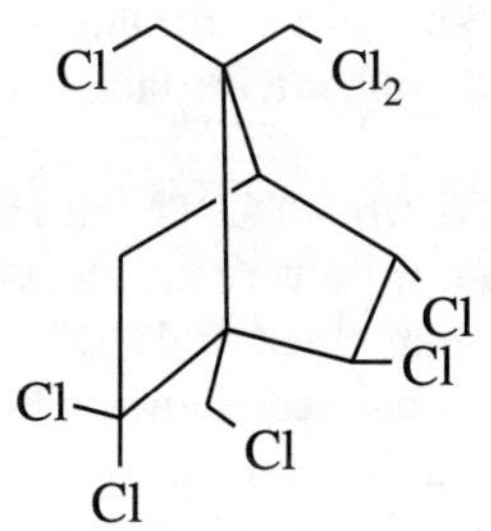

The many compounds found in formulations of toxaphene vary widely in their toxicities. One of the most toxic is 8-octachlorobornane, shown above. Toxaphene produces convulsions of an epileptic type in exposed mammals.

16.6 NONINSECTICIDAL ORGANOHALIDE PESTICIDES

The best-known noninsecticidal organohalide pesticides are the **chlorophenoxy** compounds. These consist of 2,4-dichlorophenoxyacetic acid (2,4-D), 2,4,5-trichlorophenoxyacetic acid (2,4,5-T, or Agent Orange), and a closely related compound, Silvex. These compounds, their esters, and their salts have been used as ingredients of a large number of herbicide formulations. Formulations of 2,4,5-T have become notorious largely by a manufacturing by-product, 2,3,7,8-tetrachlorodibenzo-p-dioxin (TCDD), commonly known as dioxin. The structural formulas of these compounds are shown in Figure 16.6.

16.6.1 Toxic Effects of Chlorophenoxy Herbicides

The oral toxicity rating of 2,4-dichlorophenoxyacetic acid is four, although the toxicities of its commercially marketed ester and salt forms are thought to be somewhat lower. Large doses have been shown to cause nerve damage, such as peripheral neuropathy, as well as convulsions and even brain damage. A National Cancer Institute study of Kansas farmers who had handled 2,4-D extensively showed an occurrence of non-Hodgkins lymphoma six to eight times that of comparable unexposed populations.[12] The toxicity of Silvex appears to be somewhat less than that of 2,4-D, and to a large extent, it is excreted unchanged in the urine.

Although the toxic effects of 2,4,5-T may even be somewhat less than those of 2,4-D, observations of 2,4,5-T toxicity have been complicated by the presence of manufacturing by-product TCDD. Experimental animals dosed with 2,4,5-T have exhibited mild spasticity. Some fatal poisonings of sheep have been caused by 2,4,5-T herbicide. Autopsied carcasses revealed nephritis, hepatitis, and enteritis. Humans absorb 2,4,5-T rapidly and excrete it largely unchanged through the urine.

16.6.2 Toxicity of TCDD

TCDD belongs to the class of compounds called **polychlorinated dibenzodioxins**, which have the same basic structure as TCDD, but different numbers and arrangements of chlorine atoms on the ring structure. These compounds exhibit varying degrees of toxicity. Classified as a supertoxic compound, TCDD is unquestionably extremely toxic to some animals. Its acute LD_{50} to male guinea pigs is only 0.6 µg/kg of body mass. Because of its production as a manufacturing by-product of some commercial products, such as 2,4,5-T, possible emission from municipal incineration, and widespread distribution in the environment from improper waste disposal (for example, as the infamous dioxin spread from waste oil at Times Beach, Missouri) or discharge from industrial accidents (Seveso, Italy), TCDD has become a notorious environmental pollutant. However, the degree and nature of its toxicity to humans are both rather uncertain. It is known to cause a human skin condition called chloracne.

Animal studies have shown a variety of effects from TCDD and related chemicals. As is the case with guinea pigs, some of these changes result from very low doses. Adverse effects may last for a long time. Body mass loss is the most common effect of TCDD. Tissue of the urinary tract epithelium and the gastrointestinal mucosa may be harmed, showing both abnormal cell proliferation and enlargement. The thymus often shows adverse effects, and liver cells may be killed or may undergo abnormal proliferation.[13] The half-life of TCDD in animals varies inversely with body mass and is determined by its metabolism, affinity for lipids, and binding sites in the liver.[14] These observations are consistent with a poorly metabolized lipophilic substance.

The potential carcinogenicity of TCDD in humans is uncertain and controversial.[15] It is known to be a very strong agent for the promotion of neoplasia (tumorous cell growth) in the livers of exposed rodents. The extremely long half-life of TCDD in humans of around 7 years is consistent with a carcinogen that remains in the system long enough to cause harm. Epidemiological studies of accidental human exposures have been complicated by coexposure to other chemicals.

16.6.3 Alachlor

Widely marketed as Monsanto's Lasso® herbicide, Alachlor (Figure 16.7) has become a widespread contaminant of groundwater in some corn- and soybean-producing areas. It seems to be efficiently absorbed through the skin. Allergic skin reactions and skin and eye irritation have been reported in exposed individuals. The U.S. Environmental Protection Agency regards Alachlor as a probable human carcinogen. A study of mortality and cancer rates over a more than 20-year period

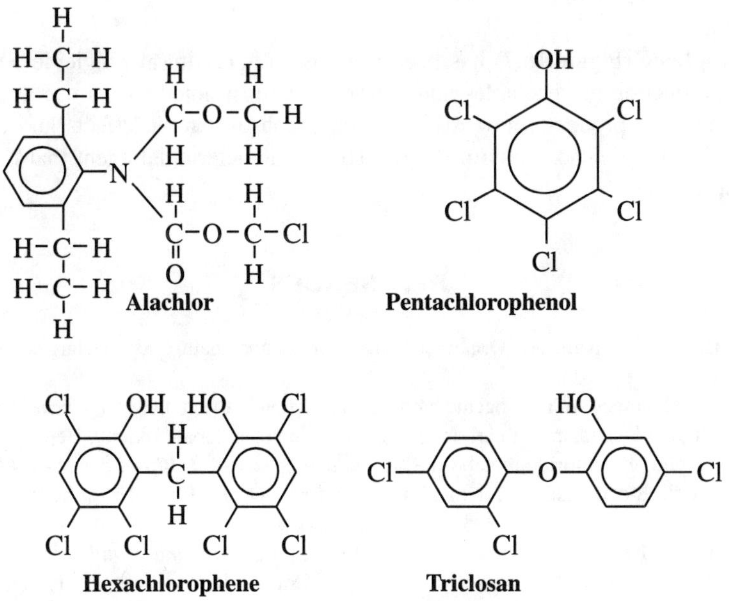

Figure 16.7 Structural formulas of Alachlor, pentachlorophenol, and microcidal hexachlorophene and triclosan.

for workers in an Alachlor manufacturing plant in the U.S. concluded that there may have been a somewhat higher incidence of colorectal cancer and myeloid leukemia among the exposed workers, suggesting a need for continued evaluation of Alachlor exposure.[16]

16.6.4 Chlorinated Phenols

The chlorinated phenols, particularly **pentachlorophenol** (Figure 16.7) and the trichlorophenol isomers, have been widely used as wood preservatives. Applied to wood, these compounds prevent wood rot through their fungicidal action and prevent termite infestation because of their insecticidal properties. Both cause liver malfunction and dermatitis. Contaminant polychlorinated dibenzodioxins may be responsible for some of the observed effects.

Studies in rodents and in human liver cells have shown that pentachlorophenol is metabolized by oxidative dechlorination to tetrachlorohydroquinone and tetrachloro-1,4-benzoquinone:

Tetrachloro-1,4-hydroquinone **Tetrachloro-1,4-benzoquinone**

Tetrachlorohydroquinone is more toxic to rats and human liver cells than its parent, pentachlorophenol.[17] Lipid peroxidation and liver damage in rats are consistent with free radical mechanisms of adverse biochemical effects from tetrachlorohydroquinone and pentachlorophenol. Liver cells exposed to tetrachlorohydroquinone and tetrachloro-1,4-benzoquinone have shown the formation of DNA adducts.[18]

16.6.5 Hexachlorophene

Hexachlorophene (Figure 16.7) has been used as an agricultural fungicide and bacteriocide, largely in the production of vegetables and cotton. It is most noted for its use as an antibacterial agent in personal care products, now discontinued because of toxic effects and possible TCDD contamination. Triclosan, also shown in Figure 16.7, is a bactericidal agent that has been used in personal care products.

REFERENCES

1. Boren, H. and Grimvall, A., Organochlorine compounds: nature as the largest producer, *Kemisk Tidskrift*, 8, 26–28, 1996.
2. Cota, G.F. and Sturges, W.T., Biogenic bromine production in the arctic, *Mar. Chem.*, 56, 181–192, 1997.
3. Boll, M. et al., Mechanisms of carbon tetrachloride-induced hepatotoxicity. Hepatocelluar damage by reactive carbon tetrachloride metabolites, *Z. Naturforsch. C-A J. Biosci.*, 56, 649–659, 2001.
4. Tong, Z. et al., Metabolism of 1,1-dichloro-1-fluoroethane (HCFC-141b) in human volunteers, *Drug Metab. Dispos.*, 26, 711–713, 1998.
5. Parkinson, A., Biotransformation of xenobiotics, in *Casarett and Doull's Toxicology: The Basic Science of Poisons*, 6th ed., Klaassen, Ed., C.D., McGraw-Hill, New York, 2001, chap. 6, pp. 133–224.
6. Kharasch, E.D. et al., Human halothane metabolism, lipid peroxidation, and cytochromes P4502A6 and P4503A4, *Eur. J. Clin. Pharmacol.*, 55, 853–859, 2000.
7. Barbin, A., Role of etheno DNA adducts in carcinogenesis induced by vinyl chloride in rats, *IARC Sci. Publ.*, 150, 303–313, 1999.
8. Green, T., Trichloroethylene and human cancer, *Hum. Ecol. Risk Assess.*, 7, 677–685, 2001.
9. Yu, K.O. et al., In vivo kinetics of trichloroacetate in male Fischer 344 rats, *Toxicol. Sci.*, 54, 302–311, 2000.
10. *PCBs: Recent Advances in Environmental Toxicology and Health Effects*, Robertson, L.W. and Hansen, L.G., Eds., University of Kentucky Press, Lexington, KY, 2001.
11. Hovander, L. et al., Identification of hydroxylated PCB metabolites and other phenolic halogenated pollutants in human blood plasma, *Arch. Environ. Contam. Toxicol.*, 42, 105–117, 2002.
12. Silberner, J., Common herbicide linked to cancer, *Sci. News*, 130, 167–174, 1986.
13. Mitrou, P.I., Dimitriadis, G., and Raptis, S.A., Toxic effects of 2,3,7,8-tetrachlorodibenzo-*p*-dioxin and related compounds, *Eur. J. Intern. Med.*, 12, 406–411, 2001.
14. Miniero, R. et al., An overview of TCDD half-life in mammals and its correlation to body weight, *Chemosphere*, 43, 839–844, 2001.
15. Pitot, H.C., III and Dragon, Y.P., Chemical carcinogenesis, in *Casarett and Doull's Toxicology: The Basic Science of Poisons*, 6th ed., Klaassen, C.D., Ed., McGraw-Hill, New York, 2001, chap. 8, pp. 241–319.
16. Leet, T. et al., Cancer incidence among Alachlor manufacturing workers, *Am. J. Ind. Med.*, 30, 300–306, 1996.
17. Wang, Y.-J. et al., Oxidative stress and liver toxicity in rats and human hepatoma cell line induced by pentachlorophenol and its major metabolite tetrachlorohydroquinone, *Toxicol. Lett.*, 122, 157–169, 2001.
18. Lin, P.-H. et al., Oxidative damage and direct adducts in calf thymus DNA induced by the pentachlorophenol metabolites tetrachlorohydroquinone and tetrachloro-1,4-benzoquinone, *Carcinogenesis*, 22, 627–634, 2001.

QUESTIONS AND PROBLEMS

1. Give an example of each of the following: alkyl halide, alkenyl halide, and aryl halide. Give an example of each of the following kinds of reactions for forming an organohalide compound: substitution halogenation, addition halogenation, and hydrohalogenation.

2. List some chemical and toxicological properties of each of the following compounds:

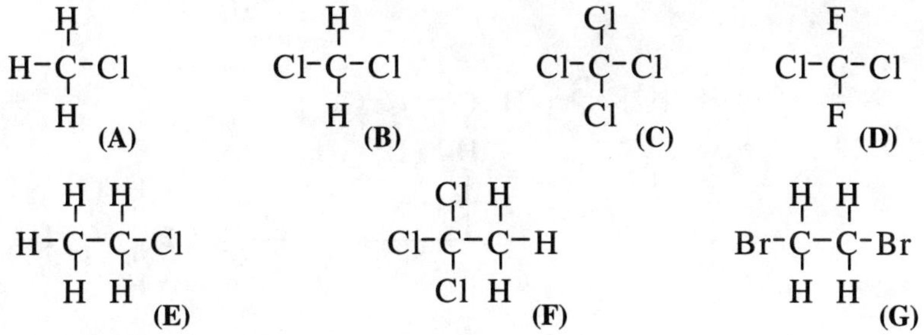

(A) CH_3Cl

(B) CH_2Cl_2

(C) $CHCl_3$

(D) CCl_2F_2

(E) CH_3-CH_2-Cl

(F) $CHCl_2-CH_2Cl$ (shown as Cl-CHCl-CH$_2$-H with Cl H / Cl H)

(G) $BrCH_2-CH_2Br$

3. Explain the following statement: carbon tetrachloride has the most notorious record of human toxicity of all organohalides especially for its toxic effects on the liver.
4. Explain the special toxicological significance of vinyl chloride.
5. Explain what is shown by the following sequence of reactions:

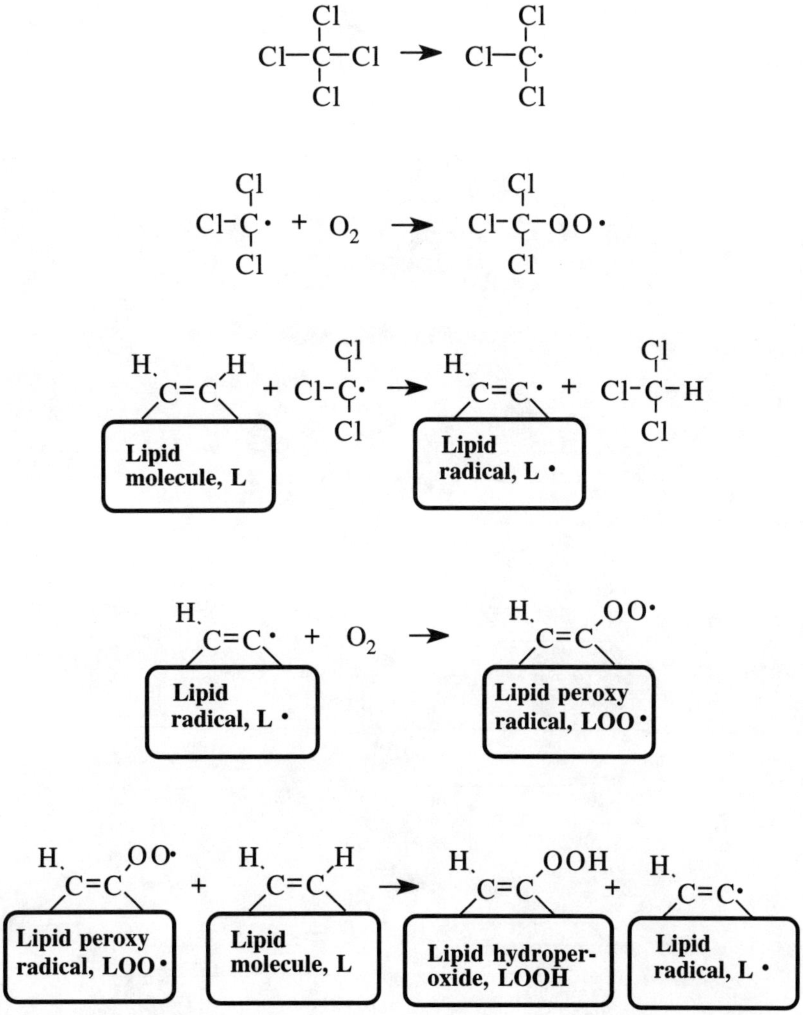

$$CCl_4 \rightarrow \cdot CCl_3$$

$$\cdot CCl_3 + O_2 \rightarrow Cl_3C\text{-}OO\cdot$$

$$\underset{\text{Lipid molecule, L}}{H_2C=CH_2} + \cdot CCl_3 \rightarrow \underset{\text{Lipid radical, L}\cdot}{HC=CH\cdot} + CHCl_3$$

$$\underset{\text{Lipid radical, L}\cdot}{HC=CH\cdot} + O_2 \rightarrow \underset{\text{Lipid peroxy radical, LOO}\cdot}{HC=CH\text{-}OO\cdot}$$

$$\underset{\text{Lipid peroxy radical, LOO}\cdot}{HC=CH\text{-}OO\cdot} + \underset{\text{Lipid molecule, L}}{H_2C=CH_2} \rightarrow \underset{\text{Lipid hydroperoxide, LOOH}}{HC=CH\text{-}OOH} + \underset{\text{Lipid radical, L}\cdot}{HC=CH\cdot}$$

6. What are the possible dichloroethylene compounds?
7. What is shown by the reaction below? What is its toxicological chemical significance?

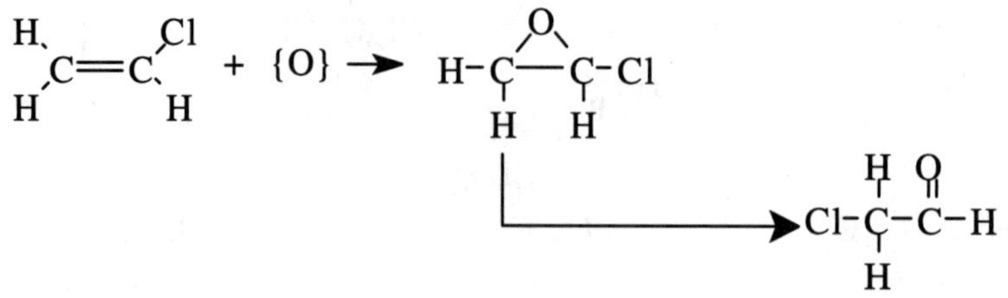

$$\underset{H}{\overset{H}{>}}C=C\underset{H}{\overset{Cl}{<}} + \{O\} \rightarrow H-\underset{H}{\overset{}{C}}\overset{O}{\underset{}{\triangle}}\underset{H}{\overset{}{C}}-Cl \longrightarrow Cl-\underset{H}{\overset{H}{C}}-\overset{O}{\underset{}{C}}-H$$

CHAPTER 17

Organosulfur Compounds

17.1 INTRODUCTION

Sulfur is directly below oxygen in the periodic table. The sulfur atom has six valence electrons, as shown in Figure 17.1, and its electron configuration is $\{Ne\}3s^23p^4$. Because of their very similar valence shell electron configurations, oxygen and sulfur behave somewhat alike chemically. However, unlike oxygen, the sulfur atom has three underlying $3d$ orbitals, and its valence shell can be expanded to more than eight electrons. This makes sulfur's chemical behavior more diverse than that of oxygen. For example, sulfur has several common oxidation states, including –2, +4, and +6, whereas most chemically combined oxygen is in the –2 oxidation state.

17.1.1 Classes of Organosulfur Compounds

The hydride of sulfur is H_2S (Figure 17.1), a highly toxic gas discussed in Section 11.9. Substitution of alkyl or aryl hydrocarbon groups such as phenyl and methyl (Figure 17.1) for H on hydrogen sulfide leads to a number of different organosulfur compounds. These include thiols (R–SH) and thioethers (R–S–R). Because of the availability of $3d$ orbitals, sulfur that is bonded to hydrocarbon moieties can also be bonded to oxygen, adding to the variety of organosulfur compounds that can exist.

Despite the high toxicity of H_2S, not all organosulfur compounds are particularly toxic. Many of the compounds have strong, offensive odors that warn of their presence, which reduces their hazard.

17.1.2 Reactions of Organic Sulfur

Organic sulfur undergoes a number of toxicological chemical reactions. These include the following:

- Oxidation of sulfur
- Reduction of sulfur
- Removal of sulfur from a molecule
- Addition of sulfur-containing groups

Examples of these kinds of reactions, some of which are very important in xenobiotic metabolism, are given below.

Oxidation of sulfur is called **S-oxidation**. Thiols can be oxidized to form disulfides:

$$2R\text{–}SH + \{O\} \rightarrow R\text{–}S\ S\text{–}R + H_2O \qquad (17.1.1)$$

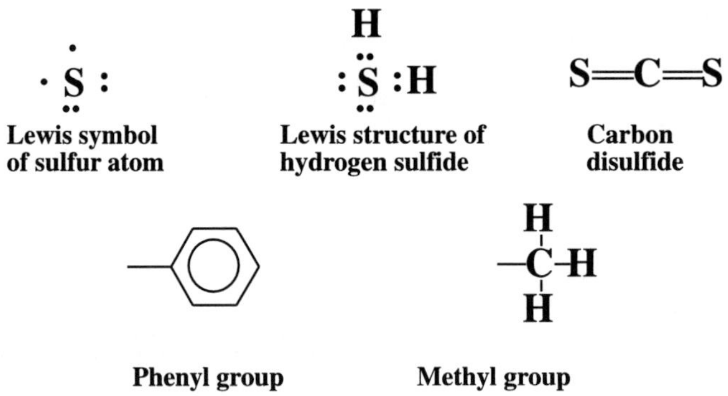

Figure 17.1 Sulfur atom, compounds, and substituent groups.

The same kind of reaction occurs with aminothiols, such as in the oxidation of cysteamine, H_2NCH_2CHSH, to cystamine, $H_2NCH_2CH_2S\ SCH_2CH_2NH_2$. S-oxidation may also involve sulfur in organosulfide compounds:

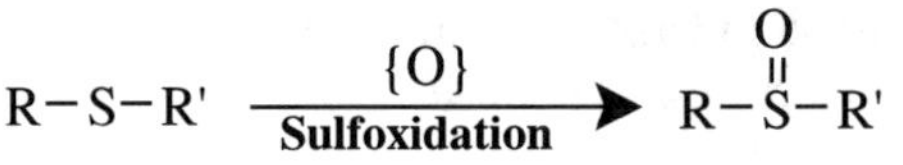

$$R-S-R' \xrightarrow[\text{Sulfoxidation}]{\{O\}} R-\overset{\overset{O}{\|}}{S}-R' \tag{17.1.2}$$

and sulfur on thioamides:

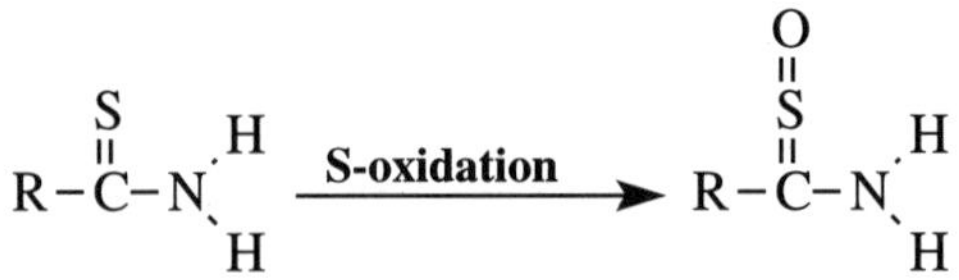

$$R-\overset{\overset{S}{\|}}{C}-N\overset{H}{\underset{H}{\diagdown}} \xrightarrow{\text{S-oxidation}} R-\overset{\overset{\overset{O}{\|}}{\underset{\|}{S}}}{C}-N\overset{H}{\underset{H}{\diagdown}} \tag{17.1.3}$$

An example of sulfur reduction is **disulfide reduction**, as shown by the following reaction:

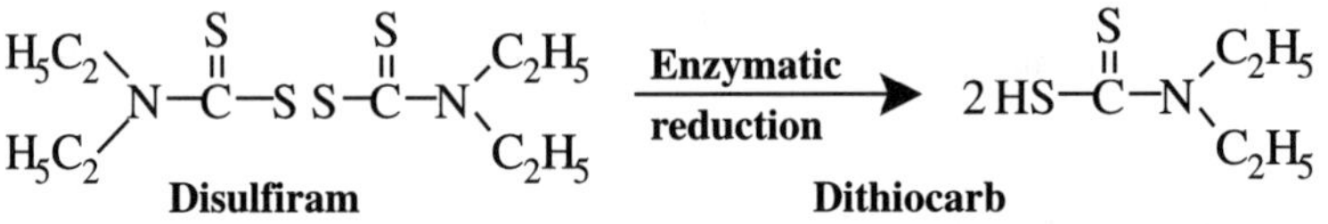

$$\underset{\text{Disulfiram}}{\overset{H_5C_2}{\underset{H_5C_2}{\diagdown}}N-\overset{\overset{S}{\|}}{C}-S\ S-\overset{\overset{S}{\|}}{C}-N\overset{C_2H_5}{\underset{C_2H_5}{\diagup}}} \xrightarrow[\text{reduction}]{\text{Enzymatic}} \underset{\text{Dithiocarb}}{2\ HS-\overset{\overset{S}{\|}}{C}-N\overset{C_2H_5}{\underset{C_2H_5}{\diagup}}} \tag{17.1.4}$$

This reaction converts disulfiram, a therapeutic agent for the treatment of alcohol abuse, discussed in Section 17.3, to dithiocarb (diethylthiocarbamate), a substance that strongly binds metals and is used for the treatment of nickel carbonyl poisoning.

Desulfuration is the term given to removal of sulfur from a molecule. One of the most common desulfuration reactions occurs with sulfur bonded to phosphorus. A common desulfuration reaction is the enzyme-mediated conversion of parathion to paraoxon (see discussion of organophosphate insecticides in Section 18.7):

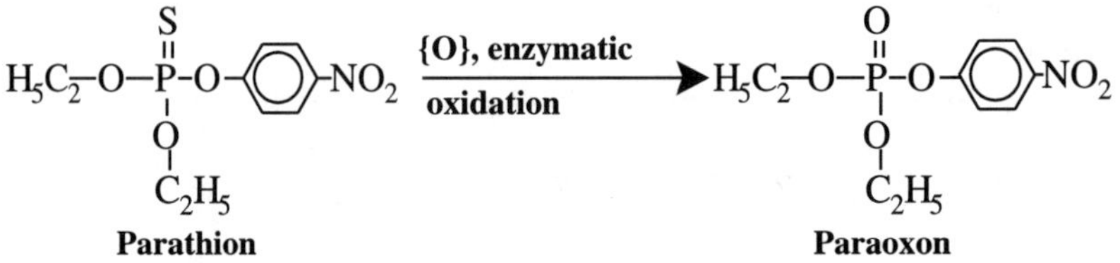

$$\underset{\text{Parathion}}{H_5C_2-O-\overset{\overset{S}{\|}}{\underset{\underset{C_2H_5}{O}}{P}}-O-\!\!\!\bigcirc\!\!\!-NO_2} \xrightarrow[\text{oxidation}]{\{O\},\ \text{enzymatic}} \underset{\text{Paraoxon}}{H_5C_2-O-\overset{\overset{O}{\|}}{\underset{\underset{C_2H_5}{O}}{P}}-O-\!\!\!\bigcirc\!\!\!-NO_2} \tag{17.1.5}$$

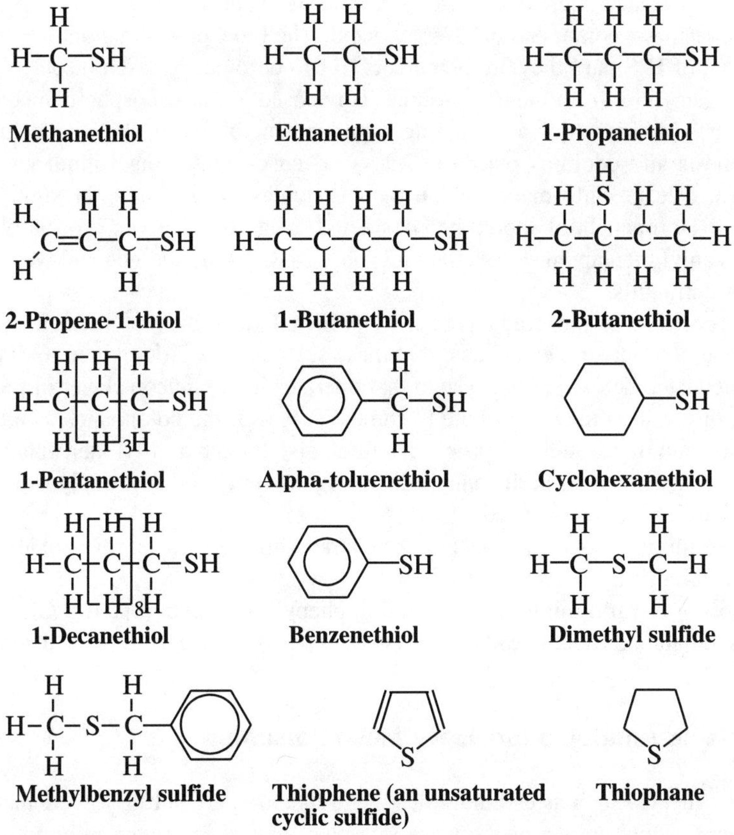

ORGANOSULFUR COMPOUNDS

Figure 17.2 Common low-molecular-mass thiols and sulfides. All are liquids at room temperature, except for methanethiol, which boils at 5.9°C.

The most significant instance of addition of a sulfur-containing group is the phase II conjugation to sulfate of a xenobiotic compound or its phase I metabolite (see Section 7.4.3) by the action of adenosine 3'-phosphate-5'-phosphosulfate, a sulfotransferase enzyme that acts as a sulfating agent:

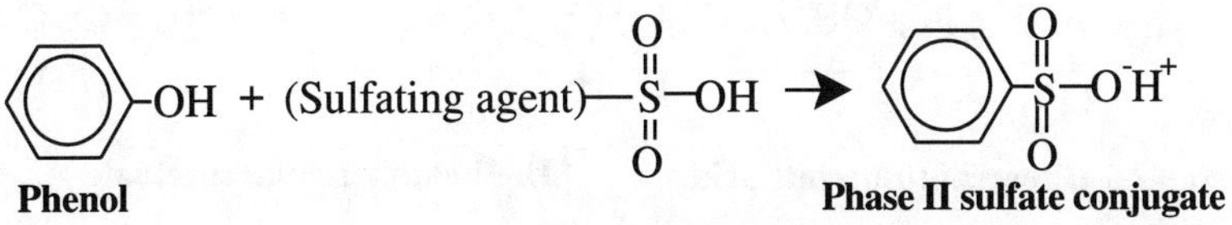

$$\text{Phenol-OH} + \text{(Sulfating agent)}-\underset{\underset{O}{\|}}{\overset{\overset{O}{\|}}{S}}-\text{OH} \rightarrow \text{Phenyl}-\underset{\underset{O}{\|}}{\overset{\overset{O}{\|}}{S}}-\text{O}^-\text{H}^+ \quad (17.1.6)$$

Phenol   Phase II sulfate conjugate

17.2 THIOLS, SULFIDES, AND DISULFIDES

Substitution of alkyl and aryl groups for H on H_2S yields **thiols** and **sulfides** (thioethers). Structural formulas of examples of these compounds are shown in Figure 17.2.

17.2.1 Thiols

Thiols are also known as mercaptans. The lighter alkyl thiols, such as methanethiol, are fairly common air pollutants with odors that may be described as ultragarlic. Inhalation of even very low concentrations of the alkyl thiols in air can be very nauseating and result in headaches. Exposure to higher levels can cause increased pulse rate, cold hands and feet, and cyanosis. With extreme cases, unconsciousness, coma, and death may occur. The biochemical action of alkyl thiols likely is similar to that of H_2S, and they are precursors to cytochrome oxidase poisons.

Volatile methanethiol from biogenic sources is released to the atmosphere in coastal and ocean upwelling areas.[1] Methanethiol and volatile liquid ethanethiol (bp, 35°C) are intermediates in pesticide synthesis and odorants placed in lines and tanks containing natural gas, propane, and butane to warn of leaks. Information about their toxicities to humans is lacking, although these compounds and 1-propanethiol should be considered dangerously toxic, especially by inhalation. Also known as amyl mercaptan, 1-pentanethiol (bp, 124°C) is an allergen and weak sensitizer that causes contact dermatitis.

Anaerobic bacteria in the colon produce significant quantities of methanethiol along with hydrogen sulfide. Rodent studies indicate that these substances are detoxified to thiosulfate by the action of a specialized detoxification system that operates in the mucous layer of the colon lining.[2] The failure of this system may contribute to some diseases of the colon, such as ulcerative colitis.

A typical alkenyl mercaptan is 2-propene-1-thiol, also known as allyl mercaptan. It is a volatile liquid (bp, 68°C) with a strong garlic odor. It has a high toxicity and is strongly irritating to mucous membranes when inhaled or ingested.

Alpha-toluenethiol, also called benzyl mercaptan (bp, 195°C) is very toxic orally. It is an experimental carcinogen.

The simplest of the aryl thiols is benzenethiol, phenyl mercaptan (bp, 168°C). It has a severely repulsive odor. Inhalation causes headache and dizziness, and skin exposure results in severe contact dermatitis.

17.2.2 Thiols as Antidotes for Heavy Metal Poisoning

Toxic heavy metals, such as cadmium, lead, and mercury, are sulfur seekers that bind strongly with thiol groups, which is one of the ways in which they interact adversely with biomolecules, including some enzymes. Advantage has been taken of this tendency to use thiols in chelation therapy in heavy metal poisoning. Among the thiols tested for this purpose are *meso*-2,3-dimercaptosuccinic acid, diethyldimercapto succinate, α-mercapto-β-(2-furyl), and α-mercapto-β-(2-thienyl) acrylic acid.[3] The structural formulas for the first two are

meso -2,3-Dimercaptosuccinic acid   **Diethyldimercapto succinate**

17.2.3 Sulfides and Disulfides

Dimethylsulfide is an alkyl sulfide or thioether. It is a volatile liquid (bp, 38°C) that is moderately toxic by ingestion. Thiophene is the most common cyclic sulfide. It is a heat-stable liquid (bp, 84°C) with a solvent action much like that of benzene. It is used in the manufacture of pharmaceuticals and dyes, as well as resins that also contain phenol or formaldehyde. Its saturated analog is tetrahydrothiophene, or thiophane.

Ruminant animals produce dimethylsulfide, a fraction of which is exhaled.[4] Therefore, cows' breath is a source of this compound in terrestrial atmospheres. Dimethylsulfide is produced in enormous quantities by marine organisms. This volatile compound is the largest source of biogenic sulfur in the atmosphere and is responsible for some of the odor emanating from coastal mud flats.

It, along with methylbenzyl sulfide and one or two other sulfides, is responsible for the nauseating stench of *Halichondria panicea*, a marine sponge.

The organic disulfides contain the –SS– functional group, as shown in the following two examples:

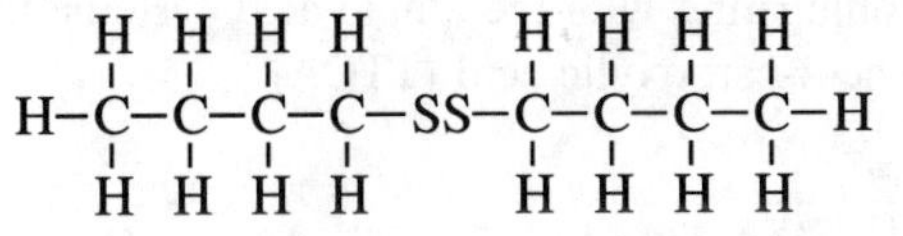

n-Butyldisulfide 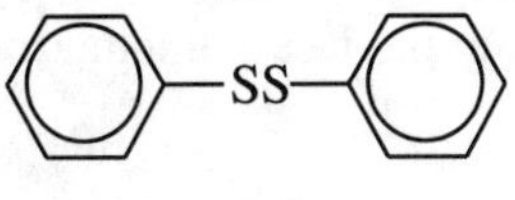   **Diphenyldisulfide**

These compounds may act as allergens that produce dermatitis in contact with skin. Not much information is available regarding their toxicities to humans, although animal studies suggest several toxic effects, including hemolytic anemia.

17.2.4 Organosulfur Compounds in Skunk Spray

Skunks are small animals that wage powerful defensive warfare in the form of spray that has a sickening offensive odor. The compounds that are responsible for skunk spray odor are organosulfur compounds. Although seven or more such compounds have been isolated from the spray of the striped skunk, *Mephitis mephitis*, it is now believed that there are only three major ones: *trans*-2-butene-1-thiol, *trans*-2-butenyl-1-thioacetate, and 3-methyl-1-butanethiol[5]:

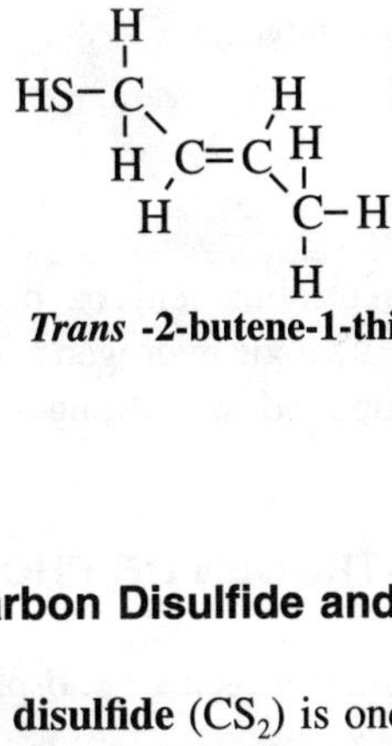

Trans-2-butene-1-thiol 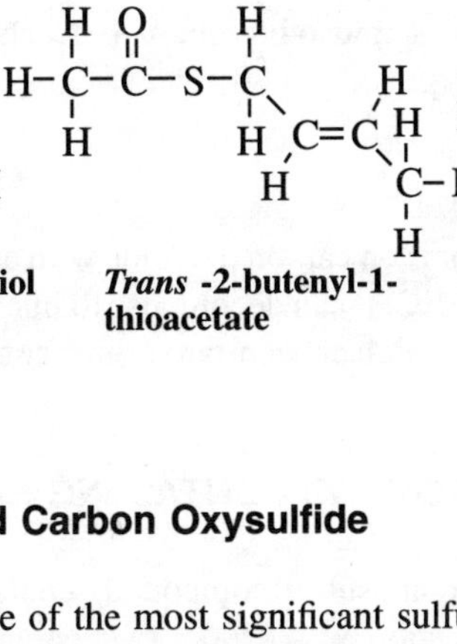   **Trans-2-butenyl-1-thioacetate** 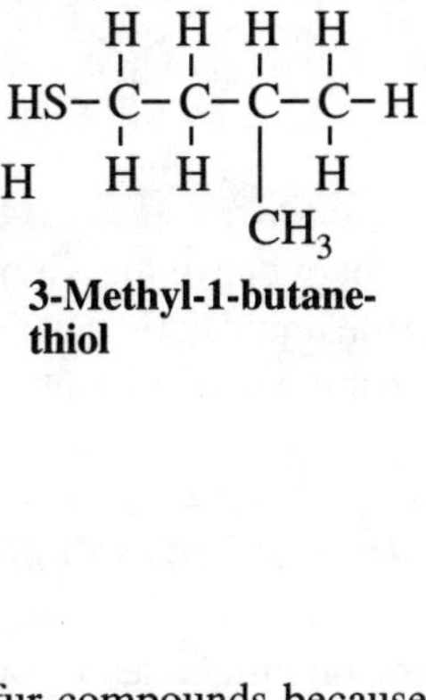   **3-Methyl-1-butanethiol**

17.2.5 Carbon Disulfide and Carbon Oxysulfide

Carbon disulfide (CS_2) is one of the most significant sulfur compounds because of its widespread use and toxicity. This compound has two sulfur atoms, each separately bonded to a carbon atom. This compound is a volatile, colorless liquid (mp, –111°C; bp, 46°C). Unlike most organosulfur compounds, it is virtually free of odor. Although its uses are declining, it has numerous applications in chemical synthesis, as a solvent to break down cellulose in viscose rayon manufacture, and in the manufacture of cellophane. It has also been used as an insecticide and fumigant.

Acute doses of carbon disulfide inhaled at 100 to 1000 ppm irritate mucous membranes and affect the central nervous system, usually causing excitation as a first noticeable effect, followed by restlessness, depression, and stupor. Carbon disulfide intoxication causes reduced conduction velocity in the peripheral nerves, which can be detected by psychomotor tests that exhibit impaired performance.[6] It is a much stronger anesthetic than chloroform (Section 16.2), causing unconsciousness and even death in cases of high exposure. Symptoms experienced during recovery from severe acute carbon disulfide poisoning resemble those that occur following intoxication from ingestion of ethanol in alcoholic beverages.

Chronic carbon disulfide poisoning by absorption through the skin or respiratory tract involves the central and peripheral nervous systems and may cause anemia. Symptoms include indistinct vision, neuritis, and a bizarre sensation of "crawling" on the skin. Psychopathological symptoms may be varied and severe, including excitation, depression, irritability, and general loss of mental capabilities to the point of insanity. Parkinsonian paralysis may result from chronic carbon disulfide poisoning.

Carbon disulfide is metabolized by conjugating with the amino acid cysteine to form adducts that rearrange to produce 2-thiothiazolidine-4-carboxylic acid (TTCA):

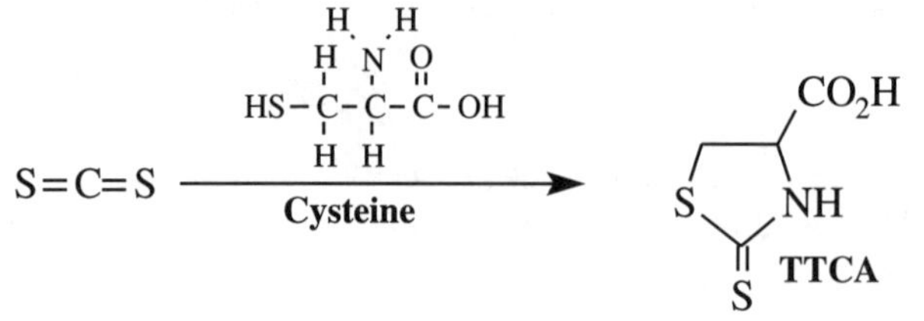

$$S=C=S \xrightarrow{\text{Cysteine}} \text{TTCA} \tag{17.2.1}$$

TTCA is commonly determined by chemical analysis of urine as a biomarker of exposure to carbon disulfide. Another metabolite of carbon disulfide that appears at levels of about 30% those of TTCA in workers exposed to carbon disulfide is 2-thioxothiazolidin-4-ylcarbonylglycine[7]:

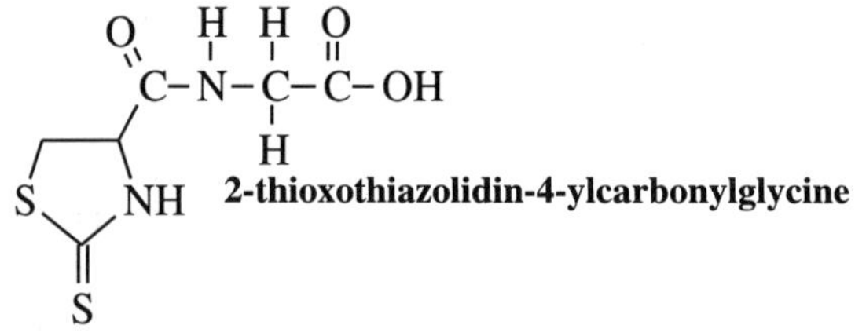

2-thioxothiazolidin-4-ylcarbonylglycine

Replacement of one of the S atoms on carbon disulfide with an O atom yields **carbon oxysulfide** (COS), a volatile liquid boiling at 50°C. It can decompose to liberate toxic hydrogen sulfide. Carbon oxysulfide vapor is a toxic irritant. At high concentrations this compound has a strong narcotic effect.

17.3 ORGANOSULFUR COMPOUNDS CONTAINING NITROGEN OR PHOSPHORUS

Several important classes of organosulfur compounds contain nitrogen or phosphorus. These compounds are discussed in this section.

17.3.1 Thiourea Compounds

Thiourea is the sulfur analog of urea. Substitution of hydrocarbon moieties on the N atoms yields various organic derivatives of thiourea, as illustrated in Figure 17.3. Thiourea has been used as a rodenticide. It has a moderate to high toxicity to humans, affecting bone marrow and causing anemia. It has been shown to cause liver and thyroid cancers in experimental animals.

Phenylthiourea is likewise a rodenticide. Its toxicity is highly selective to rodents relative to humans, although it probably is very toxic to some other animals. The compound is metabolized extensively, and some of the sulfur is excreted as sulfate in urine.

Commonly called ANTU, **1-naphthylthiourea** is a virtually tasteless rodenticide that has a very high rodent:human toxicity ratio. The lethal dose to monkeys is about 4000 mg/kg. One

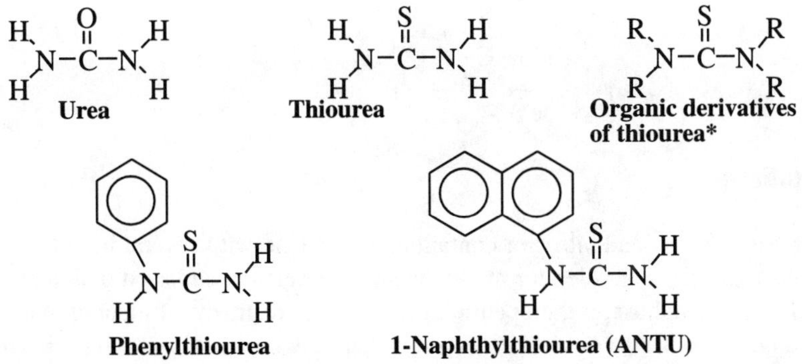

Figure 17.3 Structural formulas of urea, thiourea, and organic derivatives of thiourea. *At least one R group is an alkyl, alkenyl, or aryl substituent.

suicidal adult male human ingested about 80 g of 30% ANTU rat poison, along with a considerable amount of alcohol. He vomited soon after ingestion and survived without significant ill effects.[8] Dogs, however, are quite susceptible to ANTU poisoning.

17.3.2 Thiocyanates

Organic **thiocyanates** are derivatives of thiocyanic acid (HSCN), in which the H is replaced by hydrocarbon moieties, such as the methyl group. Dating from the 1930s and regarded as the first synthetic organic insecticides, these compounds kill insects on contact. Because of their volatilities, the lower-molecular-mass methyl, ethyl, and isopropyl thiocyanates are effective fumigants for insect control. Insecticidal lauryl thiocyanate (below) is not volatile and is used in sprays in petroleum-based solvents and in dusting powders.

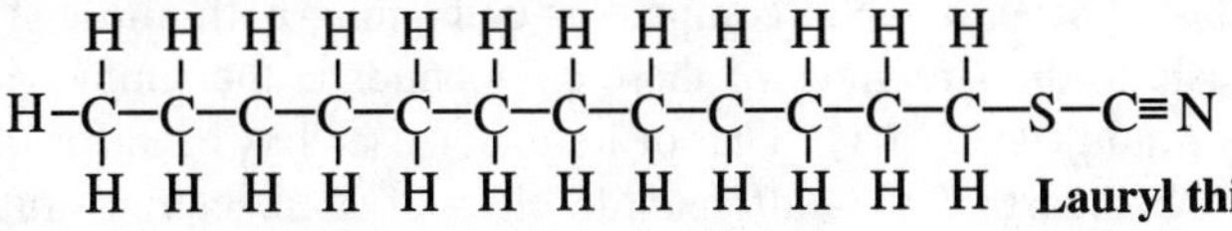

The toxicities of the thiocyanates vary widely by compound and route of administration. Some metabolic processes liberate HCN from thiocyanates. As discussed in Section 11.2.1, HCN is highly toxic, so its generation in the body can result in death. Therefore, methyl, ethyl, and isopropyl thiocyanates should be regarded as rapid-acting, potent poisons.

The **isothiocyanate** group is illustrated in the structure below:

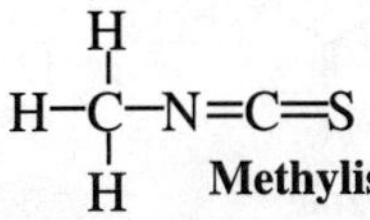

Other compounds in this class include ethyl, allyl, and phenyl isothiocyanates. Methylisothiocyanate, also known as methyl mustard oil, and its ethyl analog have been developed as military poisons. Both are powerful irritants to eyes, skin, and the respiratory tract. When decomposed by heat, these compounds emit sulfur oxides and hydrogen cyanide. Methylisothiocyanate occurs in the environment from some kinds of vegetables, as a breakdown product of some carbamate insecticides, and from deliberate addition to soil for fumigation.[9] It is a degradation product of the fungicide Vapam:

$$\text{HS-}\overset{\overset{S}{\|}}{C}\text{-}\overset{\overset{H}{|}}{N}\text{-}\overset{\overset{H}{|}}{\underset{\underset{H}{|}}{C}}\text{-H} \quad \text{Vapam}$$

17.3.3 Disulfiram

Disulfiram, is a sulfur- and nitrogen-containing compound with several industrial uses, including applications as a rubber accelerator and vulcanizer, fungicide, and seed disinfectant. It is most commonly known as antabuse, a therapeutic agent for the treatment of alcohol abuse that causes nausea, vomiting, and other adverse effects when ethanol is ingested. Disulfiram is an inhibitor of aldehyde dehydrogenase so that it allows for buildup of the acetaldehyde metabolite of ethanol, causing unpleasant effects that are a deterrent to the ingestion of alcohol. Because of the buildup of acetaldehyde, disulfiram should be given with extreme caution, especially to individuals suffering from liver cirrhosis.

[Structure of Disulfiram (antabuse)]

17.3.4 Cyclic Sulfur and Nitrogen Organic Compounds

The structural formulas of several cyclic compounds containing both nitrogen and sulfur are shown in Figure 17.4. Basic to the structures of these compounds is the simple ring structure of **thiazole**. It is a colorless liquid (bp, 117°C). One of its major uses has been for the manufacture of sulfathiazole, one of the oldest of the sulfonamide class of antibacterial drugs. The use of sulfathiazole is now confined to the practice of veterinary medicine because of its serious side effects.

Several derivatives of thiazole have commercial uses. One of these is **2-aminothiazole**, which has shown a high toxicity to experimental animals. **Benzothiazole** is another related compound used in organic synthesis. Thiazoles are used as rubber vulcanization accelerators. Benzothiazole,

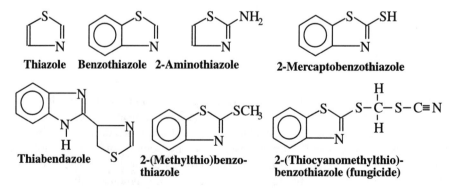

Figure 17.4 Cyclic compounds containing nitrogen and sulfur.

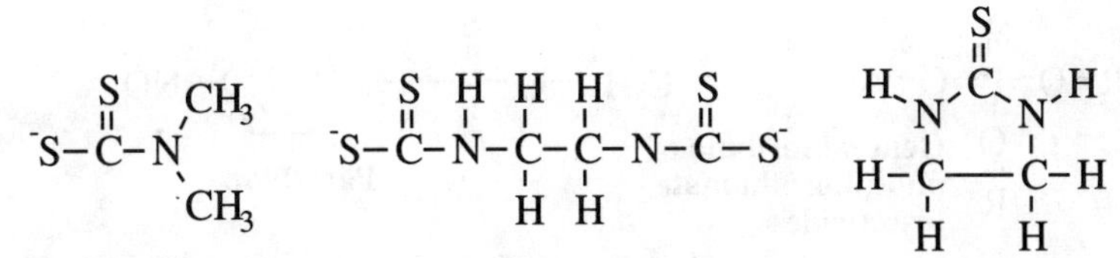

ORGANOSULFUR COMPOUNDS

Figure 17.5 Dithiocarbamate anions and ethylenethiourea.

2-mercaptobenzothiazole, and 2-(methylthio)benzothiazole are breakdown products of fungicidal 2-(thiocyanomethylthio)benzothiazole and are common constituents of tannery wastewaters. The human toxicity of 2-(thiocyanomethylthio)benzothiazole is not known, although it has a high toxicity to mice. **Thiabendazole**, 2-(4'-thiazoyl)benzimidazole, is a systemic fungicide that can be carried through a plant and onto plant leaves. Rats and dogs tolerate a relatively high dose of this chemical, although it tends to make the latter vomit. Acting as an adjuvant (a substance added to a drug or insecticide to give it a desired form and enhance its action and effectiveness), 2-mercaptobenzothiazole is mixed with dithiocarbamate fungicides (see below) to increase their potency. It is an allergen that causes type IV (cell-mediated) hypersensitivity. This condition is manifested by contact dermatitis and a delayed hypersensitive reaction that follows a latent period after exposure. This cell-mediated process results from the sensitization of T lymphocytes.

17.3.5 Dithiocarbamates

Dithiocarbamate fungicides consist of metal salts of **dimethylthiocarbamate** and **ethylenebis-dithiocarbamate** anions, as shown in Figure 17.5. These fungicides are named in accordance with the metal ion present. For example, the manganese salt of dimethyldithiocarbamate is called maneb, and the zinc and sodium salts are zineb and nabam, respectively. The iron salt of ethylenebisdithiocarbamate is called ferbam, and the zinc salt of this ion is called ziram. These salts are chelates (Section 2.3) in which two S atoms from the ethylenebisdithiocarbamate anion are bonded to the same metal ion in a ring structure.

The dithiocarbamate fungicides have been popular for agricultural use because of their effectiveness and relatively low toxicities to animals. However, there is concern over their environmental breakdown products, particularly ethylenethiourea (2-imidazolidine thione; see Figure 17.5), which is toxic to the thyroid and has been shown to be mutagenic, carcinogenic, and teratogenic in experimental animals.

17.3.6 Phosphine Sulfides

A number of toxicologically important organic compounds have sulfur bound to phosphorus. The simplest of these are the phosphine sulfides, containing only carbon, hydrogen, phosphorus, and sulfur, as illustrated by the example below:

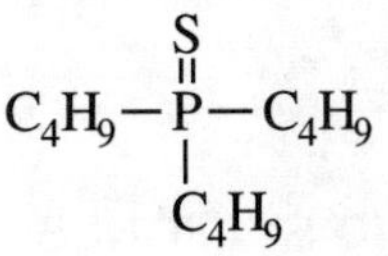

$C_4H_9-P(=S)(C_4H_9)-C_4H_9$ **Tributylphosphine sulfide**

Phosphine sulfides tend to be toxic. When burned, they give off dangerous phosphorus oxide and sulfur oxide fumes.

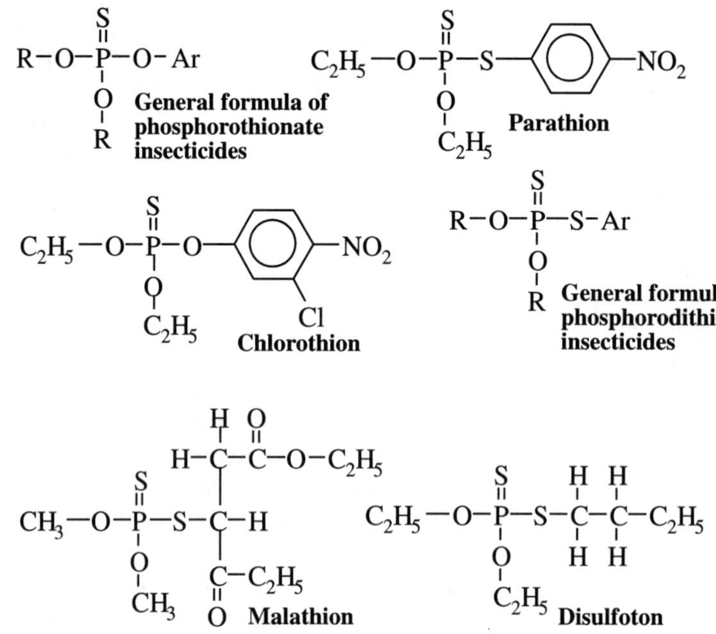

Figure 17.6 General formulas and specific examples of phosphorothionate and phosphorodithioate organophosphate insecticides.

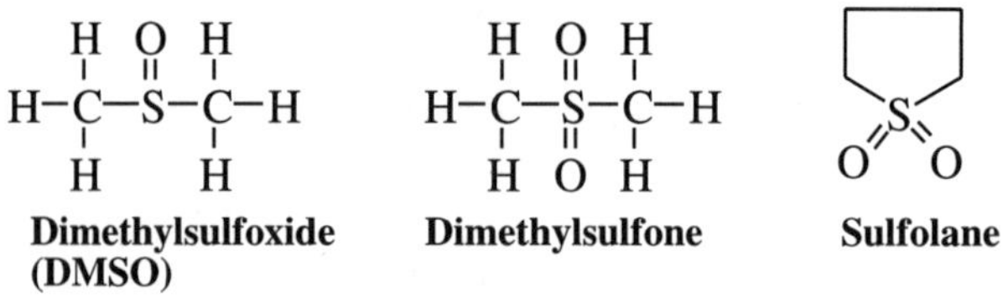

Figure 17.7 Sulfoxides and sulfones.

17.3.7 Phosphorothionate and Phosphorodithioate Esters

The most toxicologically significant organic compounds that contain both phosphorus and sulfur are the thiophosphate esters, which are used as insecticidal acetylcholinesterase inhibitors. The general formulas of insecticidal **phosphorothionate** and **phosphorodithioate** esters are shown in Figure 17.6, where R is usually a methyl (–CH$_3$) or ethyl (–C$_2$H$_5$) group, and Ar is a moiety of more complex structure, frequently aromatic. Phosphorothionate and phosphorodithioate esters contain the P=S (thiono) group, which increases their insect:mammal toxicity ratios and decreases their tendency to undergo nonenzymatic hydrolysis, compared to their analogous compounds that contain the P=O functional group. The metabolic oxidative desulfuration conversion of P=S to P=O in organisms (see Section 17.1) converts the phosphorothionate and phosphorodithioate esters to species that have insecticidal activity.

17.4 SULFOXIDES AND SULFONES

Numerous important organic compounds contain oxygen bonded to sulfur. Among these compounds are the **sulfoxides** and **sulfones**, shown by the examples in Figure 17.7.

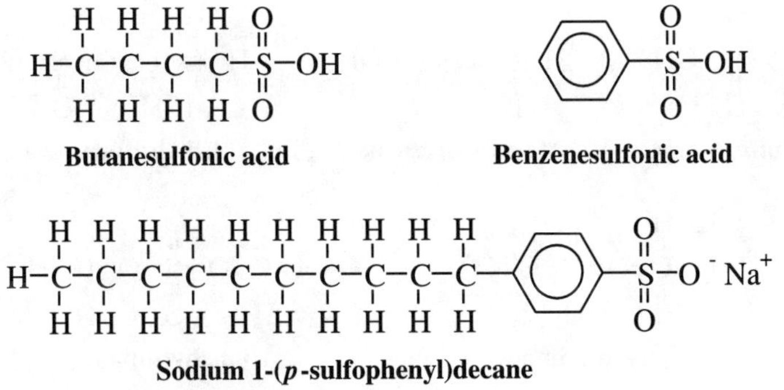

Figure 17.8 Sulfonic acids and a sulfonate salt.

Dimethylsulfoxide (DMSO) is a liquid with numerous uses and some very interesting properties. It occurs in the atmosphere over and near oceans as an oxidation product of dimethylsulfide produced by biological marine sources. Mixed with water, DMSO produces a good antifreeze solution. It is also employed to remove paint and varnish and as a hydraulic fluid. It has some potential pharmaceutical applications, for example, as an antiinflammatory and bacteriostatic agent. It has the ability to carry solutes into the skin's stratum corneum (see Figure 6.4), from which they are slowly released into the blood and lymph system. This phenomenon has some pharmaceutical potential, as well as some obvious hazards. Dimethylsulfoxide has a remarkably low acute toxicity, with an LD_{50} of 10 to 20 g/kg in several kinds of experimental animals. DMSO applied to the skin rapidly spreads throughout the body, and the subject experiences a taste in the mouth resembling that of garlic and quickly develops a garlic odor in the breath. The main acute toxic effects of DMSO are inhibition of acetylcholinesterase and breakdown of red blood cells (hemolysis). Some DMSO is excreted directly in the urine, and it also undergoes partial metabolism to dimethylsulfide and dimethylsulfone (Figures 17.2 and 17.7, respectively).

Although dimethylsulfone has some commercial uses, **sulfolane** (Figure 17.7) is the most widely used sulfone. It is a polar aprotic (no ionizable H) solvent with a relatively high dielectric constant, and it dissolves both organic and inorganic solutes. When ionic compounds are dissolved in sulfolane, the cations are solvated (bound by the solvent) rather strongly. However, the anions are left in a relatively unsolvated form, which tends to increase their reactivities substantially. The major commercial use of sulfolane is in an operation called BTX processing, in which it selectively extracts benzene, toluene, and xylene from aliphatic hydrocarbons. It is also the solvent in the Sulfinol process by which thiols (Section 17.2) and acidic compounds are removed from natural gas. Sulfolane is used as a solvent for polymerization reactions and as a polymer plasticizer. Exposure to sulfolane can cause eye and skin irritation, although its overall toxicity is relatively low.

17.5 SULFONIC ACIDS, SALTS, AND ESTERS

Sulfonic acids contain the $-SO_3H$ group attached to a hydrocarbon moiety. For many applications these acids are converted to salts, such as sodium 1-(p-sulfophenyl)decane, a biodegradable detergent surfactant. Its structural formula and those of two sulfonic acids are shown in Figure 17.8.

In general, the sulfonic acids are water soluble and are strong acids because of virtually complete loss of ionizable H^+ in aqueous solution. They have some important commercial applications, such as in the hydrolysis of fats and oils (see Section 3.5) to fatty acids and glycerol. Benzenesulfonic acid is fused with NaOH in the preparation of phenol. Dyes and some pharmaceutical compounds are manufactured from p-toluenesulfonic acid. Methanesulfonic acid has been developed as an esterification catalyst in place of sulfuric acid for the synthesis of resins in paints and coatings.

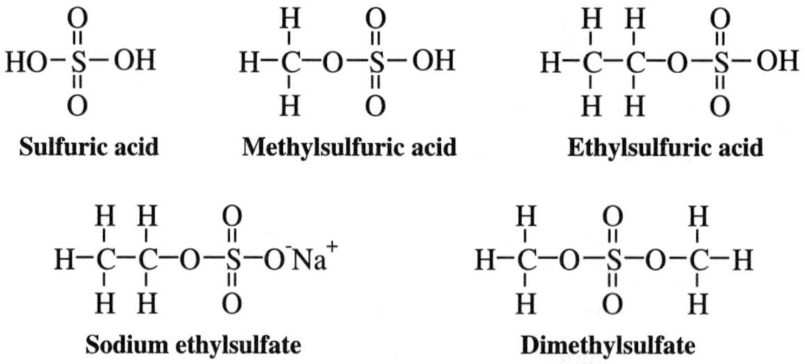

Sulfuric acid   **Methylsulfuric acid**   **Ethylsulfuric acid**

Sodium ethylsulfate   **Dimethylsulfate**

Figure 17.9 Sulfuric acid and organosulfate esters.

One of the major advantages of methanesulfonic acid over sulfuric acid is that it is not an oxidizing species.

Benzenesulfonic acid and *p*-toluenesulfonic acid are strong irritants to skin, eyes, and mucous membranes. Solutions of sulfonic acids are strongly acidic, and precautions appropriate to the handling of strong acids should be taken with them.

The methyl ester of methylsulfonic acid is methylmethane sulfonate. Its structural formula is

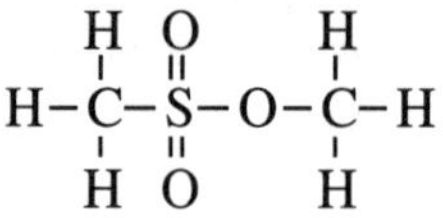

Methylmethane sulfonate

Toxicologically, it has been classified as a primary or direct-acting carcinogen that does not require metabolic conversion to act as a carcinogen.[10] Both it and ethyl methane sulfonate are strong biological alkylating agents, although they are regarded as only weak carcinogens.

17.6 ORGANIC ESTERS OF SULFURIC ACID

As shown in Figure 17.9, esters of sulfuric acid exist in which either one or both of the ionizable H atoms are replaced by hydrocarbon substituents, such as the methyl group. Replacement of one H yields an acid ester, and replacement of both yields an ester. Metabolically, acid ester sulfates are synthesized in phase II reactions to produce water-soluble products of xenobiotic compounds (such as phenol) that are readily eliminated from the body (see Section 7.4).

Sulfuric acid esters have several industrial uses, especially as alkylating agents, which act to attach alkyl groups (such as methyl) to organic molecules. Among the products made with sulfuric acid ester reactants are agricultural chemicals, dyes, and drugs.

Methylsulfuric acid is an oily water-soluble liquid. It is a strong irritant to skin, eyes, and mucous tissue. **Ethylsulfuric acid** is likewise an oily liquid and a strong tissue irritant. **Sodium ethylsulfate** is a hygroscopic white crystalline solid.

Dimethylsulfate is a liquid (bp, 188°C; fp, –32°C). It is colorless, odorless, and highly toxic. It has been classified as a primary carcinogen.[10] When skin or mucous membranes are exposed to dimethylsulfate, there is an initial latent period during which few symptoms are observed. After this period, conjunctivitis and inflammation of nasal tissue and respiratory tract mucous membranes develop. Heavier exposures damage the liver and kidney and cause pulmonary edema and cloudiness of the cornea. Death can follow in 3 or 4 days. The related compound, diethylsulfate, is an oily liquid. It reacts with water to yield sulfuric acid. Like dimethylsulfate, it is a strong irritant to tissue and has proven to be carcinogenic in experimental animals.

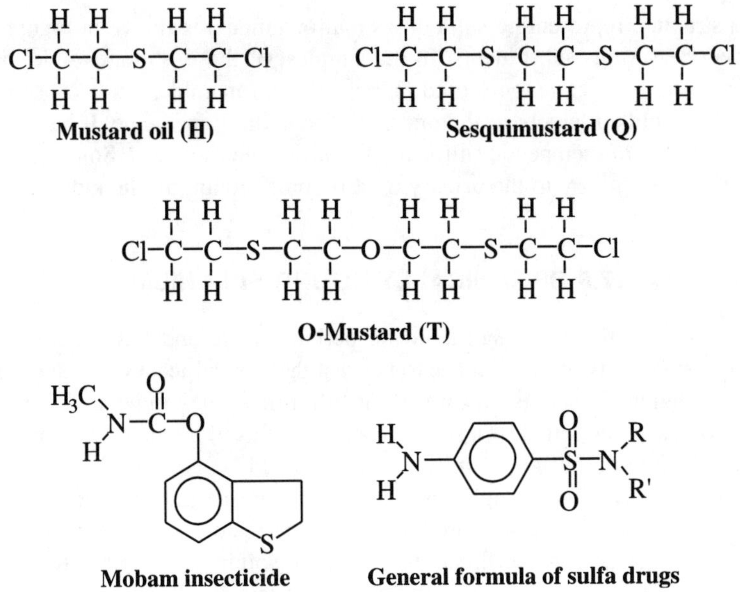

Figure 17.10 Some miscellaneous organosulfur compounds.

17.7 MISCELLANEOUS ORGANOSULFUR COMPOUNDS

A number of sulfur compounds containing other elements, such as halides, are used for various purposes. Some examples of such compounds are shown in Figure 17.10 and discussed briefly here.

17.7.1 Sulfur Mustards

The first three compounds shown in Figure 17.10 are **sulfur mustards**, which are highly toxic military poisons, or poison gases. These are mustard oil (bis(2-chloroethyl)sulfide), sesquimustard (1,2-bis(2-chloroethylthio)ethane), and O-mustard (bis(2-chloroethylthioethyl)ether). The toxic properties of mustard oil are typical of those of the sulfur mustards. As a military blistering gas poison, the vapors of this compound are very penetrating, so that it damages and destroys tissue at some depth from the point of contact. Affected tissue becomes severely inflamed, and the resulting lesions often become infected. Death can result from pulmonary lesions. Part of the hazard of mustard oil stems from the speed with which it penetrates tissue, so that efforts to remove it from the exposed area are ineffective after about 30 min. The compound is an experimental mutagen and primary carcinogen.

17.7.2 Sulfur in Pesticides

In Section 17.3, rodenticidal thioureas, insecticidal thiocyanates, and fungicidal dithiocarbamates were discussed. Sulfur is a common constituent of other classes of insecticides. These prominently include the organophosphate insecticides discussed in Chapter 18. Mobam (Figure 17.10) is a contact insecticide of the carbamate type, closely related in structure and function to the well-known carbamate insecticide carbaryl (see Figure 15.8). Mobam has been found to have a relatively high toxicity to laboratory mammals and is considerably more toxic than carbaryl.

17.7.3 Sulfa Drugs

The general structure representing sulfa drugs (sulfonamides) is shown in Figure 17.10, where the R groups may be various substituents. In the simplest of these, sulfanilamide, both R groups are H. It was once the most commonly used therapeutic sulfonamide, but because of side effects in humans, it is now limited largely to the practice of veterinary medicine. It has a toxicity rating of three. A large number of therapeutic sulfonamides have been produced. Some of the compounds have a tendency to cause injury to the urinary tract by precipitating in the kidney.

17.8 ORGANICALLY BOUND SELENIUM

Selenium, Se, is directly below sulfur in the periodic table and has a number of chemical similarities to sulfur. There is some evidence to suggest that a significant fraction of environmental selenium, such as selenium in soil, is substituted for sulfur in seleno-amino acids bound in peptides and proteins. This view is supported by a study of selenium in soil humic acid.[11] This study showed that most of the selenium in the soil samples examined could be extracted by base along with the humic and fulvic acid fractions (complex molecules that are partial biodegradation products of plant material). These humic substances are biodegradation residues of plant and animal biomass, and the results suggest that selenium originally added to the soil in the inorganic form was converted to proteinaceous selenium, which was incorporated into humic and fulvic acids. Acidic hydrolysis of these substances appeared to release selenium as seleno-amino acids. The seleno-methionine amino acid has been identified in selenium-rich soil humic matter.

REFERENCES

1. Kettle, A.J. et al., Assessing the flux of different volatile sulfur gases from the ocean to the atmosphere, *J. Geophys. Res.*, 106, 12193–12209, 2001.
2. Furne, J. et al., Oxidation of hydrogen sulfide and methanethiol to thiosulfate by rat tissues: a specialized function of the colonic mucosa, *Biochem. Pharmacol.*, 62, 255–259, 2001.
3. Tandon, S.K. and Prasad, S., Effect of thiamine on the cadmium-chelating capacity of thiol compounds, *Hum. Exp. Toxicol.*, 19, 523–528, 2000.
4. Hobbs, P. and Mottram, T., New directions: significant contributions of dimethyl sulfide from livestock to the atmosphere, *Atmos. Environ.*, 34, 3649–3650, 2000.
5. Wood, W.F., The history of skunk defensive secretion research, *Chem. Educ.*, 4, 44–50, 1999.
6. Newhook, R., Meek, M.E., and Walker, M., Carbon disulfide: hazard characterization and exposure-response analysis, *Environ. Carcinog. Ecotoxicol. Rev.*, 19, 125–160, 2001.
7. Amarnath, V. et al., *Chem. Res. Toxicol.*, 14, 1277–1283, 2001.
8. Gosselin, R.E., Smith, R.P., and Hodge, H.C., ANTU, in *Clinical Toxicology of Commercial Products*, 5th ed., Williams & Wilkins, Baltimore, 1984, pp. III-40–III-42.
9. Kassie, F. et al., Aenotoxic effects of methylisothiocyanate, *Mutat. Res.*, 490, 1–9, 2001.
10. Levi, P.E., Toxic action, in *Modern Toxicology*, Hodgson, E. and Levi, P.E., Eds., Elsevier, New York, 1987, chap. 6, pp. 133–184.
11. Kang, Y. et al., Selenium in soil humic acid, *Soil Sci. Plant Nutr.*, 37, 241–248, 1991.

QUESTIONS AND PROBLEMS

1. What is the inorganic hydride of sulfur? In what sense may organosulfur compounds be viewed as derivatives of this compound? What kind of organosulfur compound is formed by substituting alkyl or aryl hydrocarbon groups such as phenyl and methyl for H on this hydride?
2. What are the most common toxicological chemical reactions involving organosulfur? Give an example of each.

ORGANOSULFUR COMPOUNDS

3. What kind of organosulfur reaction is illustrated by the following:

$$2R\text{–}SH + \{O\} \rightarrow R\text{–}S\ S\text{–}R + H_2O$$

4. What kinds of compounds are known as mercaptans? Give an example formula of a mercaptan. How do these compounds make their presence known? How is this property put to practical use as a safety measure?

5. Despite the similarity of their names, how does carbon disulfide, CS_2, differ from the compounds shown below:

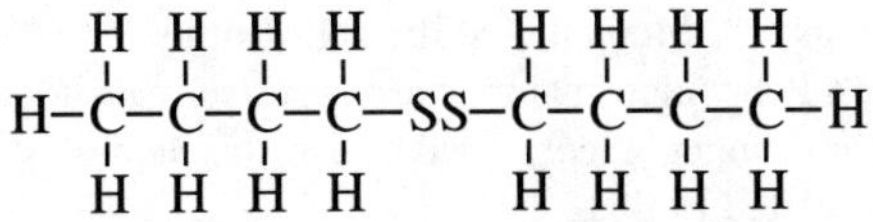

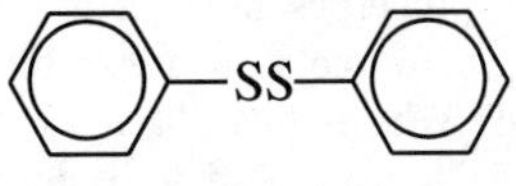

[Structures: butyl disulfide (H-C-C-C-C-SS-C-C-C-C-H with hydrogens) and diphenyl disulfide (Ph-SS-Ph)]

6. What are the symptoms of chronic carbon disulfide poisoning? How does this compound appear to affect the central nervous system?

7. To what general class of organosulfur compounds do the following belong? What are some of their uses and toxicological properties?

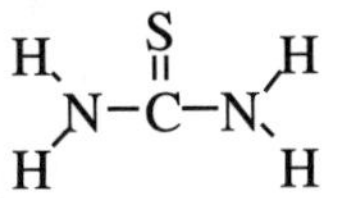

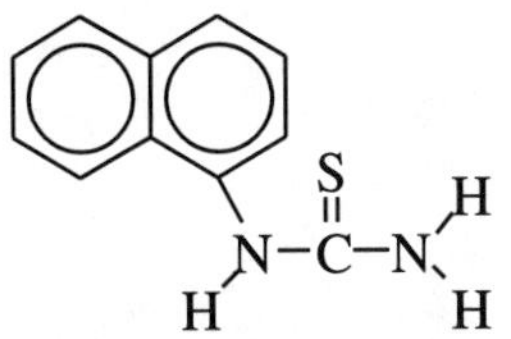

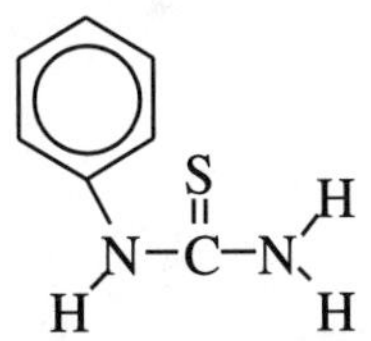

[Structures: thiourea (H₂N-C(=S)-NH₂), naphthyl thiourea, and phenyl thiourea]

8. What class of organosulfur compounds was regarded as the first synthetic organic insecticides? Give an example of one of these compounds.

9. What is an adjuvant? Explain the application of 2-mercaptobenzothiazole as an adjuvant.

10. What is the therapeutic use of the compound shown below.

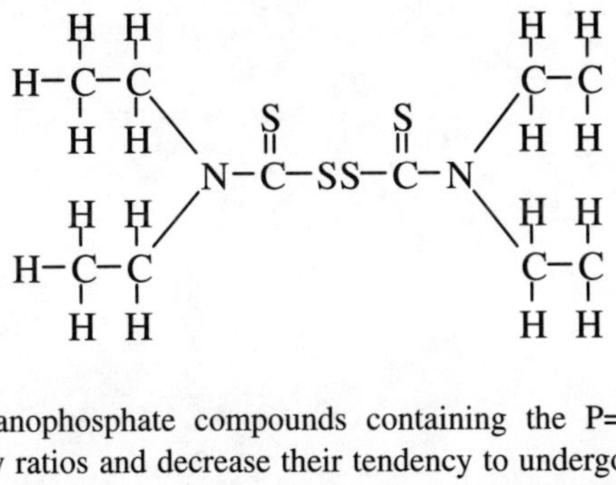

[Structure: tetraethylthiuram disulfide (disulfiram), (Et₂N-C(=S)-SS-C(=S)-NEt₂)]

11. What is done to organophosphate compounds containing the P=O group to increase their insect:mammal toxicity ratios and decrease their tendency to undergo nonenzymatic hydrolysis?

12. To what general classes of compounds do the following belong? What are some of their properties and uses?

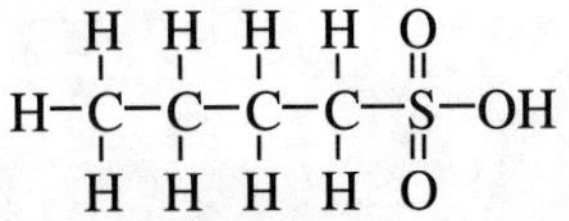

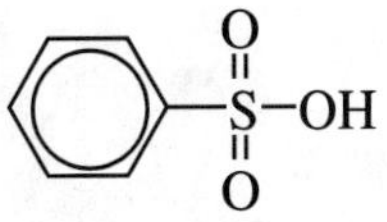

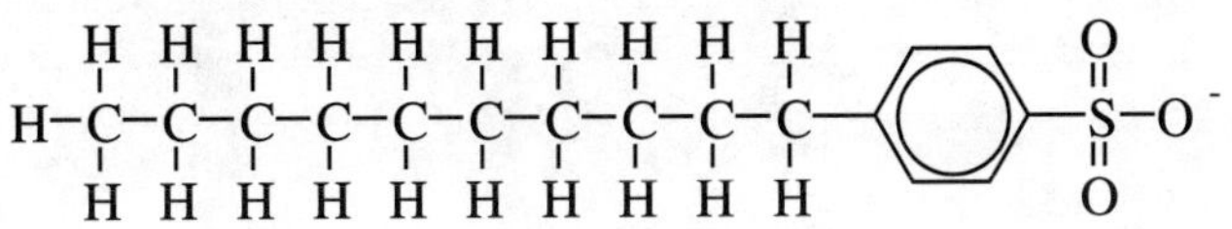

[Structures: butanesulfonic acid (H-C-C-C-C-S(=O)₂-OH), benzenesulfonic acid (Ph-S(=O)₂-OH), decylbenzenesulfonate (C₁₀H₂₁-C₆H₄-S(=O)₂-O⁻)]

13. What is the significance of the compounds below?

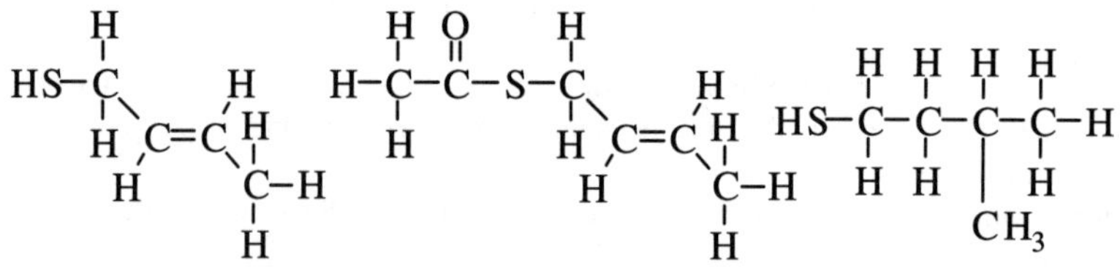

$$HS-CH_2-CH=CH-CH_2-CH_3 \qquad H-CH_2-C(=O)-S-CH_2-CH=CH-CH_3 \qquad HS-C(CH_3)_2-CH_2-CH(CH_3)-CH_3$$

14. Give the properties and uses of sesquimustard, Mobam, and the sulfonamides.
15. What is the basic structure of thiazole? For which major pharmaceutical can it serve as a raw material? Why is this pharmaceutical no longer widely used? Give the names, structures, and properties or uses of some compounds related to thiazole.
16. Give the names and structures of two dithiocarbamate fungicides. How are they used? How are zinc and iron involved with these fungicides?
17. Which are the simplest of the toxicologically important organic compounds that contain sulfur bound to phosphorus? What would be the formula of the trimethyl member of this group?

CHAPTER 18

Organophosphorus Compounds

18.1 INTRODUCTION

Phosphorus is directly below nitrogen in the periodic table. (The relationship of the chemistry of phosphorus to that of nitrogen is somewhat like the sulfur–oxygen relationship discussed in the introduction to Chapter 17.) The phosphorus atom electron configuration is $\{Ne\}3s^23p^3$, and it has five outer-shell electrons, as shown by its Lewis symbol in Figure 18.1. Because of the availability of underlying $3d$ orbitals, the valence shell of phosphorus can be expanded to more than eight electrons.

There are many kinds of organophosphorus compounds, including those with P–C bonds and those in which hydrocarbon moieties are bonded to P through an atom other than carbon, usually oxygen. These compounds have numerous industrial uses, and many of them, especially the organophosphate ester insecticides discussed later in this chapter, are economic poisons, that is, they are used to destroy pests that are harmful to crops, fruits, and vegetables. Organophosphorus compounds have varying degrees of toxicity. Some of these compounds, such as the nerve gases produced as military poisons, are deadly in minute quantities. The organophosphate esters, a class of compounds that contains the organophosphate ester insecticides and the organophosphate military poisons, are of particular toxicological interest because of their ability to inhibit acetylcholinesterase enzyme.

18.1.1 Phosphine

Phosphine (PH_3) is the hydride of phosphorus discussed as a toxic inorganic compound in Section 11.8. The formulas of many organophosphorus compounds can be derived by substituting organic groups for the H atoms in phosphine, and such an approach serves as a good starting point for the discussion of organophosphorus compounds.

18.2 ALKYL AND ARYL PHOSPHINES

Figure 18.2 gives the structural formulas of the more significant alkyl and aryl phosphine compounds. **Methylphosphine** is a colorless reactive gas that is very toxic by inhalation. **Dimethylphosphine** is a colorless, reactive, volatile liquid (bp, 25°C) that is toxic when inhaled or ingested. Both methylphosphine and dimethylphosphine have toxic effects similar to those of phosphine, a pulmonary tract irritant and central nervous system depressant that causes fatigue, vomiting, difficult breathing, and even death. **Trimethylphosphine** is a colorless volatile liquid (bp, 42°C). It is reactive enough to be spontaneously ignitable and probably has a high toxicity.

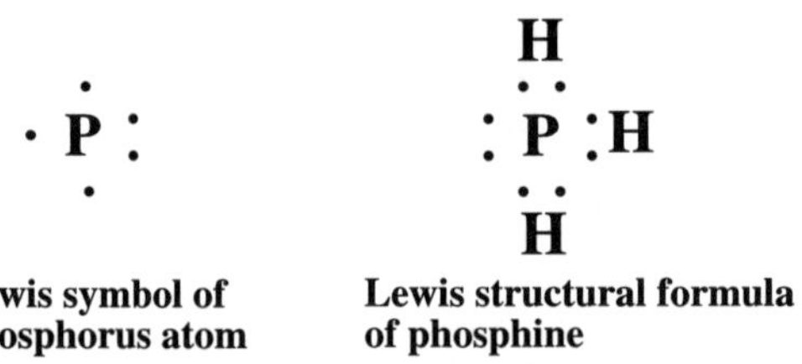

Figure 18.1 Lewis representations of the phosphorus atom and its hydride, phosphine, showing valence electrons as dots.

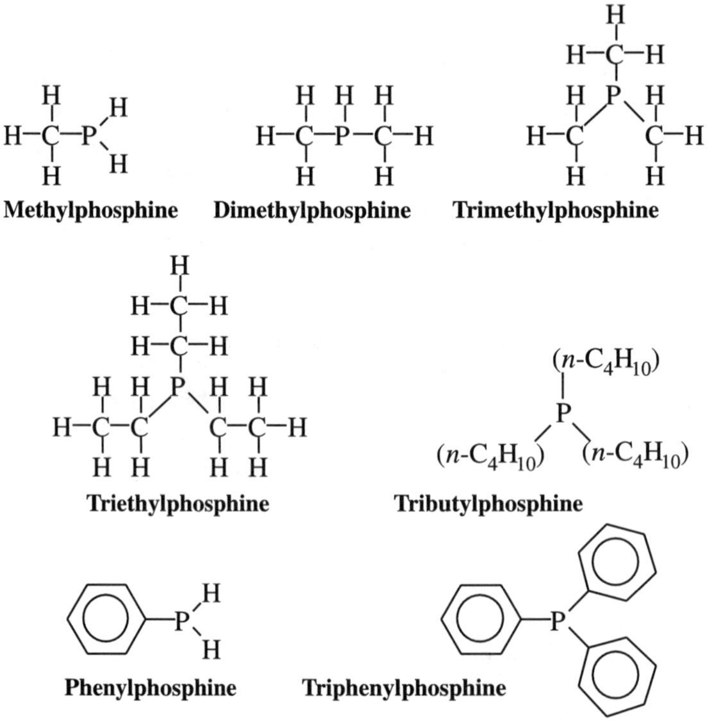

Figure 18.2 Some of the more significant alkyl and aryl phosphines.

Triethylphosphine probably has a high toxicity and tributylphosphine is a moderately toxic liquid. **Phenylphosphine** (phosphaniline) is a reactive, moderately flammable liquid (bp, 16°C) with a high toxicity by inhalation. **Triphenylphosphine** is a crystalline solid (mp, 79°C; bp > 360°C) with a low reactivity and moderate toxicity when inhaled or ingested.

The combustion of aryl and alkyl phosphines, such as trimethylphosphine, occurs as shown by the following example:

$$4C_3H_9P + 26O_2 \rightarrow 12CO_2 + 18H_2O + P_4O_{10} \tag{18.2.1}$$

Such a reaction produces P_4O_{10}, a corrosive irritant toxic substance discussed in 11.8.2, or droplets of corrosive orthophosphoric acid, H_3PO_4.

18.3 PHOSPHINE OXIDES AND SULFIDES

Phosphine oxides and sulfides have the general formulas illustrated below, where R represents hydrocarbon groups:

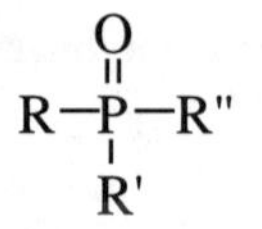

$$\begin{array}{c} O \\ \| \\ R-P-R'' \\ | \\ R' \end{array} \qquad \begin{array}{c} S \\ \| \\ R-P-R'' \\ | \\ R' \end{array}$$

Phosphine oxide 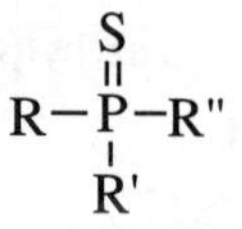   **Phosphine sulfide**

Two common phosphine oxides are **triethylphosphine** oxide (each R is a C_2H_5 group) and **tributylphosphine oxide** (each R is a C_4H_9 group). The former is a colorless, deliquescent, crystalline solid (mp, 52.9°C; bp, 243°C). The latter is a crystalline solid (mp, 94°C). Both compounds probably have high toxicities when ingested.

Triethylphosphine sulfide, $(C_2H_5)_3PS$, is a crystalline solid (mp, 94°C). Not much is known about its toxicity, which is probably high. **Tributylphosphine sulfide**, $(C_4H_9)_3PS$, is a skin irritant with a moderate toxicity hazard. When burned, both of these compounds give off dangerous fumes of phosphorus and sulfur oxides.

18.4 PHOSPHONIC AND PHOSPHOROUS ACID ESTERS

Phosphonic acid esters are derived from phosphonic acid (often erroneously called phosphorous acid), which is shown with some of its esters in Figure 18.3. Only two of the H atoms of phosphonic acid are ionizable, and hydrocarbon groups may be substituted for these atoms to give phosphonic acid esters. It is also possible to have esters in which a hydrocarbon moiety is substituted for the H atom that is bonded directly to the phosphorus atom. An example of such a compound is **dimethylmethylphosphonate**, shown in Figure 18.3. This type of compound has the same elemental formula as triesters of the hypothetical acid $P(OH)_3$, phosphorous acid. Examples of triesters of phosphorous acid, such as **trimethylphosphite**, are shown in Figure 18.3.

Trimethylphosphite is a colorless liquid (bp, 233°C). It is soluble in many organic solvents, but not in water. Little information is available regarding its toxicity or other hazards. **Tributylphosphite** is a liquid (bp, 120°C). It decomposes in water, but is probably not very toxic. **Triphenylphosphite** is a white solid or oily liquid (mp, 23°C; bp, 157°C). It is a skin irritant with a moderate oral toxicity. Although it is not soluble in water, it may hydrolyze somewhat to phenol, which adds to its toxicity. **Tris(2-ethylhexyl)phosphite**, a trialkyl phosphite in which the hydrocarbon moieties are the 2-ethylhexyl group, $-CH_2CH(C_2H_5)C_4H_9$, is a water-insoluble compound (bp, 100°C). Its toxicity is largely unknown.

Dimethylmethylphosphonate is of toxicological concern because of its widespread use; diethylethylphosphonate may be a suitable substitute for it in some applications.[1] **Methylphosphonate**, $(CH_3O)P(O)H(OH)$, has a moderate oral toxicity and is a skin and eye irritant. **Dibutylphosphonate**, $(C_4H_9O)_2P(O)H$, is a liquid boiling at 115°C at 10 mmHg pressure. Through ingestion and dermally, it has a moderately high toxicity. Like other organophosphonates and phosphites, it can decompose to evolve dangerous products when heated, burned, or exposed to reactive chemicals, such as oxidants. Thermal decomposition can result in the evolution of highly toxic phosphine, PH_3. Combustion produces corrosive orthophosphoric acid and oxides of phosphorus.

Diallylphosphonate, shown in Figure 18.3, has two alkenyl substituent groups. Information is lacking on its toxicity, although compounds with allyl groups tend to be relatively toxic. Incidents have been reported in which this compound has exploded during distillation.

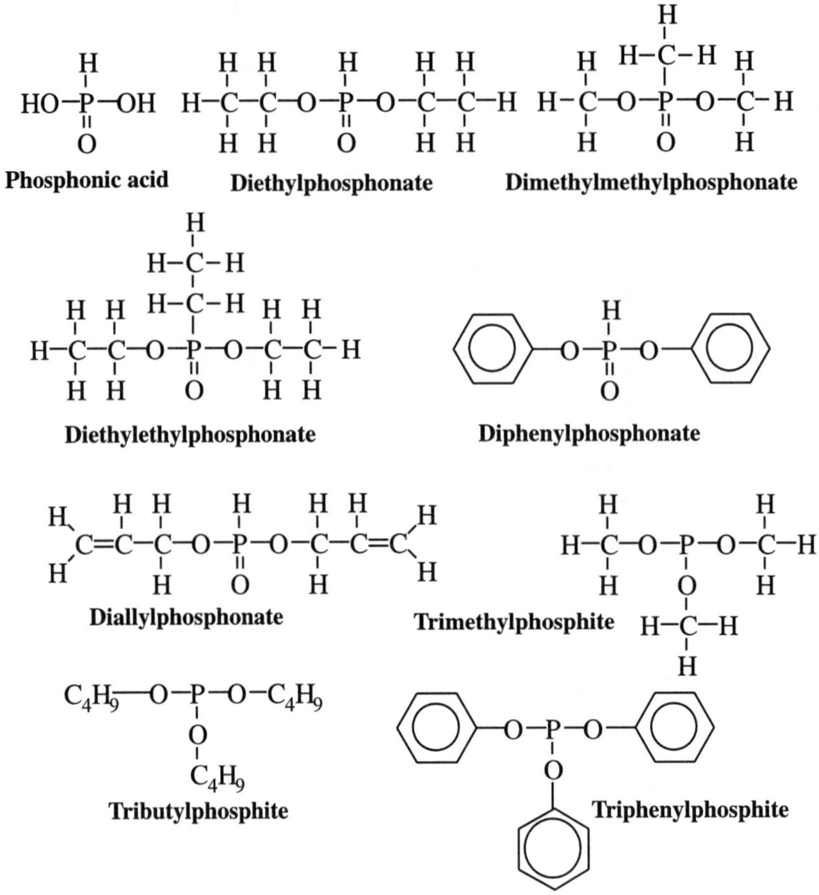

Figure 18.3 Phosphonic acid and esters of phosphonic and phosphorous acids.

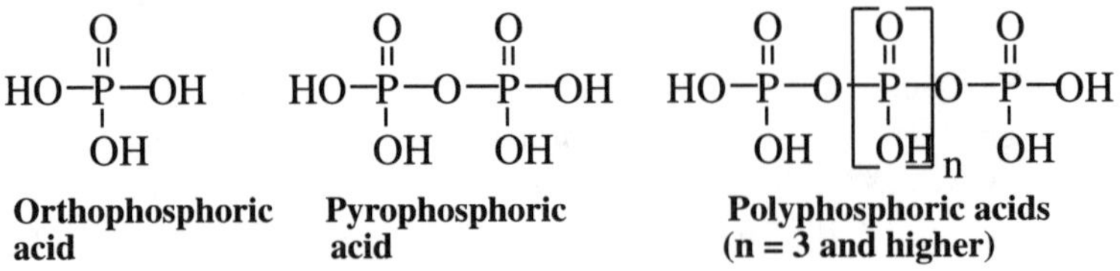

Figure 18.4 Orthophosphoric acid and acids formed by its polymerization.

18.5 ORGANOPHOSPHATE ESTERS

18.5.1 Orthophosphates and Polyphosphates

Figure 18.4 shows the structural formula of orthophosphoric acid as well as those of diphosphoric and polyphosphoric acids, produced by polymerization of orthophosphoric acid with loss of water. These compounds form esters in which alkyl, alkenyl, and aryl hydrocarbon moieties are substituted for H; most of the more common ones are esters of orthophosphoric acid. In this section,

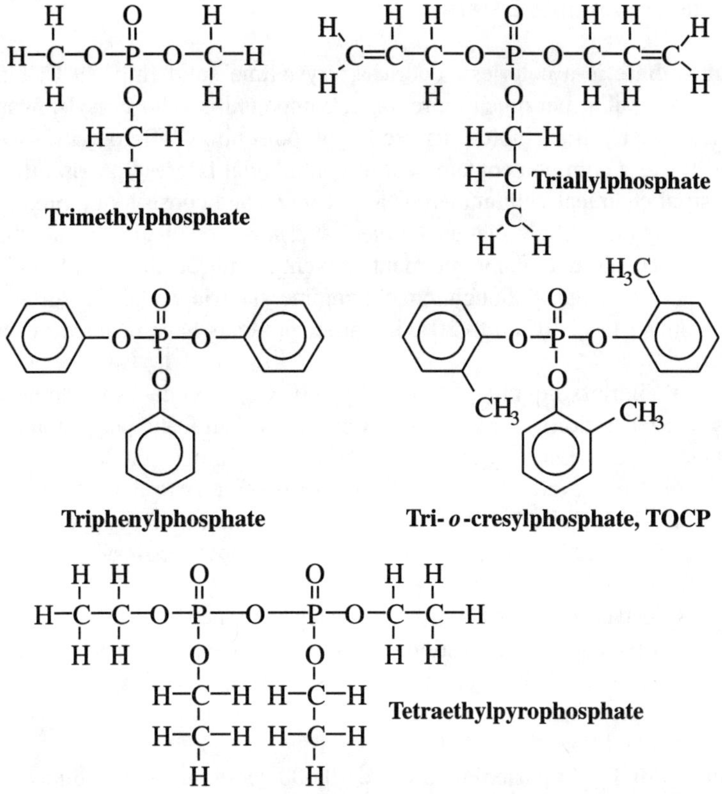

Figure 18.5 Phosphate esters.

only the relatively simple organophosphate esters are discussed. Many economic poisons — particularly insecticides — are organophosphate esters that often contain nitrogen, sulfur, or halogens. These compounds are discussed in a later section.

18.5.2 Orthophosphate Esters

Some of the more significant phosphate esters are shown in Figure 18.5. Trimethylphosphate is the simplest of the organophosphate esters; the structural formulas of the other alkyl esters of orthophosphoric acid are like those of trimethylphosphate, but with alkyl substituent groups other than methyl. Comparatively little information is available about the toxicity of trimethylphosphate, although it is probably moderately toxic orally or through skin absorption. A study of potential carcinogenicity of this compound to Wistar rats showed no evidence that it is carcinogenic to these test animals.[2]

Triethylphosphate, $(C_2H_5O)_3PO$, is a liquid (fp, –57°C; bp, 214°C). It is insoluble in water, but soluble in most organic solvents. Like other phosphate esters, it damages nerves and is a cholinesterase inhibitor. It is regarded as moderately toxic. Two other alkyl phosphates with toxicities probably similar to that of triethylphosphate are **tributylphosphate**, $(n\text{-}C_4H_9O)_3PO$, and **tris(2-ethylhexyl)-phosphate**, $(C_8H_{17}O)_3PO$.

Triallylphosphate is the phosphate triester of allyl alcohol and contains unsaturated C=C bonds in its structure. This compound is a liquid (fp, –50°C). It is regarded as having a high toxicity and produces abnormal tissue growth when administered subcutaneously. It has been known to explode during distillation.

18.5.3 Aromatic Phosphate Esters

Triphenylphosphate is a colorless, odorless, crystalline solid (mp, 49°C; bp, 245°C). It is moderately toxic. A similar, but much more toxic, compound is **tri-*o*-cresyl-phosphate** (TOCP), an aryl phosphate ester with a notorious record of poisonings.[3] Before its toxicity was fully recognized, TOCP was a common contaminant of commercial **tricresylphosphate**. Tricresylphosphate is an industrial chemical with numerous applications and consists of a mixture of phosphate esters in which the hydrocarbon moieties are *meta* and *para* cresyl substituents. It has been used as a lubricant, gasoline additive, flame retardant, solvent for nitrocellulose, plasticizer, and even a cooling fluid for machine guns. Although modern commercial tricresylphosphate contains less than 1% TOCP, contaminant levels of up to 20% in earlier products have resulted in severe poisoning incidents.

Pure TOCP is a colorless liquid (fp, −27°C; bp, 410°C). It produces pronounced neurological effects and causes degeneration of the neurons in the body's central and peripheral nervous systems, although fatalities are rare. Early symptoms of TOCP poisoning include nausea, vomiting, and diarrhea, accompanied by severe abdominal pain. Normally a 1- to 3-week latent period occurs after these symptoms have subsided, followed by manifestations of peripheral paralysis, as evidenced by "wrist drop" and "foot drop." In some cases, the slow recovery is complete, whereas in others partial paralysis remains.

The most widespread case of TOCP poisoning occurred in the U.S. in 1930 when approximately 20,000 people were affected by the ingestion of alcoholic Jamaican ginger ("Jake") adulterated by 2% TOCP. The peculiar manner in which the victims walked, including "foot drop," slapping the feet on the floor, high stepping, and unsteadiness, gave rise to the name of "jake leg" to describe the very unfortunate condition.

A major incident of TOCP poisoning affected 10,000 people in Morocco in 1959. The victims had eaten food cooked in olive oil adulterated with TOCP-contaminated lubricating oil. A number of cases of permanent paralysis resulted from ingestion of the contaminated cooking oil.

It is believed that metabolic products of TOCP inhibit acetylcholinesterase. Apparently other factors are involved in TOCP neurotoxicity. A study of tri-*o*-cresylphosphate poisoning in China has described a number of symptoms.[4] Initial pain in the lower leg muscles was followed by paralysis and lower limb nerve injury. Patients with mild poisoning recovered after several months, but more severely poisoned ones suffered permanent effects. Despite the devastating effects of TOCP, the percentage of virtually complete recovery in healthy subjects is relatively high.

18.5.4 Tetraethylpyrophosphate

Tetraethylpyrophosphate (TEPP) was the first organophosphate compound to be used as an insecticide. This compound was developed in Germany during World War II and was substituted for nicotine as an insecticide. It is a white to amber hygroscopic liquid (bp, 155°C) that readily hydrolyzes in contact with water. Because of its tendency to hydrolyze and its extremely high toxicity to mammals, TEPP was used for only a very short time as an insecticide, although it is a very effective one. It was typically applied as an insecticidal dust formulation containing 1% TEPP.

The toxicity of TEPP to humans and other mammals is very high; it has a toxicity rating of 6, supertoxic. TEPP is a very potent acetylcholinesterase inhibitor. (The inhibition of acetylcholinesterase by organophosphate insecticides is discussed in Section 18.7.)

18.6 PHOSPHOROTHIONATE AND PHOSPHORODITHIOATE ESTERS

The general formulas of **phosphorothionate** and **phosphorodithioate** esters are shown in Figure 18.6, where R represents a hydrocarbon or substituted hydrocarbon moiety. Many of the

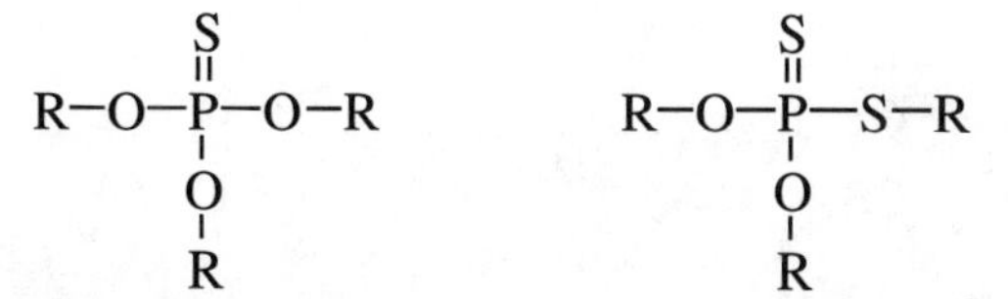

$$\text{R-O-}\underset{\underset{\text{R}}{|}}{\underset{\text{O}}{\overset{\overset{\text{S}}{\|}}{\text{P}}}}\text{-O-R} \qquad \text{R-O-}\underset{\underset{\text{R}}{|}}{\underset{\text{O}}{\overset{\overset{\text{S}}{\|}}{\text{P}}}}\text{-S-R}$$

Phosphorothionate   **Phosphorodithioate**

Figure 18.6 General formulas of phosphorothionate and phosphorodithioate esters; each R represents a hydrocarbon or substituted hydrocarbon moiety.

organophosphate insecticides are sulfur-containing esters of these general types, which often exhibit higher insect:mammal toxicity ratios than do their nonsulfur analogs. Esters containing the P=S (thiono) group are not as effective as their analogous compounds that contain the P=O functional group in inhibiting acetylcholinesterase. In addition to their lower toxicities to nontarget organisms, thiono compounds are more stable toward nonenzymatic hydrolysis. The metabolic conversion of P=S to P=O (oxidative desulfuration) in organisms is responsible for the insecticidal activity and mammalian toxicity of phosphorothionate and phosphorodithioate insecticides.

An example of a simple phosphorothionate is tributylphosphorothionate, in which the R groups (above) are n-C_4H_9 groups. It is a colorless liquid (bp, 143°C). The compound is a cholinesterase inhibitor, as are some of its metabolic products. Examples of phosphorothionate and phosphorodithioate esters with more complex formulas synthesized for their insecticidal properties are discussed in the following section.

18.7 ORGANOPHOSPHATE INSECTICIDES

The organophosphate insecticides were originally developed in Germany during the 1930s and 1940s, primarily through the efforts of Gerhard Schrader and his research group. The first of these was tetraethylpyrophosphate, discussed in Section 18.5. Its disadvantages — including high toxicity to mammals — led to the development of related compounds, starting with **parathion**, O,O-diethyl-O-p-nitrophenylphosphorothionate, which will be discussed in some detail.

18.7.1 Chemical Formulas and Properties

Many insecticidal organophosphate compounds have been synthesized. Unlike the organohalide insecticides that they largely displaced, the organophosphates readily undergo biodegradation and do not bioaccumulate. However, the neurotoxic characteristics of organophosphates pose dangers in their handling and use, so that once-popular compounds of this type have now been phased out or their uses severely curtailed.

An enormous variety of organophosphate ester compounds have been synthesized and used as pesticides. They can be categorized as phosphates, phosphorothiolates, phosphorothioates, and phosphorodithioates, depending on the number and bonding configurations of S atoms bound to the central P atom (Figure 18.7). In the generic formulas of these classes of compounds shown in Figure 18.7, the R groups are frequently methyl (–CH_3) or ethyl (–C_2H_5) groups and Ar is a moiety of a more complex structure, frequently aromatic. In some insecticides, three or even all four of the atoms bound directly to P are S atoms.

18.7.2 Phosphate Ester Insecticides

Figure 18.8 shows some organophosphate insecticides based on the phosphate esters. These compounds do not contain sulfur. One of the more significant of these compounds is paraoxon,

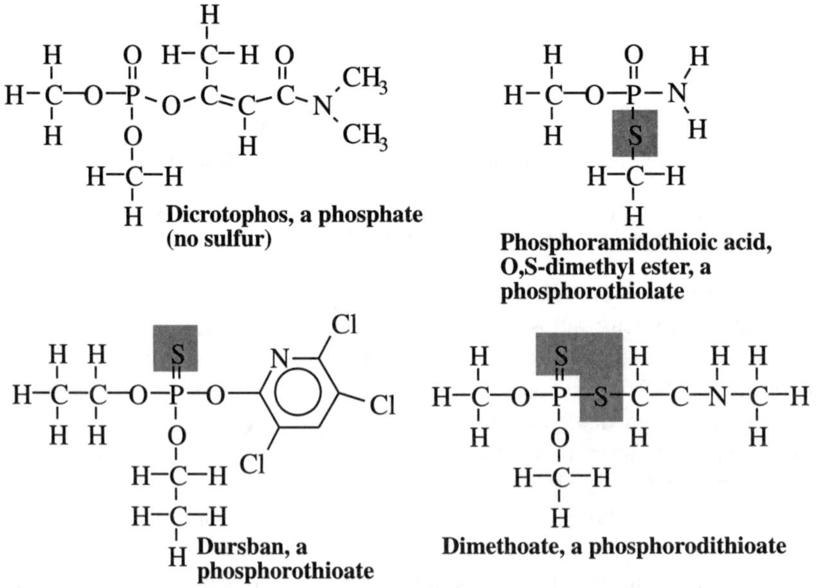

Figure 18.7 Phosphates, phosphorothiolate, phosphorothioate, and phosphorodithioate insecticides distinguished by the numbers and orientations of the sulfur atoms around the phosphorus.

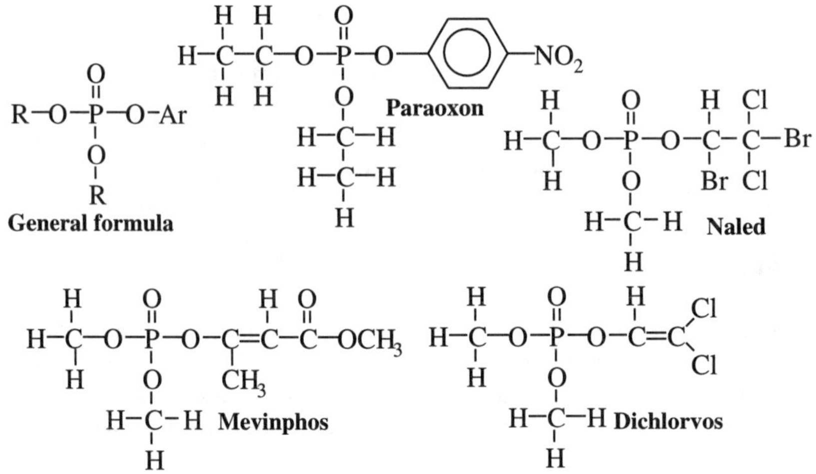

Figure 18.8 Organophosphate insecticides based on phosphate esters.

which, as noted previously, is a metabolic activation product of parathion. It has been synthesized directly and was made by Schrader in 1944 along with parathion. One of the most toxic organophosphate insecticides, paraoxon has a toxicity rating of six. It is alleged to have been provided to chemical warfare agents in South Africa's former apartheid government. **Naled** is a bromine-containing phosphate ester insecticide. **Mevinphos** is considered to be an extremely dangerous chemical. It is still used, however.[5] **Dichlorvos** has a toxicity rating of four and is deactivated by enzymes in the livers of mammals. Its tendency to vaporize has enabled its use in pest strips. In 2002, the company that had been using dichlorvos in Vapona fly killer and moth killer strips announced that it would no longer do so because of concerns over its potential carcinogenicity.[6]

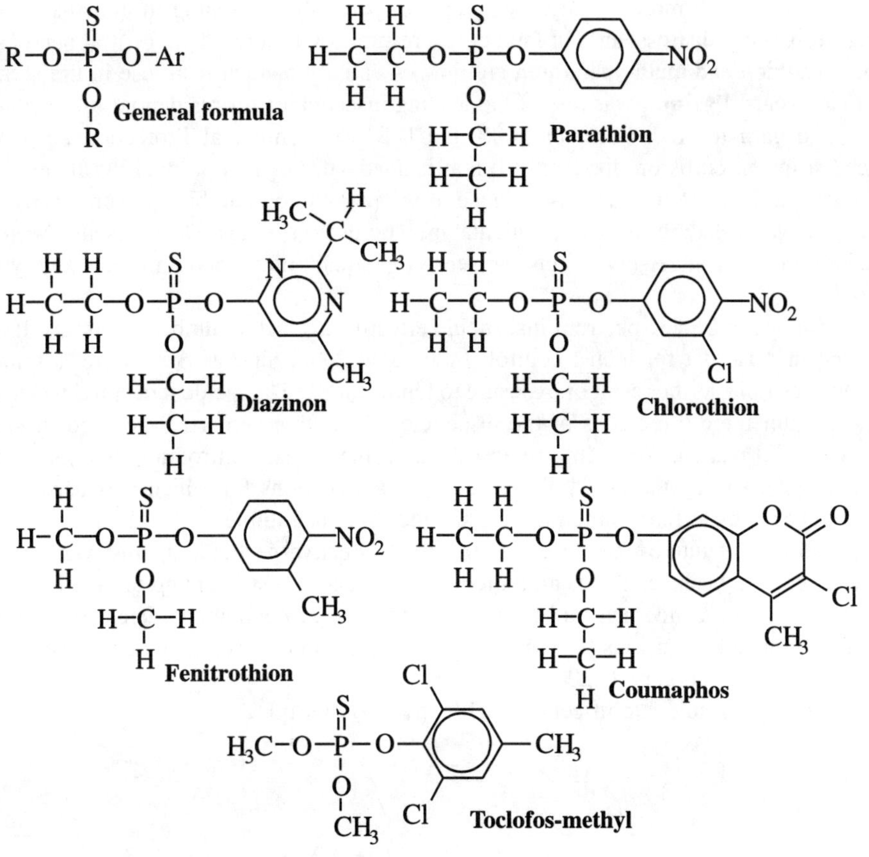

ORGANOPHOSPHORUS COMPOUNDS

Figure 18.9 Phosphorothionate organophosphate insecticides.

18.7.3 Phosphorothionate Insecticides

Figure 18.9 gives the structural formulas of some typical phosphorothionate esters and the general formula of this type of organophosphate insecticide.

Insecticidal parathion is a phosphorothionate ester first licensed for use in 1944. Pure parathion is a yellow liquid that is insoluble in kerosene and water, but stable in contact with water. Among its properties that make parathion convenient to use as an insecticide are stability in contact with neutral and somewhat basic aqueous solutions, low volatility, and toxicity to a wide range of insects. It was applied as an emulsion in water, dust, wettable powder, or aerosol. Even before it was banned for general use, it was not recommended for applications in homes or animal shelters because of its toxicity to mammals.

Parathion has a toxicity rating of six (supertoxic), and methylparathion (which has methyl groups instead of the ethyl groups shown in Figure 18.9) is regarded as extremely toxic. As little as 120 mg of parathion has been known to kill an adult human, and a dose of 2 mg has killed a child. Most accidental poisonings have occurred by absorption through the skin. Since its use began, several hundred people have been killed by parathion. One of the larger poisoning incidents occurred in Jamaica in 1976 from ingestion of parathion-contaminated flour. Of 79 people exposed, 17 died.

In the body, parathion is converted to paraoxon (structure in Figure 18.8), which is a potent inhibitor of acetylcholinesterase. Because this conversion is required for parathion to have a toxic effect, symptoms develop several hours after exposure, whereas the toxic effects of TEPP or paraoxon develop much more rapidly. Symptoms of parathion poisoning in humans include skin twitching, respiratory distress, and, in fatal cases, respiratory failure due to central nervous system paralysis. Parathion and methylparathion are now essentially banned from use in the U.S.

For many years **diazinon** was one of the leading insecticides for residential use, including use on lawns and gardens. In December 2000, the U.S. Environmental Protection Agency (EPA) announced stringent curbs on diazinon use that banned sales of this product by the end of 2004. This ban was put in place because of evidence of water pollution and bird poisonings by diazinon and because it was a leading cause of accidental insecticide poisonings. There has also been concern that diazinon in water adversely affects the sense of "smell" of salmon and their ability to avoid predators.[7]

Fenitrothion is a broad-spectrum insecticide effective against a number of insects. It has been widely used in Australia for locust control. In January 2002, nine workers were hospitalized in Melbourne, Australia, as the result of exposure to fenitrothion.[8] The incident occured when a forklift ran over and punctured three cans of the insecticide. Symptoms reported included irritated skin, stinging eyes, and nausea. According to an official at the scene, fenitrothion "works its way into the eyes, armpits, and up the nose." Because of the applications for which it is used, residential and dietary exposures to fenitrothion are considered to be negligible in the U.S.

Coumaphos is of interest because it is the most effective pesticide against varroa mites and small hive beetles in beehives and is an ingredient of insecticide strips hung in the hives to kill the mites.[9] There is some concern regarding this use because of coumaphos detected in honey. The U.S. EPA has granted extensions for the use of coumaphos in beehives, including one to run from February 2, 2002, to February 1, 2003.

Another phosphorothionate insecticide, **chlorpyrifos methyl**,

is used to protect stored grain from insects. Because of concerns about its acute, subchronic, and developmental toxicity potential, the U.S. EPA has placed a ban on sales of this insecticide after December 31, 2004. The relatively long period from announcing the ban until it takes effect was allowed to enable agricultural interests to find a suitable substitute.

18.7.4 Phosphorodithioate Insecticides

Figure 18.10 shows the general formula of phosphorodithioate insecticides and structural formulas of some examples. In 2002, **dimethoate** was canceled for residential use in the U.S., and some of the crop uses of **disulfoton** were discontinued. **Azinphos-methyl** and **phosmet** are among the older organophosphate insecticides, having first been licensed in the mid-1960s. In late 2001, the U.S. EPA canceled 28 crop uses for azinphos-methyl and announced that seven crop uses, including those on peaches, almonds, walnuts, and cotton, were to be phased out over the next 4 years. Three uses of phosmet were voluntarily withdrawn by the manufacturer, and uses on nine crops, including blueberries, grapes, pears, and plums, were allowed for five more years. These measures were taken to reduce consumer exposure and particularly to reduce hazards to workers.[10]

Malathion is the best-known phosphorodithioate insecticide. It shows how differences in structural formula can cause pronounced differences in the properties of organophosphate pesticides.

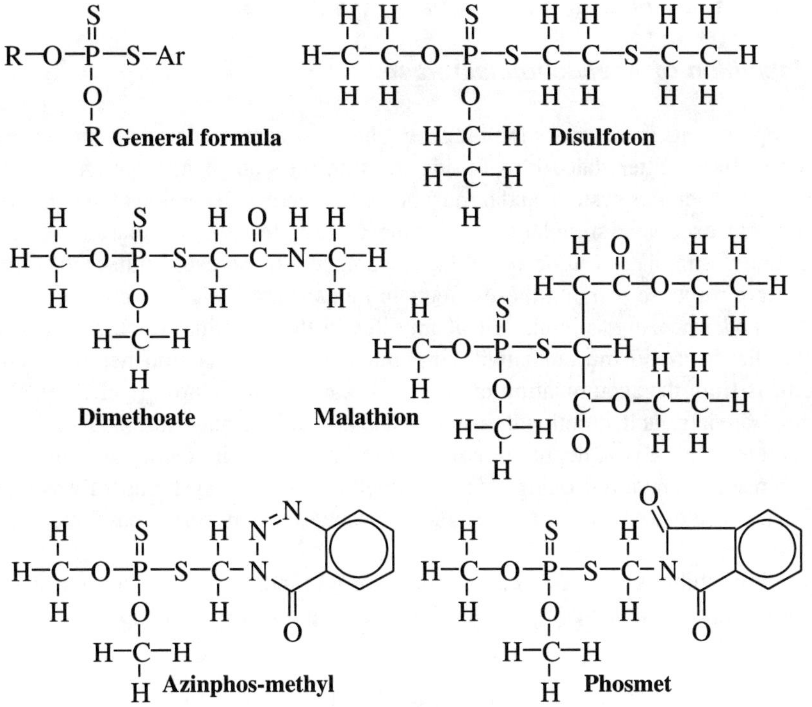

Figure 18.10 Phosphorodithioate organophosphate insecticides.

Malathion has two carboxyester linkages, which are hydrolyzable by carboxylase enzymes to relatively nontoxic products, as shown in reaction 18.7.1. The enzymes that accomplish this reaction are possessed by mammals, but not by insects, so that mammals can detoxify malathion, whereas insects cannot. The result is that malathion has selective insecticidal activity. For example, although malathion is a very effective insecticide, its LD_{50} for adult male rats is about 100 times that of parathion, reflecting the much lower mammalian toxicity of malathion than those of some of the more toxic organophosphate insecticides, such as parathion.

Carboxylase enzymes are inhibited by organophosphates other than malathion. The result of exposure of mammals to malathion plus another organophosphate is potentiation (enhancement of the action of an active substance by an otherwise inactive substance) of the toxicity of malathion.

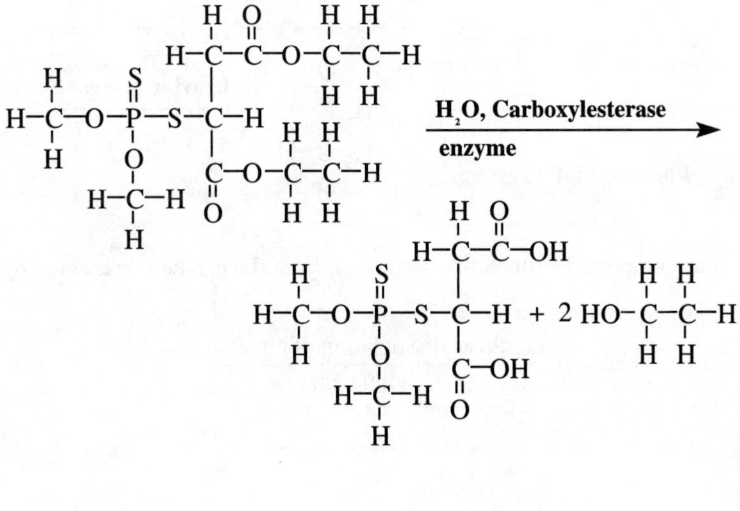

(18.7.1)

18.7.5 Toxic Actions of Organophosphate Insecticides

18.7.5.1 Inhibition of Acetylcholinesterase

The organophosphate insecticides inhibit acetylcholinesterase in mammals and insects. Acetylcholine is a neurotransmitter that forms during the transmission of nerve impulses in the body, including the central nervous system, and it must be hydrolyzed by the action of acetylcholinesterase enzyme to prevent excessive stimulation of the nerve receptors. Accumulation of acetylcholine results in continued stimulation of acetylcholine receptors and can cause numerous effects related to excessive nerve response. Among these effects in humans are bronchioconstriction, resulting in chest tightness and wheezing; stimulation of muscles in the intestinal tract, resulting in nausea, vomiting, and diarrhea; and muscular twitching and cramps. The central nervous system shows numerous effects from the accumulation of acetylcholine. These include psychological symptoms of restlessness, anxiety, and emotional instability. The subject may suffer from headache and insomnia. In more severe cases, depression of the respiratory and circulatory systems, convulsions, and coma may result. In fatal poisonings, death is due to respiratory system paralysis. Patients who recover from the acute toxic effects of organophosphates often suffer continued neurological symptoms.

Cholinesterase inhibition occurs when an inhibitor (I) binds to the cholinesterase enzyme (E) to produce an enzyme–inhibitor complex, as shown by the following reaction:

$$E + I \rightleftharpoons EI \tag{18.7.2}$$

With some inhibitors the reaction is reversible. With other kinds of compounds, such as the organophosphates, a stable, covalently bound complex (E') is formed from which it is difficult to regenerate the original enzyme, as illustrated by the reaction

$$E + I \rightleftharpoons EI \rightarrow E' \xrightarrow{\text{Slowly or not at all}} E + \text{Products} \tag{18.7.3}$$

An example of irreversible binding is that of paraoxon, which can be viewed as an organophosphate compound containing a phosphorylating group (P) and a leaving group (L), as shown below:

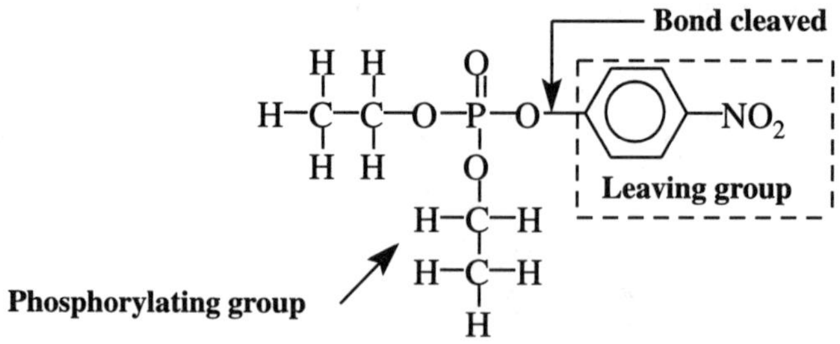

The reaction of this compound with cholinesterase enzyme (E) can be represented by the following reaction:

$$E + PL \rightarrow EP + L \xrightarrow{\text{Slow dissociation of covalently bound enzyme}} E + \text{Products} \tag{18.7.4}$$

ORGANOPHOSPHORUS COMPOUNDS

The phosphorylating group bonds to an OH group at the active site of the enzyme.

18.7.5.2 Metabolic Activation

Highly purified phosphorothionate and phosphorodithioate insecticides do not inhibit acetylcholinesterase directly. In order for these compounds to inhibit acetylcholinesterase, the following phase I metabolic conversion of P=S to P=O must occur:

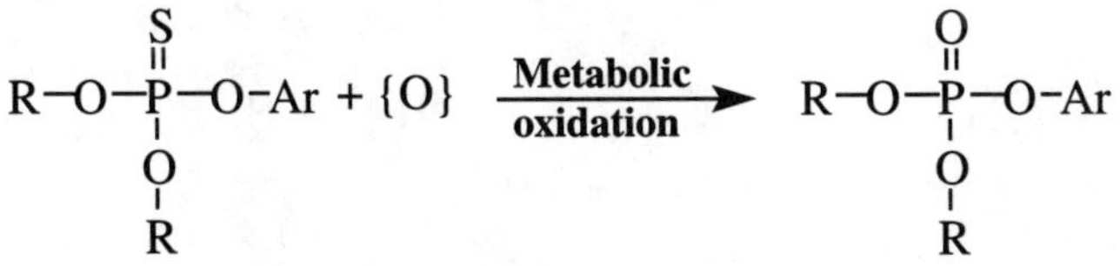

$$\begin{array}{c} S \\ \| \\ R-O-P-O-Ar \\ | \\ O \\ | \\ R \end{array} + \{O\} \xrightarrow{\text{Metabolic oxidation}} \begin{array}{c} O \\ \| \\ R-O-P-O-Ar \\ | \\ O \\ | \\ R \end{array} \quad (18.7.5)$$

A specific example of this type of reaction is the conversion of parathion to paraoxon, mentioned above.

18.7.5.3 Mammalian Toxicities

The mammalian toxicities of the organophosphate insecticides vary widely. This may be seen from the LD_{50} values to male rats of organophosphate insecticides, including some of the ones discussed above. With the approximate LD_{50} values (oral, mg/kg) given in parentheses, common organophosphate insecticides, in descending order of toxicity, are TEPP (1) > mevinphos, disulfoton (6 or 7) > parathion, methylparathion, azinphosmethyl, chlorfenvinphos (13 to 15) > dichlorvos (80) > diazinon (110) > trichlorfon (215) > chlorothion (880) > ronnel > malathion (1300).

Organophosphate poisonings are a particular problem in poorer countries, where lax controls, poor protective measures, and other problems contribute to exposure. The African country of Zimbabwe has experienced an epidemic of such poisonings, with a mortality rate from reported instances of around 8%.[11] About three fourths of the adult poisonings were the result of suicide attempts, whereas poisoning in young children was predominantly from accidental ingestion.

Treatment of organophosphate poisoning consists of (1) blocking the nerve-stimulating effects of acetylcholine that builds up when acetylcholinesterase enzyme is inhibited, and (2) binding with organophosphates to prevent them from adding to acetylcholinesterase or to break down the adducts that have formed. The standard approach to the former is administration of atropine[12]:

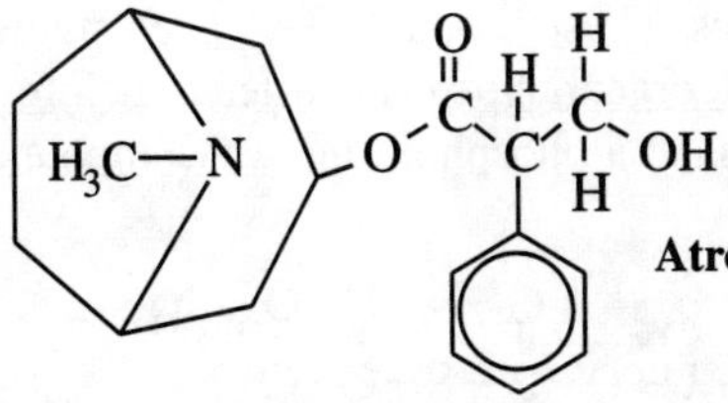

Atropine

Itself a deadly poison derived from belladonna or other nightshade plants (see Chapter 19), atropine must be administered with great care. The organophosphate residues can be bound with oximes to release acetylcholinesterase. This is shown below for the reaction of organophosphate-bound acetylcholinesterase enzyme with pralidoxime:

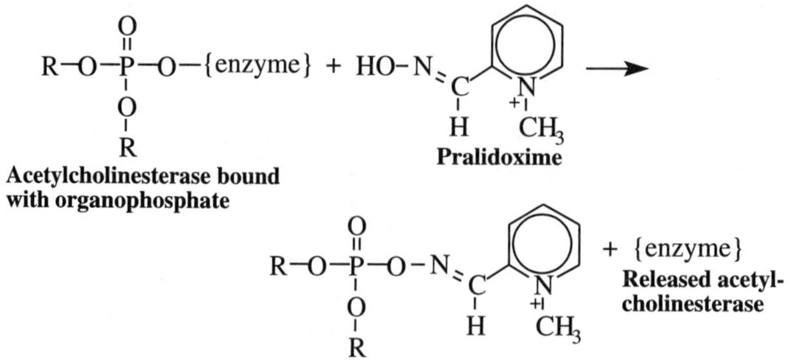

$$\text{R-O-P(=O)(O-R)-O-\{enzyme\}} + \text{HO-N=CH-C}_5\text{H}_3\text{N}^+\text{-CH}_3 \longrightarrow$$

Acetylcholinesterase bound with organophosphate   **Pralidoxime**   (18.7.6)

$$\text{R-O-P(=O)(O-R)-O-N=CH-C}_5\text{H}_3\text{N}^+\text{-CH}_3 + \text{\{enzyme\}}$$

Released acetyl-cholinesterase

18.7.5.4 Deactivation of Organophosphates

The deactivation of organophosphates is accomplished by hydrolysis catalyzed by phosphotriestrase enzymes. These enzymes are found in a wide range of organisms, from bacteria to humans.[13] Hydrolysis of organophosphates is shown by the following general reactions, where R is an alkyl group, Ar is a substituent group that is frequently aromatic, and X is either S or O:

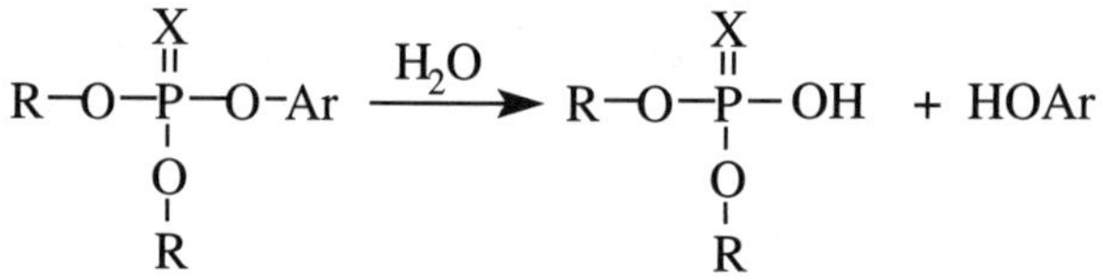

$$\text{R-O-P(=X)(O-R)-O-Ar} \xrightarrow{H_2O} \text{R-O-P(=X)(O-R)-OH} + \text{HOAr} \quad (18.7.7)$$

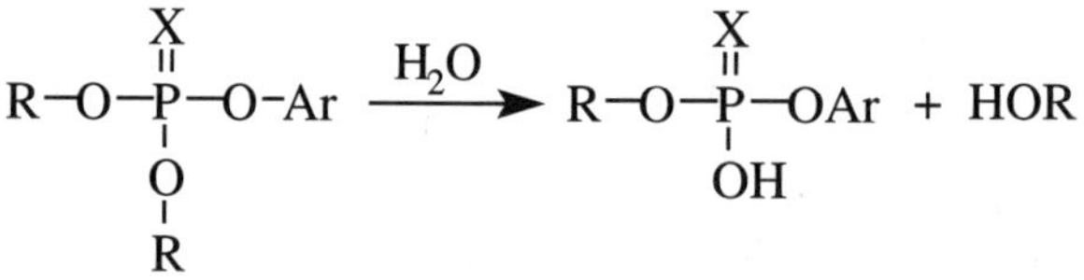

$$\text{R-O-P(=X)(O-R)-O-Ar} \xrightarrow{H_2O} \text{R-O-P(=X)(OH)-OAr} + \text{HOR} \quad (18.7.8)$$

The phosphate products may be measured in urine as bioindicators of exposure to organophosphates. The product that is observed depends on whether or not oxidative desulfuration occurs prior to esterase hydrolysis of the molecules. These products have been studied in the urine of greenhouse workers exposed to three methyl organophosphate esters: toclofos-methyl (Figure 18.9), fenitrothion (Figure 18.9), and omethoate, a phosphorothiolate compound (below)[14]:

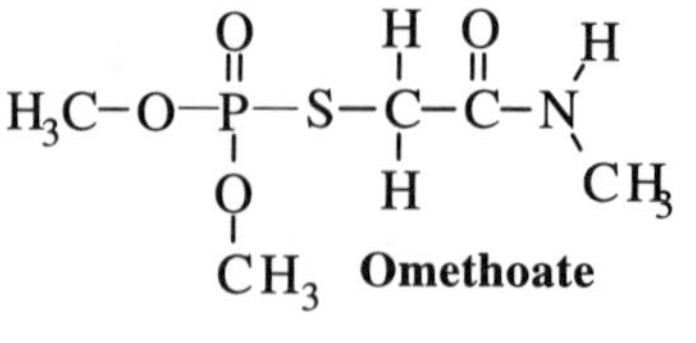

$$\text{H}_3\text{C-O-P(=O)(O-CH}_3\text{)-S-CH}_2\text{-C(=O)-NH-CH}_3$$

Omethoate

In a study of biomarkers of exposure to organophosphate insecticides, the urine of 6- and 7-year-old Italian children was analyzed for dimethyl- and diethyl- phosphates, thiophosphates, and dithiophosphates, such as the examples shown in Figure 18.11. Levels of these metabolites were found to correlate well with insecticide applications inside or outside the homes where the children

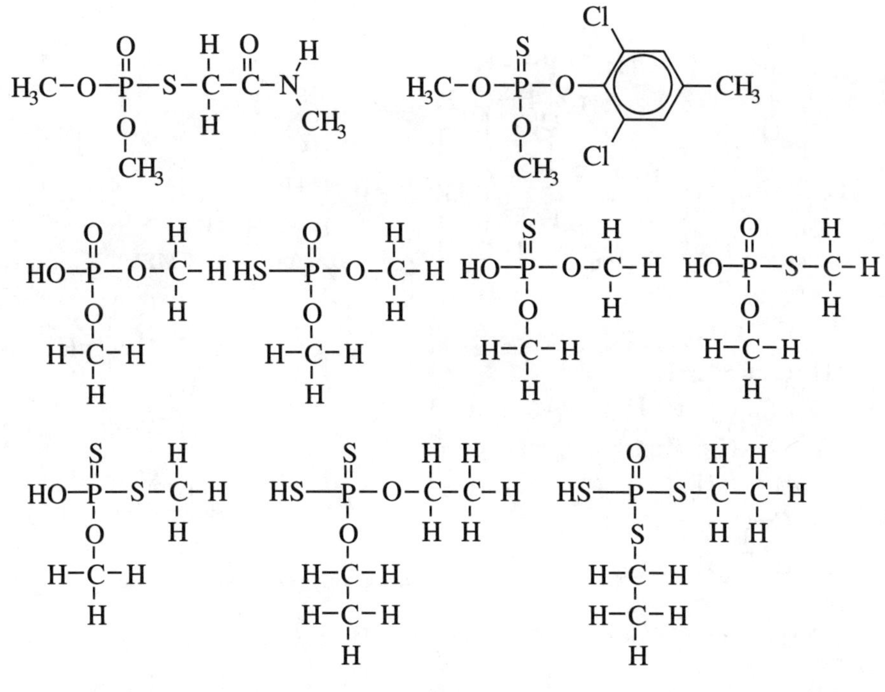

Figure 18.11 Examples of dimethyl- and diethyl- organophosphate, organothiophosphate, and organodithiophosphate esters that can be measured in urine as biomarkers of exposure to various organophosphate ester insecticides.

lived; higher levels of the metabolites were found in the urine of the children than in that of a group of adults from the same vicinity.[15]

A study of diazinon metabolites in the urine of United Kingdom sheep dippers has shown strong correlations with the kind and intensity of exposure.[16] The highest levels were found in workers who handled the concentrated dip. Exposure to dilute dip was largely through splashing (sheep object to being immersed in a solution of organophosphates and tend to splash it around). It was also observed that the sheep dippers exhibit a marked aversion to wearing protective clothing, thus increasing their exposure.

18.8 ORGANOPHOSPHORUS MILITARY POISONS

Organophosphorus compounds developed for use as military poisons — the nerve gases — are among the most toxic synthetic compounds ever made. They were first reported in 1937 by a group at the German concern Farbenfabriken Bayer AG, headed by Gerhard Schrader; development continued in Germany during World War II and in other countries afterward. Structural formulas of the most studied of these compounds are shown in Figure 18.12.

The action of **Sarin** is typical of the organophosphorus military poisons. Its lethal dose to humans may be as low as about 0.01 mg/kg. It is a systemic poison to the central nervous system that is readily absorbed as a liquid through the skin; a single drop so absorbed can kill a human. It is a colorless liquid (fp, –58°C; bp, 147°C). A compound that is chemically similar to Sarin, **diisopropylphosphorfluoridate**, a highly toxic, oily liquid that served as the basis for the development of nerve gases in Germany during World War II, acts by binding to the active site of acetylcholinesterase enzyme, thereby inhibiting the enzyme. Specifically, the reaction is thought to be with a serine side chain on the active site, as shown by the following reaction:

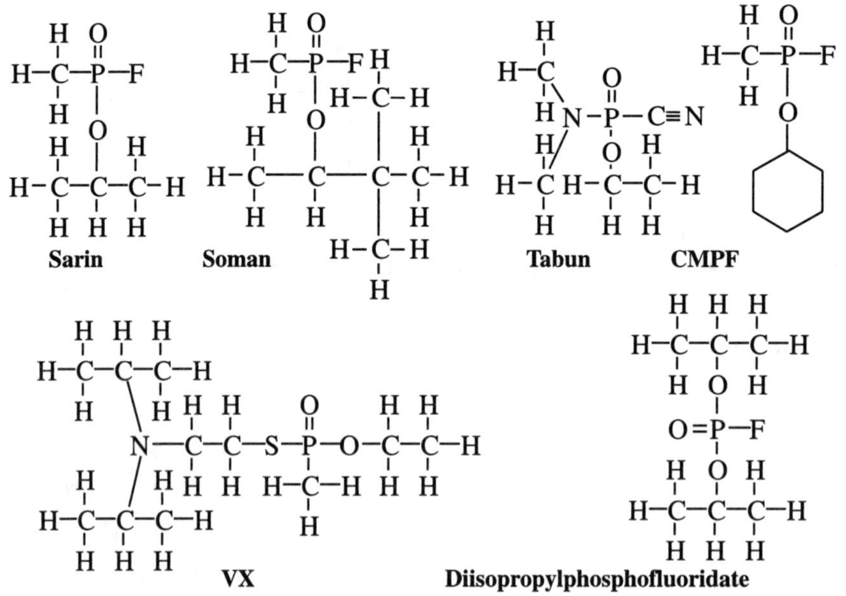

Figure 18.12 Major nerve gas organophosphate military poisons.

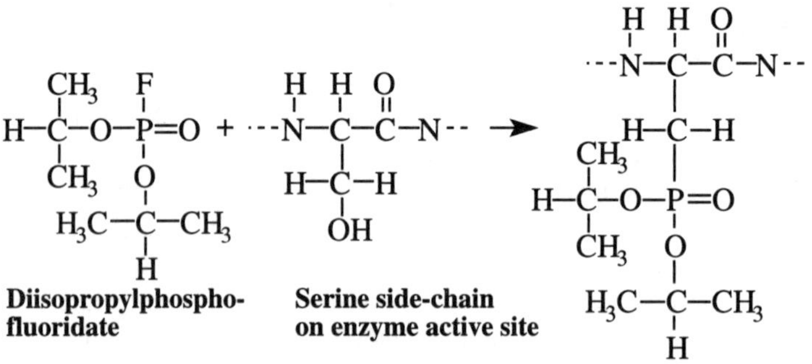

Soman is an irreversible inhibitor of acetylcholinesterase. Unlike the case with some acetylcholinesterase inhibitors, recovery from sublethal poisoning by Soman requires enzyme resynthesis, rather than reversible binding to the enzyme.

Pure Tabun is also a colorless liquid; it has a freezing point of –49°C and decomposes when heated to 238°C. Its toxicity is similar to that of Sarin. Tabun acts primarily on the sympathetic nervous system, and it has a paralytic effect on the blood vessels. Its toxic action and symptoms of poisoning are similar to those of parathion, an organophosphate insecticide for which extensive human toxicity data are available. **Diisopropylfluorophosphate**, $(i\text{-}C_3H_7O)_2P(O)F$, is a highly toxic, oily liquid that served as the basis for the development of nerve gases in Germany during World War II. The organophosphorus military poisons are powerful inhibitors of acetylcholinesterase enzyme.

REFERENCES

1. Blumbach, K. et al., Biotransformation and male rat-specific renal toxicity of diethyl ethyl- and dimethyl methylphosphonate, *Toxicol. Sci.*, 53, 24–32, 2000.

2. Bomhard, E.M. et al., Trimethyl phosphate: a 30-month chronic toxicity/carcinogenicity study in Wistar rats with administration in drinking water, *Fundam. Appl. Toxicol.*, 40, 75–89, 1997.
3. Gosselin, R.E., Smith, R.P., and Hodge, H.C., Tri-*ortho*-cresyl phosphate, in *Clinical Toxicology of Commercial Products*, 5th ed., Williams & Wilkins, Baltimore, 1984, pp. III-388–III-393.
4. Yang, J., Li, H., and Niu, Q., Four-year clinical study on fulminant intoxication of tri-*o*-cresylphosphate, *Zhonghua Laodong Weisheng Zhiyebing Zazhi*, 19, 212–214, 2001.
5. Sewald, N., Amvac rakes the pesticide fields, *Chemical Week*, May 2, 2001, pp. 41–43.
6. Vapona products in cancer scare, *Marketing Week*, Feb. 21, 2002, p. 5.
7. Bernton, H., Insecticide on sale as EPA ban nears; stores have until '04 to unload diazinon supply, *The Seattle Times*, Feb. 4, 2002, p. B1.
8. Jackson, A., Pesticide leak puts 9 workers in hospital, *The Age*, Jan. 8, 2002, p. 1.
9. Bayer's checkmite (coumaphos) has been given a specific exemption by the EPA to the California Department of Pesticide Regulation for use on beehives to control varroa mites, *Chemical Business NewsBase: Agricultural Chemical News*, May 15, 2001, p. 1.
10. EPA places new restrictions on two older pesticides, *Chemical Market Reporter*, Nov. 12, 2001, p. 28.
11. Dong, X. and Simon, M.A., The epidemiology of organophosphate poisoning in urban Zimbabwe from 1995 to 2000, *Int. J. Occup. Environ. Health*, 7, 333–338, 2001.
12. Ecobichon, D.J., Toxic effects of pesticides, in *Casarett and Doull's Toxicology: The Basic Science of Poisons*, 6th ed., Klaassen, C.D., Ed., McGraw-Hill, New York, 2001, chap. 22, pp. 763–810.
13. Sogorb, M.A. and Vilanova, E., Enzymes involved in the detoxication of organophosphorus, carbamate and pyrethroid insecticides through hydrolysis, *Toxicol. Lett.*, 128, 315–228, 2002.
14. Aprea, C. et al., Evaluation of respiratory and cutaneous doses and urinary excretion of alkylphosphates by workers in greenhouses treated with omethoate, fenitrothion, and toclofos-methyl, *Am. Ind. Hyg. J.*, 62, 87–95, 2001.
15. Aprea, C. et al., Biologic monitoring of exposure to organophosphorus pesticides in 195 Italian children, *Environ. Health Perspect.*, 108, 521–525, 2000.
16. Buchanan, D. et al., Estimation of cumulative exposure to organophosphate sheep dips in a study of chronic neurological health effects among United Kingdom sheep dippers, *Occup. Environ. Med.*, 58, 694–701, 2001.

SUPPLEMENTARY REFERENCE

Manahan, S.E., *Environmental Chemistry*, 7th ed., Lewis Publishers/CRC Press, Inc., Boca Raton, FL, 2000.

QUESTIONS AND PROBLEMS

1. What may be said about the relationship of phosphorus with nitrogen, oxygen, and sulfur in organophosphorus compounds? Give examples of organophosphorus compounds that contain N, O, or P.
2. To which kinds of nitrogen and sulfur compounds are alkyl and aryl phosphine compounds analogous?
3. What particular hazards are posed by the reaction below?

$$4C_3H_9P + 26O_2 \rightarrow 12CO_2 + 18H_2O + P_4O_{10}$$

4. What classes of compounds do the general formulas below illustrate? Give examples of each, along with their toxic effects.

$$\begin{array}{cc} \overset{\displaystyle O}{\underset{\displaystyle R'}{\overset{\|}{R-P-R''}}} & \overset{\displaystyle S}{\underset{\displaystyle R'}{\overset{\|}{R-P-R''}}} \end{array}$$

5. Discuss the toxicity characteristics of TOCP.
6. Discuss the uses and toxicity characteristics of tricresylphosphate.
7. What is phosphonic acid? How does it differ from phosphorous acid? From the structures below, pick out the esters of each of these acids.

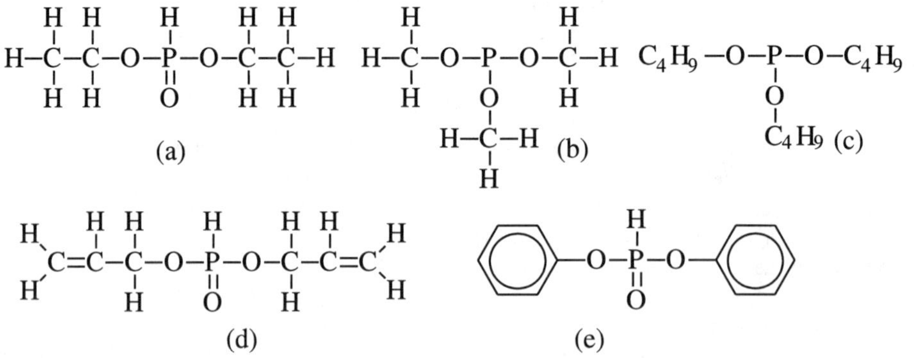

(a), (b), (c), (d), (e) [structures shown]

8. What are the general formulas of phosphorothionate and phosphorodithioate esters? How are the thiono and P=O functional groups involved in these kinds of esters? How are they used? What is oxidative desulfuration and why is it significant with these kinds of compounds?
9. For what purposes have the two compounds below been used? What are their toxicity characteristics and relative toxicities?

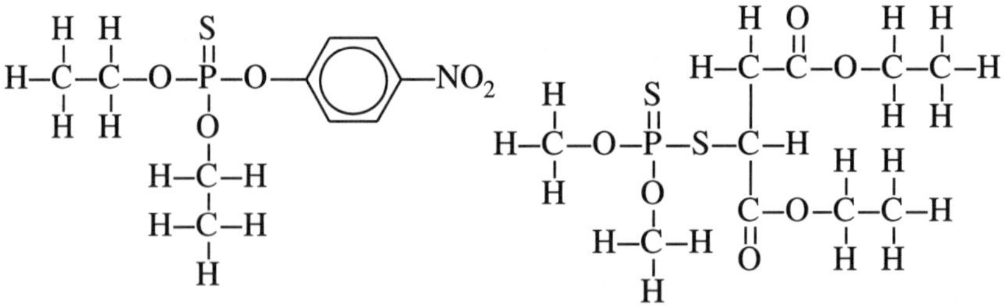

[structures shown]

10. What is parathion converted to in the body? Of what kind of reaction mentioned in Chapter 17 is this an example? What does it have to do with the uses and toxicity of parathion?
11. Designate which of the following statements is **not** true:
    (a) The organophosphate insecticides inhibit acetylcholinesterase in mammals and insects.
    (b) These insecticides are toxic because they prevent the formation of acetylcholine produced during the transmission of nerve impulses in the body.
    (c) Acetylcholine must be hydrolyzed by the action of acetylcholinesterase enzyme to prevent excessive stimulation of the nerve receptors.
    (d) Excessive accumulation of acetylcholine can cause numerous effects related to excessive nerve response.
12. Discuss what the following shows regarding the interaction of an organophosphate insecticide with cholinesterase enzyme:

$$E + PL \rightarrow EP + L \xrightarrow{\text{Slow dissociation of covalently bound enzyme}} E + \text{Products}$$

ORGANOPHOSPHORUS COMPOUNDS

13. Discuss what the following shows regarding cholinesterase inhibition:

$$E + I \rightleftarrows EI \rightarrow E' \xrightarrow{\text{Slowly or not at all}} E + \text{Products}$$

14. Explain what the following shows regarding the deactivation of organophosphates:

$$\text{R-O-}\overset{\overset{X}{\|}}{\underset{\underset{R}{|}}{\underset{|}{P}}}\text{-O-Ar} \xrightarrow{H_2O} \text{R-O-}\overset{\overset{X}{\|}}{\underset{\underset{R}{|}}{\underset{|}{P}}}\text{-OH} + \text{HOAr}$$

$$\text{R-O-}\overset{\overset{X}{\|}}{\underset{\underset{R}{|}}{\underset{|}{P}}}\text{-O-Ar} \xrightarrow{H_2O} \text{R-O-}\overset{\overset{X}{\|}}{\underset{\underset{OH}{|}}{P}}\text{-OAr} + \text{HOR}$$

15. What are the uses of the following compounds? What may be said about their toxicities?

CHAPTER 19

Toxic Natural Products

19.1 INTRODUCTION

Toxic natural products are poisons produced by organisms. They include an enormous variety of materials, including animal venoms and poisons, bacterial toxins, protozoal toxins, algal toxins, mycotoxins (from fungi), and plant toxins.[1] These substances may affect a number of organs and tissues in humans, including the skin, heart, liver, kidney, neurological system, immune system, and gastrointestinal system. Some such substances may be hypnotic or psychotropic; others are carcinogenic.

Perhaps the most acutely toxic substance known is the botulism toxin, produced by the anaerobic bacterium *Clostridium botulinum* and responsible for many food poisoning deaths, especially from improperly canned food. Mycotoxins generated by fungi (molds) can cause a number of human maladies, and some of these materials, such as the aflatoxins, are carcinogenic to some animals. Venoms from wasps, spiders, scorpions, and reptiles — consisting of an exotic variety of biomolecules, including low-molecular-mass polypeptides, proteins, enzymes, steroids, lipids, 5-HT, and glycosides — can be fatal to humans. Each year, in the Orient, tetrodotoxin from improperly prepared puffer fish makes this dish the last delicacy consumed by some unfortunate diners. The stories of Socrates' execution from being forced to drink an extract of the deadly poisonous spotted hemlock plant and Cleopatra's suicide at the fangs of a venomous asp are rooted in antiquity. Many household poisonings result from children ingesting toxic plant leaves or berries. High on this list is philodendron. Other plants that may be involved in poisonings include diefenbachia, jade plant, wandering Jew, Swedish ivy, pokeweed, string of pearls, and yew. Pollen from plants causes widespread misery from allergies, and reactions to toxins from plants such as poison ivy can be severe.

Living organisms wage chemical warfare against their potential predators and prey with a fascinating variety of chemical substances. Some organisms produce toxic metabolic by-products that have no obvious use to the organisms that make them. This chapter briefly describes some of the toxic natural products from living organisms, with emphasis on those that are toxic to humans.

A few distinctions should be made at this point that apply particularly to animals. A **poisonous organism** is one that produces toxins. A **poisonous animal** may contain toxins in its tissues that act as poisons to other animals that eat its flesh. **Venoms** are poisons that can be delivered without the need for the organism to be eaten. **Venomous animals** can deliver poisons to another animal by means such as biting (usually striking with fangs) or stinging. The puffer fish — some tissues of which are deadly when ingested — is a poisonous animal, whereas the rattlesnake is a venomous one, although its flesh may be eaten safely. Some organisms — notoriously, the skunk (see the discussion of organosulfur compounds from skunk spray in Section 17.2) — wage chemical warfare by emitting substances that are not notably toxic, but still effective in keeping predators away. Perhaps these organisms should be classified as noxious species.

19.2 TOXIC SUBSTANCES FROM BACTERIA

Bacteria are sources of a number of toxic substances, including botulinus toxin, which is arguably the most deadly substance known.[2] The two greatest concerns regarding toxic substances from bacteria are their roles in causing symptoms of bacterial disease and food poisoning. It is useful to consider as one class those bacteria that produce toxins that adversely affect a host in which the bacteria are growing, and as another class those bacteria that produce toxins to which another organism is subsequently exposed, such as by ingestion.

Bacteria are single-celled microorganisms that may grow in colonies and are shaped as spheres, rods, or spirals. They are usefully classified with respect to their need for oxygen, which accepts electrons during the metabolic oxidation of food substances, such as organic matter. **Aerobic bacteria** require molecular oxygen to survive, whereas **anaerobic bacteria** grow in the absence of oxygen, which may be toxic to them. **Facultative bacteria** can grow either aerobically or anaerobically. Anaerobic bacteria and facultative bacteria functioning anaerobically use substances other than molecular O_2 as electron acceptors (oxidants that take electrons away from other reactants in a chemical reaction). For example, sulfate takes the place of O_2 in the anaerobic degradation of organic matter (represented as $\{CH_2O\}$) by *Desulfovibrio*, yielding toxic hydrogen sulfide (H_2S) as a product, as shown by the following overall reaction:

$$SO_4^{2-} + 2\{CH_2O\} + 2H^+ \rightarrow H_2S + 2CO_2 + 2H_2O \qquad (19.2.1)$$

Toxicants such as H_2S produced by microorganisms are usually not called microbial toxins, a term that more properly refers to usually proteinaceous species of high molecular mass synthesized metabolically by microorganisms and capable of inducing a strong response in susceptible organisms at low concentrations. Bacteria and other microorganisms do produce a variety of poisonous substances, such as acetaldehyde, formaldehyde, and putrescine (see Figure 19.2). The proteinaceous toxins produced by bacteria act in a number of ways, including effects on enzymes, detrimental interactions with cell surfaces, and food poisoning.[3]

19.2.1 *In Vivo* Bacterial Toxins

Some important bacterial toxins are produced in the host and have a detrimental effect on the host. For example, such toxins are synthesized by *Clostridium tetani*, common soil bacteria that enter the body largely through puncture wounds.

The toxin from this bacterium interferes with neurotransmitters, such as acetylcholine, causing **tetanus**, commonly called lockjaw. Abnormal populations or strains of *Shigella dysenteriae* bacteria in the body can cause a severe form of dysentery, hemorrhagic colitis, and hemolytic uremic syndrome, releasing a toxin that causes intestinal hemorrhaging and gastrointestinal tract paralysis. These shiga toxins, which are also produced by some strains of the common intestinal bacteria *Escherichia coli*, possess structural groups that bind with cell surfaces and an enzymatically active component that enters the cell and inhibits protein synthesis.[4] Toxin-releasing bacteria responsible for the most common form of food poisoning are those of the genus *Salmonella*. Victims are afflicted with flu-like symptoms and may even die from the effects of the toxin. Other bacteria that cause many cases of food poisoning include *Staphylococcus aureus* and *Clostridium perfringens*.[5] Diphtheria is caused by a toxin generated by *Cornybacterium diphtheriae*. The toxin interferes with protein synthesis and is generally destructive to tissue.

19.2.1.1 Toxic Shock Syndrome

Toxic shock syndrome is a very damaging, often fatal condition caused by toxins from *Staphylococcus aureus* or *Streptococcus pyogenes*. First reported in children in 1978, it is manifested by high fever, erythroderma (a skin rash condition), and severe diarrhea.[6] Patients may exhibit confusion, hypotension, and tachycardia, and they may go into shock with failure of several organs. Survivors often suffer from skin desquamation (flaky skin).

In 1980, toxic shock syndrome in menstruating women was linked to the use of a new superabsorbent tampon that altered the vaginal environment, including sequestration of magnesium ion, in a way that greatly increased susceptibility to the infection. The tampons were quickly withdrawn from the market, and now most cases occur in surgical patients infected by the bacteria that cause the syndrome.

The bacterial toxins responsible for toxic shock syndrome are proteins that are classified as superantigens. These proteins bypass some of the steps normally involved in antigen-mediated immune response, thereby activating 5 to 30% of the T cell population, compared to 0.01 to 0.1% activated by conventional antigens. The consequence of the huge numbers of activated T cells is release of cytokines that cause capillaries to leak, resulting in many of the symptoms of the syndrome.

19.2.2 Bacterial Toxins Produced Outside the Body

The most notorious toxin produced by bacteria outside the body is that of *Clostridium botulinum*. This kind of bacteria grows naturally in soil and on vegetable material. Under anaerobic or slightly aerobic conditions, it synthesizes an almost unbelievably toxic product. The conditions for generating this toxin most commonly occur as the result of the improper canning of food, particularly vegetables. Botulinum toxin binds irreversibly to nerve terminals, preventing the release of acetylcholine; the affected muscle acts as though the nerve were disconnected. The toxin actually consists of several polypeptides in the range of 200,000 to 400,000 molecular mass. Fortunately, these proteins are inactivated by heating for a sufficient time at 80 to 100°C. Botulinum poisoning symptoms appear within 12 to 36 h after ingestion, beginning with gastrointestinal tract disorders and progressing through neurologic symptoms, paralysis of the respiratory muscles, and death by respiratory failure.

19.3 MYCOTOXINS

Mycotoxins are toxic secondary metabolites from fungi that have a wide range of structures and a variety of toxic effects. Human and animal exposure to mycotoxins usually results from ingestion of food upon which fungal molds have grown. Among the many kinds of molds that produce mycotoxins are *Aspergillus flavus*, *Fusarium*, *Trichoderma*, *Aspergillus*, and *Penicillium*. Perhaps the most well-known mycotoxins are the **aflatoxins** produced by *Aspergillus*. These molds grow on a variety of food products, including corn, cereal grains, rice, apples, peanuts, and milk. Other mycotoxins include ergot alkaloids, ochratoxins, fumonisins, trichothecenes, tremorgenic toxins, satratoxin, zearalenone, and vomitoxin.[7] Humans and other animals suffer from a variety of ill effects resulting from exposure to mycotoxins. Monogastric (single stomach) animals are relatively more susceptible to adverse effects from toxin ingestion, whereas the rumen microbiota (bacteria) in the digestive systems of ruminant animals tend to reduce the toxicity of mycotoxins to ruminants.[8] Structural formulas of some important mycotoxins are shown in Figure 19.1.

Adverse human health effects of mycotoxins have occurred during times of short food supply when substandard grain has been consumed for food. One such case occurred in the vicinity of

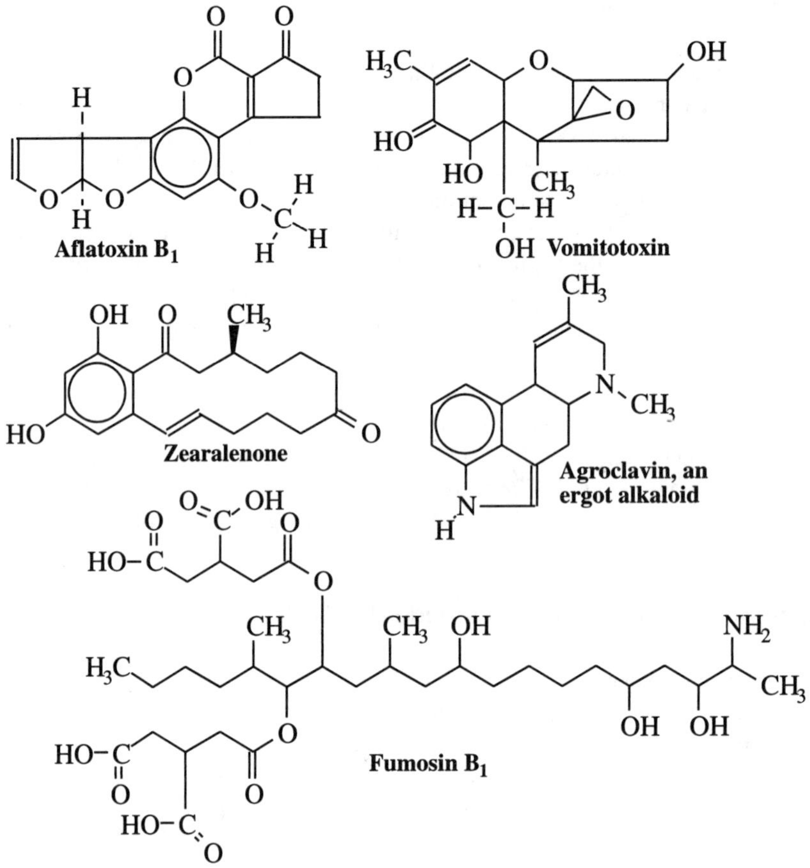

Figure 19.1 Representative compounds of the large number of mycotoxins produced as secondary metabolites by fungi.

Orenburg in Siberia in 1944 during World War II. Harvest delayed by the war resulted in contamination of barley, millet, and wheat by trichothecenes. Humans that later consumed the grain were afflicted with a number of disorders, including gastrointestinal maladies, internal hemorrhaging, and severe skin rash; about 10% of those afflicted died. Another major class of mycotoxins consists of the **ergot alkaloids** from *Claviceps*. Several genera of *fungi imperfecti* produce toxic trichothecenes.

19.3.1 Aflatoxins

The most common source of aflatoxins is moldy food, particularly nuts, some cereal grains, and oil seeds. The most notorious of the aflatoxins is aflatoxin B_1, for which the structural formula is shown in Figure 19.1. Produced by *Aspergillus niger*, it is a potent liver toxin and liver carcinogen in some species. It is metabolized in the liver to an epoxide (see Section 7.3). The product is electrophilic with a strong tendency to bond covalently to protein, DNA, and RNA. Other common aflatoxins produced by molds are those designated by the letters B_2, G_1, G_2, and M_1.

19.3.2 Other Mycotoxins

The ergot alkaloids have been associated with a number of spectacular outbreaks of central nervous system disorders, sometimes called ergotism. St. Anthony's fire is an example of convulsive ergotism. From examination of historical records, it is now known that outbreaks of this malady resulted for the most part from ingestion of moldy grain products. Although ergotism is now virtually unknown in humans, it still occurs in livestock.

Trichothecenes are composed of 40 or more structurally related compounds produced by a variety of molds, including *Cephalosporium, Fusarium, Myrothecium,* and *Trichoderma,* which grow predominantly on grains. Much of the available information on human toxicity of trichothecenes was obtained from an outbreak of poisoning in Siberia in 1944, mentioned above.

Several species of *Fusarium* that commonly grow on barley, corn, wheat, and other grains produce zearalenone (Figure 19.1). This toxic substance binds with the estrogenic receptor in animals, causing a variety of adverse estrogenic effects. These include enlarged uteri, infertility, reabsorption of the fetus, and vaginal prolapse.[9]

19.3.3 Mushroom Toxins

Mushrooms are spore-forming bodies of filamentous terrestrial fungi, some of which are considered to be food delicacies, whereas others, such as *Amanita phalloides, Amanita virosa,* and *Gyromita esculenta,* are very toxic, with reported worldwide deaths of the order of 100 per year.[10] In extreme cases, one bite of one poisonous mushroom can be fatal. Accidental mushroom poisonings are often caused by the death's head mushroom, because it is easily mistaken for edible varieties.

Some toxins in mushrooms are alkaloids that cause central nervous system effects of narcosis and convulsions. Hallucinations occur in subjects who have eaten mushrooms that contain **psilocybin**. The toxic alkaloid **muscarine** is present in some mushrooms.

Another class of toxins produced by some mushrooms consists of polypeptides, particularly amanitin and phalloidin. These substances are stable to heating (cooking). They are systemic poisons that attack cells of various organs, including the heart and liver. In early 1988, an organ transplant was performed on a woman in the U.S. to replace her liver, which was badly damaged from the ingestion of wild mushrooms that she and a companion had mistakenly identified as edible varieties and consumed.

The symptoms of mushroom poisoning vary. Typical early symptoms involve the gastrointestinal tract and include stomach pains and cramps, nausea, vomiting, and diarrhea. Victims in the second phase of severe poisoning may suffer paralysis, delirium, and coma, along with often severe liver damage.

Edible *Coprinus atramentarius* mushrooms produce an interesting ethanol-sensitizing effect similar to that of disulfiram (antabuse, see Section 17.3). Ingestion of alcohol can cause severe reactions in individuals up to several days after having eaten this kind of mushroom.

19.4 TOXINS FROM PROTOZOA

Protozoa are microscopic animals consisting of single eukaryotic cells and classified on the bases of morphology, means of locomotion, presence or absence of chloroplasts, presence or absence of shells, and ability to form cysts. Several devastating human diseases, including malaria, sleeping sickness, and some kinds of dysentery, are caused by parasitic protozoa. Parasitic protozoa can cause debilitating, even fatal, diseases in livestock and wildlife.

Toxic substances from two of the major types of unicellular protista — bacteria and fungi — were discussed in the preceding sections. Protozoans are also notable for the production of toxic substances. Most of the protozoans that produce toxins belong to the order **Dinoflagellata**, which predominantly consists of marine species. The cells of these organisms are enclosed in cellulose envelopes, which often have beautiful patterns on them. Among the effects caused by toxins from these organisms are gastrointestinal, respiratory, and skin disorders in humans; mass kills of various marine animals; and paralytic conditions caused by eating infested shellfish.

The marine growth of dinoflagellates is characterized by occasional incidents in which they multiply at such an explosive rate that they color the water yellow, olive green, or red by their vast numbers. In 1946, some sections of the Florida coast became so afflicted by "red tide" that the water became viscous, and for many miles the beaches were littered with the remains of dead fish, shellfish, turtles, and other marine organisms. The sea spray in these areas became so irritating that coastal schools and resorts were closed.

The greatest danger to humans from dinoflagellata toxins comes from the ingestion of shellfish, such as mussels and clams, that have accumulated the protozoa from sea water. In this form the toxic material is called paralytic shellfish poison. As little as 4 mg of this toxin, the amount found in several severely infested mussels or clams, can be fatal to a human. The toxin depresses respiration and affects the heart, resulting in complete cardiac arrest in extreme cases.

One of the most damaging dinoflagellates in recent years has been *Pfiestria piscicida*, an organism that thrives in polluted estuarine waters enriched in nutrients from fertilizers, feedlot runoff, and sewage. This organism has been responsible for massive fish kills in estuary regions of the Atlantic, especially North Carolina's Albermarle–Pamlico Estuary. It releases a toxin of unknown chemical structure in the presence of fish, disabling and killing fish and enabling their attack by *Pfiestria*. Humans studying this phenomenon were themselves afflicted by the toxin, through either contact with contaminated water or inhalation of spray from such water.[11] The victims suffered acute eye and respiratory irritation with concurrent fatigue and headaches. They also suffered from nausea, stomach cramps, and vomiting characteristic of gastrointestinal disorders. *Pfiestria piscicida* is a remarkable organism that takes on numerous forms during its life cycle. It is highly opportunistic, and when fish are not available for food, it will even retain chloroplasts from ingested algae (kleptochloroplasts) within a large food vacuole in the *Pfiestria piscicida* cell, where the chloroplasts can function for several days, providing nutrients to their hosts.

19.5 TOXIC SUBSTANCES FROM PLANTS

Various plants produce a wide range of toxic substances, as reflected by plant names such as deadly nightshade and poison hemlock. Although the use of "poison arrows" having tips covered with plant-derived curare has declined as the tribes that employed them have acquired the sometimes dubious traits of modern civilization, poisoning by plants is still of concern in the grazing of ruminant animals, and houseplants such as philodendron and yew are responsible for poisoning some children. Taxol from the western yew tree is a neurotoxin, but has proven to be a useful chemotherapeutic agent for the treatment of breast cancer. Plant-derived cocaine causes many deaths among those who use it or get into fatal disputes marketing it.

The toxic agents in plants have been summarized in a review chapter dealing with that topic.[12] Toxic substances from plants are discussed here in five categories: nerve poisons, internal organ poisons, skin and eye irritants, allergens, and metal (mineral) accumulators. As can be seen in Figure 19.2, plant toxins have a variety of chemical structures. Prominent among the chemical classes of toxicants synthesized by plants are nitrogen-containing alkaloids (see Section 15.10), which usually occur in plants as salts. Some harmful plant compounds undergo metabolic reactions to form toxic substances. For example, amygdalin, present in the meats of fruit seeds such as those of apples and peaches, undergoes acid hydrolysis in the stomach or enzymatic hydrolysis elsewhere in the body to yield toxic HCN.

Extracts from the castor bean can be used to isolate ricin, an extremely potent poison that is one of the leading candidates for use by terrorists. Ricin is a large-molecule, heterogeneous protein. A chemical marker for ricin is the hydrolysis product alkaloid ricinine, shown in Figure 19.2.

Pyrrolizidine alkaloids in the plant *Senecio vernalis* have been implicated in the poisoning of cattle.[13] The toxic agents in this plant include three closely related alkaloids — senecionine, senkirkin, and seneciphyllin. The structural formula of senecionine is shown in Figure 19.2.

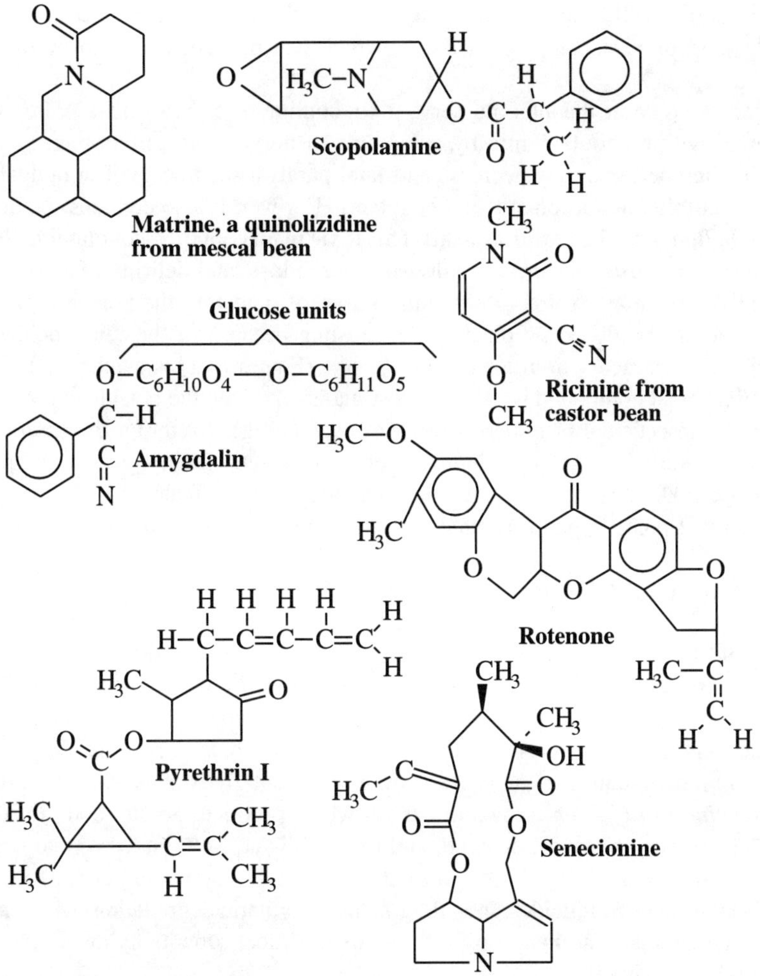

Figure 19.2 Representative toxic substances from plants.

19.5.1 Nerve Toxins from Plants

Nerve toxins from plants cause a variety of central nervous and peripheral nervous system effects. Several examples are cited here.

Plant-derived **neurotoxic psychodysleptics** affect peripheral neural functions and motor coordination, sometimes accompanied by delirium, stupor, trance states, and vomiting. Prominent among these toxins are the pyrollizidines from peyote. Also included are erythrionones from the coral tree and quinolizidines from the mescal bean (see Figure 19.2).

A plant neurotoxin that is receiving much current publicity because of its effectiveness in the chemotherapeutic treatment of at least one form of cancer is **taxol**, a complex molecule that belongs to the class of taxine alkaloids. Taxol occurs in most tissues of *Taxus breviofolia*, the western yew tree, and is isolated from the bark of that tree (once considered a nuisance tree in forestry, but in short supply following discovery of the therapeutic value of taxol until alternate sources were developed). Ingestion of taxol causes a number of neurotoxic effects, including sensory neuropathy, nausea and gastrointestinal disturbances, and impaired respiration and cardiac function. It also causes blood disorders (leukopenia and thrombocytopenia). The mechanism of taxol neurotoxicity involves binding to tubulin, a protein involved in the assembly of microtubules, which assemble and dissociate as part of cell function. This binding of tubulin in nerve cell microtubules stabilizes the microtubules and prevents their dissociation, which can be detrimental to normal nerve cell function.

Spotted hemlock contains the alkaloid nerve toxin coniine (see Figure 15.10). Ingestion of this poison is followed within about 15 min by symptoms of nervousness, trembling, arrythmia, and bradycardia. Body temperature may decrease and fatal paralysis can occur. The nightshade family of plants contains edible potato, tomato, and eggplant. However, it also contains the deadly nightshade, or *Atropa Belladonna* (beautiful woman). This toxic plant contains scopolamine (Figure 19.2) and atropine. Ingestion causes dizziness, mydriasis, speech loss, and delirium. Paralysis can occur. Fatally poisoned victims may expire within half an hour of ingesting the poison.

Several nerve toxins produced by plants are interesting because of their insecticidal properties. Insecticidal nicotine is extracted from tobacco. Rotenone (Figure 19.2) is synthesized by almost 70 legumes. This insecticidal compound is safe for most mammals, with the notable exception of swine. The most significant insecticidal plant derivatives, however, are the pyrethrins, discussed below.

Tall larkspur, a plant of the genus *Delphinium* of the buttercup family growing in the mountain ranges of the western U.S., produces more than 40 alkaloids. The alkaloids in these plants are the largest source of plant poisonings of livestock on mountain rangelands in the western U.S.[14]

19.5.1.1 Pyrethrins and Pyrethroids

Pyrethrins and their synthetic analogs represent both the oldest and newest of insecticides. Extracts of dried chrysanthemum or pyrethrum flowers, which contain pyrethrin I (Figure 19.2) and II, jasmolin I and II, and cinerin I and II, have been known for their insecticidal properties for a long time, and may have even been used as botanical insecticides in China almost 2000 years ago. The most important commerical sources of insecticidal pyrethrins are pyrethrum flowers (*Chrysanthemum cinerariaefolium*) grown in Kenya, which produces about 7000 tons of pyrethrin each year. Production does not meet demand, and research is actively under way to biosynthesize natural pyrethrins from plant cell cultures and with genetically engineered organisms. Pyrethrins have several advantages as insecticides, including facile enzymatic degradation, which makes them relatively safe for mammals, ability to rapidly paralyze (knock down) flying insects, and good biodegradability characteristics.

Structural formulas of three major natural pyrethrins, which are chemically classified as monoterpene esters, are given in Figure 19.3. The shaded part of the first structural formula is common to natural pyrethrins, the most characteristic structural feature of which is the cyclopropane group.

Synthetic analogs of the pyrethrins, **pyrethroids**, have been widely produced as insecticides during recent years. The first of these was allethrin (Figure 19.4). Three pyrethroids that are relatively safe and can be used on cereal crops are fluvalinate, zeta-cypermethrin, and deltametrin, structural formulas of which are shown in Figure 19.4.[15]

Pyrethroids act as neurotoxins that cause excitation of the nervous system. Organisms may become hypersensitive to stimuli and experience paresthesia (abnormal sensations of burning or prickling of the skin). Tremors and salivation are common symptoms of mammals exposed to toxic levels of pyrethroids.

19.5.2 Internal Organ Plant Toxins

Toxins from plants may affect internal organs, such as the heart, kidney, liver, and stomach. Because of their very different digestive systems involving multiple stomachs, ruminant animals may react differently to these toxins than do monogastric animals. Some major plant toxins that affect internal organs are summarized in Table 19.1.

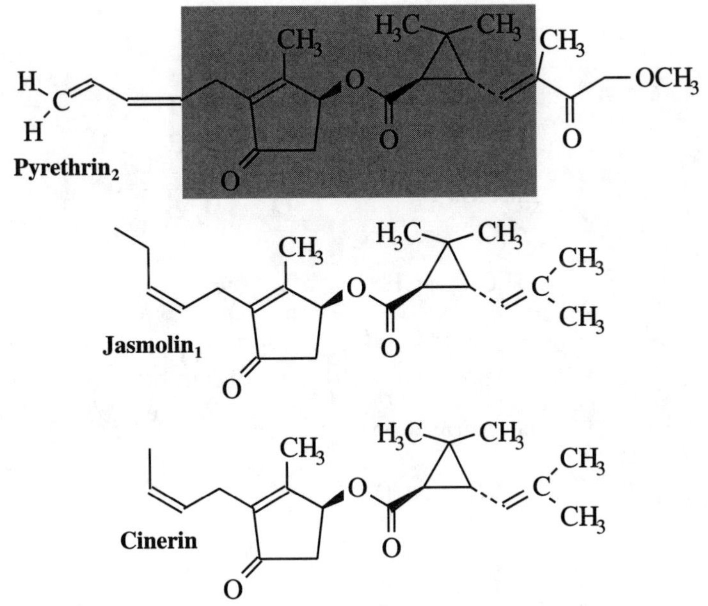

Figure 19.3 Three major naturally occurring pyrethrins. The shaded structural segment in the pyrethrin II formula is common to pyrethrins.

Table 19.1 Plant Toxins That Affect Internal Organs of Animals

Toxin	Source	Target Organ
Saponins	Alfalfa, cockles, English ivy	Noncardioactive steroid glycosides that cause gastric upset
Pyrrolizidine alkaloids such as festucine	Fescue hay	Liver: obstructs veins
Hypercin	St. John's wort, horsebrush	Liver: releases pigmented molecules to the bloodstream, causing photosensitization
Digitoxin	Foxglove	Heart: overdose causes heart to stop; strengthen heartbeat and eliminate fluids present in congestive heart failure
Oxalates	Oak tannin	Kidney: precipitation of CaC_2O_4

19.5.3 Eye and Skin Irritants

Anyone who has been afflicted by poison ivy, poison oak, or poison sumac appreciates the high potential that some plant toxins have to irritate skin and eyes. The toxic agents in the plants just mentioned are catechol compounds, such as urishikiol in poison ivy. Contact with the poison causes a characteristic skin rash that may be disabling and very persistent in heavily exposed, sensitive individuals. Lungs may be affected — often by inhalation of smoke from the burning plants — to the extent that hospitalization is required.

Photosensitizers constitute a class of systemic plant poisons capable of affecting areas other than those exposed. These pigmented substances may pass through the liver without being conjugated and collect in skin capillaries. When these areas of the skin are subsequently exposed to light, the capillaries leak. (Since light is required, the phenomenon is called a photosensitized condition.) In severe cases, tissue and hair are sloughed off. St. John's wort or horsebrush causes this kind of condition in farm animals.

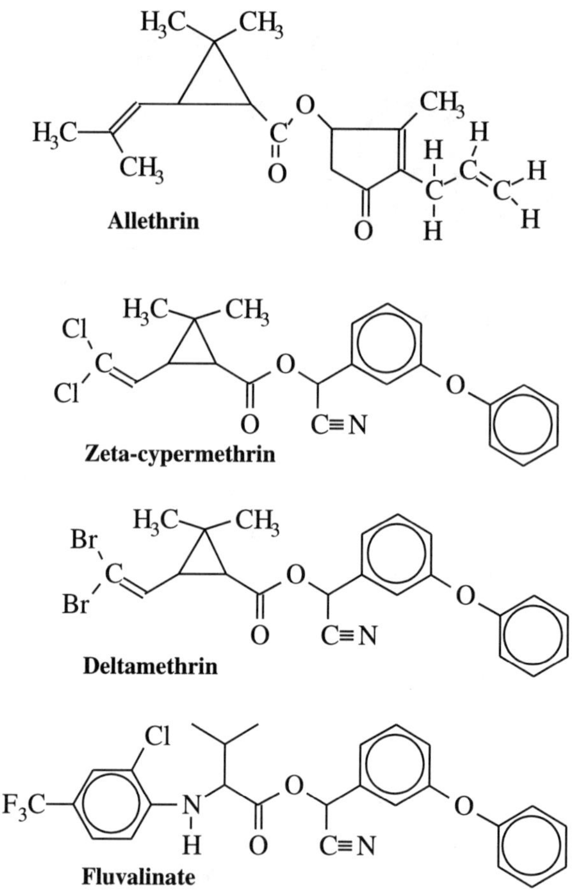

Figure 19.4 Four synthetic pyrethroids with insecticidal properties similar to those of the natural pyrethrins.

19.5.4 Allergens

Many plants are notorious for producing allergens that cause allergic reactions in sensitized individuals. The most common plant allergens consist of pollen. The process that leads to an allergic reaction starts when the allergen, acting as a hapten, combines with an endogenous protein in the body to form an antigen. Antibodies are generated that react with the antigen and produce histamine, resulting in an allergic reaction (see Section 9.6, "Immune System"). The severity of the symptoms varies with the amount of histamine produced. These symptoms can include skin rash, watery eyes, and runny nose. In severe cases, victims suffer fatal anaphylactic shock. An example of an allergic reaction to a plant product — all too familiar to many of its victims — is hay fever induced by the pollen of ragweed or goldenrod.

19.5.5 Mineral Accumulators

Some plants classified as mineral accumulators become toxic because of the inorganic materials that they absorb from soil and water and retain in the plant biomass. An important example of such a plant is *Astragalus*, sometimes called locoweed. This plant causes serious problems in some western U.S. grazing areas because it accumulates selenium. Animals that eat too much of it get selenium poisoning, characterized by anemia and a condition known descriptively as "blind staggers."

Nitrate accumulation may occur in plants growing on soil fertilized with nitrate under moisture-deficient conditions. In the stomachs of ruminant animals, nitrate (NO_3^-) ingested with plant material is reduced to nitrite (NO_2^-). The nitrite product enters the bloodstream and oxidizes the iron(II) in hemoglobin to iron(III). The condition that results is methemoglobinemia, which was discussed in Section 15.3 in connection with aniline poisoning.

Another toxicological problem that can result from excessive nitrate in plant material is the generation of toxic nitrogen dioxide gas (see Section 11.3) during the fermentation of ensilage composed of chopped plant matter contaminated with nitrate. The toxic effect of NO_2 from this source has been called silo-filler's disease.

19.5.6 Toxic Algae

Photosynthetic algae that live in water can produce toxic substances. These are usually manifested during so-called algae blooms characterized by a rapid increase in the numbers of algae. Typical of such outbreaks was a February and March 2002 infestation of *Dinophysis acuminata* in the Potomac River and its Virginia tributaries. Toxic algae may accumulate in shellfish and cause diarrhetic shellfish poisoning in people who eat the contaminated shellfish. The toxins that cause this condition consist of fiendishly complicated molecules. They include okadaic acid ($C_{44}H_{68}O_{13}$), pectenotoxin ($C_{47}H_{68}O_{16}$), dinophysistoxin ($C_{45}H_{70}O_{13}$), and yessotoxin ($C_{55}H_{82}O_{21}S_2$).[16]

Reef barrier fish, including barracuda, grouper, red snapper, and sea bass, may accumulate toxic levels of ciguatoxin produced by ciguetera algae. Problems from this toxin have occurred in the Caribbean and in some areas of the Indian and Pacific Oceans. The toxin is insidious and gives no warning by taste or odor. Cooking does not destroy it. Symptoms of ciguatoxin poisoning include gastrointestinal effects of diarrhea, vomiting, and abdominal pain; tingling sensation in the mouth; pain, tingling, and weakness in the legs; and hot–cold sensation reversal.

19.6 INSECT TOXINS

Although relatively few insect species produce enough toxin to endanger humans, insects cause more fatal poisonings in the U.S. each year than do all other venomous animals combined. Most venomous insects are from the order *Hymenoptera*, which includes ants, bees, hornets, wasps, and yellow jackets. These insects deliver their toxins by a stinging mechanism.

Chemically, the toxic substances produced by insects are variable and have been incompletely characterized. In general, hymenopteran venoms are composed of water-soluble, nitrogen-containing chemical species in concentrated mixtures. Although they contain chemical compounds in common, the compositions of insect venoms from different species are variable. The three major types of chemical species are biologically synthesized (biogenic) amines, peptides and small proteins, and enzymes. Of the biogenic amines, the most common is histamine, which is found in the venoms of bees, wasps, and hornets. Wasp and hornet venoms contain serotonin, and hornet venom contains the biogenic amine acetylcholine. Among the peptides and low-molecular-mass proteins in insect venoms are apamin, mellitin, and mast cell degranulating peptide in bee venom, and kinin in wasp and hornet venoms. Enzymes contained in bee, wasp, and hornet venoms are phospholipase A and hyaluronidase. Phospholipase B occurs in wasp and hornet venoms.

19.6.1 Bee Venom

Bee venom contains a greater variety of proteinaceous materials than do wasp and hornet venoms. Apamin in bee venom is a polypeptide containing 18 amino acids and having three disulfide (–SS–) bridges in its structure. Because of these bridges and its small size, the apamine molecule is able to traverse the blood–brain barrier and function as a central nervous system poison. Mellitin in bee venom consists of a chain of 27 amino acids. It can be a direct cause of erythrocyte hemolysis. Symptoms of bradycardia and arrythmia can be caused by mellitin. Mast cell degranulating peptide in bee venom acts on mast cells. These are a type of white blood cell believed to be involved in the production of heparin, a key participant in the blood-clotting process. The degranulating peptide causes mast cells to disperse, with an accompanying release of histamine into the system.

19.6.2 Wasp and Hornet Venoms

Wasp and hornet venoms are distinguished from bee venoms by their lower content of peptides. They contain kinin peptide, which may cause smooth muscle contraction and lowered blood pressure. Two biogenic amines in wasp and hornet venoms (serotonin and acetylcholine) lower blood pressure and cause pain. Acetylcholine may cause malfunction of heart and skeletal muscles.

19.6.3 Toxicities of Insect Venoms

The toxicities of insect venoms are low to most people. Despite this, relatively large numbers of fatalities occur each year from insect stings because of allergic reactions in sensitized individuals. These reactions can lead to potentially fatal anaphylactic shock, which affects the nervous system, cardiovascular function, and respiratory function. The agents in bee venom that are responsible for severe allergic reactions are mellitin and two enzymes of high molecular mass — hyaluronidase and phospholipase A-2.

19.7 SPIDER TOXINS

There are about 30,000 species of spiders, virtually all of which produce venom. Fortunately, most lack dangerous quantities of venom or the means to deliver it. Nevertheless, about 200 species of spiders are significantly poisonous to humans. Many of these have colorful common names, such as tarantula, trap-door spider, black widow, giant crab spider, poison lady, and deadly spider. Space permits only a brief discussion of spider venoms here.

19.7.1 Brown Recluse Spiders

Brown recluse spiders (*Loxosceles*) are of concern because of their common occurrence in households in temperate regions. Many people are bitten by this spider despite its nonaggressive nature. A brown spider bite can cause severe damage at the site of the injury. When this occurs, the tissue and underlying muscle around the bite undergo severe necrosis, leaving a gaping wound up to 10 cm across. Plastic surgery is often required in an attempt to repair the damage. In addition, *Loxosceles* venom may cause systemic effects, such as fever, vomiting, and nausea. In rare cases, death results. The venom of *Loxosceles* contains protein and includes enzymes. The mechanisms by which the venom produces lesions are not completely understood.

19.7.2 Widow Spiders

The widow spiders are *Latrodectus* species. Unlike the *Loxosceles* species described above, the bite sites from widow spiders show virtually no damage. The symptoms of widow spider poisoning are many and varied. They include pain, cramps, sweating, headache, dizziness, tremor, nausea, vomiting, and elevated blood pressure. The venom contains several proteins, including a proteinaceous neurotoxin with a molecular mass of about 130,000.

19.7.3 Other Spiders

Several other types of venomous spiders should be mentioned here. Running spiders (*Chiranthium* species) are noted for the tenacity with which they cling to the bite area, causing a sharply painful wound. The bites of cobweb spiders (*Steatoda* species) cause localized pain and tissue damage. Venomous jumping spiders (*Phidippus* species) produce a wheal (raised area) up to 5 cm across in the bite area.

19.8 REPTILE TOXINS

Snakes are the most notorious of the venomous animals. The names of venomous snakes suggest danger — Eastern diamondback rattlesnake, king cobra, black mamba, fer-de-lance, horned puff adder, *Crotalus horridus horridus*. About 10% of the approximately 3500 snake species are sufficiently venomous to be hazardous to humans. These may be divided among *Crotilidae* (including rattlesnakes, bushmaster, and fer-de-lance), *Elipidae* (including cobras, mambas, and coral snakes), *Hydrophidae* (true sea snakes), *Laticaudae* (sea kraits), and *Colubridae* (including the boomslang and Australian death adder). Although snake bites in the U.S. are generally regarded as more of a problem from former times in rural areas, they can be a very serious danger as human dwellings and activities intrude into the snakes' normal territories. In 2001, rattlesnake bites in Arizona approached record levels. Incidents were described in which victims had to be placed in intensive care, and one victim walking to the pool in his apartment complex was bitten on the foot by a baby rattlesnake.[17]

19.8.1 Chemical Composition of Snake Venoms

Snake venoms are complex mixtures that may contain biogenic amines, carbohydrates, glycoproteins, lipids, and metal ions. (These materials are too complex to discuss in detail here; the reader is referred to an excellent review dealing with animal venoms and poisons.[18]) The most important snake venom constituents, however, are proteins, including numerous enzymes. Approximately 25 different enzymes have been identified in various snake venoms. The most prominent of the enzymes in snake venom are the proteolytic enzymes, which bring about the breakdown of proteins, thereby causing tissue to deteriorate. Some proteolytic enzymes are associated with hemorrhaging. Collagen (connective tissue in tendons, skin, and bones) is broken down by collagenase enzyme contained in some snake venoms. Among the other kinds of enzymes that occur in snake venoms are hyaluronidase, arginine ester hydrolase, lactate dehydrogenase, DNase, L-amino acid oxidase, nucleotidase enzymes, RNase, phospholipase enzymes, phosphoesterase enzymes, and acetylcholinesterase.

Numerous nonenzyme polypeptides occur in snake venom. Some of these polypeptides, though by no means all, are neurotoxins.

19.8.2 Toxic Effects of Snake Venom

The effects of snake bite can range from relatively minor discomfort to almost instant death. The latter is often associated with drastically lowered blood pressure and shock. The predominant effects of snake venoms can be divided into two major categories: cardiotoxic and neurotoxic effects. Blood-clotting mechanisms may be affected by enzymes in snake venom, and blood vessels may be damaged as well. Almost all organs in the victims of poisoning by *Crotilidae* have exhibited adverse effects, many of which appear to be associated with changes in the blood and with alterations in the lung. Clumped blood cells and clots in blood vessels have been observed in the lungs of victims. These effects are caused in part by the action of thrombin-like enzymes, which are constituents of *Crotilidae* (e.g., copperheads, rattlesnakes, Chinese habu) and *Viperidae* (e.g., puff adders, European viper, Sahara sand viper) venoms. Thrombin-like enzymes cause the release of fibrinopeptides that result in fibrinogen clot formation. Agents in cobra toxin break down the blood–brain barrier by disrupting capillaries and cell membranes. So altered, the barrier loses its effectiveness in preventing the entry of other brain-damaging toxic agents.

19.9 NONREPTILE ANIMAL TOXINS

Several major types of animals that produce poisonous substances have been considered so far in this chapter. With the exception of birds, all classes of the animal kingdom contain members that produce toxic substances. It has now been demonstrated that there are even birds that are "toxic." It is believed that such birds do not produce toxins but accumulate toxic alkaloids, including andromedotoxin, batrachotoxins, and cantheridin, from their diets and deposit these poisonous materials in their skin and feathers.[19] Toxic animals not covered so far in this chapter are summarized here.

Numerous kinds of fish contain poisons in their organs and flesh. The most notorious of the poisonous fish are puffers and puffer-like fish that produce tetrodotoxin. This supertoxic substance is present in the liver and ovary of the fish. It acts on nerve cell membranes by affecting the passage of sodium ions, a process involved in generating nerve impulses. The fatality rate for persons developing clinical symptoms of tetrodotoxin poisoning is about 40%. Usually associated with Japan, puffer fish poisoning kills about 100 people per year globally. Some of these poisonings are self-inflicted by suicidal individuals.

Some fish are venomous and have means of delivering venom to other animals. This is accomplished, for example, by spines on weever fish. The infamous stingray has a serrated spine on its tail that can be used to inflict severe wounds, while depositing venom from specialized cells along the spine. The venom increases the pain from the wound and has systemic effects, especially on the cardiovascular system.

Numerous species of amphibia (frogs, toads, newts, salamanders) produce poisons, such as bufotenin, in specialized skin secretory glands. Most of these animals pose no hazard to humans. However, some of the toxins are extremely poisonous. For example, Central American Indian hunters have used hunting arrows tipped with poison from the golden arrow frog.

Bufotenin, a compound isolated from some amphibian toxins

In addition to the poisonous fish and sea snakes mentioned previously in this chapter, several other forms of marine life produce toxins. Among these are *Porifera*, or sponges, consisting of colonies of unicellular animals. The sponges release poisons to keep predators away. They may have sharp spicules that can injure human skin, while simultaneously exposing it to poison. Various species of the *Coelenterates*, including corals, jellyfish, and sea anenomes, are capable of delivering venom by stinging. Some of these venoms have highly neurotoxic effects. *Echinoderms*, exemplified by starfish, sea cucumbers, and sea urchins, may possess spines capable of delivering toxins. People injured by these spines often experience severe pain and other symptoms of poisoning. Various mollusks produce poisons, such as the poison contained in the liver of the abalone, *Haliotis*. Some mollusk poisons, such as those of the genus *Conus*, are delivered as venoms by a stinging mechanism.

Arthropods, which consist of a vast variety of invertebrate animals that have jointed legs and a segmented body, are notable for their production of toxins. Of the arthropods, insects and spiders were discussed earlier in this chapter. Some **scorpions**, arachnidal arthropods with nipper-equipped front claws and stinger-equipped long, curved, segmented tails, are notably venomous animals. The stings of scorpions, some of which reach a length of 8 in., are a very serious hazard, especially to children; most fatalities occur in children under age 3. In Mexico, the particularly dangerous scorpion *Centruroides suffusus* attains a length up to 9 cm. Mexico has had a particularly serious problem with fatal scorpion bites. During the two decades following 1940, it is estimated that over 20,000 deaths occurred from venomous scorpion stings. Some **centipedes** are capable of delivering venom by biting. The site becomes swollen, inflamed, and painful. Some millipedes secrete a toxic skin irritant when touched.

Although the greater hazard from ticks is their ability to carry human diseases, such as Rocky Mountain spotted fever or Lyme disease (a debilitating condition that is of great concern in New England and some of the upper midwestern states of the U.S.), some species discharge a venom that causes a condition called tick paralysis, characterized by weakness and lack of coordination. An infamous mite larva, the chigger, causes inflamed spots on the skin that itch badly. The chigger is so small that most people require a magnifying glass to see it, but a large number of chigger bites can cause intense misery in a victim.

REFERENCES

1. Westendorf, J., Natural compounds, in *Toxicology*, Academic Press, San Diego, CA, 1999, pp. 959–1007.
2. Rappuoli, R. and Pizza, M., Bacterial toxins, in *Cell Microbiology*, Pascale Cossart, Ed., ASMPress, Herndon, VA, 2000, pp. 193–220.
3. Alouf, J.E., Bacterial protein toxins, *Methods Mol. Biol.*, 145, 1–26, 2000.
4. Sandvig, K., Shig toxins, *Toxicon*, 39, 1629–1635, 2001.
5. Atanassova, V., Meindl, A., and Ring, C., Prevalence of *Staphylococcus aureus* and staphylococcal enterotoxins in raw pork and uncooked smoked ham: a comparison of classical culturing detection and RFLP-PCR, *Int. J. Food Microbiol.*, 68, 105–113, 2001.
6. Issa, N.C. and Thompson, R.L., Staphylococcal toxic shock syndrome: suspicion and prevention are keys to control, *Postgraduate Medicine*, Oct. 1, 2001, p. 55.
7. Etzel, R.A., Mycotoxins, *J. Am. Med. Assoc.*, 287, 425–427, 2002.
8. Hussein, H.S. and Brasel, J.M., Toxicity, metabolism, and impact of mycotoxins on humans and animals, *Toxicology*, 167, 101–134, 2001.
9. Hisashi, K., Toxicological approaches of zearalenone, *Mycotoxins*, 50, 111–117, 2000.
10. Dreisbach, R.H. and Robertson, W.O., *Handbook of Poisoning*, 12th ed., Appleton and Lange, Norwalk, CT, 1987.
11. Grattan, L.M., Oldach, D., and Morris, J.G., Human health risks of exposure to *Pfiesteria piscicida*, *Bioscience*, October 1, 2001, p. 853.
12. Norton, S., Toxic effects of plants, in *Casarett and Doull's Toxicology: The Basic Science of Poisons*, 6th ed., Klaassen, C.D., Ed., McGraw-Hill, New York, 2001, chap. 27, pp. 965–976.
13. Skaanild, M.T., Friis, C., and Brimer, L., Interplant alkaloid variation and *Senecion vernalis* toxicity in cattle, *Vet. Hum. Toxicol.*, 43, 147–151, 2001.
14. Panter, K.E. et al., Larkspur poisoning: toxicology and alkaloid structure-activity relationships, *Biochem. Syst. Ecol.*, 30, 113–128, 2002.
15. Moreby, S.J. et al., A comparison of the effect of new and established insecticides on nontarget invertebrates of winter wheat fields, *Environ. Toxicol. Chem.*, 20, 2243–2254, 2001.
16. Ito, S. and Tsukada, K., Matrix effect and correlation by standard addition in quantitative liquid chromatographic spectrometric analysis of diarrhetic shellfish poisoning toxins, *J. Chromatogr.*, 943, 39–46, 2002.

17. McClain, C., Snake bites rise: antivenom supply remains spotty, *Arizona Daily Star*, Sept. 9, 2001, p. A1.
18. Russell, F.E., Toxic effects of terrestrial animal venoms and poisons, in *Casarett and Doull's Toxicology: The Basic Science of Poisons*, 6th ed., Klaassen, C.D., Ed., McGraw-Hill, New York, 2001, chap. 26, pp. 945–964.
19. Bartram, S. and Boland, W., Chemistry and ecology of toxic birds, *Chem. Biochem.*, 2, 809–811, 2001.

SUPPLEMENTARY REFERENCES

Aktories, K., Ed., *Bacterial Toxins: Tools in Cell Biology and Pharmacology*, Chapman & Hall, London, 1997.
Alper, K.R. and Glick, S.D., Eds., *The Alkaloids. Chemistry and Biology*, Academic Press, San Diego, 2001.
Dvorácková, I., *Aflatoxins and Human Health*, CRC Press, Boca Raton, FL, 1990.
Gilles, H.M., Ed., *Protozoal Diseases*, Oxford University Press, London, 1999.
Holst, O., Ed., *Bacterial Toxins: Methods and Protocols*, Humana Press, Totowa, NJ, 2000.
Keen, N.T., Ed., *Delivery and Perception of Pathogen Signals in Plants*, American Phytopathological Society, St. Paul, Minnesota, 2001.
Mara, W.P., *Venomous Snakes of the World*, T.F.H. Publications, Neptune, NJ, 1993.
Mebs, D., *Venomous and Poisonous Animals*, CRC Press, Boca Raton, FL, 2002.
Newlands, G., *Venomous Creatures*, Struik Publishers, Cape Town, 1997.
Richard, J., Ed., *Mycotoxins — An Overview*, Romer Labs, Union, MO, 2000.
Tu, A.T., *Insect Poisons, Allergens, and Other Invertebrate Venoms*, Marcel Dekker, New York, 1984.

QUESTIONS AND PROBLEMS

1. Distinguish between poisonous organisms and venomous organisms. Give an example of each.
2. What does the following reaction show about toxic natural products?

$$SO_4^{2-} + 2\{CH_2O\} + 2H^+ \rightarrow H_2S + 2CO_2 + 2H_2O$$

3. What two kinds of bacterial toxins are illustrated by *Shigella dysenteriae*, compared to *Clostridium botulinum*?
4. What kinds of toxins are produced by *Aspergillus flavus, Fusarium, Trichoderma, Aspergillus*, and *Penicillium*? Give some examples of these toxins.
5. What kind of toxic substance is illustrated by the following? What produces it? What are some of its effects?

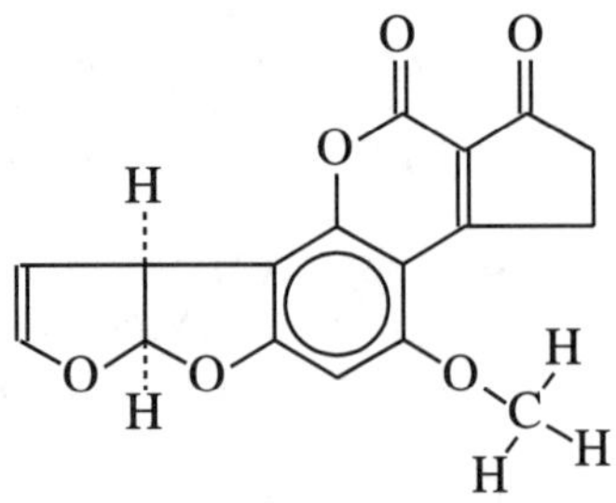

6. What kind of organism causes "red tide"? What are some of the symptoms of poisoning by "red tide" toxins?
7. What is the greatest danger to humans from dinoflagellata toxins? What are the toxic effects of these poisons?
8. A very large number of plant toxins are classified in a diverse group of natural products. What is this group? What are some of the toxic effects of substances belonging to it?

9. Taxol has both harmful and potentially beneficial effects. List and discuss both of these. What is the source of taxol?
10. List and discuss several of the prominent nerve toxins from plants. From which plants do they come? What are some of their important effects?
11. What are the most common plant allergens consisting of pollen? What are some of the major allergic conditions caused by these substances?
12. What is the toxicological significance of *Amanita phalloides*, *Amanita virosa*, and *Gyromita esculenta*? What are psilocybin and muscarine?
13. To which order do most venomous insects belong? What is the greatest danger from insect stings?
14. What is contained in the venom of *Loxosceles*? What are some of the major toxic effects of this venom?
15. What are the major constituents of snake venoms? What are the toxicological effects of snake venoms?
16. What is bufotenin? What kind of organism produces it?
17. What venomous animal is an "arachnidal arthropod" with nipper-equipped front claws and stinger-equipped long, curved, segmented tails? What are some significant aspects of its hazards?

CHAPTER 20

Analysis of Xenobiotics

20.1 INTRODUCTION

As defined in Section 6.8, a xenobiotic species is one that is foreign to living systems. Common examples include heavy metals, such as lead, which serve no physiologic function, and synthetic organic compounds, which are not made in nature. Exposure of organisms to xenobiotic materials is a very important consideration in environmental and toxicological chemistry. Therefore, the determination of exposure by various analytical techniques is one of the more crucial aspects of environmental chemistry.

This chapter deals with the determination of xenobiotic substances in biological materials. Although such substances can be measured in a variety of tissues, the greatest concern is their presence in human tissues and other samples of human origin. Therefore, the methods described in this chapter apply primarily to exposed human subjects. They are essentially identical to methods used on other animals, and in fact, most were developed through animal studies. Significantly different techniques may be required for plant or microbiological samples.

The measurement of xenobiotic substances and their metabolites in blood, urine, breath, and other samples of biological origin to determine exposure to toxic substances is called **biological monitoring**. Comparison of the levels of analytes measured with the degree and type of exposure to foreign substances is a crucial aspect of toxicological chemistry. It is an area in which rapid advances are being made. For current information regarding this area, refer to reviews of the topic;[1,2] books on biological monitoring, such as those by Angerer, Draper, Baselt, and Kneip and coauthors, listed at the end of this chapter under "Supplementary References," are available as well.

The two main approaches to monitoring of toxic chemicals are workplace monitoring, using samplers that sample xenobiotic substances from workplace air, and biological monitoring. Although the analyses are generally much more difficult, biological monitoring is a much better indicator of exposure because it measures exposure to all routes — oral, dermal, and inhalation — and it gives an integrated value of exposure. Furthermore, biological monitoring is very useful in determining the effectiveness of measures taken to prevent exposure, such as protective clothing and hygienic measures.

20.2 INDICATORS OF EXPOSURE TO XENOBIOTICS

The two major considerations in determining exposure to xenobiotics are the type of sample and the type of analyte. Both of these are influenced by what happens to a xenobiotic material when it gets into the body. For some exposures, the entry site composes the sample. This is the case, for example, in exposure to asbestos fibers in the air, which is manifested by lesions to the lung. More commonly, the analyte may appear at some distance from the site of exposure, such as lead in bone that was originally taken in by the respiratory route. In other cases, the original xenobiotic is not even present in the analyte. An example of this is methemoglobin in blood, the result of exposure to aniline absorbed through the skin.

The two major kinds of samples analyzed for xenobiotics exposure are blood and urine. Both of these sample types are analyzed for systemic xenobiotics, which are those that are transported in the body and metabolized in various tissues. Xenobiotic substances, their metabolites, and their adducts are absorbed into the body and transported through it in the bloodstream. Therefore, blood is of unique importance as a sample for biological monitoring. Blood is not a simple sample to process, and subjects often object to the process of taking it. Upon collection, blood may be treated with an anticoagulant, usually a salt of ethylenediaminetetraacetic acid (EDTA), and processed for analysis as whole blood. It may also be allowed to clot and be centrifuged to remove solids; the liquid remaining is blood serum.

Recall from Chapter 7 that as the result of phase I and phase II reactions, xenobiotics tend to be converted to more polar and water-soluble metabolites. These are eliminated with the urine, making urine a good sample to analyze as evidence of exposure to xenobiotic substances. Urine has the advantage of being a simpler matrix than blood and one that subjects more readily give for analysis. Other kinds of samples that may be analyzed include breath (for volatile xenobiotics and volatile metabolites), hair or nails (for trace elements, such as selenium), adipose tissue (fat), and milk (obviously limited to lactating females). Various kinds of organ tissue can be analyzed in cadavers, which can be useful in trying to determine cause of death by poisoning.

The choice of the analyte actually measured varies with the xenobiotic substance to which the subject has been exposed. Therefore, it is convenient to divide xenobiotic analysis on the basis of the type of chemical species determined. The most straightforward analyte is, of course, the xenobiotic itself. This applies to elemental xenobiotics, especially metals, which are almost always determined in the elemental form. In a few cases, organic xenobiotics can also be determined as the parent compound. However, organic xenobiotics are commonly metabolized to other products by phase I and phase II reactions. Commonly, the phase I reaction product is measured, often after it is hydrolyzed from the phase II conjugate, using enzymes or acid hydrolysis procedures. Thus, for example, *trans,trans*-muconic acid can be measured as evidence of exposure to the parent compound benzene. In other cases, a phase II reaction product is measured, for example, hippuric acid determined as evidence of exposure to toluene. Some xenobiotics or their metabolites form adducts with endogenous materials in the body, which are then measured as evidence of exposure. A simple example is the adduct formed between carbon monoxide and hemoglobin, carboxyhemoglobin. More complicated examples are the adducts formed by the carcinogenic phase I reaction products of polycyclic aromatic hydrocarbons with DNA or hemoglobin. Another class of analytes consists of endogenous substances produced upon exposure to a xenobiotic material. Methemoglobin formed as a result of exposure to nitrobenzene, aniline, and related compounds is an example of such a substance that does not contain any of the original xenobiotic material. Another class of substance causes measurable alterations in enzyme activity. The most common example of this is the inhibition of acetylcholinesterase enzyme by organophosphates or carbamate insecticides.

20.3 DETERMINATION OF METALS

20.3.1 Direct Analysis of Metals

Several biologically important metals can be determined directly in body fluids, especially urine, by atomic absorption. In the simplest cases, the urine is diluted with water or acid and a portion analyzed directly by graphite furnace atomic absorption, taking advantage of the very high sensitivity of that technique for some metals. Metals that can be determined directly in urine by this approach include chromium, copper, lead, lithium, and zinc. Very low levels of metals can be measured using a graphite furnace atomic absorption technique, and Zeeman background correction with a graphite furnace enables measurement of metals in samples that contain enough biological material to cause significant amounts of "smoke" during the atomization process, so that ashing the samples is less necessary.

A method has been published for the determination of a variety of metals in diluted blood and serum using inductively coupled plasma atomization with mass spectrometric detection.[3] Blood was diluted tenfold and serum fivefold with a solution containing ammonia, Triton X-100 surfactant, and EDTA. Detection limits adequate for measurement in blood or serum were found for cadmium, cobalt, copper, lead, rubidium, and zinc.

20.3.2 Metals in Wet-Ashed Blood and Urine

Several toxicologically important metals are readily determined from wet-ashed blood or urine using atomic spectroscopic techniques. The ashing procedure may vary, but always entails heating the sample with strong acid and oxidant to dryness and redissolving the residue in acid. A typical procedure is digestion of blood or urine for cadmium analysis, which consists of mixing the sample with a comparable volume of concentrated nitric acid, heating to a reduced volume, adding 30% hydrogen peroxide oxidant, heating to dryness, and dissolving in nitric acid prior to measurement by atomic absorption or emission. Mixtures of nitric, sulfuric, and perchloric acid are effective though somewhat hazardous media for digesting blood, urine, or tissue. Wet ashing followed by atomic absorption analysis can be used for the determination in blood or urine of cadmium, chromium, copper, lead, manganese, and zinc, among other metals. Although atomic absorption, especially highly sensitive graphite furnace atomic absorption, has long been favored for measuring metals in biological samples, the multielement capability and other advantages of inductively coupled plasma atomic spectroscopy have led to its use for determining metals in blood and urine samples.[4]

20.3.3 Extraction of Metals for Atomic Absorption Analysis

A number of procedures for the determination of metals and biological samples call for the extraction of the metal with an organic chelating agent in order to remove interferences and concentrate the metal to enable detection of low levels. The urine or blood sample may be first subjected to wet ashing to enable extraction of the metal. Beryllium from an acid-digested blood or urine sample may be extracted by acetylacetone into methylisobutyl ketone prior to atomic absorption analysis. Virtually all of the common metals can be determined by this approach using appropriate extractants.

The availability of strongly chelating extractant reagents for a number of metals has lead to the development of procedures in which the metal is extracted from minimally treated blood or urine and then quantified by atomic absorption analysis. The metals for which such extractions can be used include cobalt, lead, and thallium extracted into organic solvent as the dithiocarbamate chelate, and nickel extracted into methylisobutyl ketone as a chelate formed with ammonium pyrrolidinedithiocarbamate.

Methods for several metals or metalloids involve conversion to a volatile form. Arsenic, antimony, and selenium can be reduced to their volatile hydrides, AsH_3, SbH_3, and H_2Se, repectively, which can be determined by atomic absorption or other means. Mercury is reduced to volatile mercury metal, which is evolved from solution and measured by cold vapor atomic absorption.

20.4 DETERMINATION OF NONMETALS AND INORGANIC COMPOUNDS

Relatively few nonmetals require determination in biological samples. One important example is fluoride, which occurs in biological fluids as the fluoride ion, F^-. In some cases of occupational exposure or exposure through food or drinking water, excessive levels of fluoride in the body can be a health concern. Fluoride is readily determined potentiometrically with a fluoride ion-selective electrode. The sample is diluted with an appropriate buffer and the potential of the fluoride electrode measured very accurately against a reference electrode, with the concentration calculated from a calibration plot. Even more accurate values can be obtained by the use of standard addition, in which the potential of the electrode system in a known volume of sample is read, a measured amount of standard fluoride is added, and the shift in potential is used to calculate the unknown concentration of fluoride.

Another nonmetal for which a method of determining biological exposure would be useful is white phosphorus, the most common and relatively toxic elemental form. Unfortunately, there is not a chemical method suitable for the determination of exposure to white phosphorus that would distinguish such exposure from relatively high background levels of organic and inorganic phosphorus in body fluids and tissues.

Toxic cyanide can be isolated in a special device called a Conway microdiffusion cell by treatment with acid, followed by collection of the weakly acidic HCN gas that is evolved in a base solution. The cyanide released can be measured spectrophotometrically by formation of a colored species.

Carbon monoxide is readily determined in blood by virtue of the colored carboxyhemoglogin that it forms with hemoglobin. The procedure consists of measuring the absorbances at wavelengths of 414, 421, and 428 nm of the blood sample, a sample through which oxygen has been bubbled to change all the hemoglobin to the oxyhemoglobin form, and a sample through which carbon monoxide has been bubbled to change all the hemoglobin to carboxyhemoglobin. With the appropriate calculations, a percentage conversion to carboxyhemoglobin can be obtained.

20.5 DETERMINATION OF PARENT ORGANIC COMPOUNDS

A number of organic compounds can be measured as the unmetabolized compound in blood, urine, and breath. In some cases, the sample can be injected along with its water content directly into a gas chromatograph. Direct injection is used for the measurement of acetone, n-butanol, dimethylformamide, cyclopropane, halothane, methoxyflurane, diethyl ether, isopropanol, methanol, methyl-n-butyl ketone, methyl chloride, methylethyl ketone, toluene, trichloroethane, and trichloroethylene.

For the determination of volatile compounds in blood or urine, a straightforward approach is to liberate the analyte at an elevated temperature, allowing the volatile compound to accumulate in headspace above the sample, followed by direct injection of headspace gas into a gas chromatograph. A reagent such as perchloric acid may be added to deproteinize the blood or urine sample and facilitate release of the volatile xenobiotic compound. Among the compounds determined by this approach are acetaldehyde, dichloromethane, chloroform, carbon tetrachloride, benzene, trichloroethylene, toluene, cyclohexane, and ethylene oxide. The use of multiple detectors for the gas chromatographic determination of analytes in headspace increases the versatility of this technique and enables the determination of a variety of physiologically important volatile organic compounds.[5]

Purge-and-trap techniques in which volatile analytes are evolved from blood or urine in a gas stream and collected on a trap for subsequent chromatographic analysis have been developed. Such a technique employing gas chromatographic separation and Fourier transform infrared detection has been described for a number of volatile organic compounds in blood.[6]

20.6 MEASUREMENT OF PHASE I AND PHASE II REACTION PRODUCTS

20.6.1 Phase I Reaction Products

For a number of organic compounds, the most accurate indication of exposure is to be obtained by determining their phase I reaction products. This is because many compounds are metabolized in the body and do not show up as the parent compound. And those fractions of volatile organic compounds that are not metabolized may be readily eliminated with expired air from the lungs and may thus be missed. In cases where a significant fraction of the xenobiotic compound has undergone a phase II reaction, the phase I product may be regenerated by acid hydrolysis.

One of the compounds commonly determined as its phase I metabolite is benzene,[7] which undergoes the following reactions in the body (see Section 13.5):

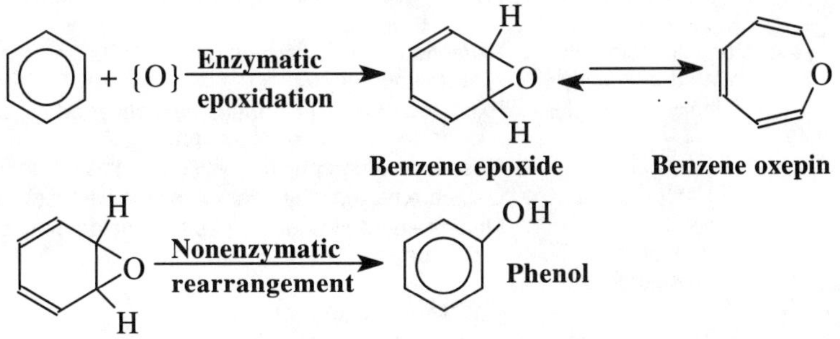

$$\text{Benzene} + \{O\} \xrightarrow{\text{Enzymatic epoxidation}} \text{Benzene epoxide} \rightleftharpoons \text{Benzene oxepin} \quad (20.6.1)$$

$$\text{Benzene epoxide} \xrightarrow{\text{Nonenzymatic rearrangement}} \text{Phenol}$$

Therefore, exposure to benzene can be determined by analysis of urine for phenol. Although a very sensitive colorimetric method for phenol involving diazotized *p*-nitroaniline has long been available, gas chromatographic analysis is now favored. The urine sample is treated with perchloric acid to hydrolyze phenol conjugates, and the phenol is extracted into diisopropyl ether for chromatographic analysis. Two other metabolic products of benzene, *trans,trans*-muconic acid[8] and S-phenyl mercapturic acid[9] are now commonly measured as more specific biomarkers of benzene exposure.

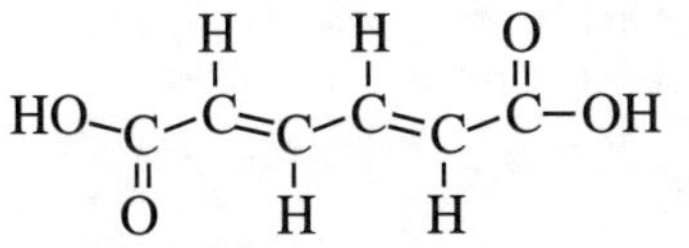

$$\text{HO–C–C=C–C=C–C–OH} \quad \textit{Trans,trans}\text{-muconic acid}$$

Insecticidal carbaryl undergoes the following metabolic reaction:

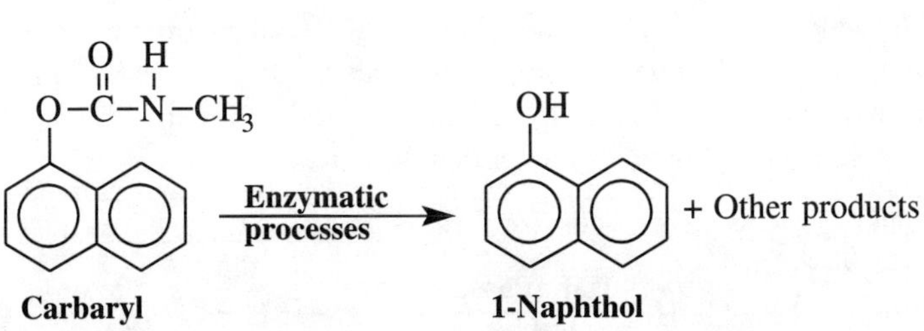

$$\text{Carbaryl} \xrightarrow{\text{Enzymatic processes}} \text{1-Naphthol} + \text{Other products} \quad (20.6.2)$$

Therefore, the analysis of 1-naphthol in urine indicates exposure to carbaryl. The 1-naphthol that is conjugated by a phase II reaction is liberated by acid hydrolysis, and then determined spectrophotometrically or by chromatography.

In addition to the examples discussed above, a number of other xenobiotics are measured by their phase I reaction products. These compounds and their metabolites are listed in Table 20.1. These methods are for metabolites in urine. Normally, the urine sample is acidified to release the phase I metabolites from phase II conjugates that they might have formed, and except where direct sample injection is employed, the analyte is collected as vapor or extracted into an organic solvent. In some cases, the analyte is reacted with a reagent that produces a volatile derivative that is readily separated and detected by gas chromatography.

Table 20.1 Phase I Reaction Products of Xenobiotics Determined

Parent Compound	Metabolite	Method of Analysis
Cyclohexane	Cyclohexanol	Extraction of acidified, hydrolyzed urine with dichloromethane, followed by gas chromatography
Diazinon	Organic phosphates	Colorimetric determination of phosphates
p-Dichlorobenzene	2,5-Dichlorophenol	Extraction into benzene; gas chromatographic analysis
Dimethylformamide	Methylformamide	Gas chromatography with direct sample introduction
Dioxane	β-hydroxyethoxy-acetic acid	Formation of volatile methyl ester; gas chromatography
Ethylbenzene	Mandelic acid and related aryl acids	Extraction of acids; formation of volatile derivatives; gas chromatography
Ethylene glycol monomethyl ether	Methoxyacetic acid	Extracted with dichloromethane; converted to volatile methyl derivative; gas chromatography
Formaldehyde	Formic acid	Gas chromatography of volatile formic acid derivative
Hexane	2,5-Hexanedione	Gas chromatography after extraction with dichloromethane
n-Heptane	2-Heptanone, valerolactone, 2,5-heptanedione	Measurement in urine by gas chromatography or mass spectrometry
Isopropanol	Acetone	Gas chromatography following extraction with methylethyl ketone
Malathion	Organic phosphates	Colorimetric determination of phosphates
Methanol	Formic acid	Gas chromatography of volatile formic acid derivative
Methyl bromide	Bromide ion	Formation of volatile organobromine compounds; gas chromatography
Nitrobenzene	p-Nitrophenol	Gas chromatography of volatile derivative
Parathion	p-Nitrophenol	Gas chromatography of volatile derivative
Polycyclic aryl hydrocarbons	1-Hydroxypyrene	HPLC of urine
Styrene	Mandelic acid	Extraction of acids; formation of volatile derivatives; gas chromatography
Tetrachloroethylene	Trichloroacetic acid	Extracted into pyridine and measured colorimetrically
Trichloroethane	Trichloroacetic acid	Extracted into pyridine and measured colorimetrically
Trichloroethylene	Trichloroacetic acid	Extracted into pyridine and measured colorimetrically

20.6.2 Phase II Reaction Products

Hippuric acids, which are formed as phase II metabolic products from toluene, xylenes, benzoic acid, ethylbenzene, and closely related compounds, can be determined as biological markers of exposure. The formation of hippuric acid from toluene is shown Figure 13.9, and the formation of 4-methylhippuric acid from p-xylene is shown below:

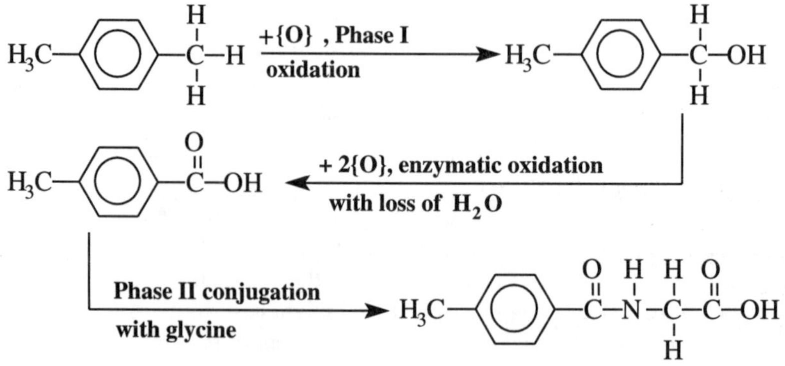

$$H_3C-C_6H_4-CH_3 \xrightarrow{+\{O\},\ \text{Phase I oxidation}} H_3C-C_6H_4-CH_2OH$$

$$H_3C-C_6H_4-COOH \xleftarrow{+2\{O\},\ \text{enzymatic oxidation with loss of } H_2O}$$

$$\xrightarrow{\text{Phase II conjugation with glycine}} H_3C-C_6H_4-C(O)-NH-CH_2-COOH \quad (20.6.3)$$

Other metabolites that may be formed from aryl solvent precursors include mandelic acid and phenylgloxylic acid.

Exposure to toluene can be detected by extracting hippuric acid from acidified urine into diethyl ether or isopropanol and direct ultraviolet absorbance measurement of the extracted acid at 230 nm. When the analysis is designed to detect xylenes, ethylbenzene, and related compounds, several metabolites related to hippuric acid may be formed and the ultraviolet spectrometric method does not give the required specificity. However, the various acids produced from these compounds can be extracted from acidified urine into ethyl acetate, derivatized to produce volatile species, and quantified by gas chromatography.

A disadvantage to measuring toluene exposure by hippuric acid is the production of this metabolite from natural sources, and the determination of toluylmercapturic acid is now favored as a biomarker of toluene exposure.[10] An interesting sidelight is that dietary habits can cause uncertainties in the measurement of xenobiotic metabolites. An example of this is the measurement of worker exposure to 3-chloropropene by the production of allylmercapturic acid.[11] This metabolite is also produced by garlic, and garlic consumption by workers was found to be a confounding factor in the method. Thiocyanate monitored as evidence of exposure to cyanide is increased markedly by the consumption of cooked cassava.

20.6.3 Mercapturates

Mercapturates are proving to be very useful phase II reaction products for measuring exposure to xenobiotics, especially because of the sensitive determination of these substances by high-performance liquid chromatography (HPLC) separation, and fluorescence detection of their *o*-phthaldialdehyde derivatives. In addition to toluene, the xenobiotics for which mercapturates may be monitored include styrene, structurally similar to toluene; acrylonitrile; allyl chloride; atrazine; butadiene; and epichlorohydrin.

The formation of mercapturates or mercapturic acid derivatives by metabolism of xenobiotics is the result of a phase II conjugation by glutathione. **Glutathione** (commonly abbreviated GSH) is a crucial conjugating agent in the body. This compound is a tripeptide, meaning that it is composed of three amino acids linked together. These amino acids and their abbreviations are glutamic acid (Glu), cysteine (Cys), and glycine (Gly). The formula of glutathione may be represented as illustrated in Figure 20.1, where the SH is shown specifically because of its crucial role in forming the covalent link to a xenobiotic compound. Glutathione conjugate may be excreted directly, although this is rare. More commonly, the GSH conjugate undergoes further biochemical reactions that produce mercapturic acids (compounds with N-acetylcysteine attached) or other species. The specific mercapturic acids can be monitored as biological markers of exposure to the xenobiotic species that result in their formation. The overall process for the production of mercapturic acids as applied to a generic xenobiotic species, HX–R (see previous discussion), is illustrated in Figure 20.1.

20.7 DETERMINATION OF ADDUCTS

Determination of adducts is often a useful and elegant means of measuring exposure to xenobiotics. Adducts, as the name implies, are substances produced when xenobiotic substances add to endogenous chemical species. The measurement of carbon monoxide from its hemoglobin adduct is discussed in Section 20.4. In general, adducts are produced when a relatively simple xenobiotic molecule adds to a large macromolecular biomolecule that is naturally present in the body. The fact that adduct formation is a mode of toxic action, such as occurs in the methylation of DNA during carcinogenesis (Section 7.8.1), makes adduct measurement as a means of biological monitoring even more pertinent.

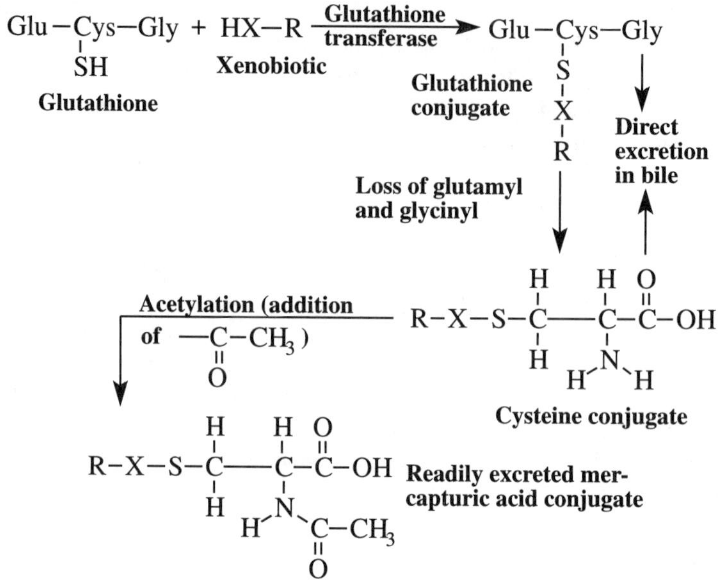

$$\text{Glu}-\underset{\underset{\text{SH}}{|}}{\text{Cys}}-\text{Gly} + \text{HX}-\text{R} \xrightarrow{\text{Glutathione transferase}} \text{Glu}-\underset{\underset{\underset{\underset{R}{|}}{\underset{X}{|}}}{\underset{S}{|}}}{\text{Cys}}-\text{Gly}$$

Glutathione — Xenobiotic — Glutathione conjugate → Direct excretion in bile

Loss of glutamyl and glycinyl ↓

Cysteine conjugate: R–X–S–CH(H)–C(H)(NH₂)–C(=O)–OH

Acetylation (addition of —C(=O)—CH₃) ↓

R–X–S–CH(H)–C(H)(NH–C(=O)–CH₃)–C(=O)–OH   Readily excreted mercapturic acid conjugate

Figure 20.1 Glutathione conjugate of a xenobiotic species (HX–R) followed by formation of glutathione and cysteine conjugate intermediates (which may be excreted in bile) and acetylation to form readily excreted mercapturic acid conjugate.

Adducts to hemoglobin are perhaps the most useful means of biological monitoring by adduct formation. Hemoglobin is, of course, present in blood, which is the most accurate type of sample for biological monitoring. Adducts to blood plasma albumin are also useful monitors and have been applied to the determination of exposure to toluene diisocyanate, benzo(a)pyrene, styrene, styrene oxide, and aflatoxin B_1. The DNA adduct of styrene oxide has been measured to indicate exposure to carcinogenic styrene oxide.[12]

One disadvantage of biological monitoring by adduct formation can be the relatively complicated procedures and expensive, specialized instruments required. Lysing red blood cells may be required to release the hemoglobin adducts, derivatization may be necessary, and the measurements of the final analyte species can require relatively sophisticated instrumental techniques. Despite these complexities, the measurement of hemoglobin adducts is emerging as a method of choice for a number of xenobiotics, including acrylamide, acrylonitrile, 1,3-butadiene, 3,3' dichlorobenzidine, ethylene oxide, and hexahydrophthalic anhydride.

20.8 THE PROMISE OF IMMUNOLOGICAL METHODS

Immunoassay methods based upon biologically synthesized antibodies to specific molecules offer distinct advantages in specificity, selectivity, simplicity, and costs. Although used in simple test kits for blood glucose and pregnancy testing, immunoassay methods have been limited in biological monitoring of xenobiotics, in part because of interferences in complex biological systems. Because of their inherent advantages, however, it can be anticipated that immunoassays will grow in importance for biological monitoring of xenobiotics.[13] As an example of such an application, polychlorinated biphenyls (PCBs) have been measured in blood plasma by immunoassay.[14]

In addition to immunoassay measurement of xenobiotics and their metabolites, immunological techniques can be used for the separation of analytes from complex biological samples employing immobilized antibodies. This approach has been used to isolate aflatoxicol from urine and enable its determination, along with aflatoxins B_1, B_2, G_1, G_2, M_1, and Q_1, using HPLC and postcolumn derivatization and fluorescence detection.[15] A monoclonal antibody reactive with S-phenylmercapturic acid, an important phase II reaction product of benzene resulting from glutathione conjugation, has been generated from an appropriate hapten–protein conjugate. The immobilized antibody has been used in a column to enrich S-phenylmercapturic acid from the urine of workers exposed to benzene.[16] Many more such applications can be anticipated in future years.

REFERENCES

1. Draper, W. et al., Industrial hygiene chemistry: keeping pace with rapid change in the workplace, *Anal. Chem.*, 71, 33R–60R, 1999. (A comprehensive review of this topic is published every two years in *Analytical Chemistry*.)
2. Atio, A., Special issue: biological monitoring in occupational and environmental health, *Sci. Total Environ.*, 199, 1–226, 1997.
3. Barany, E. et al., Inductively coupled plasma mass spectrometry for direct multielement analysis of diluted human blood and serum, *J. Anal. At. Spectrosc.*, 12, 1005–1009, 1997.
4. Paschal, D.C. et al., Trace metals in urine of United States residents: reference range concentrations, *Environ. Res.*, 76, 53–59, 1998.
5. Schroers, H.-J. and Jermann, E., Determination of physiological levels of volatile organic compounds in blood using static headspace capillary gas chromatography with serial triple detection, *Analyst*, 123, 715–720, 1998.
6. Ojanpera, I., Pihlainen, K., and Vuori, E., Identification limits for volatile organic compounds in the blood by purge-and-trap GC-FTIR, *J. Anal. Toxicol.*, 22, 290–295, 1998.
7. Agency for Toxic Substances and Disease Registry, U.S. Department of Health and Human Services, *Toxicological Profile for Benzene*, CD/ROM Version, CRC Press/Lewis Publishers, Boca Raton, FL, 1999.
8. Scherer, G., Renner, T., and Meger, M., Analysis and evaluation of *trans,trans*-muconic acid as a biomarker for benzene exposure, *J. Chromatogr. B Biomed. Sci. Appl.*, 717, 179–199, 1998.
9. Boogaard, P.J. and Van Sittert, N.J., Suitability of S-phenyl mercapturic acid and *trans-trans*-muconic acid as biomarkers for exposure to low concentrations of benzene, *Environ. Health Perspect. Suppl.*, 104, 1151–1157, 1996.
10. Angerer, J., Schildbach, M., and Kramer, A., S-toluylmercapturic acid in the urine of workers exposed to toluene: a new biomarker for toluene exposure, *Arch. Toxicol.*, 72, 119–123, 1998.
11. De Ruij, B.M. et al., Allylmercapturic acid as urinary biomarker of human exposure to allyl chloride, *Occup. Environ. Med.*, 54, 653–661, 1997.
12. Rappaport, S.M. et al., An Investigation of multiple biomarkers among workers exposed to styrene and styrene-7,8-oxide, *Cancer Res.*, 56, 5410–5416, 1996.
13. Wengatz, I. et al., Recent developments in immunoassays and related methods for the detection of xenobiotics, *ACS Symposium Series 646 (Environmental Immunochemical Methods)*, American Chemical Society, Washington, D.C., 1996, pp. 110–126.
14. Griffin, P., Jones, K., and Cocker, J., Biological monitoring of polychlorinated biphenyls in plasma: a comparison of enzyme-linked immunosorbent assay and gas chromatography detection methods, *Biomarkers*, 2, 193–195, 1997.
15. Kussak, A. et al., Determination of aflatoxicol in human urine by immunoaffinity column clean-up and liquid chromatography, *Chemosphere*, 36, 1841–1848, 1998.
16. Ball, L. et al., Immunoenrichment of urinary S-phenylmercapturic acid, *Biomarkers*, 2, 29–33, 1997.

SUPPLEMENTARY REFERENCES

Angerer, J.K. and Schaller, K.H., *Analyses of Hazardous Substances in Biological Materials*, Vol. 1, VCH, Weinheim, Germany, 1985.

Angerer, J.K. and Schaller, K.H., *Analyses of Hazardous Substances in Biological Materials*, Vol. 2, VCH, Weinheim, Germany, 1988.

Angerer, J.K. and Schaller, K.H., *Analyses of Hazardous Substances in Biological Materials*, Vol. 3, VCH, Weinheim, Germany, 1991.

Angerer, J.K. and Schaller, K.H., *Analyses of Hazardous Substances in Biological Materials*, Vol. 4, VCH, Weinheim, Germany, 1994.

Angerer, J.K. and Schaller, K.H., *Analyses of Hazardous Substances in Biological Materials*, Vol. 5, John Wiley & Sons, New York, 1996.

Angerer, J.K. and Schaller, K.H., *Analyses of Hazardous Substances in Biological Materials*, Vol. 6, John Wiley & Sons, New York, 1999.

Baselt, R.C., *Biological Monitoring Methods for Industrial Chemicals*, 2nd ed., PSG Publishing Company, Inc., Littleton, MA, 1988.

Committee on National Monitoring of Human Tissues, Board on Environmental Studies and Toxicology, Commission on Life Sciences, *Monitoring Human Tissues for Toxic Substances*, National Academy Press, Washington, D.C., 1991.

Draper, W.M. et al., Industrial hygiene chemistry: keeping pace with rapid change in the workplace, *Anal. Chem.*, 71, 33R–60R, 1999. (A comprehensive review of this topic is published every two years in *Analytical Chemistry*.)

Ellenberg, H., *Biological Monitoring: Signals from the Environment*, Braunschweig, Vieweg, Germany, 1991.

Hee, S.Q., *Biological Monitoring: An Introduction*, Van Nostrand Reinhold, New York, 1993.

Ioannides, C., Ed., *Cytochromes P450: Metabolic and Toxicological Aspects*, CRC Press, Boca Raton, FL, 1996.

Kneip, T.J. and Crable, J.V., *Methods for Biological Monitoring*, American Public Health Association, Washington, D.C., 1988.

Lauwerys, R.R. and Hoet, P., *Industrial Chemical Exposure: Guidelines for Biological Monitoring*, 2nd ed., CRC Press/Lewis Publishers, Boca Raton, FL, 1993.

Mendelsohn, M.L., Peeters, J.P., and Normandy, M.J., Eds., *Biomarkers and Occupational Health: Progress and Perspectives*, Joseph Henry Press, Washington, D.C., 1995.

Minear, R.A. et al., Eds., *Applications of Molecular Biology in Environmental Chemistry*, CRC Press/Lewis Publishers, Boca Raton, FL, 1995.

Richardson, M., Ed., *Environmental Xenobiotics*, Taylor & Francis, London, 1996.

Saleh, M.A., Blancato, J.N., and Nauman, C.H., *Biomarkers of Human Exposure to Pesticides*, American Chemical Society, Washington, D.C., 1994.

Singh,V.P., Ed., *Biotransformations: Microbial Degradation of Health-Risk Compounds*, Elsevier, Amsterdam, 1995.

Travis, C.C., Ed., *Use of Biomarkers in Assessing Health and Environmental Impacts of Chemical Pollutants*, Plenum Press, New York, 1993.

Williams, W.P., *Human Exposure to Pollutants: Report on the Pilot Phase of the Human Exposure Assessment Locations Programme*, United Nations Environment Programme, New York, 1992.

World Health Organization, *Biological Monitoring of Chemical Exposure in the Workplace*, World Health Organization, Geneva, Switzerland, 1996.

QUESTIONS AND PROBLEMS

1. Personnel monitoring in the workplace is commonly practiced with vapor samplers that workers carry around. How does this differ from biological monitoring? In what respects is biological monitoring superior?
2. Why is blood arguably the best kind of sample for biological monitoring? What are some of the disadvantages of blood in terms of sampling and sample processing? What are some disadvantages of blood as a matrix for analysis? What are the advantages of urine? Discuss why urine might be the kind of sample most likely to show metabolites and least likely to show parent species.
3. Distinguish among the following kinds of analytes measured for biological monitoring: parent compound, phase I reaction product, phase II reaction product, adducts.
4. What is wet ashing? For what kinds of analytes is wet ashing of blood commonly performed? What kinds of reagents are used for wet ashing, and what are some of the special safety precautions that should be taken with the use of these kinds of reagents for wet ashing?

5. What species is commonly measured potentiometrically in biological monitoring?
6. Compare the analysis of phase I and phase II metabolic products for biological monitoring. How are phase II products converted back to phase I metabolites for analysis?
7. Which biomolecule is most commonly involved in the formation of adducts for biological monitoring? What is a problem with measuring adducts for biological monitoring?
8. What are two general uses of immunology in biological monitoring? What is a disadvantage of immunological techniques? Discuss the likelihood that immunological techniques will find increasing use in the future as a means of biological monitoring.
9. The determination of DNA adducts is a favored means of measuring exposure to carcinogens. Based on what is known about the mechanism of carcinogenicity, why would this method be favored? What might be some limitations of measuring DNA adducts as evidence of exposure to carcinogens?
10. How are mercapturic acid conjugates formed? What special role do they play in biological monitoring? What advantage do they afford in terms of measurement?
11. For what kinds of xenobiotics is trichloroacetic acid measured? Suggest the pathways by which these compounds might form trichloroacetic acid metabolically.
12. Match each xenobiotic species from the left column with the analyte that is measured in its biological monitoring from the right column.

    1. Methanol   (a) Mandelic acid
    2. Malathion   (b) A diketone
    3. Styrene   (c) Organic phosphates
    4. Nitrobenzene   (d) Formic acid
    5. n-Heptane   (e) p-Nitrophenol

Index

A

Abiotic, 40
Accutane, 206
Acetaldehyde, 301
Acetaminophen, toxicity to liver, 180
Acetic acid, 303
Acetone cyanohydrin, 315
Acetone, 25, 301
Acetonitrile, 314
Acetylaminofluorene, 2-, 146
Acetylation reaction, 154
Acetylcholine, 130
Acetylcholinesterase inhibition by organophosphates, 374
Acetylcholinesterase, 130, 159
Acetylene, 20, 283
Acid, 12
Acid anhydrides, 304
Acidic solutions, 13
Acrolein, 25, 301
Acrylic acid, 303
Acrylonitrile, 35, 314
Activation of toxic substances, 140
Activity of electron, 44
Acute local exposure, 119
Acute systemic exposure, 119
Acute toxicity, 116
Addition reactions, 20, 275
Additive effects, 132
Adducts, determination of, 407
Adenine, 73, 168
Adenosine diphosphate (ADP), 87
Adenosine triphosphate (ATP), 79
Adipic acid, 303
Adjuvants and toxic substances, 118
ADP (adenosine diphosphate), 87
Aerobic (water), 43
Aerobic bacteria, 384
Aerobic respiration, 46, 76, 91
Aerosols, 54
Aflatoxins, 385
Agroclavin, 386

Alachlor, 342
Alcohol dehydrogenase, 145
Alcohol dehydrogenation, 145
Alcohols, 293
Alcohols, higher, 296
Aldehydes, 300
Aldicarb, 146, 320
Aldrin, 339
Alevoli of lungs, 122
Algae, toxic, 393
Aliphatic hydrocarbons, 15
Alkaloids, 322
Alkanes, 15, 273
Alkenes, 20, 275
Alkenyl halides, 332
Alkyl halides, 27, 328
Alkyl polyamines, 310
Alkylating agents in carcinogenesis, 163
Alkylating agents, 172
Alkylation of DNA, 160
Alkyne, 20
Allergens, 392
Allergy, 197
Allethrin, 392
Allyl chloride, 333
Alpha helix (in protein structure), 64
Alpha-olefins, 282
Amidases, 148
Amine compounds, 26
Amines, 309
Amino acids, 61, 62, 85
Aminothiazole, 354
Ammonia, 238
Ammonium ion, 8
Amygdalin, 236, 389
Amyl acetate, 305
Anabolic metabolic processes, 92
Anabolism, 76, 106
Anaerobic (water), 43
Anaerobic bacteria, 384
Anaerobic respiration, 76
Analysis of xenobiotics, 401

Anemia, aplastic, 195
Aneuploidy, 170
Aniline, 311
Aniline metabolites, 313
Animal toxins, nonreptile, 396
Anions, 8
Antagonism, 132
Antagonist action, 158
Anthropocene, 56
Anthrosphere, 39, 55
Anticodon, 169
Antigen, 197
ANTU (1-naphthylthiourea), 32, 353
Aplastic anemia, 195
Apoptosis, 183
Apoptotic bodies, 183
Aquatic chemistry, 44
Aquatic ecosystems, 110
Aquatic ecotoxicology, 111
Aromatic amines, 311
Aromatic compounds, 275
Aromatic hydrocarbons, 21
Aromatic hydrocarbons, 21
Aromaticity, 22, 275
Arrhythmia, 132, 195
Arsanilic acid, 269
Arsenic toxicity, 223
Arsenic, drinking water standards, 224
Arsenic, organometalloid compounds, 267
Arsenobetaine, 268
Arsenocholine, 268
Arsenosugar, 268
Arsine, 245
Arteriosclerosis, 196
Arthropods, 397
Aryl halides, 336
Aryl hydrocarbons, 21
Arylating agents in carcinogenesis, 163
Asbestos, 244
Asphyxiant, chemical, cyanide as, 236
Asphyxiant, simple, 276
Ataxia, 135
Atherosclerosis, 196
Atmosphere, 39, 40, 51
Atmospheric chemistry, 52
Atomic absorption analysis, 403
Atomic mass, 2
Atomic mass unit, 2
Atomic weight, 2
Atoms, 2
Atoxyl, 268
ATP (adenosine triphosphate), 79
ATP, 87
Atropine, 375
Autoimmunity, 197
Autotrophic, 77
Auxotrophs, 175
Axon, 201
Axonopathies, 202
Azides, 243
Azinphos-methyl, 372

B

Back mutation (reversion), 175
Bacterial toxins, 385
BAL (British anti-Lewisite), antidote to lead poisoning, 222
Balanced chemical equation, 10
Base, 12
Base excision, 170
Base-pair substitution, 170
Basic solutions, 13
Baycol, toxic effects, 184
Bee venom, 393
Behavioral response, 131, 132
Benzene, 23, 283
Benzene epoxide, 405
Benzene metabolism, 284
Benzene oxepin, 284, 405
Benzene oxide, 284
Benzene-1,2-oxide, 299
Benzidine, 313
Benzo(a)pyrene, 23, 145, 289
Benzo(a)pyrene 7,8-diol-9,10-epoxide, 145
Benzoic acid, 303
Benzothiazole, 354
Berylliosis, 215
Beryllium, toxicity, 215
Bile, 80
Bile salts, 81
Binding species, toxic, 140
Bioavailability, 102, 103
Biochemical effect, 131
Biochemical mechanisms of toxicity, 157
Biochemical transformations of toxic substances, 140
Biochemistry, 1, 59
Bioconcentration, 102
Bioconcentration by vegetation, 105
Bioconcentration factor, 104
Biodegradability, 108
Biodegradation, 105
Biogeochemical cycles, 40
Biological monitoring, 401
Biomagnification, 101
Biomarker of exposure, 109
Biomarkers of effect, 109
Biomarkers of susceptibility, 109
Biomarkers, 108
Biomolecules, 59
Biorefractory substances, 108
Biosphere, 40, 55
Biotic, 40
Biotransfer factor, 105
Biotransformation, 106, 141
Bladder, toxic effects to, 206
Blastocyst, 204
Blocker, 132
Blood for metals analysis, 403
Blood pressure, 132
Blood vessels, 192
Blood, 192
Blood-brain barrier, 201
Bradycardia, 132, 195

INDEX

Branched-chain alkanes, 16
British anti-Lewisite(BAL), antidote to lead poisoning, 222
Bromine toxicity, 229
Bronchitis, 185
Brown recluse spiders, 394
Bruce Ames test, 163, 175
Bufotenin, 396
Bulky group, large, and DNA, 172
Butadiene, 1,3-, 281
Butane, 277
Butyl acetate, 305
Butylene compounds, 282
Butyric acid, 303

C

Cadmium organometallic compounds, 263
Cadmium toxicity, 217
Caffeine, 322
Calomel, 219
Canalicular choleostasis, 191
Carbamates, 320
Carbamic acid, 320
Carbanions, 254
Carbarsone, 269
Carbaryl, 320, 405
Carbofuran, 320
Carbohydrate digestion, 81
Carbohydrate metabolism, 85
Carbohydrates, 65
Carbon disulfide, 351
Carbon monoxide, 53
Carbon oxysulfide, 352
Carbon tetrachloride, 28, 328
Carbonyl compounds, 254, 256
Carbonyl group, 25
Carboxyhemoglobin, 194
Carboxylic acids, 302
Carcinogenesis, 161
Carcinogenic agents, 161
Carcinogenic process, 162
Cardiotoxicants, 195
Cardiovascular system, 192, 193
Carrying capacity, 110
Castor bean, toxins from, 389
Catabolic metabolism, 84
Catabolism, 76, 83, 106
Cation exchanger, 35
Cations, 8
Cell membrane, 60
Cell necrosis, 183
Cell nucleus, 60
Cell walls, 61
Cell, 60
Cellular respiration, 61, 87
Cellulose, 66
Centipedes, 397
Central metal ion, 212
Central nervous system, 200

Central nervous system, effects from toxic substance exposure, 135
Cerebral cortex, 200
Cerebrum, 200
Cervix, 203
CFCs (chlorofluorocarbons), 29
Chain reactions, 19, 54
Chelate, metal, 212
Chelating agent, 212
Chemical allergy, 197
Chemical antagonism, 132
Chemical bond, 7
Chemical carcinogenesis, 161
Chemical compound, 7
Chemical equations, 10
Chemical formula, 7
Chemical idiosyncrasy, 116
Chemical reactions, 10
Chemical symbol, 2
Chemical weathering, 50
Chemistry, 1
Chemodynamics, 99
Chiral, 34
Chloracetaldehyde, 334
Chloracne, 189
Chloramines, 243
Chlordane, 339
Chlordecone (Kepone), 339
Chlorine toxicity, 229
Chlorobenzene (monochlorobenzene), 336
Chlorofluorocarbons (CFCs), 29, 330
Chloromethane, 328
Chlorophenoxy insecticides, 341
Chloroplasts, 61
Chloroprene, 333
Chlorothion, 356, 371
Chlorpyrifos methyl, 372
Chromate toxicity, 216
Chromium toxicity, 216
Chromosomal structural alterations, 170
Chromosomes, 60, 167
Chronic local exposure, 119
Chronic systemic exposure, 119
Chronic toxicity, 116
Cilia, respiratory tract, 122
Cinerin, 391
Cirrhosis, 191
Cis-trans isomers, 21
Citric acid cycle, 84, 89
Clinical toxicology, 116
Cloning vehicles, 76
Clostridium botulinum, 383, 385
Coagulation, 14
Cobalt toxicity, 216
Cocaine, 322
Codon, 169
Coenzymes, 70
Colloidal particles, 14
Colloidal suspension, 14
Colloids, 45

Coma, 135
Combined available chlorine, 243
Cometabolism, 107
Common names, 18
Community tolerance, pollution-induced, 112
Complementary strands of DNA, 74
Complex, 45, 212
Complex ion, 14
Complexation, 14
Condensation aerosols, 55
Condensation nuclei, 54
Condensed strucutal formulas, 21
Coniine, 322
Conjugating agent, endogenous, 150
Conjugation product, 142
Conjugation reactions, 149
Conjunctivitis, 135
Contact dermatitis, 187
Convulsions, 135
Corrosive substances, 13
Cotinine, 322
Coumaphos, 372
Covalent bond, 7
Cresol, 297
Crick, Francis, 74
Crotonaldehyde, 301
Crust (Earth), 46
Crust (of earth), 40
Crystalline lattice, 8
Cumulative poison, 218
Cyamic acid, 238
Cyanamide, 238
Cyanide, 235
Cyanide analysis, 404
Cyanoethylene oxide, 315
Cyanogen, 238
Cyanogens, 236
Cyanotic appearance, 134
Cycloalkanes, 16
Cyclohexane analysis, 406
Cyclohexane, 278
Cyclopentadiene, 282
Cytogenetic assays, 175
Cytoplasm, 61
Cytosine, 73, 168

D

Dalton, 2
Dative bond, 256
DDT, 99, 339
Dealkylation, 148
Dehalogenation, 149
Deltamethrin, 392
Denaturation of proteins, 65
Dendrites, 201
Deoxyribonucleic acid (DNA), 72
Deoxyribose, 72, 73, 168
Dermatitis, contact, 187
Dermatitis, irritant, 187
Detoxication, 140

Deuterium, 2
Developmental toxic effects, 136
Developmental toxicology, 205
Diazinon, 372
Diazinon analysis, 406
Dibromoethane, 1,2-, 28
Dichloramine, 243
Dichlorobenzene analysis, 406
Dichlorodifluoromethane, 328
Dichloromethane, 27, 328
Dichlorophenoxyacetic acid, 2,4- (2,4-D), 341
Dichlorvos, 370
Dicrotophos, 370
Dicyclopentadiene, 282
Dieldrin, 339
Diesel fuel, 278
Diethyl ether, 304
Digestion, 80
Digitoxin, 391
Di-isopropyl ether, 304
Diisopropylphosphorfluoridate, 378
Dimercaptosuccinic acid, meso-2,3, 350
Dimethoate, 370, 372
Dimethylarsine, 267
Dimethylarsinic acid, 267
Dimethylformamide analysis, 406
Dimethylmercury, 264
Dimethylnitrosamine, 26, 317
Dimethylsulfate, 32, 358
Dimethylsulfide, 350
Dimethylsufone, 356
Dimethylsulfoxide (DMSO), 32, 357
Dinoseb, 316
Dioxane analysis, 406
Diquat, 321
Direct-acting carcinogens, 162
Disaccharides, 65
Dispersion aerosols, 54
Dispositional antagonism, 132
Distribution of toxic substances in the body, 120
Disulfides (organic), 350
Disulfiram, 348, 354
Disulfoton, 356, 372
Dithiocarb, 348
Dithiocarbamates, 355
DMSO (dimethylsulfoxide), 357
DNA (deoxyribonucleic acid), 72
DNA adducts, 172
DNA, modified, 75
Dose of toxic substance, 119, 124
Dose-response, 124
Double helix (in DNA), 74
Dunet, 320
Duration of toxic substance exposure, 119
Dursban, 370

E

Ecology, 39, 97
Economic form, 126
Ecosystems, 97

Ecotoxicology, 97, 116
Ecotoxicology, aquatic, 111
Ecotoxicology, terrestrial, 111
EDTA (ethylenediaminetetraacetic acid), antidote to lead poisoning, 222
Eggs, 203
Eggshell thinning from DDT exposure, 111
Electron activity, 44
Electron carriers, 91
Electron donor, 212
Electron receptor, 90
Electron transfer chain, 90
Electronegativities of elements, 253
Electronegativity, 10
Electrons, 2
Electrophilic species, 157
Elements, 1
Elements, essential, 212
Elements, toxic, 211
Elimination of toxic substances from the body, 120
Emphysema, 185
Enantiomers, 34
Encephelopathy, 201
Endocrine cells, 198
Endocrine disruptors, 109, 200
Endocrine glands, 68, 69, 198
Endocrine system, 198
Endogenous conjugating agent, 150
Endogenous substances, 128
Endogenous substrate, 141
Endoplasmic reticulum, 61, 143
Enterohepatic circulation, 152
Enterohepatic circulation system, 123
Environmental biochemistry, 59
Environmental chemistry, 41
Environmental factors in toxic substance exposure, 119
Environmental geochemistry, 49
Environmental science, 39
Environmental toxicology, 116
Enzyme induction, 158
Enzyme inhibitors, 158
Enzyme-substrate complex, 69
Epidermis, 186
Epigenetic carcinogen, 161
Epilimnion, 42
Epoxidation, 143
Epoxide hydration, 144
Ergot alkaloids, 386
Erythema, 187
Erythropoieten, 192
Erytrhocytes, 192, 193
Escherichia coli, 384
Essential amino acids, 85
Essential elements, 212
Esterase, 148
Esters, 305
Ethanol, 295
Ethene, 20
Ethers, 303
Ethybenzene analysis, 406
Ethyl acetate, 305

Ethyl group, 17
Ethylene glycol monomethyl ether analysis, 406
Ethylene glycol, 296
Ethylene oxide, 24, 279
Ethylene, 20, 35, 279
Ethylenediamine, 311
Ethylenediaminetetraacetic acid (EDTA), antidote to lead poisoning, 222
Ethylsufuric acid, 32, 358
Eukaryotic cells, 60
Excipients and toxic substances, 118
Excision (of DNA), 170
Excited state, 53
Exons, 169
Exposure to toxic substances, 120
Extractions of metals for analysis, 403
Eye irritants, 391
Eyes, effects from toxic substance exposure, 135

F

Facultative bacteria, 384
Fat digestion, 81
Fat metabolism, 85
Fatty acid cycle, 85
Fatty amines, 310
Fenitrothion, 372
Fermentation, 76, 87, 92
Fertilization, 203
Fetal alcohol syndrome, 206, 295
Fibrous proteins, 64
Fission products, toxic radionuclides, 231
Flocculation, 14
Fluoride analysis, 403
Fluorine toxicity, 229
Fluorosis, 240
Fluvalinate, 392
Fly ash, 55
Forensic toxicology, 116
Formaldehyde analysis, 406
Formaldehyde, 299
Formalin, 300
Formed elements, 192
Formic acid, 303
Forward mutations, 175
Frameshift mutation, 170
Free radical species, 19
Free radicals, 54, 157, 226
Freon 12, 328
Frequency of exposure to toxic substance, 119
Fructose, 81
Fuel oil, 278
Fumosin B_1, 386
Functional antagonism, 132
Functional groups, 23, 24
Furfural, 277
Furfural, 301

G

Gametes, 203

Gaseous toxicants, 99
Gastrointestinal tract, 79, 123
Gene mutation, 170
Gene, 75
Genes, 169
Genetic engineering, 76
Genetic status, 119
Genome, 169
Genomics, 169
Genotoxic carcinogens, 161
Genotykpe, 169
Geochemical cycle, 49
Geochemistry, 49
Geometrical isomerism, 34
Geosphere, 39, 46
Germ cell, 167
Globular protein, 64
Glucose, 81
Glucuronides, 150
Glutathione, 152, 407
Glycoaldehyde, 296
Glycogen, 65
Glycolic acid, 296
Glycolysis, 87
Glycoproteins, 66
Glyoxylic acid, 287, 296
Golgi bodies, 61
Granulomatous inflammation, 189
Green chemistry, 56
Grignard reagents, 260
Ground state, 53
Groundwater, 99
Groups of elements, 4
Guanine, 73, 168
Gyneomastia, 204

H

Hallucinattions, 135
Halogen oxides, 241
Halogens toxicity, 228
Halons, 29
Halothane, 331
Haptens, 197
HCFC-123, 331
HCFC-141b, 331
Heavy metal poisoning, defenses against, 222
Heavy metal toxicants, 98
Heavy metals, toxic, 140
Hemangiosarcoma, 191
Hematology, 192
Heme group, 91, 194
Hemoglobin, 192, 194
Hepatotoxicity, 179, 190
Hepatotoxins, 190
Heptachlor, 339
Heterotrophic cells, 77
Hexachlorobenzene, 336
Hexachlorobutadiene, 28, 333
Hexachlorocyclohexane (lindane), 340
Hexachlorocyclopentadiene, 335

Hexachlorophene, 202, 343
Hexane analysis, 406
Hexane, n- analysis, 406
HFC-134a, 30
Hippuric acid analysis, 406
Hippuric acid, 155, 286
Horizons (soil), 47
Hormones, 68, 198
Hornet venoms, 394
Host defense mechanisms, 197
Human chorionic gonadotropin, 204
Humus, soil, 51
Hydorchloric acid, 240
Hydration, 12
Hydrazine, 238
Hydrocarbon toxicities, 273
Hydrocarbons, 14, 274
Hydrochlorofluorocarbons, 330
Hydrofluoric acid, 240
Hydrogen, 2
Hydrogen bond, 12
Hydrogen bromide, 240
Hydrogen chloride, 240
Hydrogen cyanide, 235
Hydrogen fluoride, 240
Hydrogen halides, 240
Hydrogen iodide, 240
Hydrogen sulfide, 246, 248
Hydrogenation reaction, 20
Hydrolase enzymes, 148
Hydrologic cycle, 42, 43
Hydrolysis, 148
Hydrolyzing enzymes, 71
Hydroperoxyl radical, 164
Hydrophobicity model, 103
Hydroquinone, 285
Hydrosphere, 39, 42
Hydroxyl radical, 164, 226
Hydroxylation, 144
Hypercin, 391
Hyperpigmentation, 188
Hypersensitivity, 127, 197
Hypochlorites, 242
Hypochlorous acid, 242
Hypolimnion, 42
Hypopigmentation, 188
Hyposensitivity, 127
Hypoxia, 194

I

Igneous rock, 46
Immune system, 196
Immunogen, 197
Immunogenesis, 197
Immunoglobin, 197
Immunological methods of analysis, 408
Immunological status, 119
Immunosuppression, 197
Indicators of exposure to xenobiotics, 401
Induration, 187

Industrial ecology, 56
Infectious agents, 196
Inhalation toxicology, 184
Inorganic compounds analysis, 403
Inorganic compounds toxicity, 235
Inorganic toxicants, 98
Insect toxins, 393
Intercalation, 173
Interhalogen compounds, 241
Intermediary xenobiotic metabolism, 141
Intestines, exposure to toxic substances, 123
Introns, 169
Iodine toxicity, 230
Ionic bonds, 8
Ionizing radiation, toxic effects, 163
Ions, 8
Irreversible effects of toxic substances, 127
Irritant dermatitis, 187
Isocyanates, 318
Isomerases, 71
Isopropanol analysis, 406
Isotopes, 3
Itai, itai disease, 218

J

Jasmolin$_1$, 391
Jaundiced appearance, 134

K

Karotype, 175
Kepone, 339
Keratin, 186
Kerosene, 278
Ketones, 300
Kidney, toxic effects to, 206
Kinetic phase, 130
Kinetic toxicology, 129
K_{ow}, octanol-water partition coefficient, 104
Krebs cycle, 90
Kupffer cells, 191

L

Lactic acid fermentation, 92
Lactose, 81
Lauryl thiocyanate, 353
LD_{50}, 124
Lead palsy, 221
Lead toxicity, 220
Leaving group, 374
Leukemia, 195
Leukocytes, 192, 195, 197
Lewis formula, 7
Lewis symbols, 6
Ligand, 14, 45, 157, 212
Ligase enzymes, 71
Lindane, 340
Lipid peroxidation, 226, 239, 329
Lipids, 66

Lipid-soluble compounds, toxic, 140
Lipobay, toxic effects, 184
Lipophilic, 101
Lithium organometallic compounds, 258
Lithium, toxicity, 214
Lithosphere, 39
Liver, 123, 189
Liver damage by pharmaceuticals, 179
Liver tumor, 191
Lung cancer, 185
Lung exposure to toxic substances, 122
Lyase enzymes, 71
Lymph, 192
Lysososme, 61

M

Magma, 46
Magnesium organometallic compounds, 260
Main group elements, 4
Malathion analysis, 406
Malathion, 356, 372
Mandelic acid, 287
Margin of safety, 126
Matrine, 389
Matrix (of toxic substance), 118
Melanin, 176, 188
Mercaptan, 30
Mercapturates, 407
Mercapturic acid conjugate, 153
Mercury organometallic compounds, 263
Mercury toxicity, 218
Messenger RNA (mRNA), 75, 169
Metabolic activation of insecticides, 375
Metabolic adaptation, 107
Metabolic processes, 79
Metabolic reductions, 147
Metabolism, 76, 79, 106
Metagenesis, 159
Metal chelate, 212
Metal fume fever, 262
Metal toxicity, 213
Metalloenzymes, 159
Metalloids toxicity, 223
Metallothionein, 223
Metals analysis, 402
Metals in an organism, 212
Metals, 6
Metamorphic rock, 47
Methanethiol, 349
Methanol, 25, 293
Methanol analysis, 406
Methemoglobinema, 194, 312
Methoxychlor, 339
Methyl acetate, 305
Methyl bromide analysis, 406
Methyl chloroform, 328
Methyl formate, 305
Methyl group, 17
Methyl methacrylate, 305
Methyl phosphonate, 365

Methylamine, 26, 310
Methylation, 156, 213
Methylethyl ketone, 301
Methylisocyanate, 318
Methylisothiocyanate (methyl mustard oil), 353
Methylmethane sulfonate, 358
Methylphosphine, 32, 363
Methylsulfuric acid, 32, 358
Methyl*tert*butyl ether (MTBE),,25, 304
Mevinphos, 370
Military poisons, 377
Mineral, 46
Mineral accumulators, 392
Mineral oil, 278
Mineralization, 40, 106
Minimata Bay, mercury contamination, 220
Miosis, eye, 135
Mitochondria, 61
Mixed organometallic compounds, 257
Mobam, 359
Molarity, 11
Molecular formulas, 16
Molecular geometry, 8, 15
Molecular mass, 9
Molecular weight, 9
Molecules, 7
Monochloramine, 243
Monomers, 34
Monomethylmercury ion, 264
Monsaccharides, 65
MPTP (1,2,3,6-Tetrahydro-1-methyl-4-phenylpyridine), 314
mRNA (messenger RNA), 75
MTBE (methyltertiarybutyl ether), 25
Muconic acid, 285, 405
Muscarine, 387
Mushroom toxins, 387
Mustard oil, methyl, 353
Mutagenesis, 136
Mutagens, 159
Mutation, 75
Mycotoxins, 385
Mydriasis, 135
Myelin sheathing, 201
Myelinopathies, 203
Myocardium, 192

N

NAD (nicotinamide adenine dinucleotide), 88
NADP (nicotinamide adenine dinucleotide phosphate), 88
Naled, 370
Naphthalene, 23, 287
Naphthol, 297
Naphthylamines, 313
Naphthylthiourea (ANTU), 32, 353
Natural history, 39
Necrosis, cell, 183
Neoplastic cells, 197
Nephritis, from lead poisoning, 221
Nephron, 206

Nephrotoxic, 207
Nervous system, 200
Neurons, 201
Neuropathy, 201
Neuropharmacology, 203
Neurotoxic psychodysleptics, 389
Neurotoxins, 200
Neurotransmission, 203
Neurotransmitters, 201
Neutral solution, 13
Neutralization reaction, 12
Neutrons, 3
Nickel toxicity, 217
Nicotinamide adenine dinucleotide (NAD), 88
Nicotinamide adenine dinucleotide phosphate (NADP), 88
Nicotine, 203, 322
Nitric oxide, 53, 239
Nitriles, 314
Nitrilotriacetate (NTA), 26
Nitro alcohols, 316
Nitro compounds, 315
Nitrobenzene analysis, 406
Nitrobenzene, 316
Nitrogen compounds, organic, 309
Nitrogen cycle, 41
Nitrogen dioxide, 53, 239
Nitrogen halides, 242
Nitrogen oxides, 239
Nitromethane, 316
Nitrophenol, *p*-, 316
Nitrosamines, 317
Nitrous oxide, 239
N-nitroso compounds, 317
Noble gases, 4
Nonkinetic toxicology, 129
Nonmetals analysis, 403
Nonmetals, 7
Nonmetals, toxicity, 225
Normals, and toxic response, 128
NTA (nitrilotriacetate), 26
Nucleophilic species, 157
Nucleotide, 73, 168
Nucleotide excision, 170
Nucleus, atom, 3
Nucleus, cell, 60
Nystagmus, 135

O

Octachlorobornane, 341
Octanol-water partition coefficient, Kow, 104
Octet rule, 6
Octet, 6
Odors from toxic substance exposure, 134
Olefins, 20, 275
Omethoate, 376
O-mustard, 359
Oncogenes, 171, 176
Optical isomerism, 34
Orbitals, 6
Orfila, M.J.B., 115

Organelles (cell), 61
Organic chemistry, 14
Organoarsenic compounds, 267
Organochlorine compounds, 19
Organogermanium compounds, 266
Organohalide compounds, 27, 327
Organohalide compounds, biogenic, 327
Organohalide insecticides, 338
Organohalides, 27
Organolead compounds, 266
Organometal compounds, 258
Organometallic compound toxicity, 258
Organometallic compounds, 45, 213, 253
Organometalloid compound, 253
Organonitrogen compounds, 25, 309
Organooxygen compounds, 14, 24
Organophosphate esters, 366
Organophosphate insecticides, 369
Organophosphates, 366
Organophosphorus compounds, 363
Organophosphorus military poisons, 377
Organoselenium compounds, 270
Organosulfur compounds, 347
Organotellurium compounds, 270
Organotin compounds, 264
Orthophosphoric acid, 366
Outer electrons, 6
Ovarian cycle, 203
Oviducts, 203
Ovulation, 203
Oxalate, calcium, 296
Oxalates, 391
Oxidases, 71
Oxidation, 10
Oxidation number, 10
Oxidation of hydrocarbons, 274
Oxidation state, 10
Oxidation-reduction, 44
Oxidative alteration (of DNA), 172
Oxidative burden, 185
Oxidative respiration, 83
Oxidative stress, 226
Oxides of hydrocarbons, 298
Oxidizer, 10
Oxidoreductase enzymes, 71
Oxygen toxicity, 225
Oxygen-containing organic compounds, 293
Oxyhemoglobin, 194
Ozone toxicity, 227

P

PAH (polycyclic aromatic hydrocarbons), 23, 107
PAPS, 153
Paraffins, 15
Paraoxon, 348, 370
Paraquat, 321
Parathion analysis, 406
Parathion, 33, 348, 356
Parent compound analysis, 404
Particles (atmospheric), 54

Particulate matter, 54
PBBs (polybromineated biphenyls), 28
PCBs (polychlorinated biphenyls), 28
Pentachlorophenol, 30, 343
Pepsin, 80
Peptide linkage, 63
Perchlorates, 242
Perchloroethylene, 333
Percutaneous absorption, 186
Percutaneous exposure, 121
Periodic table, 4, 5
Peripheral nervous system, 201
Peripheral neuropathy, 201
Peroxidation, lipid, 239
Peroxide from diethyl ether, 304
Persistent organic toxicants, 98
Petroleum jelly, 278
Pfiesteria piscicida, 388
Phagocytosis, 191
Phanerochaete chrysoporium, 107
Phase I reaction, 142, 143
Phase I reaction product analysis, 404
Phase II reaction, 142
Phase II reaction product analysis, 406
Phase II reactions, 149
Phases of toxicity, 130
Phenol, 25, 297
Phenols, 297
Phenotype, 169
Phenyl glucuronide, 298
Phenyl groups, 276
Phenylthiourea, 352
Phosgene, 318, 327
Phosmet, 372
Phosphate (in DNA and RNA), 73
Phosphate ester insecticides, 369
Phosphine, 245, 363
Phosphine oxides, 365
Phosphine sulfides, 355, 365
Phosphoglycerides, 67
Phosphonic acid esters, 365
Phosphonic acid, 366
Phosphordithioate insecticides, 356
Phosphorodithioate esters, 33, 368, 369
Phosphorothionate insecticides, 356, 371
Phosphorus acid esters, 365
Phosphorus halides, 246
Phosphorus oxyhalides, 246
Phosphorus pentoxide, 245
Phosphorus toxicity, 228
Phosphorylating group, 374
Phossy jaw, 228
Photoallergy, 188
Photochemical dissociation, 53
Photochemical process, 19
Photochemical reactions, 52
Photochemical smog, 54
Photosensitivity, 177, 188
Photosynthesis, 76
Phototoxic responses of skin, 188
Phototoxicity, 188

Phthalate, dimethyl, 305
Phthalic acid, 303
Phthalic anhydride, 287
Physiological response, 131, 132
Phytochelatins, 223
Phytoxic substances, 228
Pi electrons, 21
Pi-bonded organometallic compounds, 254, 257
Piperazine, 312
Piperidine, 312, 314
Pirimicarb, 320
Plants, toxic substances from, 388
Plasma, 192
Plasticizer, 35
Platlets, blood, 192
Pneumonitis, acute chemical, 215
Point mutation, 170
Poison, 115
Poisonous animals, 383
Poisonous organism, 383
Poisons, 97
Polar covalent bonds in organometallic compounds, 255
Polar molecule, 12
Pollutants, water, 45
Pollution-induced community tolerance, 112
Polybrominated biphenyls (PBBs), 28
Polychlorinated biphenyls (PCBs), 28, 336
Polychlorinated dibenzodioxins, 342
Polycyclic aromatic hydrocarbons (PAH), 23, 107, 288
Polycyclic aromatic hydrocarbons analysis, 406
Polyethylene, 279
Polymers, 34
Polyneuropathy, 277
Polypeptides, 61
Polyphosphates, 366
Polysaccharides, 65
Polyvinylchloride polymer, 34
Population (in ecology), 110
Porphobilinogen, 221
Porphyria, 177, 188
Porphyria cutanea tarda, 337
Portal circulatory system, 123
Potassium organometallic compounds, 260
Potent, as related to toxic substances, 125
Potentiation, 132
Precarcinogens, 161
Primary carcinogens, 162
Primary protein structure, 64
Primary reaction, 131
Primary water treatment, 46
Procarcinogens, 161
Product, 10
Prokaryotic cells, 60
Proliferation, uncontrolled, of cells, 197
Promoters in carcinogensis, 163
Propane, 277
Propionic acid, 25, 303
Propylene, 35, 280
Protein digestion, 83
Protein metabolism, 86
Protein structure, 64

Proteins, 61
Proteins, types, 63
Protons, 3
Proto-oncogenes, 171
Protoxicant, 130, 141
Protozoa, toxins from, 387
Proximate carcinogens, 161
Psilocybin, 387
Psychodysleptics, neurotoxic, 389
Pulmonary exposure, 122
Pulmonary fibrosis, 185
Pulmonary, 184
Pumonary edema, 185
Putrescine, 311
Pyrene, 290
Pyrethrin I, 389
Pyrethrins, 390
Pyrethroids, 390
Pyridine, 314
Pyrophosphoric acid, 366
Pyrrole, 312
Pyrrolidine, 312
Pyrrolizidine, 391

Q

QSARs (quantitative structure-activity relationships, 139
Quantitative structure-activity relationships (QSARs), 139
Quaternary protein structure, 64

R

Racemic mixture, 34
Radioactive isotopes, 3
Radioactivity, 3
Radionuclides, 3
Radionuclides, toxicities, 230
Radium toxicity, 230
Radon toxicity, 230
Rate of exposure to toxic substance, 119
Rattlesnake, 395
Reactant, 10
Reaction rates, 11
Reactive substances, toxic, 140
Recalcitrant substances, 108
Receptor, 129, 157
Receptor antagonism, 132
Recombinant DNA, 76
Redox-reactive reagents, 157
Reducer, 10
Reducing agent, 10
Reductase enzymes, 147
Reductive dehalogenation, 149
Renal failure, 207
Replication of DNA, 75
Reproductive system, 203
Reproductive toxicology, 136
Reptile toxins, 395
Residue, amino acid, 61
Resonance stabilization, 22
Respiration, 76

Respiration, oxidative, 83
Respiratory rate, 132
Respiratory tract toxicology, 184
Response, 124
Reversal (of DNA damage), 170
Reversible effects, 125
Reversible effects of toxic substances, 127
Reversion (back mutation), 175
Ribonucleic acid (RNA), 72
Ribose, 72, 73, 168
Ribosomes, 61, 169
Ricinine, 389
RNA (ribonucleic acid), 72
Rock, 46
Rotenone, 389
Route of exposure to toxic substance, 119, 120
Roxarsone, 269

S

S-adenosylmethionine (SAM), 156
Saffrole, 155
Salmonella typhimurium (in Bruce Ames test), 175
Salt, 13
Salvarsan, 268
SAM, S-adenosylmethionine, 156
Saponins, 391
Sarin, 378
Scaling (skin), 187
Scleroderma, 244
Scopolamine, 389
Scorpions, 397
Screening organ, 122
Seaweed, organoarsenic in, 268
Secondary water treatment, 46
Sedimentary rocks, 47
Selenium, organically bound, 360
Senecionine, 389
Sensitivity to toxic substances, 127
Serum, blood, 194
Sesquimustard, 359
Shigella dysenteriae, 384
Silanes, 244
Silica, 243
Silicates, 47
Silicon halides, 245
Silicon, inorganc compounds, 243
Silvex, 341
Simple sugars, 65
Site of exposure to toxic substance, 119, 120
Skin, 186
Skin cancer, 189
Skin exposure to toxic substances, 121
Skin irritants, 391
Skin permeability, 121
Skin symptoms of toxic substance exposure, 134
Skunk spray, organosulfur compounds in, 351
Snake venoms, 395
Sodium ethylsulfate, 358
Sodium organometallic compounds, 260
Soil, 47
Soil chemistry, 50
Soil humus, 51
Soil profile, 47
Soil solution, 50
Solute, 11
Solution, 11
Solution concentration, 11
Solvation, 12
Solvent, 11
Soman, 378
Somatic cells, 167
S-oxidation, 348
Sperm, 203
Spider toxins, 394
Spinal cord, 200
Spirogermanium, 266
Spleen, 192
Starch, 65
Stearic acid, 303
Steatosis, 190
Stem cells, 192
Stereochemical molecular configuration, 129
Steroids, 69
Stomach, exposure to toxic substances, 123
Straight-chain alkanes, 16
Stratification of a body of water, 43
Stratosphere, 52
Stratum corneum, 121, 186
Structural formulas, 16
Structural isomers, 16
Structure-activity relationships, 173
Strychnine, 322
Styrene analysis, 406
Styrene, 35, 286
Styrene-7,8-oxide, 286
Subatomic particles, 2
Sublethal effects, 125
Sublethal effects of toxic substances, 111
Substitution reactions, 19
Substitution reactions, 274
Substrates (enzyme), 69
Sucrose, 81
Sugars, simple, 65
Sulfa drugs, 360
Sulfate conjugate, 349
Sulfides, 30
Sulfides (organic), 350
Sulfites, 247
Sulfolane, 32, 357
Sulfones, 32, 356
Sulfonic acids, 32, 357
Sulfoxidation, 348
Sulfoxides, 32, 356
Sulfur dioxide, 247
Sulfur mustards, 359
Sulfur, inorganic compounds, 246, 249
Sulfuric acid, 248
Sulfurous acid, 248
Superoxide radical ion, 226
Superoxide, 164
Surface water, 99

Synergistic effects, 132
Systematic names, 17
Systemic poisons, 129
Systemic toxicology, 129

T

Tabun, 378
Tachycardia, 132, 195
Target cells, 198
Taxol, 389
Taxonic classification, 119
TCDD (2,3,7,8-tetrachlorodibenzo-*p*-dioxin), 99, 189, 342, 406
Temik, 146
Temperature (body), 132
Teratogens, 205
Teratology, 136, 205
Terrestrial ecosystems, 110
Terrestrial ecotoxicology, 111
Tertiary protein structure, 64
Testes, 203
Testosterone, 203
Tetanus, 384
Tetrachlorodibenzo-*p*-dioxin, 2,3,7,8- (TCDD), 99, 189, 342, 406
Tetrachloroethylene, 28, 333
Tetrachloroethylene analysis, 406
Tetrachloro-p-dioxin, 2,3,7,8- (TCDD) 341
Tetraethylpyrophosphate, 33, 368
Tetrafluoroethylene, 35
Tetrahydrofuran, 304
Tetrahydro-1-methyl-4-phenylpyridine, 1,2,3,6- (MPTP), 314
Thalidomide, 206
Thiabendazole, 355
Thiazole, 354
Thiocyanates, 353
Thiols, 30, 349
Thiophane, 349
Thiourea, 32
Thiourea compounds, 352
Threshold dose, 125
Threshold limit values (TLVs), 276
Thymine, 73, 168
Time-weighted average exposure, 276
Tiron, antidote for beryllium poisoning, 215
TNT (trinitrotoluene), 26, 316
Toclofos-methyl, 371
TOCP (tri-*o*-cresylphosphate), 33, 367
Tolerance, 128
Toluene, 285
Topsoil, 47
Toxaphene, 340
Toxic epidermal necrolysis, 189
Toxic shock dyndrome, 385
Toxic substances, 97
Toxicant concentration, 119
Toxicants, 97, 115, 117
Toxication, 140
Toxicity ratings, 125

Toxicity, acute, 116
Toxicity, chronic, 116
Toxicity, metal, 213
Toxicity-influencing factors, 117
Toxicogenomics, 177
Toxicokinetics, 141
Toxicological chemistry 1, 116, 139
Toxicologist, 115
Toxicology, 115
Toxicology, clinical, 116
Toxicology, environmental, 116
Toxicology, forensic, 116
Toxicometrics, 129
Toxins, 117
Transcription in protein synthesis, 75
Transcription, 169
Transfer RNA, 169
Transferase enzymes, 71
Transition (in DNA), 170
Transition elements, 4
Translation, 169
Translation in protein synthesis, 75
Transpiration, 48
Transversion (in DNA), 170
Tributyl tin (TBT), 265
Trichloroacetate, 335
Trichloroethane analysis, 406
Trichloroethylene, 28, 333
Trichlorophenoxyacetic acid, 2,4,5- (2,4,5-T), 341
Trifluoroacetic acid, 332
Triglycerides, 67
Trimethylphosphate, 33, 365, 367
Trimethylphosphite, 365
Trinitrotoluene (TNT), 26, 316
Tri-*o*-cresylphosphate (TOCP), 33, 367
Triphenylphosphine, 32
Trophic level, 101
Troposphere, 51
Tumor suppressor genes, 171, 176

U

UDPGA (uridine-5'-diphospho-a-D-glucuronic acid), 151
Ultimate carcinogen, 161
Uneconomic form, 126
Unsaturated compound, 20
Unsaturated hydrocarbons, 275
Uracil, 73, 168
Urea, 86
Uridine-5'-diphospho-a-D-glucuronic acid (UDPGA), 151
Urine for metals analysis, 403
Urticaria, 187

V

Vacular toxicants, 196
Vacuoles, 61
Valence electrons, 6
Vanadium toxicity, 215
Vapam, 354
Vasodilation, 194

Vehicle (of toxic substance), 118
Venomous animals, 383
Venoms, 383
Vesiculation, 187
Viagra, 196
Viagra, toxic effects, 184
Vinyl acetate, 305
Vinyl chloride, 20, 28, 333
Vital signs, 132
Vomitoxins, 386
VX, 378

W

Wasp venoms, 394
Water as a solvent, 12
Water pollutants, 45
Water treatment, 46
Watson, James B., 74
Weathering, 46, 49

Weathering, chemical, 50
Wet-ashed blood and urine, 403
Widow spiders, 394

X

Xenobiotic compounds, 141
Xenobiotic molecules, 106
Xenobiotics analysis, 401
Xylene analysis, 407
Xylenes, 283, 286

Z

Zearalenone, 386
Zeta-cypermethrin, 392
Zinc organometallic compounds, 261
Zwitterion, 63
Zygote, 203